# The Neurosciences: Paths of Discovery, I

# The Neurosciences: Paths of Discovery, I

Frederic G. Worden
Judith P. Swazey
George Adelman
Editors

**Birkhäuser**
Boston • Basel • Berlin

Library of Congress Cataloging-in-Publication Data

The Neurosciences: Paths of Discovery I/Frederic G. Worden, Judith P. Swazey,
   George Adelman, editors
        p.    cm.
     Originally published: Cambridge, Mass.: MIT Press, 1975.
     Proceedings of a symposium held at the Massachusetts Institute of Technology,
   Oct. 29-30, 1973.
     Includes bibliographical references and indexes.

     1. Neurobiology—Congresses. I. Worden, Frederic G.    II. Swazey, Judith P.
III. Adelman, George, 1926-
     [DNLM: 1. Neurology—congresses.    2. Neurosciences—congresses.
WL 100 N4954 1975a]
QP361.N482    1992
612.8—dc20
DNLM/DLC
for Library of Congress                                              91-33450
                                                                          CIP

Printed on acid-free paper.

ISBN-13: 978-0-8176-3621-0                    e-ISBN-13: 978-1-4684-6817-5
DOI: 10.1007/978-1-4684-6817-5

9 8 7 6 5 4 3 2 1

To the memory of
Barbara Hecker Schmitt

The figure on the cover is based on Figure 2, "Medial surface of the right half of the brain in the bisected head", from *The Human Central Nervous System* (3rd revised edition, 1988) by R. Nieuwenhuys, J. Voogd, and Chr. van Huijzen; Springer-Verlag Berlin, Heidelberg, New York. We gratefully acknowledge permission to use this figure.

# Contents

## The Neuroscience Community: Some Organizational Developments

## On the Nature of Research in Neuroscience

# Preface

To commemorate properly the 70th birthday of a man who, by his very nature, is too busy to pause for any kind of ceremonial event unless it has a concomitant functional output was a difficult problem for the Staff and Associates of the Neurosciences Research Program. Frank (F.O.S.) has always dreaded the prospect that sometime it might be appropriate for his colleagues to present him a *Festschrift*. In fact, "Fest me no Schriften" became his battle cry, expressing his feeling that the idea of testimonials clustered into a book was *anathema*. So the "breakthrough" idea for the planners was to organize a symposium around the theme of discovery in neuroscience that would be valuable scientifically and, in its demonstration of interdisciplinary interaction, would support that emphasis in Frank's career.

After much planning a program was developed, beginning with a birthday party the evening before, followed by the two-day symposium, and closing with the first F.O. Schmitt Lecture in Neuroscience. We hope that publication of the scientific proceedings in this volume will be of interest not only to the neuroscience community, but also to a broad general readership interested in discovery, understanding, and the creative processes in scientific work.

An organizing committee, chaired by Fred Worden, collected advice and guidance leading to the selection of speakers whose scientific careers have played an important part in the recent history of modern neuroscience. The choice of speakers was not easy; many individuals other than those selected might well have been asked to participate. A few who had accepted invitations had to cancel at the last moment. Heinrich Klüver, one of Frank's closest friends, and Yngve Zotterman were two who were unable to participate, and one, the great Polish psychologist Jerzy Konorski, passed away before the symposium; they were all sorely missed. We were fortunate in having had the benefit of Ralph Gerard's presence and contributed paper, and hope that its inclusion in this volume can be something of a memorial to this pioneer.

The group whose papers appear in this volume illustrate many of the various paths of discovery in neuroscience; their work truly epitomizes the emphasis that Frank has always placed on the importance of interdisciplinary interaction in science. Their enormously productive and diverse careers are perhaps best summed up by the two phrases that Frank has always used when he wants to urge his colleagues on to one more go around at solving a problem in the lab or at a conference—"You ain't seen nothin' yet" and "It's later than you think."

The symposium at which most of these papers were presented was held at MIT's Kresge Auditorium on 29 and 30 October 1973. The first F.O. Schmitt Lecture in Neuroscience, "Sources of discovery in neuroscience," was presented by J.Z. Young on the evening of 30 October 1973. The symposium was open to the public and was simultaneously broadcast on FM radio by the MIT radio station. Approximately 2000 people attended some or all of the presentations.

Special thanks go to the administrative staff of the Neurosciences Research Program, under the direction of L.E. Johnson and Kay Cusick, whose arrangements made the program so successful. The help and cooperation of the MIT staff of Kresge Auditorium is also deeply appreciated. We are also grateful to Neurosciences Research Program Associates Ross Adey, Robert Galambos, Melvin Calvin, Richard Sidman, Robert Livingston, Donald MacKay, Gardner Quarton, and Vernon Mountcastle, who served as session chairmen during the symposium; to Walter Rosenblith, provost of MIT, who introduced the first F.O. Schmitt Lecture in Neuroscience, and to Irwin Sizer, dean of the graduate school at MIT, who was chairman of arrangements for the birthday party at Boston's Museum of Science and a source of wise counsel and help throughout these events.

Boston, Massachusetts                                                    Frederic G. Worden
28 October 1974                                                             Judith P. Swazey
                                                                                    George Adelman

Francis O. Schmitt

# Francis O. Schmitt—A Profile

## The Biophysical Years
Jerome Gross and Carl F. Cori

Francis Otto Schmitt, a native of Saint Louis, received his A.B. degree at Washington University in 1924, and his Ph.D. degree in physiology (medical sciences) in 1927. Two of his teachers in the physiology department at the Medical School were Joseph Erlanger and Herbert Gasser, eminent neurophysiologists who in 1944 shared the Nobel Prize in physiology for their work on single nerve fibers. Schmitt remained on intimate terms with these two men, who undoubtedly had a great effect on his subsequent development.

Armed with a National Research Council Fellowship, Schmitt did postdoctoral work with G.N. Lewis at the University of California, with J.G. Drummond at University College, London, and with O. Warburg and O. Meyerhof at the Kaiser Wilhelm Institute in Germany. On his return, he joined the zoology department of Washington University and rose quickly through the academic ranks, becoming full professor in 1938. On the retirement in 1940 of Caswell Grave, a kindly man of the old biology school, Schmitt became head of the department, only to leave a year later for the Massachusetts Institute of Technology, where he has remained until now.

## The Saint Louis period (1924–1942)
First papers are not necessarily significant in the widespread scientific activity of a man like F.O. Schmitt. But, for the record, his first two papers as an undergraduate, published with Caswell Grave, dealt with the coordination and regulation of ciliary movement in a mollusk. His third paper, published in 1925 on "Fluid crystals and meristematic growth," is significant because it marks the beginning of Schmitt's lifelong interest in molecular biology. Other early publications included several papers on kidney function in *Necturus maculosus* (with H.L. White) and on conduction of an impulse through heart muscle (with Joseph Erlanger) that formed part of his Ph.D. thesis.

Studies of the effect of high-frequency sound waves on protoplasm show one characteristic of F.O. even during his early postdoctoral period—his fascination with new apparatus and techniques. A paper with Otto Meyerhof on the respiratory quotient of nerve ushered in a long series of studies dealing with the respiratory activity of nerve and the effect of respiratory inhibitors on the conduction of the nerve impulse. A paper with C.F. Cori dealt with lactic-acid formation in stimulated frog sciatic nerve.

F.O. has always been a great stimulator of work by promoting collaborative efforts. A particularly productive early collaboration developed with R.S. Bear and led to a study of the ultrastructure of nerve sheath as revealed through the use of x-ray diffraction and polarized light. Later these studies were aided by J.Z.

Young, G.L. Clark, and others. Characteristically, Schmitt expanded his studies on ultrastructure horizontally by studying other biological objects, such as the red-cell membrane and the granules in spores of *Tradescantia*. Several papers dealt with x-ray diffraction of lipids, particularly those occurring in nerve, and with lipid-protein complexes. This series of papers contributed greatly to our current understanding of membrane structure. Another series of papers, also begun with Bear, dealt with the electrolytes, inorganic and organic, in the axoplasm of different nerves.

Schmitt was an excellent teacher at Washington University, and he took a great personal interest in his students. He also organized an interdepartmental seminar that met informally in the evening and became known as the "Schmittie Verein." Attendance was by invitation only, and failure to attend without excuse could lead to your being dropped. This sounds worse than it was—Schmitt was a most amiable tyrant.

Mention should also be made here of F.O.'s younger brother Otto, who was a Ph.D. student in physics and is now a professor of biophysics at the University of Minnesota. Brother Otto had a special aptitude for putting together electronic devices, and reports on several of these for temperature regulation and stimulation of nerve were published by the two brothers.

By the end of his highly productive Saint Louis period, Schmitt and his collaborators had published some one hundred papers dealing with a variety of problems, among which the structural analysis of tissues using physical and physicochemical methods seemed to emerge as the predominant interest. His emphasis was clearly shifting toward the molecular level. Schmitt's attention may have been attracted to collagen for the first time during the early thirties, when he and his colleagues discovered with chagrin that the diffraction and thermal-shrinkage properties of neurons could not be ascribed to axon constituents, but rather were due to that ubiquitous extracellular fiber associated with nerve. Others have had similar painful experiences! In any event, the time was ripe for analysis of the structure of collagen and other fibrous proteins by electron microscopy, a subject which was to occupy Schmitt's interest for many years.

**The Cambridge period (1941 to present)**

Frank left Washington University in 1941 to become professor of biology and head of the Department of Biology and Public Health at the Massachusetts Institute of Technology. This department, one of the first of its kind in the United States, had, a few years before his arrival, begun to expand into the new area of physical biology under the direction of Professor John T. Bunker. Schmitt was the logical choice to direct this new field at MIT with vigor and inspiration. He brought with him Richard Bear and David Waugh, and Cecil Hall also joined the department at this time. It was during this period that Frank began his pioneering studies with the electron microscope, with Hall and M.A. Jakus, on the ultrastructure of fibrous proteins, collagen and paramyosin, striated muscle, and cilia and flagella.

During most of his professional life, F.O. has been active in stimulating biological research at the molecular level. Those who know him, however, know that he could never be accused of being a reductionist in philosophy. His approach has been to learn the structure and behavior of the "tinker toys" in order to reconstruct the functional tissue component. Schmitt and his colleagues were working with molecular self-assembly mechanisms long before this phrase was invented.

During the exciting days of the forties and fifties, "a discovery a day" was being made in the cramped biology-department laboratories under the massive dome of Building 10 at MIT. The team of Hall, Jakus, and Schmitt published some of the earliest electron micrographs of the supramolecular fabric of striated muscle fibers, and with Bear they correlated the new structures with those arrived at by x-ray diffraction. These investigators also pioneered the use of heavy-metal electron stains in these and other studies. Jakus and Hall went on to examine by electron microscopy the structure of isolated and purified myosin and actin molecules and their mode of polymerization and interaction. During their stay in Schmitt's laboratory, Jean Hanson and Hugh Huxley developed their now famous "sliding-filament" hypothesis of the mechanism of muscle-fiber contraction. C.E. Hall, working at resolutions of 15 Å in the early fifties, published the first electron micrographs of replicas of the globular protein molecules, catalase and edstin, assembled in microcrystals, and also high-resolution electron micrographs of fibrinogen molecules in the act of polymerizing to fibrin. Early observations on the structure of the myelin sheath, the "jelly-roll" hypothesis of myelin synthesis by B. Geren Uzman, Schmitt and Geren's discovery and characterization of the axon filaments of nerve, high-resolution electron micrographs of the structure of cell membranes, and the "unit-membrane" model of J.D. Robertson, all these were solid pioneering contributions. The glass knife, still a mainstay of thin-sectioning technique for electron microscopy, was invented one tired evening around midnight in Schmitt's laboratory by H. Latta and F. Hartman using glass fragments from broken milk bottles.

In 1942, using one of the first electron microscopes made commercially in this country, Schmitt, Hall, and Jakus discovered the periodic cross-striations in collagen fibrils, and they were soon using heavy-metal electron stains for the first time to reveal greater fine structural detail. Schmitt, J. Gross, and J.H. Highberger then embarked upon a systematic analysis of the in vitro reconstitution of cross-striated fibrils from solutions of purified collagen molecules. They provided the first indications that structural macromolecules in random dispersion could, under appropriate environmental conditions, be brought together to reconstitute a variety of highly ordered supramolecular fabrics, including striated fibrils identical with the native form. In the early fifties, Highberger, Gross, and Schmitt described the unusual long-spacing structures into which they had forced collagen molecules to self-assemble in vitro. These experiments led them to propose the correct dimensions of the collagen molecule (later visualized directly by C.E. Hall), which they called "tropocollagen." From these studies they developed the "quarter-stagger hypothesis," later elucidated in greater detail by Hodge and Schmitt, describing the manner in which these molecules align in the fibril to give rise to the characteristic 640 Å axial periodicity. Their methods for the isolation and purification of collagen and the long-spacing and reconstituted fibrillar forms they developed are still in general use in studies of metabolic and structural collagen in normal and diseased tissues. Early in the studies on collagen fibrogenesis, Schmitt, with characteristic prescience, predicted that the molecules would have short non-collagen-like peptides at either end, and that these "telopeptides" would be important in fibrillogenesis and perhaps even in human disease. Several years later, this prediction was proved correct by Schmitt and his colleagues A.L. Rubin, D. Pfahl, P.T. Speakman, and P.F. Davison. In recent years the telopeptides have been isolated and sequenced; they have been determined to be critical in the formation of intra- and intermolecular cross-links and to be important immunogenic determinants for collagen. They are probably also involved in several human birth defects of connective tissue.

Although Francis O. Schmitt turned away from clinical medicine after two years of preclinical training to follow a career in the basic biological sciences, his service as trustee of the Massachusetts General Hospital and as adviser, mentor, and colleague to numerous M.D. postdoctoral fellows at MIT provides strong evidence of his continuing interest in medicine. There is no doubt that his contributions to fundamental biology have had a powerful impact on medical science and have amply justified his early career decision. His major influence has been the powerful stimulus he has given to the field of tissue fine structure (or, as he would phrase it, "analytical cytology") and the model he has set for the large number of students and colleagues who have been trying to follow in his footsteps.

## The Neuroscience Years
Walter A. Rosenblith

It was the year France fell and Wendell Willkie got the Republican nomination. Thanks to a Rockefeller grant, I was spending the summer at the Cold Spring Harbor Biological Laboratory experimenting with a flock (or pack) of electric eels. I also attended the 1940 Symposium on Quantitative Biology, which was being held at the laboratory and whose subject was permeability.

In that symposium, Frank Schmitt was the senior author of a paper on x-ray diffraction studies of biological materials. I was a budding physicist who—under the tutelage of R.T. Cox from NYU's uptown physics department—had spent the previous year becoming acquainted with electric organs, biological impedances, etc. Having talked to Frank after his paper, I was led to read some of his earlier papers, an experience that came in handy the next year at UCLA, where I spent a fair amount of time looking into the biological effects of ultrasonics.

My next encounter with Frank was eleven years later. I was moving from Harvard's Psycho-Acoustic Laboratory to MIT's Department of Electrical Engineering. Frank had come a decade earlier to MIT. He had transformed a Department of Biology and Public Health into a Department of Biology and Biological Engineering and later (1944–1945) into a Department of Biology whose teaching and research programs were defined along the general lines of physical and chemical biology, with particular reference to those aspects that were susceptible to quantitative treatment. The department's interest in molecular structure was symbolized by the role of the electron microscope, which was beginning to take its place next to the more established methods of biochemistry and those of physical biology or biophysics.

Frank's dynamic transformation of the department's image at the Institute had not been lost on his colleagues in other MIT departments. Thus Professor Hazen, the head of the Department of Electrical Engineering, suggested that I go discuss my requested title as Associate Professor of Communications Biophysics with Frank in order to avoid either confusion or jurisdictional conflict. Frank was very gracious and not at all concerned. As a matter of fact, he was interested in the research I proposed to do. When I told him a bit naively that I was hoping to study sensory information processing somewhat in the spirit of cybernetics and communication theory, with the ultimate purpose of trying to understand how the brain functioned, he remarked jokingly, "Aren't you 200 years early?"

But, given Frank's concept of speed, two centuries went by awfully fast, and it was less than a chronological decade before he—who had long been concerned with the structure of nerve—was to bring together scientists from a broad range of disciplines in order to crystallize the neurosciences. But let us not jump too far ahead in our story.

In 1955 MIT's faculty bestowed upon Frank the highest honor that one's peers in the Institute can bestow: they made him an Institute Professor. (He was, to be precise, the second Institute Professor, the solid-state physicist John C. Slater having been chosen a few years earlier.) This distinguished post not only recognizes outstanding achievement, but also allows the incumbent to be freed from administrative duties so as to be able to concentrate on advanced teaching and research. This was the period during which revolutionary advances in molecular biology were changing the structure of the life sciences. It was the time during which the Biophysical Society was conceived and then born (in 1957). It was the time during which the special NIH Study Section in Biophysics and Biophysical Chemistry, which Frank chaired, was trying to arrange adequate research support and to provide academic institutional arrangements to attract young people from throughout the physical and life sciences to the new fields.

To those who were inclined to scoff "Biophysics, what's that?" Frank, in collaboration with Larry Oncley, Robley Williams, Murray Rosenberg, Dick Bolt, and others, offered a substantive answer. Under the auspices of the Study Section, a monthlong program was organized in Boulder, Colorado. About 120 senior researchers and selected younger scientists took part in this program, which was held on the campus of the University of Colorado between mid-July and mid-August 1958. The setting was particularly conducive to an informal interchange of ideas; as a matter of fact, well-meaning tongues referred to the program variously as a "summer kibbutz," a "semieternal Gordon conference," or simply a "Symposium" with an emphasis on the meaning of the Greek root word, namely "to drink together." Even though some of us were not familiar with either the name or the concept, it is now clear to us that another "Schmittie Verein" had seen the light of day in the clear air of the Rockies.

To Frank, the general purpose of the Boulder experiment was the exchange of information among representatives of the physical and the life sciences. He anticipated that the blending of concepts and methods from the various fields would lead to significant advances in the investigation of biological problems. As he has since written, "The core of the study program was a carefully integrated series of about sixty lectures, constituting compact summaries of certain key problems and critical evaluation of recent advances. The lecture material provided a framework for other activities such as planned workshop sessions, spontaneous discussion groups, and library study."

The suitably edited text of the sixty-odd lectures appeared first as the January and April 1959 issues of *Reviews of Modern Physics* and later that year as *Biophysical Science—A Study Program* (New York: Wiley, 1959), which stands to this day as the most ambitious survey of the range of biophysical science.

The Boulder study program was a watershed in the history of biophysics. It produced a key book to which researchers, teachers, students, and even administrators could refer. It also created an understanding among a group of influential scientists from different fields and different universities that would help them in establishing a wide variety of biophysical institutions: from a subdepartment in a medical school to the IUPAB, the International Union of Pure and Applied Biophysics, which was founded in the early 1960s.

Frank gave three lectures at Boulder, one of which was entitled "Molecular organization of the nerve fiber." But approximately one-sixth of the lectures were related to neural structures and functioning, and the book's prefatory material assures us that "future conferences and publications of this kind will undoubtedly emphasize alternative subjects such as, for example, those at higher levels of biological complexity." Of course, it was Frank's own Boulder in Neuroscience, the monthlong conference in the summer of 1966, that did just that; and in many people's opinion the Intensive Study Program and its resulting book (*The Neurosciences: A Study Program*, New York: Rockefeller University Press, 1967) served to crystallize the new field, the neurosciences.

It was because of these antecedents that I asked Frank, a few months ago, whether he felt that he had conceived what he calls the "NRP idiom" at the 1966 Boulder conference. The letter Frank wrote in answer to this question concludes with the following paragraph: "It seems that in my early professional career I was concerned with a direct attack on the physical basis of my mind, because years later, after the establishment of NRP, one of my former students from Washington University in the early 1930s said I was 'talking that way' to the students in my classes on general physiology (the word used in those days for molecular biology)."

Dr. Swazey's chapter deals with the origin and early history of the NRP. All that is left here is to provide a bit more of the international zeitgeist of the period during which the NRP was born.

During the so-called thaw that followed Stalin's death many attempts were made to bring together Western brain and behavioral scientists with Soviet students of "Higher Nervous Activity" à la Pavlov. The neuroscientifically speaking relevant historic event was a symposium held in Moscow in 1958. Subsequent to it, efforts were made to organize an International Brain Research Organization under the auspices of UNESCO. These efforts were successful, and IBRO was launched in Paris in 1960. The organization was initially made up of scientists grouped in seven panels: neuroanatomy, neurochemistry, neuroendocrinology, neuropharmacology, neurophysiology, the behavioral sciences, and neurocommunications and biophysics. Later on an eighth panel on brain pathology was added. Frank Schmitt was, of course, quickly elected to membership in the Neurocommunications and Biophysics Panel.

In spite of valiant efforts by Herbert Jasper, the first secretary-general, and his successors, IBRO, though extraordinarily useful, has never had available to it the necessary resources to conduct more than a few relatively low-cost programs: international research teams, fellowships, symposia, etc. IBRO has, however, led to the formation of various national Brain Research Committees (such as the NRC committee to which Frank belonged) and ultimately to the formation of large interdisciplinary societies such as the Society for Neuroscience in the United States. Frank Schmitt has always aided and abetted these incarnations of the zeitgeist, but they have not satisfied his restless desire for intensive personal activity and deep involvement. Thus we are fortunate that he has complemented these worthwhile organizations with that particular *Ortsgeist* called the Neurosciences Research Program, which has inhabited Brandegee House from the moment Frank brought it into being.

# The Neurosciences: Paths of Discovery

George Adelman (b. 1926, Boston, Massachusetts) is managing editor and librarian of the Neurosciences Research Program. He is editor of the *NRP Bulletin*, an editor of the series *Neurosciences Research Symposium Summaries*, and associate editor of the three volumes in the Intensive Study Program series *The Neurosciences*.

Frederic G. Worden (b. 1918, Syracuse, New York) has been director of the Neurosciences Research Program since 1974. His principal research interests have been the biology and psychology of mental illness and the neurobiology of brain function and behavior. He is professor of psychiatry at MIT.

# 1
# Introduction and Overview

## George Adelman and Frederic G. Worden

Exploratory activity is a fascinating attribute of any animal, but especially of man, who has relentlessly extended his exploration from his neighborhood to his world, to the reaches of outer space, and even to that ultimate mystery, his own brain, at once the source and object of his curiosity. The challenge of the brain has led gifted scientists to explore, and indeed to create, diverse intellectual pathways toward an understanding of how this marvelous organ mediates behavior and subjective experience. This book does not attempt to describe all the pathways in neuroscientific research; even less does it attempt to catalogue all the knowledge already available. Rather, it focuses on how scientific discovery and understanding come about through the interaction of individual human beings with each other, with ideas, with technological developments, and above all else, with those peculiarly human social and organizational influences that create a shared experience—the neuroscience community.

The contributors to this volume are only a small sample of the total neuroscience community. In selecting them the goal was, as far as possible, to have the main avenues of neuroscientific research represented by highly productive scientists, but to make no attempt at representative coverage of all the scientific domains involved. The charge to the contributors was deliberately broad so that the personality of each author could be reflected in the way he responded to the task. As a result, the reader will find a diversity of writing styles, a wide range of scientific domains, and, occasionally, highly personal opinions, value judgments, and descriptions of people, places, and ideas. The reader can discover many different pathways, along each of which he sees, through the eyes of a gifted scientist, something of the high adventure of research, discovery, and the construction of scientific models of reality.

Organizing such a book posed some perplexing problems. We have arranged the papers to some extent by subject area and emphasis; in some cases, though, the work is grouped according to such arbitrary, but intuitively appealing, categories as areas of specialization, kinds of preparations studied, or the fact that the main emphasis in one group is on unexpected observations and how they can lead to new understanding.

The rest of this introduction is a synoptic outline of what the book is about. We hope it will give the reader an overview of the book and of the multiplicity of problems that make up the neurosciences.

## The F. O. Schmitt Lecture in Neuroscience

In the first F. O. Schmitt Lecture in Neuroscience, J. Z. Young looks back over some forty years of his experience in neuroscience and forward to the new paths that he sees opening up for future exploration. His career, like that of F. O. Schmitt whose birthday he was helping to celebrate, has spanned so many core fields of research on the nervous system that it epitomizes neuroscience. In studies in anatomy, physiology, and zoology, he has always sought to develop explanations for the major questions of brain and behavior, whether working on the axon of a squid or the dual brain of an octopus. His paper is a personalized overview of the neuroscience community of people, methods, and ideas, and it provides an eminently appropriate beginning for the book.

## Comparative Approaches to the Nervous System

Many different kinds of nervous systems have emerged during the evolutionary development of animals, each one shaped by selective pressures to meet the adaptive demands confronting the organism, whether it be a clam or a snake, a bird or a man. As Donald Kennedy, with a zoologist's deep awareness of and "respect for diversity," suggests, the phylogenetic diversity of nervous systems offers both opportunities and dangers to the investigator. Generalizing from one nervous system to another may be hazardous, but comparisons of different neural solutions to the same problem may illuminate underlying principles, and the variety of animal forms offers an opportunity to find a particular nervous system suited for research on a particular problem. Kennedy's own work with simple systems, notably mollusk vision and the crayfish stretch receptor, is marked by a focus on the scrupulous analysis of a tiny bit of behavior as a means to understanding general principles.

Another approach seeks insights about the nervous system by comparisons with some quite different system. Gerald Edelman examines the comparison of "memory" in the immune system with memory in the brain. From his research on the immune system, perhaps the best understood molecular recognition system, he derives certain criteria and conceptual models that may elucidate how the brain functions as a selective-recognition system. Edelman, besides succinctly reviewing how these models from research on the immune system might have application in neuroscience, also discusses the importance of a strong guiding personality in the launching of a scientific program.

The phenomenon of specificity in nerve cells, their ability to make specific functional connections required by the "wiring diagram" of the normal nervous system, both in development and in regrowth after injury, provides another type of comparative approach, a comparison across stages of development. Paul Weiss describes certain general principles that he believes apply to all nervous-system development, and in his fascinating review of the classic reinnervation studies stemming from his own discovery of the "myotypic" response in limb regeneration in amphibia, he shows how these principles bear on future neuroscience research goals. Weiss has deliberately chosen this one path to describe, from among the many he has followed, because it is the longest continuous path with which he has been involved and because it best illustrates the unity of neuroscience even while it reflects its broad ramifications.

## Nerve Cells and Brain Circuits

Are the neurons that make up the brain separate and individual, like cells in other tissues, or is there a continuity of cytoplasm amounting to a continuous reticulum or syncytium? The history of controversy about this issue illustrates especially well the interaction between technological advances and progress in understanding. John Szentágothai describes his part in this controversy and shows how his research progressed, favored by a fertile imagination and the ready use of newly developed techniques, to the modern study of the functional organization of neural networks, as in the cerebellum. He emphasizes the importance of good concepts even when the supporting evidence may be scanty.

The functional approach to the structural anatomy of the nervous system is the central theme of Alf Brodal's research, and expresses his conviction that a detailed knowledge of structure is prerequisite for understanding the functions of the nervous system. He restates here his careerlong insistence that, since there are extremely subtle and probably unrecognized interactions and influences in the brain, attempts at model making without knowledge of the functional anatomy are futile or misleading. He and Szentágothai both hint at the need to reassess the modern neuron doctrine in the light of such phenomena as gap junctions, dendrodendritic interaction, nonspike interaction, etc.

## Membrane Excitability and Synaptic Transmission

The excitability properties of nerve cells underlie their function in communication systems capable of processing information and controlling behavior; in Kenneth Cole's opinion, "fast, reliable, and unlimited communication by axons" is one of the "most superb achievements of the nervous system." His research on the fundamental biophysics of nerve-membrane permeability has helped to clarify the complex ionic flows and molecular membrane changes that constitute the nerve implulse, or action potential, which propagates along axons to carry signals between neurons. The marvelous differentiation between sodium and potassium ions made by ionic channels in the nerve membrane is further discussed in Cole's description of some of the tribulations, triumphs, and tantalizing mysteries in this area of research.

The excitable responses of one nerve cell are communicated to other nerve cells by several different mechanisms, the best known of which is the specialized connection called a synapse. Another of the great controversies of modern neurophysiology raged over the question of whether transmission across the synapse was electrical or chemical. John Eccles, one of the earlier proponents of the electrical hypothesis, describes the intracellular recording experiments that persuaded him in 1951 to abandon the electrical in favor of the chemical hypothesis. Influenced by Karl Popper, who held that the business of science consists of creating hypotheses that are sufficiently specific to be tested, Eccles takes his success in disproving his own hypotheses as a triumph rather than a defeat. He describes how it inaugurated his further explorations of synaptic mechanisms and eventually led him to the conclusion that even a complete understanding of the synapse would not wholly solve the brain-mind problem.

The study of chemical substances liberated by presynaptic membranes has become an important pathway in neuroscience, perhaps more a highway in recent years with the greatly increased interest in finding new "certified" neurotransmitters. Ulf von Euler describes his own pioneering role, stimulated in 1926 by Loewi's experiments demonstrating neurochemical transmission. In particular, he describes his and other work on the adrenergic system that established noradrenaline as the "sympathetic" transmitter. But as for the main goal of future research, von Euler seems to turn to Cole as he suggests that further progress in basic neuroscience requires an understanding of membrane ion channels.

## Neurotransmitters and Brain Function

Further information about the neurotransmitters followed from two developments. One, described by Julius Axelrod in relation to his early experiments, was the development of the field of biochemical pharmacology, based on the notion that the physiological action of chemical agents depends upon the products created as they are metabolically transformed within the body. Using this approach, Axelrod clarified mechanisms terminating the action of neurotransmitters that have been released. He showed that noradrenaline is inactivated, not only by monoamine oxidase, but also by another enzyme, catechol-O-methyltransferase, and that, even without enzymatic action, it is inactivated by reuptake into the terminals of sympathetic nerves.

In the other approach, neurotransmitter action has been investigated at the level of the individual nerve cell. Floyd Bloom describes two technological advances that bridged the gap between ultrastructure and microphysiology. One was the introduction of multibarrel micropipettes, which release minute amounts of chemicals in the immediate vicinity of a single neuron while simultaneously recording its electrical response. This permits the direct testing of chemicals for properties qualifying them as putative neurotransmitters. The second was the introduction of the electron microscope, which, given selective stains that identify neurotransmitter substances, can demonstrate concentrations of such substances in nerve terminals and even in individual synapses.

Many new neurotransmitters have been thus identified, and brain circuits, which were unsuspected by the classical neuroanatomists, have been defined by the particular chemical that is characteristic for transmission throughout the system. These "chemically coded" circuits figure importantly in the current picture of brain function.

## Adventures with Unexpected Observations

Occasionally in scientific research an unexpected phenomenon will be observed that will become the central theme for a lifetime of research.

Berta Scharrer has followed such a leitmotiv throughout her career: the phenomenon of neurosecretion. In her paper she tells how she and her husband Ernst followed up his unexpected discovery that some neurons (in the hypothalamus) secrete, not neurotransmitters affecting other neurons, but rather hormones, discharged directly into the blood to affect distant targets. Before this concept of an endocrinelike category of neurons was scientifically accepted, the Scharrers made many lonely morphological and physiological explorations in Germany and the United States to demonstrate neurosecretory systems in vertebrates and invertebrates and to explain how these mechanisms supplement, and hold an intermediary position between, the neural and endocrine systems. Her description of the long path that led from German institutes to American universities is inspiring as well as instructive.

The name of Rita Levi-Montalcini has become identified with the phenomenon of the "nerve growth factor," an unexpected finding that stemmed from a report by Bueker, in Hamburger's laboratory, that a fast-growing tumor implanted in chick embryo was rapidly invaded by nerve fibers from the host. Subsequently, she and her colleagues in Italy, the United States, and Brazil developed research programs involving the NGF, now recognized as a substance *released* by action of the tumor, which has the remarkable property of stimulating the growth of nerve fibers and causing the proliferation of fibrous proteins, neurotubules, and neurofilaments within the neuron. Here also is the story of a valiant scientist who continued to do research despite the fact that as a Jew in wartime Italy she was in continual danger of arrest.

Frédéric Bremer describes how a particular experimental procedure, the isolation of the forebrain (the *cerveau isolé* preparation), unexpectedly focused his research goals and enabled him and others to study many aspects of the regulation of cortical excitability. When the forebrain is isolated from all inputs except those from the eyes, it manifests the functional and EEG condition of "deep sleep." It can be "aroused" by direct electrical stimulation of the internal capsule. Thus this preparation seemed to make available a path to a basic understanding of sleep and arousal. But sleep and wakefulness have turned out to be more complicated, as shown by subsequent investigation of the role of the ascending reticular formation and new methods of studying the chemistry of sleep.

**Sensory and Motor Systems**

Information about the outer world, and the inner world of the body, reaches the brain over sensory systems and interacts with ongoing brain processes, leading to outputs over the motor system that constitute the actions of an animal. William Rushton's title, "From nerves to eyes," refers to the movement of his career, from his early experiments on nerve excitation and its relation to muscle response to his later research on the photochemistry and physiology of vision. Utilizing a variety of physical measures, he demonstrated quantitative relationships between the bleaching and regeneration of visual pigments and the psychophysics of vision in normal and in so-called color-blind subjects. His tour de force career is an outstanding example of how a change of research path in midlife may very well prove to be both reinvigorating and rewarding.

The other great sensory system for perception at a distance is the auditory system, and Hallowell Davis describes his participation in the pioneering physiological investigations of the auditory receptor and the central auditory pathway. He focuses on the effects of technological advances, such as the replacement of the string galvanometer by the cathode-ray oscilloscope, and of early experimental observations, such as the discovery by Wever and Bray that the electrical responses evoked by spoken language and picked up by an electrode in the auditory nerve of a cat could be reproduced through a loudspeaker as intelligible speech. This posed a classic problem for auditory research because it meant that a neural response was occurring at a frequency greater than the 1000 Hz limit of individual auditory nerve fibers. Davis's career in auditory research was supplemented by an equally important contribution to the use of EEG in research and in clinical diagnosis.

At higher levels of neural organization, questions arise concerning mechanisms for hierarchically arranged systems of coordination and integration. How are quadripedal walking movements, that is, the coordinated responses of lower and upper limbs, organized by a spinal cord experimentally isolated from higher brain centers? How is it that the sensory receptive area of each spinal dorsal root is regulated by adjacent spinal roots even without any higher center for spinal-root coordination? Derek Denny-Brown, whose career spans physiology, neuropathology, histology, and neurology, describes experimental attacks on these kinds of questions and suggests that polysynaptic chains of short-axon cells, constituting an intercalated diffuse neuropil, store response patterns and regulate the background tone of the nervous system upon which differentiated response patterns are superimposed.

Ragnar Granit's career, moving in a direction opposite to that of Rushton, started with an early interest in psychology, visual psychophysics, and neurophysiology and moved on to studies of the motor system, with special reference to sensory feedback operating in the control of muscular action. He raises questions involving methodological and conceptual issues relevant to the different techniques for investigating the nervous system, giving particular consideration to electrophysiological recording techniques (gross and microelectrode). Do single-unit studies provide a good window into multicellular organization? He suggests that ablation studies, in which effects are observed following the removal or destruction of a part of the nervous system, may become a more powerful experimental approach through the use of improved anatomical and behavioral techniques.

## Behavior and Brain

The papers in this section are concerned with the areas of research that to most people *are* the neurosciences: behavior and brain function. Neuropsychology, correlating human behavior with its substrate brain function, tries to discover the role of different parts of the brain in higher-level psychological functions, such as cognitive processes and language.

Alexander Luria has, throughout his career, been closely associated with the development of neuropsychology, and he is especially noted for his studies of disturbances of cognitive functions in brain-injured people. He reviews the history of the field, differentiating it from other aspects of neuroscience in that it focuses on the role of whole brain systems, rather than on the analytical dissection of factors at lower levels of organization, such as molecular, biochemical, or neurophysiological subsystems. Historically, two opposing models of brain function developed from studies of the brain-injured: one postulated that complex functions such as language are localized in highly specialized "centers" in the brain, while the other stressed that psychological functions depend more on the total amount of undamaged brain tissue available for "mass action." Luria traces the development of the new concept that higher psychological functions are produced by the interaction of many different brain systems and that, after damage to neural systems, the end result (goal) of a function such as locomotion may be achieved in a new way involving new cooperative interactions among the available brain systems.

Physiological psychology attacks the same problems of neuropsychology with experimental approaches that use the data and concepts of the neurological disciplines but emphasize animal experiments rather than human observations. Eliot Stellar is an exemplar of this field, having been involved with its first great growth period in the forties. He describes his own interdisciplinary research on cognitive functions and the classic affective functions—instinct, emotion, motivation, drive—and points out that the goal of understanding the brain's role in subjective experience may be much closer than we realize, especially in relation to hedonic experience.

Working at the interface between classic learning-theory psychology and physiological psychology, James Olds developed an experimental approach in which the animal delivers electrical stimulation to its own brain by pressing a lever. He describes how the mapping of "reward" and "punishment" centers in the brain has led to a wide range of behavioral, anatomical, electrophysiological, and pharmacological analyses of the neural mechanisms mediating motivational and drive systems. He also discusses his current research program, in which he hopes to trace each stage of the learning process through the brain from incoming sensation to outgoing response.

## The Brain: Neural Object and Conscious Subject

Herbert Jasper reviews electrophysiological approaches to the study of complex brain systems and the conscious phenomena related to them. He starts by describing the early history of the scalp-recorded electroencephalogram, including the discoveries of how it reflects states of consciousness, alterations of brain metabolism, and pathological conditions, and its early application in clinical diagnosis. The technological advance from scalp electrodes to implanted electrodes in unanesthetized animals and man to the microelectrode analysis of individual neuronal responses, provided the basis for the development of conceptual models of the neural mechanisms regulating sleep and wakefulness, and even of such complex human functions as speech, perception, and memory. Jasper also describes his own work on functional specificity in local neuron assemblies and the discovery of thalamic neurons that detect "novelty." He then discusses two competing models of brain function: the mosaic concept, emphasizing the specific connectivity of neuron assemblies and their role in complex functions, and the more dynamic view that neural response patterns are variable and can be modulated by neuro-behavioral states of the organism. This modulation of specifically connected systems is mediated especially by chemically coded neural systems, transmission in which depends upon a specific neurotransmitter substance.

Wilder Penfield was a pioneer in the use of electrical stimulation and recording to explore the waking brains of neurosurgical patients. He gives an overview of the conclusions he has derived from neurosurgical observations on over 1000 human patients, stressing that in electrical stimulation of the brain, as well as in epileptic seizures, the local effect is transmitted over nerve pathways to activate distantly located gray matter. For example, local epileptic discharge in motor or sensory cortex can spread to automatic sensorimotor mechanisms in higher brainstem centers and culminate in a major epileptic convulsion. In contrast, local seizure activity in prefrontal or temporal gray matter can spread to a neural mechanism in the brainstem, leading to the automatonlike behavior manifested by patients with petit mal epilepsy. These two types of epileptic spread suggest a functional interaction between brain mechanisms serving consciousness and those serving automatic sensorimotor mechanisms coordinating behavioral output. Penfield conceives of three separate brain mechanisms, one being nothing less than the neural basis of conscious experience, another being the mechanism for automatic sensorimotor coordination, and the third being that for recall of past experience.

Roger Sperry develops a different conceptual model of the nervous system. His career as an experimentalist began with relatively "simple" problems concerning the controversy over whether brain function should be conceptualized in terms of equipotentiality and plasticity or in terms of "hard-wired" connectionistic specificity. One of the strongest arguments for the equipotentiality position was the apparent fact that cutting the corpus callosum, a massive set of fibers connecting the two cerebral hemispheres, failed to produce any disturbance in human mental or behavioral function. In a series of brilliant experiments, first in animals then in man, Sperry established that profound disturbances of function do exist in such patients with a "split brain," but that sophisticated tests are necessary to reveal them. Studies of animals and humans with brain bisections led Sperry to conclude that consciousness is a special and selectively localized property of the brain, emergent from the activities of specific brain systems, and that it is not an epiphenomenon, but acts as a causal determinant upon brain mechanisms. Mentalistic phenomena such as beliefs and value systems are thus drawn into the world of scientific causality, bringing science to grips with problems hitherto left to the domains of humanistic and philosophical studies.

**The Neuroscience Community: People and Ideas**

Many chapters in this book give highly personal accounts, but Ralph Gerard and Richard Jung provide a special view of persons who are scientists and of the influence on them cf family, friends, colleagues, students, and organizational settings.

Ralph Gerard, one of the acknowledged deans of American neuroscience and honorary president of the Society for Neuroscience, died shortly after he participated in this symposium. This paper is thus his last publication, and its heavily autobiographical nature should make it especially valuable to historians of science. His long career followed many diverse and branching pathways of neuroscience research, and his intimate account thus reveals interactions with many of the outstanding neuroscientists of the past four or more decades. Many of them were friends as well as scientific collaborators, and at the symposium he renewed many old acquaintances. The message in his paper is one that he exemplified in his own life: the neuroscientist, to be truly effective, should be a specialist in the minute experiment; he should be an expert in one of the neurosciences, but at the same time he should try to see and understand and be guided by "the large picture." Gerard's abounding talents and curiosity brought him into close involvement with a broad range of methods and techniques involving not simply the microelectrode, for which he is best known, but also neurochemical, neurophysiological, biophysical, and metabolic approaches and methods. He was one of the small group of pioneers who created neuroscience.

Richard Jung writes about the many European scientists whose teaching and influence made him one of Germany's leading neuroscientists. He focuses on the years 1930–1953, and he emphasizes the international interaction which, despite the war, was critical in forming his own scientific consciousness. His work has always had a base in human clinical work, and he points out that pure science may well be meaningless unless it has some relevance to real human problems. He suggests that interaction with scientists from other traditions and with other points of view, including philosophy, is an essential component of creativity. He also suggests that even failure can be valuable, and in a fascinating section he describes two research failures in his laboratory, both coming in part from an overscrupulous insistence on quantitative scientific vigor and adherence to theoretical convention.

**The Neuroscience Community: Some Organizational Developments**

The two papers in this section discuss specific organized research "institutes" in neuroscience and their influence on the development of the field. Horace Magoun, who was a founder of the Brain Research Institute of UCLA, reviews the work and influence of Ranson's Institute of Neurology at Northwestern, a research organization in the great tradition of the nineteenth-century German university institutes. Ranson's institute flourished under his direction from 1928 to 1942, and from it flowed a wealth of research findings and fundamental discoveries, especially involving the use of stereotaxic apparatus, that have continued to influence neuroscience research to the present day. But, as Magoun points out, behind the productivity of such institutes as that of Ranson, and of Bailey at the University of Illinois, there must be a catalytic individual whose highly directed genius and dynamism provides the motive power to shape the organization.

Judith Swazey provides a science historian's expertise in her review of the history of the Neurosciences Research Program, the "invisible organized research unit" (in Magoun's terms) that was founded and, through its first twelve years, directed by Frank Schmitt, the exemplar of the catalytic "impresario of science" (in Edelman's terms). Her paper provides an appropriate tribute to Schmitt, whose efforts in creating an interdisciplinary institute also helped to create an interdisciplinary neuroscience. Her carefully documented short history not only provides the NRP with its first objective look backward but also gives a strong reminder of the level of effort that will be required to keep the program moving forward.

## On the Nature of Research in Neuroscience

This last section consists of two papers that consider the nature of the processes and interactions that make up neuroscience research.

Paul Dell considers procedural and philosophical themes also touched upon by many of the other participants: the importance of interaction with people and ideas, the unexpected nature of some findings, the difference between discovery and invention, etc. He formulates an encompassing rubric, "creative dialogues," describing the influence on the scientist's mind of his "dialogues" with the external environment, with the sociocultural environment, and with the object of his research activities. He illustrates this with a review of research on neural substrates of sleep and wakefulness. The use of the *cerveau isolé* preparation, the new understanding of the role of the ascending reticular formation and of the monoamine systems in the sleep-waking cycle, and the discovery of the phenomenon of REM (rapid eye movement) and slow-wave sleep all show how disparate lines of research can come together to deepen understanding of a scientific problem.

Finally, Swazey and Worden look at neuroscience through the window of the ideas of Thomas Kuhn, whose influential and controversial book *The Structure of Scientific Revolutions* has helped to clarify the nature of scientific research. From an examination largely of the history of the physical sciences, Kuhn developed a model of research that conflicts with the traditional view of science as the construction of an edifice of truth through the incremental accumulation of knowledge gained by testing hypotheses against reality. Swazey and Worden apply Kuhnian concepts to some of the examples of neuroscience research described in this volume and discuss the special difficulties that arise from the multidisciplinary nature of the field.

# The F.O. Schmitt Lecture

John Z. Young (b. 1907, Bristol, England) is professor emeritus of anatomy at University College, London. His studies have ranged widely in the neurosciences, from peripheral to central nervous systems in both vertebrates and invertebrates and from electrical transmission in single nerves to the memory systems of the entire brain. He was awarded the 1973 Francis O. Schmitt Medal and Prize in Neuroscience, with the following citation:

John Zachary Young has provided generations of students and researchers with both living material and conceptual propositions at levels from the single nerve fiber to memory. Prolific generalist, exacting microscopist, literate essayist, imaginative experimentalist, he is not above using common sense, underlining qualitative and subjective science, or posing unanswerable questions. He introduced the squid giant axon to physiology, pioneered with his co-workers in measuring intracellular potassium concentration, birefringence of axon sheaths (now known to be stacked cell membrane), the dependence of axon velocity on diameter, and the regeneration of nerves, and in localizing the separable assets of learning and of memory within the brain of an advanced invertebrate. Perpetual optimist while emphasizing the complexities, he constantly spurs us on by his zest and his talent.

# 2
# Sources of Discovery in Neuroscience

## John Z. Young

### The Aims and Methods of Neuroscience

Frank Schmitt has been associated with so many aspects of research on the nervous system that it may be appropriate to do honour to him by trying to look at what has been discovered in neuroscience over the last forty years. This is a vast and indeed impossible task, and I shall make it even larger by trying to examine the general framework within which our science has grown. What have we been trying to do as we researched into the nervous system? What motives led each of us to study this part of the body? Is our work an academic exercise (whatever that may be), or is it to be thought of as an aid to humanity, say in medicine or education? What determines the methods that we use? Clearly the social milieu has an influence on each of us. Are we then slaves of bourgeois capitalism, or marxist or maoist socialism, or of some other ism? Does the system we belong to influence the research that we do?

We should also ask what logical methods have been used. Are there differences in the neuroscience of empiricists, holists or existentialists? Do we proceed by induction from many instances? Have we been trying to disconfirm hypotheses, as Popper believes scientists do (1972)? Have new paradigms for normal science been established, as Kuhn would have it (1962)? If I had to answer such questions, I should say that it is a naive oversimplification to suppose that we know enough to make any general statement about how biological science evolves. It uses many of the subtle operations that are adopted by the human brain and which as yet we hardly begin to understand. Our job as neuroscientists is to try to find ways to specify what these operations are. Inevitably we make our observations and specifications in terms of the technology available to us and the hypotheses and languages that go with it.

Speaking of technology may remind us that a special capacity of the human brain is that it enables us to discuss how to make tools to help with the work of our senses and of our hands. Having devised these tools we use them to study the body itself—say by dissection, by chemistry or by electrical recording. Then we apply the principles of the machines we have made to the biological facts we have found with them and so provide for explanation and control of the body's own working. We may therefore gain insight into the development of neuroscience in recent years by noting how new technological developments have influenced the experiments that have been made and the explanatory models that have been developed from the results.

Whatever may be the method by which science advances, it is as a product of human brains. Evidence now begins to suggest that at least some of the special capacities of the brain are based upon inherited organization: for instance, the capacities for social interaction and ethical behaviour, language, logical induction, perhaps the making of anthropomorphic comparisons and possibly even religious observance. No one can pretend that we yet understand much about this, but I shall follow the clue that the human brain and memory are specialized for certain functions appropriate to life in society. As the bee's brain is organized for social action to collect honey, or the lion's or the wolf's to catch prey, so ours is organized to enable men to get a living by social life, in ever more diverse ways.

The historian of neuroscience therefore has a doubly difficult task. Not only must he look at the discoveries that have been made about the brain, but he must also look behind the scenes at the way the human brain investigates the human brain. Here we are in deep waters indeed and shall inevitably be swept around in many philosophical and linguistic whirlpools. Let us not be alarmed by this, but humbly proud that perhaps our science is indeed in process of competing with theology as the Queen of Sciences, the basis for all the rest.

## Some Types of Scientific Thinkers

In looking at neuroscience I think we shall find that it has proceeded mainly in the tradition of what may be called Anglo-American logical empiricism. We like to operate with clear definitions. We have what has been called by Jameson (1972) "a stubborn will to isolate the object in question whether it be a material thing, an event or a word." This tendency is often traced back to John Locke, and Jameson comments that it has a social and political background. It allows "a turning away of the eyes and avoiding observation of those larger wholes and totalities which would force the mind into uncomfortable social and political conclusions."

Has neuroscience been guilty of such turning away in the pursuit of isolable entities, even of minutiae? Where do we stand in the current tendency to depart from analysis and take to what existentialists believe to be a more truly synthetic view of the whole man? You may feel uncomfortable at the very thought of such an enquiry and I share your alarm. But perhaps there may be something for us in such a view. If we are to have a model of the whole brain at work, we may need new, large concepts. Their value may be hinted at in a parallel with linguistics. De Saussure long ago (1916) contrasted words with language *(la parole et le langage)*: words can be defined, but language, like society or an engram in the brain, is a different sort of entity; it is nowhere present all at once. Such entities do not take the form of objects or substances, and yet their presence is felt in every moment of our thought or our speech. Perhaps we need such more general concepts in trying to speak about the brain, and we may find more help than we at present expect from the ideas of thinkers such as Sartre and Levi-Strauss. But let us not forget that we are planning also to look behind the scenes, trying to discover ways to describe how the brains of these more romantic or holistic workers function, as well as our own presumedly more analytical ones.

## Saint Louis, 1935

How then can we see our scientific lifetimes, those of Frank Schmitt, myself and the rest of you, in this framework? I must ask your indulgence as I try to do my best and tell you of some of the landmarks that we have passed in the last forty years. To avoid covering only old ground I shall try to show how some of these phases have led to modern work, illustrating partly from our own laboratories. Fortunately there is also a much better way to form a conspectus of neuroscience over recent years, by using the *NRP Bulletins*. They have been so cleverly devised and guided by Frank and others that they provide a means of following progress in neuroscience in a manner that I have not seen equalled for any other part of science.

A good place for us to begin, however, might be long before the days of neuroscience or NRP, when Frank and I first met in the spring of 1935 in Saint Louis. I remember very well the impression of entering that laboratory, where Frank was stimulating a frog's nerve repetitively to study its metabolism. What struck me was that the room was totally silent and quite uncluttered by wires. I came from Oxford, where in Sherrington's laboratory Denny-Brown, Eccles, Granit and others were beginning their careers (for descriptions in depth see Denny-Brown, 1966; Creed et al., 1932; Granit, 1972). The equipment that they used was of course wonderfully effective for its time, but it was not silent, and each cubicle was a maze of wires so dense that you had to stoop to enter. The electronic revolution had not reached Oxford, and stimulation was by means of pendulums or rotating drums, which made a tremendous noise.

Those were great days in Saint Louis. Herbert Gasser and Joseph Erlanger were the laboratory chiefs, and very amiable they were to a young man in their quite different ways. Their book marked a great step toward an understanding that the nervous system operates as a set of channels, each highly differentiated to carry a distinct item of information in the form of all-or-nothing signals (Erlanger and Gasser, 1937). This conception had of course long been developing, from Müller's "law of specific nerve energies" (Müller, 1848) on to the work of Keith Lucas (1909, 1912), Adrian (1914) and others. But the use of cathode-ray amplifiers and oscilloscopes to demonstrate the distinct properties of fibres of different diameter really began the modern analysis of the diversity of pathways. The presence of many separate lines of transmission is of course distinctive of nervous functioning, a method of working very different from the means of communication that the engineer uses when he passes many messages along a single channel. I found this emphasis by Gasser and Erlanger on the variety of nerve fibres fascinating, and at the same time wondered whether the classes of fibre revealed as a function of diameter (A, B and C, etc.) were really exclusive and distinct. They have stood up well, however, even until today. Perhaps too well, a critic might say; it may be time for an advance to a more rational terminology.

At Saint Louis at that time there were many others who have made neuroscience what it is today. Besides Frank and his brother Otto, there were George Bishop, Jim O'Leary, Peter Heinbecker and of course Lorente de Nó. Being interested in the sympathetic system I made a special pilgrimage to see Kuntz at the Saint Louis University. I considered that his book (Kuntz, 1934) had made the best synthesis about the autonomic nervous system since Langley (1903), and I was surprised to find that he was little regarded by the pundits at Washington University. He was largely a histologist and did not use the new electrophysiological methods.

## First Studies of Squid Giant Axons

That summer a group of us went to Cold Spring Harbor and Woods Hole for what was the main point of my visit to the States: to show that the giant nerve fibres I had found in the squid really were nerve fibres and could be of use to physiologists. We carried squids across Long Island in milk cans, but made little progress until we went on to Woods Hole. Several groups there quickly showed what could be done with these wonderful fibres. The proof that they are nerve fibres was done quite simply, with a pin. Stimulate near the ganglion and watch the muscle twitch. Prick the fibre and the twitch stops. Stimulate again *below* the prick and you've proved your point (Young, 1938b).

At first we had less luck with action potentials. Det Bronk, Ralph Gerard and Keffer Hartline all helped me with this, but every time we passed a stimulus to the nerve the oscillograph trace went sailing away, and there was no clear evidence of an action potential. So one day when Det and Ralph were out I said to Keffer, "Why don't we do away with all this equipment and stimulate it with sodium citrate?" So we hooked a fibre up to an amplifier and loudspeaker and put some citrate on the end. Out came "buzzzzzzzzzz"—one of the best sounds I have ever heard (Figure 2.1).

But why all this interest in giant fibres? It had become clear already from Lucas and Adrian and many others that the signal in the nervous system is an all-or-nothing event. Following the empiricist tradition of isolation physiologists in-

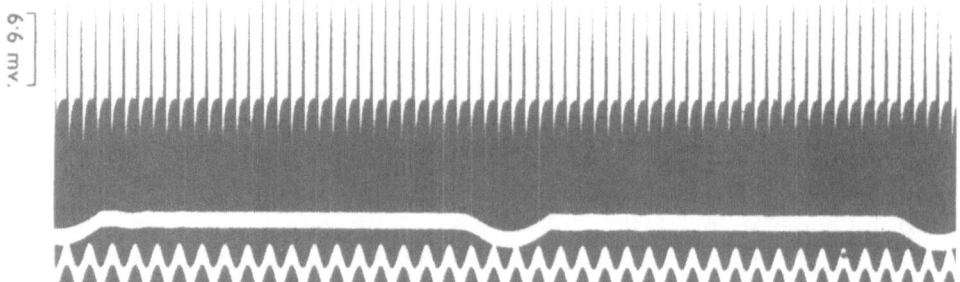

Figure 2.1   One of the first records of action-potential discharge in a giant fibre of a squid, after the application of a solution of sodium citrate (1%) to one end. The time signals are 10 and 200 msec. From Young (1944).

deed felt they had hit upon a true unit of function, and they wanted to study it hard. For this single fibres are needed, and they were first studied by Adrian and Bronk in 1928. But they are difficult to produce by dissection, whether of vertebrate or invertebrate nerves, so we were all looking for big fibres.

Incidentally, at Oxford Eccles and Granit and I had been exploring the possibility of using the giant fibres of the earthworm, and we published a short note on their action potentials in 1932. I did the dissection, Eccles the recording, while Granit sat in a deck chair. We were not quite sure of his function. Perhaps he was deciding what logical methods we were to use, though I doubt whether scientists really proceed in the way that philosophers of science seem to suppose. It is a banal truism that all scientific workers operate with some hypotheses, but this alone does not adequately describe the motivation or process of their activities. Eccles, Granit and I were certainly not doing the work on earthworms to try to disprove the hypothesis that nerve fibres conduct. We were groping our way, trying to find new material for study. Disparagers can say this is not science, but we three seem to have been moderately successful scientists.

Looking behind the scenes we may ask what it is about the human brain that makes people search for new phenomena or new ideas. This may be the most important scientific activity of all. It is very relevant to ask how new things are found, and surely it is not simply by trying to disprove old hypotheses. The faculty of curiosity or enquiry can be traced back to our half-human ancestors. It involves of course the observation that a situation is unfamiliar and the hypothesis that it is worth investigating. Incidentally, there is evidence that the hippocampus is involved in this curiosity (see below). So perhaps the brain has built-in systems for enquiring and also for forming maps, models or hypotheses incorporating the information it obtains. All these capacities for curiosity, as well as hypothesis, are essential sources of new discovery.

Several contributors to this symposium mention the importance of chance or accident in scientific discovery. The finding of the giant nerve fibres of the squid was itself due to a combination of accident, curiosity and an interest in nervous communication. At the hind end of the stellate ganglion of an octopus there is a yellow spot (Young, 1929). There were no hypotheses about it at all. Moved by simple curiosity I cut sections to see what it was and found a vesicle in which ended processes arising from what seemed to be nerve cells (Figure 2. 2). I thought it was a sort of adrenal gland, and Ernst Scharrer suggested it as one of the first neurosecretory organs. We were both wrong, for thirty years later Howard Bern and his colleagues showed that the processes entering this cavity are rhabdomes and it is a sort of internal eye (Nishioka, Hagadorn and Bern, 1962). Looking in squids for this epistellar body, as we had called it, I found instead the giant nerve fibres, though for several years I was foolish enough to think that they were veins and neglected to examine them closely (Young, 1936a).

But to return to Woods Hole (Young, 1936b), it was in structural and biophysical studies of the giant fibres with Frank Schmitt and Richard Bear that we made in some ways the best progress (Bear, Schmitt and Young, 1937a,b,c). For

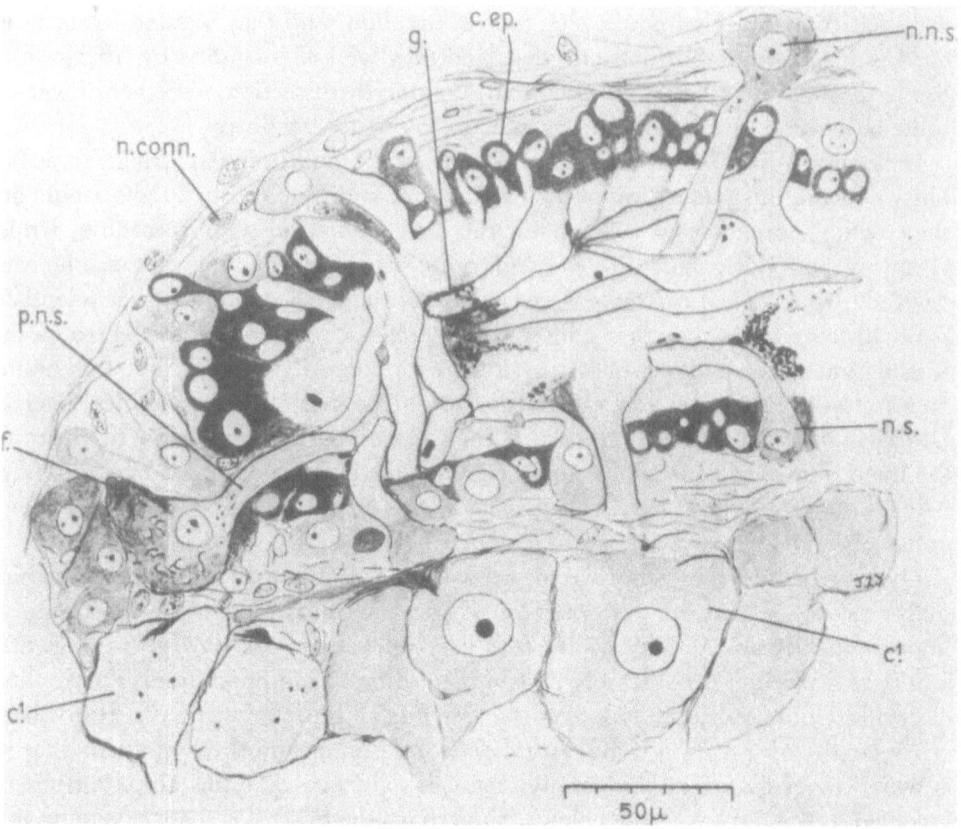

Figure 2.2   Drawing of a sagittal section of the epistellar body of the octopod *Ocythöe*: C¹.,
neurons; c.ep., epithelial cells; f., connective-tissue fibrils; g., osmiophilic granules; n.s., "neurosecretory cell" (now considered the cell body of the photoreceptive cell); n.conn., nucleus of connective tissue; n.n.s., nucleus of "neurosecretory cell"; p.n.s., process of "neurosecretory cell"
(now known to be a rhabdome). Fixed in Flemming's fluid and stained with iron haematoxylin.
From Young (1936a).

me it was a revelation of what a biophysical approach could mean. We dissected
out single fibres and soaked them in fluids of different refractive index (Figure
2.3). I learned about metatropic sheaths and x-ray diffraction and all the mysteries of fine structure, which I admit our crude morphology at Oxford had passed
by.

Before we go on to other themes, let us think for a moment what has sprung
from this introduction by Frank and others of biophysical concepts to the study of
nerve structure. A great part of what we now know about the lamellar structure
of myelin comes from his work with Richard Bear in those early days (see Mokrasch, Bear and Schmitt, 1971). This provided the background for the interpretation of early electron micrographs of myelin by Sjöstrand (1953) and Dave Robertson (1955) and for the revelation of its development from the Schwann-cell
surface by Betty Geren Uzman, who was of course also a pupil of Frank Schmitt
(Geren, 1954). From this, in time, it became apparent that in studying myelin

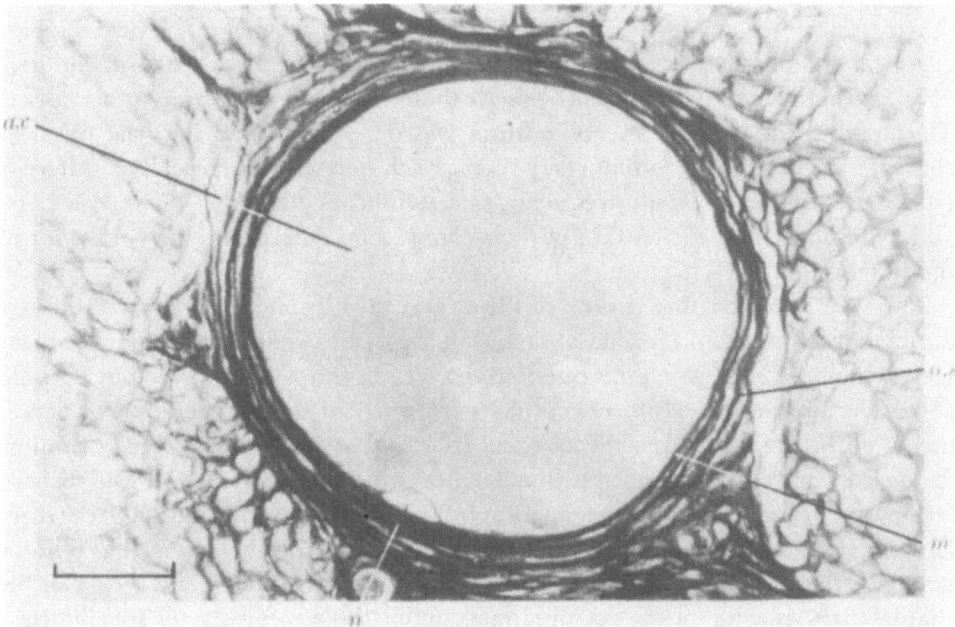

Figure 2.3   Transverse section of the hindmost stellar nerve of *Loligo pealeii* in the region of the
giant fibre: ax, axoplasm; m, region of metatropic reaction; n, one of the nuclei; o.s., outer sheath.
The scale represents 50 $\mu$m. Fixed in Flemming's fluid and stained with Mallory's stain. From
Bear, Schmitt and Young (1937a).

one is studying a model of the cell surface, and indeed, as Dave Robertson would
say, a model of a unit membrane that is typical of all surfaces. This is therefore a
concept that permeates the whole of biology, and much of it springs from the
biology department at MIT.

Squid axons did indeed prove valuable for an enormous variety of studies. The
concepts of the structure of their axoplasm and sheaths were basic to much of the
work on the nature of the action potential, culminating in the Hodgkin-Huxley
equations (1952) and the sodium pump. Some characteristics of the membrane,
such as the impedance changes, were determined even in those early days at
Woods Hole by K.S. Cole and H.J. Curtis (1936; see Cole, 1968). A little later
Webb and I (1940) used these fibres to make what I believe were the first direct
determinations of the internal potassium content of a nerve cell. Pumphrey and I
(1938) also showed that conduction velocity follows approximately the square
root of the diameter, as was predicted by the equations of Frank Offner and his
colleagues (1940) and later by Rushton (1951) and Hodgkin (1954).

**Development of Connections in the Nervous System**

So in this phase of neuroscience many very precise biophysical facts were established by the application to suitably chosen material of the new technologies of
electronics, electron microscopy, x-ray diffraction, microchemical estimation and

others. These technologies enabled investigators to provide a picture in biochemical and physical terms of the functioning of the peripheral parts of the nervous system that was perhaps more precise than was available for any other tissue.
The nature of the signals in nerve fibres was determined and also the physical
characteristics of the fibres that carry them. Each fibre carries basically a different
type of information, though frequency is a significant variable. There is a huge
variety of fibres, and the pattern of their connections determines the behaviour of
the system.

The way in which this variety of fibres and their connections are established
during development and regeneration has therefore been one of the major problem
areas of neuroscience. With this question too, Frank and I have played our humble
parts. During the war I was concerned with study of the regeneration of nerve,
then unfortunately a topic of all too great concern to surgeons. The direct stimulus
for me to work on this topic was thus medical, and much detail was established,
for instance about rates of regeneration and conditions promoting recovery.
However, the really profound discoveries in this field came from work that did not
have, so far as I know, a direct social or medical motivation. Paul Weiss, in his
chapter, tells of some of the startling facts about the appearance of specific connections in the regeneration of nerves of amphibia. Roger Sperry (1944) awoke
us to the strength of the factors that somehow lead nerve fibres to remake connections after injury. We have recently been able to show one example of this
capacity by interrupting the nerves to the skin of the back of an octopus (Figure
2.4; see Sanders and Young, 1974). After about two months the full range of
display of colour patterns by the chromatophores returns. This must surely mean
that correct individual connections have been remade between the nerves and the

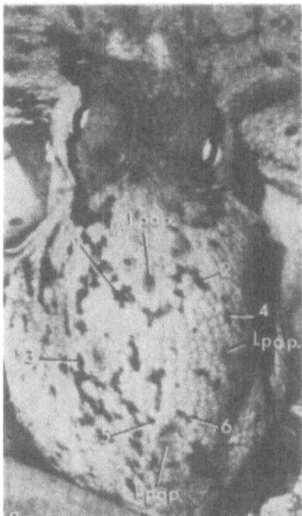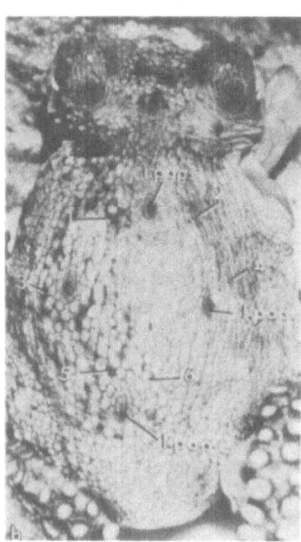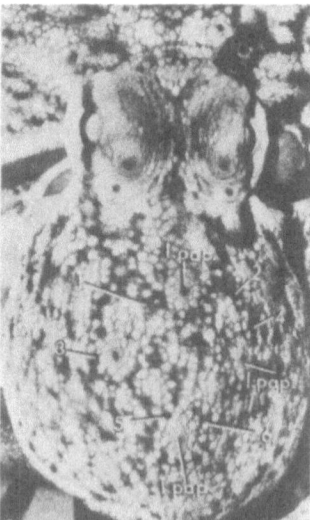

Figure 2.4 *Octopus vulgaris* (a) before the pallial nerve was crushed, (b) immediately after the
operation, and (c) 62 days after the operation. The numbers indicate identical spots on the skin.
The even-numbered spots on the right disappear immediately after the operation, but reappear
again later. From Sanders and Young (1974).

muscles of the chromatophores, so that the colour cells of a given region show light or dark in the many combinations that are indicated by the commands from the brain centres. This is an example of the fact that the very precise connections made during development can in some animals be remade after injury (Gaze, 1970; Gaze, Keating and Chung, 1974). It would be very hard for an historian to trace the sequence of natural history, curiosity, hypothesis and experiment by which this knowledge has been produced. Even now we have only guesses as to how the fibres are guided back to the old connections, but I venture the forecast that over the next decades we shall discover the nature of the labelling by which nerve fibres recognize each other and how each learns its proper name. Levinthal and his colleagues have recently shown how a so-called lead axon in a developing water flea carries a terminal bulb, which probes for recognition. When it has found its mark, other fibres in the bundle follow (Lopresti, Macagno and Levinthal, 1973).

Whatever the nature of these relations proves to be, it is clear that the maintenance of the integrity of the nerve fibre and of its connections must depend upon some form of communication up from the periphery as well as down from the cell. The first nerve fibres formed, whether in development or regeneration, are very thin. They grow to their proper size (and their proper conduction velocity, etc.) under the influence of their connections at the periphery as well as the centre (Aitken, Sharman and Young, 1947). The most obvious means of communication both down and up is by actual axoplasmic flow, which we now know indeed occurs, probably in both directions. Frank Schmitt has suggested a physical mechanism for it on a basis of transport along neurofilaments or neurotubules by an actomyosin type of system (Schmitt, 1967).

## Problems of the Brain: Method and Model

So far we have been dealing mainly with the peripheral nervous system. But of course it is in the connections between nerve cells, and especially in the brain, that we have to look for the aspects of neuroscience that interest us most. In the brain the problems are vastly more difficult. For example, the mechanism of the sodium pump of a peripheral nerve fibre can be regarded as a physical problem, but it is difficult to know even what is meant by asking to understand the cerebral cortex. Yet the problems of the brain are so challenging intellectually, and so demanding of human attention for their social and medical importance, that we cannot afford to say that they are too difficult for us to attack at present. We *must* have some model for talking about the central nervous system, unless we are to be content to ask our children, our students and our patients to be satisfied with the age-old explanations of those whose brain models operate with ideas about spirits, even if they are dressed up as ids, egos and superegos.

The problem is how to formulate an adequate model. Whereas physical and chemical models will allow some progress in attacks on systems of relatively simple units such as synapses, they fail when we try to think about the operation of whole sections of the nervous system, say the hippocampus, cerebellum or cerebral cortex. Have we given enough attention to this question of what we hope to achieve when we study higher nervous centres? The experimenter must often proceed with the vaguest of probes, stimulating this part of the brain or excising that, with no more hypothesis than a belief that all parts of it contribute something to the behaviour of the animal or man.

## Simpler Models

It would be impossible within this paper to follow the sequence of discovery about how nerve cells stimulate each other. I will call attention only to some aspects bearing on our general problem of method. There has been a marked tendency to concentrate upon study of relatively simple parts of the central nervous system and to try to isolate units both of structure and of function. Sherrington himself re ognized the need to identify "reflexes," though he was aware of the dangers of isolation (Sherrington, 1906). He confined himself mainly, but not of course absolutely, to the spinal cord. Pavlov (1927), working with the brain, showed much the same tendencies to simplification. Notice also the limitations imposed upon both of these great workers by the technology then available, which was mainly mechanical recordings of movements or of droplets of secretion. As electronics developed, more subtle investigations became possible. Attention went first to the synapses of relatively simple units such as muscle end-plates (Katz, 1966), to sympathetic ganglia, to the spinal cord (Eccles, 1953, and many others) or to invertebrate preparations (Wiersma, 1952; Bullock, 1948). It is from such admittedly simple situations that our knowledge of the fundamentals of synaptic action has been derived. Still, today our understanding of the various ways that the activities of unit cells combine is expanding largely from study of simple ganglia of crustacea, insects and mollusks, as by Miledi (1967), Kennedy, Selverston and Remler (1969), Horridge (1962), Tauc (1967), Kandel and Kupfermann (1970), Hoyle (1970), Kerkut (1969) and many others.

Investigations of the electrical changes occurring during synaptic stimulation have given us a rather complete picture of the processes of summation that form the base of the decision whether or not a neuron shall fire. Sherrington (1906) could only postulate hypothetical central excitatory and inhibitory states. His successors have been able to measure these directly as physical parameters, excitatory and inhibitory postsynaptic potentials.

The model of the assemblies of single neurons as on-off units has of course become the theme of an immense amount of experimental and theoretical investigation. Much of this originated at MIT under the influence of Warren McCulloch, Walter Pitts and others. Indeed, in some sense the whole development of cybernetics has flowed from here through Rosenblueth and Norbert Wiener (1948). It is interesting also to follow this thread back to more conventional physiology, as Drs. Magoun and H. Davis remind us in their chapters, through Cannon (1932) and his concept of homeostasis, Henderson (1913) of Harvard and back to Claude Bernard (1878–1879) and his concept of the internal environment.

In spite of all this work the neuron model is far from complete and has not yet succeeded in giving us a really clear basis for understanding higher nervous functions. Perhaps these facts may be connected. The classical neuron model does not properly allow for interactions between the dendrites of one cell or between those of different cells. There is no doubt that action potentials can be set up in the dendrites of some cells, for instance those of the hippocampus (Spencer and Kandel, 1961). It may be that such impulses propagate past branching points only if they arrive in certain temporal patterns. Wall (1965) has provided evidence that natural stimulation of the central part of the skin field that activates a spinal-cord neuron produces large potentials, whereas small ones, presumably in dendrites, are set up from the periphery of the field. Certainly the firing pattern of a neuron can be influenced at various points along its length, so that it is not a simple on-off unit but has computational functions as a cascaded array of filters (see Waxman, 1974).

## Microneurons

Furthermore, the classical neuron model, implying an all-or-nothing on-off function, takes no account at all of the actions of the cells that are most numerous in many parts of nervous systems, the small cells, microneurons or amacrine cells. Many of these have no axons, and, as Dowling (1970) has shown, in the retina they produce only graded potentials, often hyperpolarizing.

The highest lobes of the brain of an octopus contain some 25 millions of such minute cells, with axons limited to the lobe (Figure 2.5). Boycott and I found that animals deprived of this part of the brain show obvious defects only when one tries to teach them (Boycott and Young, 1957). Then they have great difficulty in learning, especially not to attack objects that have yielded pain. Such quite simple techniques of excision followed by study of behaviour can tell us a lot about how brains work, but of course not nearly enough.

How do these small cells produce this effect? Presumably chemical interactions are involved, and the amacrine cells are packed with synaptic vesicles and show serial synapses (Gray and Young, 1968). It may be that this large mass of small cells serves as a sort of reservoir of inhibition, which may well be necessary for the functioning of a learning system. If neurons that have received particular sets of stimuli are to exert control over behaviour, it is important that the others be restrained. It must be significant that such neural components with very many small cells are found in both the visual and tactile learning systems of the octopus and also in the higher nervous centres of the more complex annelids and arthropods. In any system using the multichannel principle of coding there must be special devices to ensure that the relevant channels have control, changing from moment to moment. The great mass of small cells in brains probably serve this function, perhaps largely by inhibition. We badly need more information about them, but progress is held back because it is difficult to record from the small cells electrically.

This is a very clear case where technical considerations are polarizing investigation into the properties of large cells and leading us to forget that the small ones

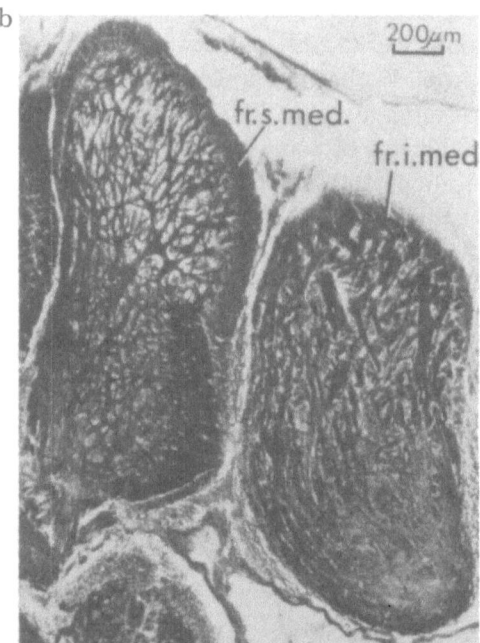

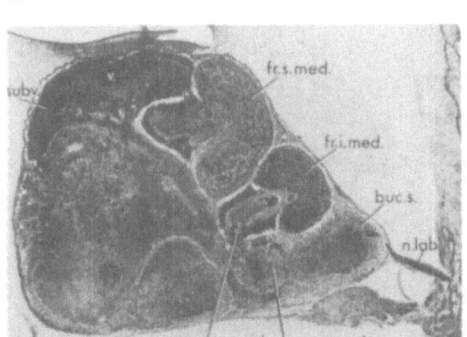

Figure 2.5    (a) Sagittal section of the supra-oesophageal lobes of the nervous system of *Octopus vulgaris* (Cajal stain): buc.p., posterior buccal lobe; buc.s., superior buccal lobe; fr.i.med., median inferior frontal lobe; fr.s.med., median superior frontal lobe; n.lab., labial nerve; subfr., subfrontal lobe; subv., subvertical lobe; v., vertical lobe. (b) Sagittal section of the median inferior frontal (fr.i.med.) and the median superior frontal (fr.s.med.) lobes of *Octopus vulgaris* (Cajal stain) to show the branching and interweaving fibres responsible for generalization.

may be equally important. If current methods do not permit study of them, surely we should look around for new methods. Once again this suggests the importance of curiosity and probing for novelty rather than either proof or disproof of existing hypotheses.

## Chemical Signalling

Certainly there are communications at work in the nervous system other than by all-or-none signals. And this brings me to a topic I should have introduced long ago, namely, chemical signalling, discussed elsewhere in this symposium so beautifully by Drs. von Euler, Scharrer, Bloom and others. Concepts involving humours have been popular for centuries, but it is only with the developments of analytic and synthetic chemistry, centrifugation and other methods in recent decades that they have become precise and respectable. The micromethods and immunological techniques now becoming available allow a spatial chemical resolution that is revolutionizing knowledge of the detailed distribution of active substances around neurons (see Schmitt and Samson, 1969).

It is interesting that in the twenties and thirties the chemical and pharmacological approach to the nervous system was quite strongly resisted. Many people, certainly in Oxford, felt that the electrical properties of nerve told most or all that we needed to know about synaptic transmission, and there was distinct unwilling-ness (to say the least) by some people to admit that synaptic transmission was chemical. In his chapter Eccles describes the controversy in detail, including his own unwillingness to accept chemical transmission and how he finally had to give up the electrical concept. This is a good example of how a climate of opinion or fashion of thought, based on previous technical advances, can influence progress. The very fact of increasing capacity to understand the electrical properties of nerve inhibited consideration of the chemical properties. As Henry Dale put it when Eccles in 1952 announced his conversion, "his acceptance now of the principle of such transmission for central synapses is the more significant in that, even a few years ago, he was still most tenaciously sceptical of cholinergic transmission at the synapses in autonomic ganglia, where to many of us the evidence for its occurrence had appeared to be very strong, and much more direct than any to be deduced from a change of electrical polarization" (Dale, 1953).

In studies of the microstructure and micropharmacology of the nervous system, chemistry and histology are now converging, as Bloom shows in his chapter. We can now recognize the sites of cholinesterases and monoamine transmitters with light and perhaps electron microscopy (Wurtman, 1971). Study of the "morphol-ogy" of the synapse has long been controversial, first about whether there was continuity and now about the means of transmission (Robertson, 1965; Bloom, Iversen and Schmitt, 1970). There is still very much to be found out, and not only about the detailed forms of single endings. Equally or more serious is our lack of knowledge of the distribution of the branches of each incoming fibre. Are the boutons attached to any one afferent fibre distributed close together on a den-drite? Are endings from different fibres intermixed? We know about these things only for a very few large neurons. These and similar questions will be answered only when people find suitable methods and suitable parts of the nervous system to which they can be applied. Once again these are matters of technology and of curiosity about what elements there are to be studied before we can begin making and testing hypotheses. We need to know much more about the numbers of cells and the disposition of their axons and dendrites. Electron microscopy can help with fine details, but is not by itself altogether suitable for study of the interrela-tions of many neurons.

The methods of staining by Procion yellow or cobalt are a great technical ad-vance with these problems. It may be noticed that what they provide is not a finer analysis of the minute "ultrastructural" components of the nervous system but the possibility of study of the *relations* of the various parts. These fundamental relational problems require methods of study with the light microscope as well as the electron microscope.

**How Far Can Neuroscience Go?**

This brings us to discuss the threshold between studies of the action of units and of the interactions of many of them. How far can neuroscience go? If it is to continue to insist on exactness, must it be confined to the study of single units or a few of them? To put the question in another way, what are the respective limits of the techniques of biophysical science, physiology and psychology? I don't think that any of us knows the answer that the future will give. It is clear that the powerful and exact concepts of physical science can produce immensely rich rewards in biology. One has only to think of DNA. Can similar methods solve problems such as perception, memory, decision-making or even consciousness? Many people would say "obviously not," because these subjects lie outside the domain of material science, but in his chapter Sperry proposes a very different view. I believe this is, at least partly, a matter of convention about the use of words and should not be dismissed as "semantic gymnastics" as Sperry disparagingly suggests. We may recall that at the beginning of this century many embryologists could not believe that the processes of development could be studied as physical events. The orderliness of development seemed to require the intervention of an organizing vital force.

Again, I remember that in the thirties many of our seniors could not understand the belief we young biologists had that our biochemical colleagues would solve the problem of heredity. What a wonderful experience it has been to live through the decades when they have begun to do so. I believe that the same will happen in neuroscience. But it will be partly by an evolution of physical science itself, allowing it to develop new concepts, for example, of the performance of organized molecular aggregates such as membranes, mitochondria and vesicles. This is happening already, and mathematicians and physicists are looking for ways to handle the activities of such aggregates as the cells of the nervous system. We should perhaps humbly remember that we still do not know how the specific ordering of nucleotides that ensures self-maintenance first arose at the origin of life, so in that sense physical science has not yet solved the problem of heredity.

## The Many Channels of the Nervous System

It would be absurd now to try to summarize the vast range of studies of cerebral functions. It may be interesting, however, to mention some work that suggests directions in which exact neuroscience may expand. The combined actions of many neurons challenge the technical powers of the investigator even more than single ones. First we need to know how many there are in any group and how they are interconnected. There is a wonderful field here for the neuroanatomist. As he looks at a section with thousands of neurons he has an opportunity denied to the physiologist. He can literally *see* the plan or blueprint upon which a part of the brain is built, *if he has the wit to understand it*. He has excellent resolution and display of information about the spatial distributions of the system, though very poor resolution of its actions in time.

An example of this kind of work can again be seen in the octopus. This animal has two memory systems, one for vision and one for touch (Figures 2.5 and 2.6; see Young, 1964). Each is composed of several lobes forming a circuit. In both systems the first lobe is occupied by an interlacing web of fibres, coming in the one case from the optic lobe and the other from the arms. What does this arrangement mean? Experiments by Martin Wells (1965) showed that after removal of this lobe from the touch system, an octopus cannot generalize from what it has learned with one arm to the others. How inefficient it would be to have to learn

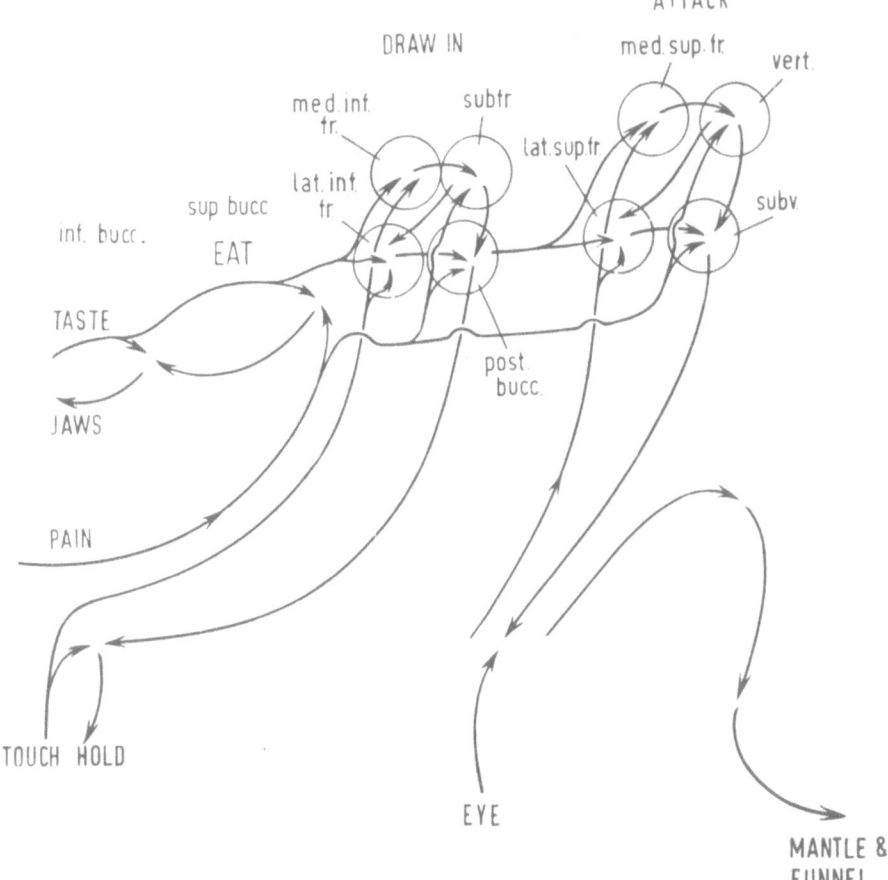

Figure 2.6   A diagram showing the two sets of pairs of centres involved in the visual and tactile memory system of *Octopus vulgaris*: inf.bucc., inferior buccal lobe; lat.inf.fr., lateral inferior frontal lobe; lat.sup.fr., lateral superior frontal lobe; med.inf.fr., median inferior frontal lobe; med.sup. fr., median superior frontal lobe; post.bucc., posterior buccal lobe; subfr., subfrontal lobe; subv., subvertical lobe; sup.bucc., superior buccal lobe; vert., vertical lobe. From Young (1964).

everything eight times over. Psychologists have long insisted that a mechanism for generalization must be included in a memory system. Here then we can literally *see* such a mechanism, and we can see how it makes use of this web of crossing fibres, allowing the spread of information. And there are two separate lobes with the same plan, one for touch and one for vision.

    In interpreting the plan of nervous tissues that he sees, the neuroanatomist can take advantage of the fact that the demands imposed by the world have produced independent systems using quite surprisingly similar principles of organization in the brains of very different animals. Perhaps we should not really be surprised at this. One of the chief lessons we have learned from an exact biology is that life is essentially uniform (for example, all living things use one genetic code). Neural transmitters seem to be remarkably uniform, and we shall probably find that nervous systems all share the same plans of organization, including that of their memory mechanisms. We expect that vision will involve a lens—should we be surprised that retinas are all similar? Indeed the deep retina of a cephalopod is astonishingly like that of a vertebrate, with plexiform and granule-cell layers

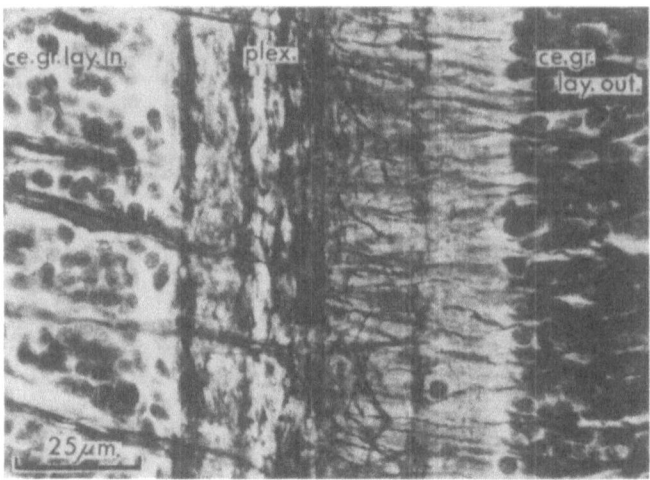

Figure 2.7 (a) A section through the deep retina of the optic lobe of *Octopus* (Cajal stain): ce.gr. lay.in., inner granule-cell layer; ce.gr.lay.out., outer granule-cell layer; plex., plexiform zone. From Young (1971).

providing amacrine and multipolar cells (Figure 2.7). Moreover analysis of the visual information goes on in a system of columns rather like a cerebral cortex (Figure 2.8). These are some of the most valuable clues we have for understanding the nervous system. If we know how to look, we can learn from nature's own experiments and then exploit them to make further experiments ourselves.

By studying the plan of a part of the brain the neuroanatomist gives essential clues to the physiologist, who badly needs them as he sets about studying activity. With electrical recording the physiologist has excellent resolution in time, but very poor resolution in space. Recording the combined interactions of nerve cells is not easy. Multiple electrodes can still sample only a fraction. More diffuse electrical recording, as with the EEG, has been a disappointment, if indeed much of the record comes from the brain at all rather than from the eye muscles as Lippold suggests (see Adey, 1969). It may be relevant that Ralph Gerard and I showed long ago that a frog's brain still continues to show rhythmic activity after it has been taken out of the head (Gerard and Young, 1937). Perhaps more could be discovered by following this line.

**Feature Detectors**

More promising at present is use of electrodes that we *do* understand, as probes into the organization of coding in the cortex. As Dr. Jung relates in his chapter, biophysicists made only limited progress with the study of units in the brain so long as they followed physical and quantitative theory, rather than nature, and probed with spots of monochromatic light or pure tones. But Lettvin, Maturana, McCulloch and Pitts (1959) asked "what the frog's eye tells the frog's brain." Vernon Mountcastle (1957) showed that units in the cat's cortex respond to particular types of skin stimulation, and Hubel and Wiesel (1959) began to show how coding is accomplished by the columns of cells. Using such techniques we can now begin to see what these individual feature detectors or classifying cells are

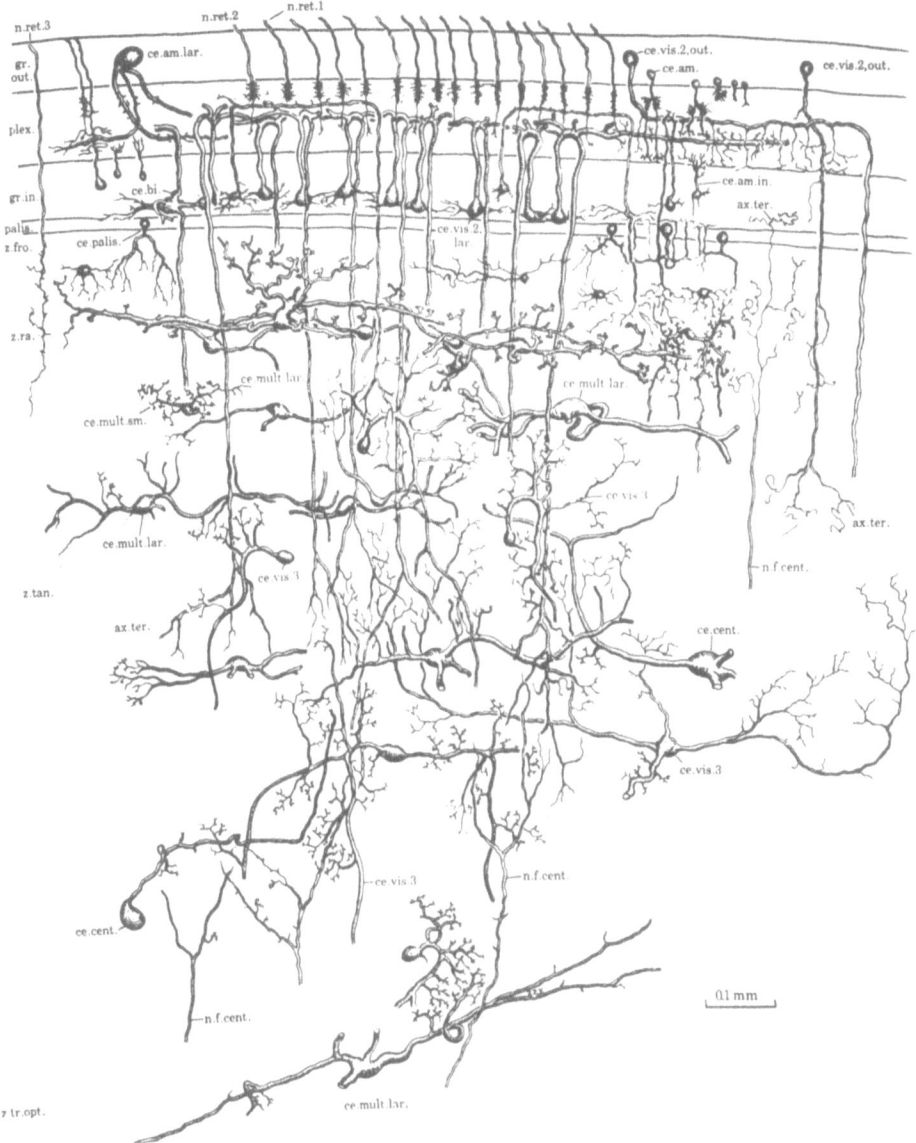

Figure 2.7 (b) A drawing of cells seen in the sagittal section of the optic lobe of *Loligo*, showing the different cell types, particularly the large multipolar cells of the medulla and the third-order cells of the lobe: ax.ter., axon terminal; ce.am., amacrine cell (microneuron); ce.am.in., inner amacrine cell; ce.bi., bipolar cell; ce.cent., cell with centrifugal nerve fibres; ce.mult.lar., large multipolar cell; ce.mult.sm., small multipolar cell; ce.palis., cell of palisade layer; ce.vis.2, second-order visual cell; ce.vis.2,lar., large second-order visual cell; ce.vis.2, out., second-order visual cell of outer granule-cell layer; ce.vis.3., third-order visual cell; gr.in., inner granule-cell layer; gr.out., outer granule-cell layer; n.f.cent., centrifugal nerve fibre; n.ret.1–3, retinal nerve fibre, types 1–3; palis., palisade layer; plex., plexiform zone; z.fro., frontier zone of optic lobe; z.ra., zone of radial columns of medulla; z.tan., zone of tangential bundles; z.tr.opt., zone of optic-tract bundles in optic lobe. From Young (1974).

doing in the sequences of the cortical hierarchies. For instance, in our laboratory Zeki has recently shown that in the fourth visual area of the monkey all the cells of one column are coded for a single colour (Zeki, 1973). My ancestor Thomas

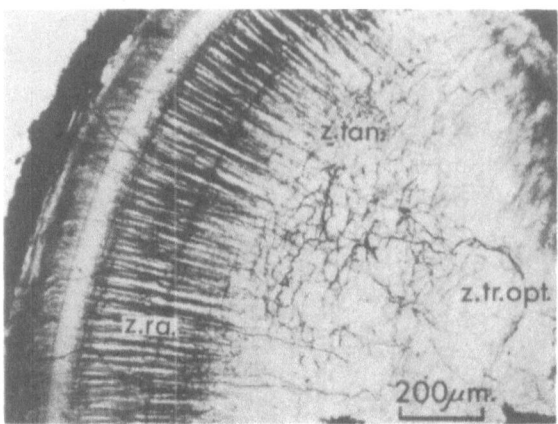

**Figure 2.8**  Sagittal section of optic lobe of *Loligo* to show the division of the medulla into an outer region of radial columns and an inner one with mainly tangential layers (Golgi stain): z.ra., zone of radial columns of medulla; z.ta., zone of tangential bundles; z.tr.opt., zone of optic-tract bundles. From Young (1974).

Young, whose bicentenary we celebrate this year, would have been intrigued to know that the cells Zeki has found so far signal either for white, red, green, blue or purple. Incidentally it was Thomas Young in his Bakerian Lecture (T. Young, 1802), long before Müller (1848), who first suggested that the nervous system operates by putting distinct items of information into separate channels.

These recent discoveries about the brain thus, in the main, confirm the concept of specific neuronal channels, which we have already stressed in dealing with the peripheral nervous system. Those of us who work with the brain often feel a sense of despair at the complexity of these channels and seek refuge in some idea of a "general" activity of whole masses of nervous tissue. Addiction to holistic concepts is indeed an occupational disease of neuroscientists (and especially of psychologists). But it is a curable disease from which one recovers by patient therapy with microscopy, microanalysis or microelectrodes. When one is most discouraged some piece of information comes along which tells that each single cell has its own function—say in the process of detection of a red square or in moving a flexor finger muscle (Evarts, 1968). It is even claimed that in the inferotemporal region of the cortex there are neurons that respond only when the stimulus is a monkey's hand (Gross, Rocha-Miranda and Bender, 1972). So one can return refreshed to the real task of finding how each such cell acts, not only when alone but also in conjunction with many others, to produce the overall behavioural phenomena that are so rightly emphasized by Gestalt psychologists and others of a "holistic" temperament.

We already have some information about the master controls of action that are at work. For example the reticular system activates and, as it were, challenges the whole brain. The hypothalamus and its marvellous sets of centres, discovered originally by the genius of Hess (1956), sets the brain and whole organism upon whatever path the needs of its body dictate.

But we have such a vast amount of information about what may be called the executive side of the brain that it would be ridiculous to try to summarize it here.

Single cells of the motor cortex are implicated in particular muscle movements, though this should not mislead us into thinking that the structuring and sequencing of commands is a simple matter limited to the motor cortex (see Evarts et al., 1971). There is a yawning gap between what we know about the sensory and the motor sides of the cortex. Surely the filling of this should lead to further understanding of what may be called decision making by the brain. Sperry in his chapter claims that we are already filling the gaps between input and output with a science of conscious experience based largely on split-brain studies.

If an animal is to produce adequate schedules of behaviour it must have equipment that recognizes familiar situations and suggests responses. There is evidence that this is a function of the hippocampus and the associated "Papez circuit of emotion" (Papez, 1937). Rats with lesions to the fornix show defects of recognition of familiarity, and this may be the basis for the conspicuous defects of immediate memory seen in humans after bilateral hippocampal lesions (Gaffan, 1972, 1973; Warrington, 1971). It may be that the hippocampus contains cells that respond when an animal reaches a particular place in a program of action. At University College, O'Keefe and Nadel have recorded with indwelling electrodes in the hippocampus of rats (O'Keefe and Dostrovsky, 1971; O'Keefe and Nadel, in preparation). The animals are trained to take food in one of three arms of a maze, water in another and dried milk in the third. Individual cells have then been found to fire when the rat goes to any one of the arms and looks in a particular direction. The experimenters believe that the stimulus is not the food or water or milk to be found there but the *place* as such, and that the hippocampus functions in some sense as what they call a "cognitive map." The activity of a particular cell indicates the position and direction of the animal and may activate a further set of cells giving anticipations of the likely results of subsequent movements. So we can begin to suggest how the brain builds up sequences of instructions to produce effective life patterns.

**Copies of Efferent Commands**

The carrying out of what we loosely call commands is of course an extremely complicated business. Indeed the command itself is often directed back into the receptor areas before, or as well as, producing motor action. The operations of the cerebellum and other centres in producing an "efference copy" or "corollary discharge" (Sperry, 1950) allow the organism to monitor what it selects from the incoming information (von Holst, 1973). We have found that centres with a cerebellum-like structure are prominent in cephalopods (Figure 2.9). They have large numbers of very long, thin, parallel fibres like those of the granule cells. They send fibres to motor centres and a large output back to the primary visual centres. Such a plan suggests devices for producing the correct timing and sequencing of selected patterns of movement, as has been suggested by Braitenberg (1967). The outputs of these lobes go partly to eye-muscle centres and partly to those controlling the direction of movement of the whole animal, as well as back to the optic lobes where vision is analyzed. In order for a squid to follow its prey there must be a combination of the information coming from the eye with information as to the

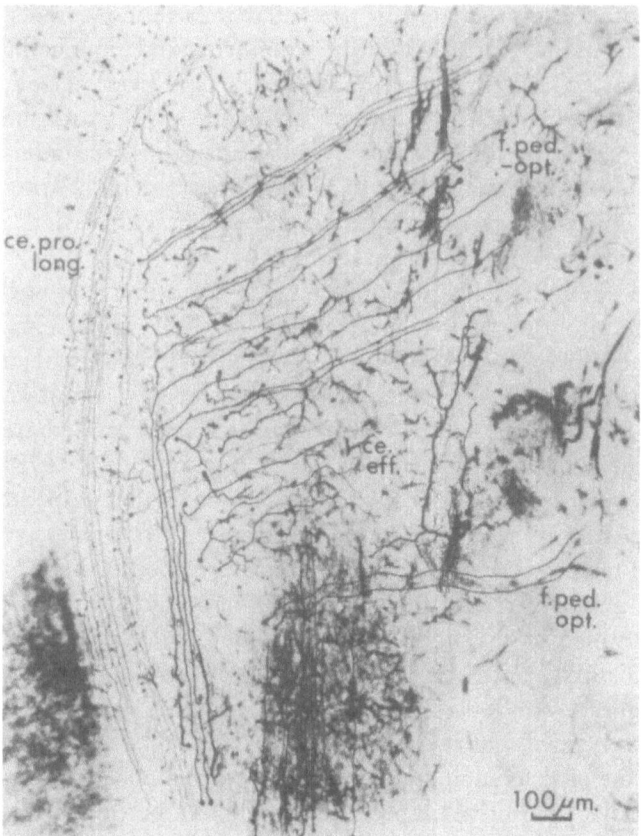

Figure 2.9   Retouched photograph of Golgi-stained section of the visuostatic (peduncle) lobe of the squid (*Loligo*) to show features resembling a cerebellum. The input fibres are not stained, but they activate two sorts of output fibres: large ones to eye muscles and other motor centres (ce.eff.), and smaller ones going back to the optic lobe (f.ped.opt.), which can thus produce an "efference copy." There are also very numerous T-shaped cells with long parallel fibres and collaterals, like the cerebellar granule cells. From Hobbs and Young (1973).

commands that the animal has issued to its own muscles. These cerebellum-like centres provide exactly those connections, and the long thin fibres may serve as delay lines to allow the appropriate timing.

## Memory

So far we have considered only the linking of afferents with the executive side of behaviour. But how is decision reached as to what the animal or man should do? This brings us to the problem that is often considered the most challenging of all, namely that of memory. Some nervous systems, conspicuously our own, have this power to change as a result of experience. The search for the nature of the change was the subject of the very first and of several of the earliest NRP Work Sessions in 1963 and 1964 (see *Neurosciences Research Program Bulletin* 1(1), 1963; 2(1), 2(3) and 2(4), 1964), but of fewer later. We must admit that the search has been only partly successful. Here too we are still guessing and looking for materials rather than testing hypotheses. True, the hypothesis of synthesis of specific memory substances has attracted many biochemists—and in general resulted in disproof. But we may notice that this disproof has not led to any advance in our understanding of memory. Disproving as such does not provide new ideas.

In the subject of memory there has been some advance by study of simple preparations. The ingenious technique invented by Horridge in which cells of the cockroach nerve cord can show a form of learning, even without the head, has allowed some study of changes in enzymes and protein synthesis in the cells (Horridge, 1973). We can notice again the combination of use of a suitable material and microchemical technique. But knowledge of these changes does not tell us how the new connections have been formed. The trouble is that, in spite of the vast effort expended on behavioural learning study, especially by psychologists, we still do not really know what we are looking for. What is the nature of the entity we call memory? Are we right to look upon it as a specific "function" distinct from others in the brain? What change takes place on each single occasion of learning? And where in the brain does it happen?

Yet perhaps even here we know a little more than thirty years ago. In an octopus, as has been said already, the visual and tactile memories are located in distinct identifiable lobules (Young, 1964, 1965). So the thesis put forward by some people that the record is wholly distributed is shown to be untrue, and this removes one barrier to progress. Of course the record within each of these systems may be distributed in some way, as in a hologram, but at least we know more precisely where to look for it. The two systems, visual and tactile, each contain several distinct parts, and to find these parts is a further substantial advance. As we have seen, one part, the vertical lobe, is necessary for learning (at least of some tasks) but is not itself the whole seat of the memory.

In 1938 I suggested that in the cephalopod brain there are self-re-excitatory circuits such as were first postulated by Alex Forbes (1922) in the spinal cord and

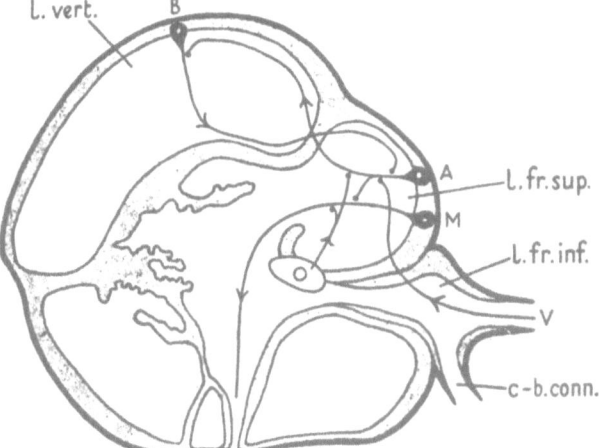

Figure 2.10   Diagram showing the memory circuit suggested in 1938 in the vertical lobe system of the cuttlefish (*Sepia*). Visual impulses from the optic lobe (O) cannot alone excite the motoneuron (M). If accompanied by impulses of taste (V) they start the circuit A → B → A whose collaterals lower the threshold of M and allow O to excite it. From Young (1938b).

studied by Lorente de Nó (1938) in the oculomotor system (Figure 2.10; see Young, 1938b). I proposed then that they could at least provide a basis for what we now call short-term memory, a suggestion later independently developed by Hebb (1949). Evidence shows that in cephalopods memory formation is indeed impaired when the circuit is broken. Electrical recording has recently shown that events occur in these circuits with delays of some seconds, and there is evidence of reverberation (Stephens, 1974). It may be, though it is not proved, that continued circulation serves to maintain for a short time a representation of external events, later to be printed if a suitably interesting "reward" arrives.

It is surprising how little attention has until recently been devoted to a search for the anatomy and physiology of the pathways by which "rewards" reach to the higher centres and are combined there during learning with inputs from the special centres. It is interesting that the subject is raised several times in this symposium. The lobes present in the octopus provide precisely this facility for spreading and recombining signals. Several of my fellow contributors refer to the locus coeruleus, the blue place in the medulla of mammals, which provides just this source of reward. Noradrenergic fibres ascend from the locus coeruleus in the median forebrain bundle ultimately to the septum, hippocampus and cortex (Figure 2.11). Crow and his colleagues have shown that rats with lesions of the locus coeruleus show gross impairment of capacity to learn the correct pathway to food in a maze (Anlezark, Crow and Greenway, 1973). Self-stimulation behaviour, as is noted in the chapter by Bloom, can be obtained by electrodes placed near the locus coeruleus (Ritter and Stein, 1973). Crow suggests that this is the basic positive-reinforcement pathway, and perhaps it is activated by the taste signals entering the medulla via the VII, IX and X nerves. He suggests that there is a second, different pathway responsible for motivational incentive or "drive," involving dopaminergic fibres from the interpeduncular nucleus to the corpus striatum and activated by the distance receptors of olfaction. This attractive

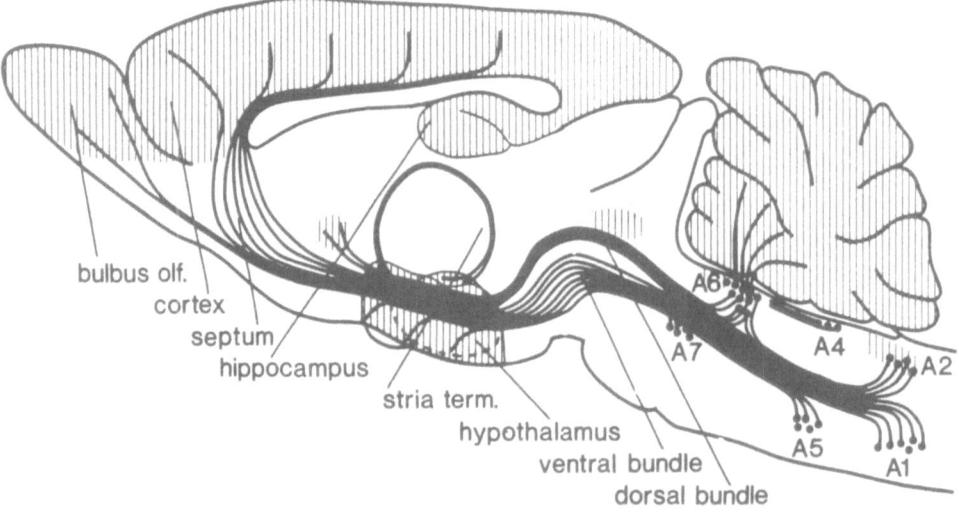

Figure 2.11   A sagittal projection of the noradrenergic fibres that ascend from the locus coeruleus (A6) and other centres in the mid-reticular formation to the hypothalamus and cortex. From Ungerstedt (1971).

hypothesis would have very interesting connections with the finding that affective disorders are connected with catecholamine metabolism. Obviously the situation is not simple, but more knowledge in this field holds out great hope for the treatment of some of the problems of human psychology.

There may be even more to it than this. In man the "rewards" by which we learn are not only the basic ones of food, sex, etc., but the secondary reinforcements of more sophisticated pleasures of being socially "right" or of making correct solutions to problems. We may speculate that there is some pathway by which the successful solution of our social or other problems is linked with whatever constitutes the pleasure reward centres, and of course vice versa. One of the most powerful determinants of human behaviour is the need for satisfaction of what we may call our ethical hunger—even at our most wicked we yearn to do what is "right" by the standards of those around us. Understanding of reward centres may even help us to be "better" in many senses, or at least less bad.

But knowledge about rewards does not tell us what change in the nervous system produces a registration in the long-term memory. My particular mnemon hypothesis is that the record is made at least partly by the switching off of unwanted channels (Figure 2.12). Learning would thus be a selective process (see Young, 1973) in some respects similar to the production of antibodies as suggested by Dr. Edelman (see Chapter 4, this volume). Heredity certainly provides an array of feature detectors, and these, or combinations of these, have more than one possible output since the essence of learning is that two or more distinct responses can be given initially to a given input, whereas after learning a particular response is always performed. Learning thus consists at least partly of a reduction of this great initial redundancy of the brain. A somewhat similar selective theory of memory has been set forth by Mark (1973). I suggest that the physical process involved may be the induction of the appropriate enzyme synthesis in the small cells that are so numerous in the very parts of the brain that are concerned with

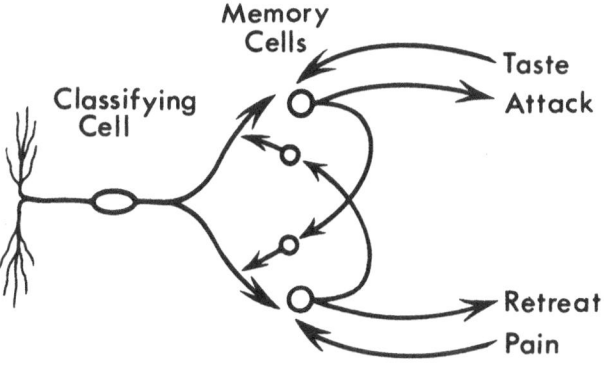

## MNEMON

Figure 2.12   A mnemon (the components of a single memory unit). The classifying cell will record the occurrence of a particular type of event; it has two outputs allowing for alternative motor actions. The system has a slight bias to one of these actions, say attack. Following this action, signals will arrive indicating the results of the action and either reinforce what was done or produce the opposite action. Collaterals of higher motor cells will then activate the small (amacrine) cells. These produce an inhibitory transmitter to close the unused pathway. After Young (1965).

learning. This agrees with the evidence that inhibition of protein synthesis interrupts the process.

This mnemon model may be proved to be wrong in detail, but as expressing the logic of the situation it can hardly be wrong. We know that there are feature detectors and that alone or as combinations they initially have the power to produce more than one output. This is the essence of parts of the brain that incorporate memory rather than fixed reflexes. We also know that what is stored in memory is a record of which actions produce "correct" results for survival; certainly, therefore, receptors indicating results, rewards or "reinforcement" are involved. The mnemon thus provides a diagram or schema of the elements that must be included in a unit of memory. We can say that each mnemon stores one bit of information, namely, the record that a given action following stimulation of a feature detector (or set of them) was good or bad for survival of the animal.

It is hard to relate these general concepts to what is known about changes that have been shown to take place in the physiological behaviour of units in the nervous system during the process of learning (John, 1967). A recent example of long-lasting change has been shown with indwelling electrodes in the hippocampus of the rabbit (Bliss and Gardner-Medwin, 1973). Electrical stimulation of the fibres of the perforant pathway produces characteristic responses of the cells of the dentate gyrus. After repetitive stimulation for fifteen seconds the response pattern changes and the new pattern remains for as long as nine days (Figure 2.13). This is a very dramatic finding, though we do not know how such changes are related to the types of memory storage we have discussed. Such experiments with indwelling electrodes are very laborious, but like those of O'Keefe and Nadel already discussed they provide a further example of the new stage of technical development that allows continuous monitoring of cell activity in the awake normal animal and provides the possibility of relating physiological activity to behaviour.

**The Model in the Brain**

Which brings us back to the question of how the brain is organized to produce the marvellous complex actions, especially of a man. I believe we can see the outlines. As usual in biology heredity and experience both play a part, and, as so often before, investigators are quarrelling over the extent of the contributions (Wiesel and Hubel, 1963; Pettigrew, 1974). There seems no doubt, however, that both kittens and monkeys are provided at birth with sets of feature detectors that are ready to respond to particular events once the opportunity has been provided by the environment (Blakemore, 1974). Heredity probably also provides the outlines of ways in which the detectors can readily come to be combined to produce complex outputs, such as language. Given an appropriate environment, choices between alternative possibilities will be made in such a way as to provide an action system suitable for survival in the light of the past history of the species and of the individual. This process I call building a model in the brain, and we can see in outline the units from which it is built and the restrictions within which it operates. The form of the model in the brain will be partly determined by heredity to conform to the needs of the species. There is some evidence that our own memories are biased to store information relevant to human communication.

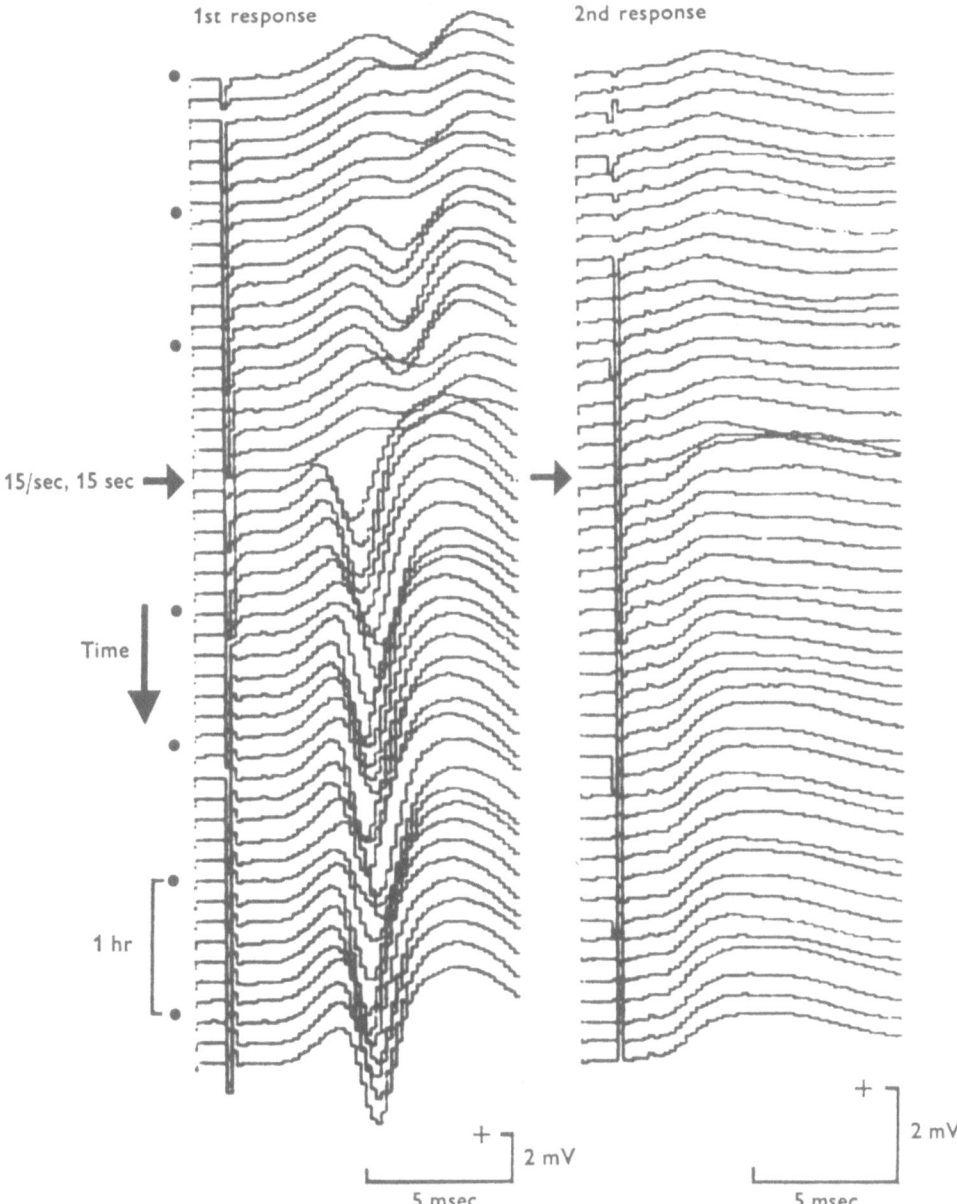

Figure 2.13  Average responses recorded from the dentate gyrus of an awake rabbit. Stimuli
were given to the perforant pathway every 9 minutes and responses are plotted down the page. The
responses are to pairs of equal stimuli of 30V given 22 msec apart. In the middle a train of stimuli
at 15/sec was given for 15 sec. This produced a change in the pattern which persisted for hours or
days. From Bliss and Gardner-Medwin (1973).

Perhaps the models in our brains are constructed around representations of
other people and their actions, and this would explain much of our tendency to
anthropomorphism. Some people may say that this is no longer neuroscience but
fantasy. But units with complex properties are already known to exist. What we
need is a technology that will show us how they come to interact to give the brain
an action system or model capable of producing appropriate behaviour.

We certainly do not know enough to be able to answer our initial question of how the brain investigates the brain. But I suggest that we do begin to see the outlines of a solution. We have even been able to say a little about possible mechanisms for such fundamental features of our lives as motivation, reinforcement or reward, curiosity and exploration, and even a little about memory.

It is sometimes said that knowledge is independent of the psychology and other characteristics of those who produce it (Popper, 1972). It seems to me that our biological studies do not agree with this. Knowledge is the product of brains that seek it out as part of their life systems. It is true that in our complex civilization they do not seek knowledge simply as part of their search for food, sex or shelter, but they are activated by a motivation system nonetheless. And the knowledge they produce is useful to themselves and to others. Not equally of course and not always, but broadly, I believe that it is. Moreover the use itself stimulates the search for further knowledge and further techniques. With the knowledge we have of the brain we can already give some help to medicine in alleviating human suffering. Have we perhaps considered too little how we might help with other human affairs, for instance with learning in education or avoidance of conflict? Neuroscience has not felt unified or strong enough to venture into such fields. In the past we have not been sure whether we were anatomists, physiologists or biochemists, psychologists or pharmacologists. But now we know our identity. Largely thanks to Frank Schmitt's initiative, we are Neuroscientists. We are united in the belief that analysis of how the brain works is good and helpful to mankind. It can proceed alongside other methods of study and in no sense "reduces" human beings. Quite the contrary, its aim and result is to expand them. Let us hope that over the next few years neuroscience will come to show much more about how all the marvellously intricate circuits and units of the brain operate together to give us those capacities upon which all human life and happiness depends.

## References

Adey, W. R. (1969): Slow electrical phenomena in the central nervous system. *Neurosci. Res. Program Bull.* 7:75–180.

Adrian, E. D. (1914): The all-or-none principle in nerve. *J. Physiol.* 47:460–474.

Adrian, E. D., and Bronk, D. W. (1928): The discharge of impulses in motor nerve fibres. Part I. Impulses in single fibres of the phrenic nerve. *J. Physiol.* 66:81–101.

Aitken, J. T., Sharman, M., and Young, J. Z. (1947): Maturation of regenerating nerve fibres with various peripheral connexions. *J. Anat.* 81:1–22.

Anlezark, G. M., Crow, T. J., and Greenway, A. P. (1973): Evidence that the noradrenergic innervation of the cerebral cortex is necessary for learning. *J. Physiol.* 231:119P–120P.

Bear, R. S., Schmitt, F. O., and Young, J. Z. (1937a): The sheath components of the giant nerve fibres of the squid. *Proc. R. Soc. B.* 123:496–504.

Bear, R. S., Schmitt, F. O., and Young, J. Z. (1937b): The ultrastructure of nerve axoplasm. *Proc. R. Soc. B.* 123:505–519.

Bear, R. S., Schmitt, F. O., and Young, J. Z. (1937c): Investigations on the protein constituents of nerve axoplasm. *Proc. R. Soc. B.* 123:520–529.

Bernard, C. (1878–1879): *Leçons sur les Phénomènes de la Vie Communs aux Animaux et aux Végétaux* (Cours de Physiologie Générale du Muséum d'Historie Naturelle). Vol. I and II. Paris: Ballière.

Blakemore, C. (1974): Developmental factors in the formation of feature extracting neurons. *In: The Neurosciences: Third Study Program.* Schmitt, F. O., and Worden, F. G., eds. Cambridge, Mass.: The MIT Press, pp. 105–113.

Bliss, T. V. P., and Gardner-Medwin, A. R. (1973): Long-lasting potentiation of synaptic transmission in the dentate area of the unanaesthetized rabbit following stimulation of the perforant path. *J. Physiol.* 232:357–374.

Bloom, F. E., Iversen, L. L., and Schmitt, F. O. (1970): Macromolecules in synaptic function. *Neurosci. Res. Program Bull.* 8:325–455.

Boycott, B. B., and Young, J. Z. (1957): Effects of interference with the vertical lobe on visual discriminations in *Octopus vulgaris* Lamarck. *Proc. R. Soc. B.* 146:439–459.

Braitenberg, V. (1967): Is the cerebellar cortex a biological clock in the millisecond range? *Prog. Brain Res.* 25:334–346.

Bullock, T. H. (1948): Properties of a single synapse in the stellate ganglion of squid. *J. Neurophysiol.* 11:343–364.

Cannon, W. B. (1932): *The Wisdom of the Body.* New York: W. W. Norton.

Cole, K. S. (1968): *Membranes, Ions and Impulses. A Chapter of Classical Biophysics.* Berkeley: University of California Press.

Cole, K. S., and Curtis, H. J. (1936): Electric impedance of nerve and muscle. *Cold Spring Harbor Symp. Quant. Biol.* 4:73–89.

Creed, R. S., Denny-Brown, D., Eccles, J. C., Liddell, E. G. T., and Sherrington, C. S. (1932): *Reflex Activity of the Spinal Cord.* London: Oxford University Press.

Dale, H. H. (1953): Acetylcholine as a chemical transmitter of the effects of nerve impulses. *In: Adventures in Physiology. With Excursions into Autopharmacology.* London: Pergamon Press, pp. 611–637.

Denny-Brown, D. (1966): *The Cerebral Control of Movement* (The Sherrington Lectures VIII). Liverpool, Eng.: Liverpool University Press.

Dowling, J. E. (1970): Organization of vertebrate retinas. The Jonas M. Friedenwald Memorial Lecture. *Invest. Ophthalmol.* 9:655–680.

Eccles, J. C. (1953): *The Neurophysiological Basis of Mind. The Principles of Neurophysiology.* Oxford: Clarendon Press.

Eccles, J. C., Granit, R., and Young, J. Z. (1932): Impulses in the giant nerve fibres of earthworms. *J. Physiol.* 77:23P–25P.

Erlanger, J., and Gasser, H. S. (1937): *Electrical Signs of Nervous Activity*. Philadelphia: University of Pennsylvania Press.

Evarts, E. V. (1968): Relation of pyramidal tract activity to force exerted during voluntary movement. *J. Neurophysiol.* 31:14–27.

Evarts, E. V., Bizzi, E., Burke, R. E., De Long, M., and Thach, W. T., Jr. (1971): Central control of movement. *Neurosci. Res. Program Bull.* 9:2–170.

Forbes, A. (1922): The interpretation of spinal reflexes in terms of present knowledge of nerve conduction. *Physiol. Rev.* 2:361–414.

Gaffan, D. (1972): Loss of recognition memory in rats with lesions of the fornix. *Neuropsychologia* 10:327–341.

Gaffan, D. (1973): Inhibitory gradients and behavioural contrast in rats with lesions of the fornix. *Physiol. Behav.* 11:215–220.

Gaze, R. M. (1970): *The Formation of Nerve Connections. A Consideration of Neural Specificity Modulation and Comparable Phenomena*. New York: Academic Press.

Gaze, R. M., Keating, M. J., and Chung, S. H. (1974): The evolution of the retinotectal map during development in *Xenopus*. *Proc. R. Soc. B.* 185:301–330.

Gerard, R. W., and Young, J. Z. (1937): Electrical activity of the central nervous system of the frog. *Proc. R. Soc. B.* 122:343–352.

Geren, B. B. (1954): The formation from the Schwann cell surface of myelin in the peripheral nerves of chick embryos. *Exp. Cell Res.* 7:558–562.

Granit, R. (1972): *Mechanisms Regulating the Discharge of Motoneurons* (The Sherrington Lectures XI). Springfield, Ill.: C. C Thomas.

Gray, E. G., and Young, J. Z. (1968): The electron microscopy of experimental degeneration in the octopus brain. *In: Cell Structure and Its Interpretation. Essays Presented to John Randal Baker F. R. S.* McGee-Russell, S. M., and Ross, K. F. A., eds. London: Edward Arnold, pp. 371–380.

Gross, C. G., Rocha-Miranda, C. E., and Bender, D. B. (1972): Visual properties of neurons in inferotemporal cortex of the macaque. *J. Neurophysiol.* 35:96–111.

Hebb, D. O. (1949): *The Organization of Behavior. A Neuropsychological Theory*. New York: John Wiley.

Henderson, L. J. (1913): *The Fitness of the Environment. An Inquiry into the Biological Significance of the Properties of Matter*. New York: Crowell-Collier and MacMillan.

Hess, W. R. (1956): *Hypothalamus and Thalamus. Experimental Documentation*. Stuttgart: Georg Thieme Verlag.

Hobbs, M. J., and Young, J. Z. (1973): A cephalopod cerebellum. *Brain Res.* 55:424–430.

Hodgkin, A. L. (1954): A note on conduction velocity. *J. Physiol.* 125:221–224.

Hodgkin, A. L., and Huxley, A. F. (1952): A quantitative description of membrane current and its application to conduction and excitation in nerve. *J. Physiol.* 117:500–544.

Horridge, G. A. (1962): Learning of leg position by the ventral nerve cord in headless insects. *Proc. R. Soc. B.* 157:33–52.

Horridge, G. A. (1973): Summary. *In: Australian Academy of Science Report: Symposium on Biological Memory.*

Hoyle, G. (1970): Cellular mechanisms underlying behavior—neuroethology. *Adv. Insect Physiol.* 7:349–444.

Hubel, D. H., and Wiesel, T. N. (1959): Receptive fields of single neurones in the cat's striate cortex. *J. Physiol.* 148:574–591.

Jameson, F. (1972): *The Prison-House of Language: A Critical Account of Structuralism and Russian Formalism* (Princeton Essays in European and Comparative Literature). Princeton, N. J.: Princeton University Press.

John, E. R. (1967): *Mechanisms of Memory.* New York: Academic Press.

Kandel, E. R., and Kupfermann, I. (1970): The functional organization of invertebrate ganglia. *Annu. Rev. Physiol.* 32:193–258.

Katz, B. (1966): *Nerve, Muscle and Synapse.* New York: McGraw-Hill.

Kennedy, D., Selverston, A. I., and Remler, M. P. (1969): Analysis of restricted neural networks. *Science* 164:1488–1496.

Kerkut, G. A. (1969): The use of snail neurons in neurophysiological studies. *Endeavour* 28:22–26.

Kuhn, T. S. (1962): *The Structure of Scientific Revolutions.* Chicago: University of Chicago Press.

Kuntz, A. (1934): *The Autonomic Nervous System.* Philadelphia: Lea and Febiger.

Langley, J. N. (1903): The autonomic nervous system. *Brain* 26:1–26.

Lettvin, J. Y., Maturana, H. R., McCulloch, W. S., and Pitts, W. H. (1959): What the frog's eye tells the frog's brain. *Proc. IRE* 47:1940–1951.

Lopresti, V., Macagno, E. R., and Levinthal, C. (1973): Structure and development of neuronal connections in isogenic organisms: cellular interactions in the development of the optic lamina of *Daphnia. Proc. Natl. Acad. Sci. USA* 70:433–437.

Lorente de Nó, R. (1938): Analysis of the activity of the chains of internuncial neurons. *J. Neurophysiol.* 1:207–244.

Lucas, K. (1909): The "all or none" contraction of the amphibian skeletal muscle fibre. *J. Physiol.* 38:113–133.

Lucas, K. (1912): Croonian Lecture: The process of excitation in nerve and muscle. *Proc. R. Soc. B.* 85:495–524.

Mark, R. (1973): Cellular mechanisms of neural memory. *In: Australian Academy of Science Report: Symposium on Biological Memory.*

Miledi, R. (1967): Spontaneous synaptic potentials and quantal release of transmitter in the stellate ganglion of the squid. *J. Physiol.* 192:379–406.

Mokrasch, L. C., Bear, R. S., and Schmitt, F. O. (1971): Myelin. *Neurosci. Res. Program Bull.* 9:440–598.

Mountcastle, V. B. (1957): Modality and topographic properties of single neurons of cat's somatic sensory cortex. *J. Neurophysiol.* 20:408–434.

Müller, J. (1848): *The Physiology of the Senses, Voice, and Muscular Motion, with the Mental Facilities.* Baly, W., trans. London: Taylor, Walton and Maberly.

Nishioka, R. S., Hagadorn, I. R., and Bern, H. A. (1962): Ultrastructure of the epistellar body of the Octopus. *Z. Zellforsch. Mikrosk. Anat.* 57:406–421.

Offner, F., Weinberg, A., and Young, G. (1940): Nerve conduction theory: Some mathematical consequences of Bernstein's model. *Bull. Math. Biophys.* 2:89–103.

O'Keefe, J., and Dostrovsky, J. (1971): The hippocampus as a spatial map. Preliminary evidence from unit activity in the freely-moving rat. *Brain Res.* 34:171–175.

Papez, J. W. (1937): A proposed mechanism of emotion. *Arch. Neurol. Psychiatr.* 38:725–743.

Pavlov, I. P. (1927): *Conditioned Reflexes: An Investigation of the Physiological Activity of the Cerebral Cortex.* Anrep, G. V., trans. and ed. Oxford: Oxford University Press.

Pettigrew, J. D. (1974): The effect of visual experience on the development of stimulus specificity by kitten cortical neurones. *J. Physiol.* 237:49–74.

Popper, K. R. (1972): *Objective Knowledge: An Evolutionary Approach.* Oxford: Clarendon Press.

Pumphrey, R. J., and Young, J. Z. (1938): The rates of conduction of nerve fibres of various diameters in cephalopods. *J. Exp. Biol.* 15:453–466.

Ritter, S., and Stein, L. (1973): Self-stimulation of noradrenergic cell group (A6) in *Locus coeruleus* of rats. *J. Comp. Physiol. Psychol.* 85:443–452.

Robertson, J. D. (1955): The ultrastructure of adult vertebrate peripheral myelinated nerve fibers in relation to myelinogenesis. *J. Biophys. Biochem. Cytol.* 1:271–278.

Robertson, J. D. (1965): The synapse: Morphological and chemical correlates of function. *Neurosci. Res. Program Bull.* 3(4):1–79.

Rushton, W. A. H. (1951): A theory of the effects of fibre size in medullated nerve. *J. Physiol.* 115:101–122.

Sanders, G. D., and Young, J. Z. (1974): Reappearance of specific colour patterns after nerve regeneration in *Octopus. Proc. R. Soc. B.* 186:1–11.

Saussure, F. de (1916): *Cours de Linguistique Générale.* Paris: Payot.

Schmitt, F. O. (1967): Molecular parameters in brain function. *In: The Human Mind.* Roslansky, J. D., ed. Amsterdam: North-Holland, pp. 109–138.

Schmitt, F. O., and Samson, F. E., Jr. (1969): Brain cell microenvironment. *Neurosci. Res. Program Bull.* 7:277–417.

Sherrington, C. S. (1906): *The Integrative Action of the Nervous System.* New York: Scribner.

Sjöstrand, F. S. (1953): The lamellated structure of the nerve myelin sheath as revealed by high resolution electron microscopy. *Experientia* 9:68–69.

Spencer, W. A., and Kandel, E. R. (1961): Electrophysiology of hippocampal neurons. IV. Fast prepotentials. *J. Neurophysiol.* 24:272–285.

Sperry, R. W. (1944): Optic nerve regeneration with return of vision in anurans. *J. Neurophysiol.* 7:57–69.

Sperry, R.W. (1950): Neural basis of the spontaneous optokinetic response, *J. Comp, Physiol. Psychol.* 4:482–489.

Stephens, R. (1974): Electrophysiological studies of the brain of *Octopus vulgaris. J. Physiol.* 240:19 pp.

Tauc, L. (1967): Transmission in invertebrate and vertebrate ganglia. *Physiol. Rev.* 47:521–593.

Ungerstedt, U. (1971): Stereotaxic mapping of the monoamine pathways in the rat brain. *Acta Physiol. Scand.* 367 (Suppl.):1–48.

Von Holst, E. (1973): *The Selected Papers of Eric von Holst.* Vol. 1. *The Behavioural Physiology of Animals and Man.* Martin, R., trans. London: Methuen.

Wall, P. D. (1965): Functional specificity. *Neurosci. Res. Program Bull.* 3:55–56, 61. Also *In: Neurosciences Research Symposium Summaries.* Vol. 1. Schmitt, F. O., et al., eds. Cambridge, Mass.: The MIT Press (1966) pp. 229–230, 235.

Warrington, E. K. (1971): Neurological disorders of memory. *Br. Med. Bull.* 27:243–247.

Waxman, S. G. (1974): Ultrastructural differentiation of the axon membrane at synaptic and non-synaptic central nodes of *Ranvier. Brain Res.* 65:338–342.

Webb, D. A., and Young, J. Z. (1940): Electrolyte content and action potential of the giant nerve fibres of *Loligo. J. Physiol.* 98:299–313.

Wells, M. J. (1965): The vertical lobe and touch learning in the octopus. *J. Exp. Biol.* 42:233–255.

Wiener, N. (1948): *Cybernetics, or Control and Communication in the Animal and the Machine.* Cambridge, Mass.: The MIT Press.

Wiersma, C. A. G. (1952): The neuron soma. Neurons of arthropods. *Cold Spring Harbor Symp. Quant. Biol.* 17:155–163.

Wiesel, T. N., and Hubel, D. H. (1963): Effects of visual deprivation on morphology and physiology of cells in the cat's lateral geniculate body. *J. Neurophysiol.* 26:978–993.

Wurtman, R. J. (1971): Brain monoamines and endocrine function. *Neurosci. Res. Program Bull.* 9:172–297.

Young, J. Z. (1929): Sopra un nuovo organo dei cefalopodi. *Boll. Soc. Ital. Biol. Sper.* 4:1022–1024.

Young, J. Z. (1936a): The giant nerve fibres and epistellar body of cephalopods. *Q. J. Microsc. Sci.* 78:367–386.

Young, J. Z. (1936b): Structure of nerve fibres and synapses in some invertebrates. *Cold Spring Harbor Symp. Quant. Biol.* 4:1–6.

Young, J. Z. (1938a): The evolution of the nervous system and of the relationship of organism and environment. *In: Evolution. Essays on Aspects of Evolutionary Biology Presented to Prof. E. S. Goodrich on his 70th Birthday.* De Beer, G. R., ed. Oxford: Clarendon Press, pp. 179–204.

Young, J. Z. (1938b): The functioning of the giant nerve fibres of the squid. *J. Exp. Biol.* 15:170–185.

Young, J. Z. (1944): Giant nerve-fibres. *Endeavour* 3:108–113.

Young, J. Z. (1964): *A Model of the Brain.* Oxford: Clarendon Press.

Young, J. Z. (1965): The Croonian Lecture, 1965: The organization of a memory system. *Proc. R. Soc. B.* 163:285–320.

Young, J. Z. (1973): Memory as a selective process. *In: Australian Academy of Science Report: Symposium on Biological Memory*, pp. 25–45.

Young, J. Z. (1974): The central nervous system of *Loligo*. I. The optic lobe. *Phil. Trans. R. Soc. B.* 267:263–302.

Young, T. (1802): The Bakerian Lecture: On the theory of light and colours. *Phil. Trans. R. Soc. B.* 95:12–48.

Zeki, S. M. (1973): Colour coding in rhesus monkey prestriate cortex. *Brain Res.* 53:422–427.

# Comparative Approaches to the Nervous System

Donald Kennedy (b. 1931, New York, New York) is professor in the Department of Biological Sciences at Stanford University. His research has dealt with invertebrate vision, with synaptic transmission and motor control in arthropod central nervous systems, and with reflex physiology.

# 3

# Behavior to Neurobiology: A Zoologist's Approach to Nervous Systems

## Donald Kennedy

This symposium was described as an exercise in contemporary history, and the participants were invited to discuss their work. That is a dangerous offer: the temptation is overwhelming to respond with the kind of history that sacrifices realism for plausibility. This dreadful urge overtakes many of us in the introductions to papers, wherein accident is readily transduced to the more elegant processes of reasoned choice and expected outcome. "The experimental system was chosen," we write, "because it offered an opportunity to test the efficacy of alpha in the absence of competing inputs." (Translation: when the work was nearly done, we realized that the input was pure alpha.) How often are such motivations remembered conveniently just about the time the introduction is being composed? Yet how comforting they are for the historians of science, whose trade is not much helped by the revelation that scientific progress occasionally resembles the blind staggers more than the measured tread of rationality. A similar conspiracy is engaged in by coaches and sportswriters, who, having a vested interest in the proposition that games are complex and intellectually demanding, insist on a level of retrospective analysis that participants find a little funny. But why should the coach, or even the players, tell on the writer who analyzes the play selection during the winning drive in the Rose Bowl? *They* know that half the plays were argued out in the huddle, but saying so wouldn't really help the image of the game.

This analogy may suggest that I do not believe much that has been written about how science works, Perhaps that simply indicates that I do science in a different way. I suspect, though, that there may be a lot of other closet violators—people who, like me, develop sweaty palms when the word *paradigm* is mentioned, and discover on sitting down at the typewriter that they have embarrassingly little to say about how they do experiments. This cynicism, on the other hand, could be a little self-protective. As I consider my own work, I must admit that it is more scattered, and less dictated by a reasoned approach to major problems, than that of most of the people I admire in neurobiology. So in the remainder of this discussion—in the true spirit of intellectual retrospection—I will make the briskest defense I can muster of the virtues of messing around. As the title suggests, I intend to emphasize that a focus on comparative behavior (and, equally, on evolution) yields some investigative styles that differ usefully from those borrowed from the more medical tradition of physiology.

### Early Training

My convictions on this matter are, as the preamble suggests, wholly self-serving. As a Harvard undergraduate I majored in biology, but I took most of the required courses with a sullenness that stayed just on the borderline of rebellion. This was only partly because they interfered with my skiing career: the late forties and early fifties represented the first flush of biochemistry in undergraduate curricula, and it seemed to me that Jeffries Wyman and his friends were doing their best to keep me from being a forest ranger or a museum collector. I was an authentic zoologist, captivated by butterflies and birds, and hopelessly hooked on fish. I didn't take organic chemistry until I wandered into graduate school, and one of the great reliefs of my life was experienced at a Harvard cocktail party last fall when I realized that Gerald Holton doesn't recall almost failing me in physics.

Two events rescued me from the inevitable conclusion to this beginning. The first was John Welsh's course in comparative physiology, which I took in my senior year. It was everything an academic experience should be: demanding, exciting, an opening-up of totally new prospects. The second happened when Harvard persuaded Don Griffin to move to Cambridge from Cornell. Griffin soon convinced me that you could, after all, have it both ways, that it was possible to retain a focus on the whole animal in its natural state and yet do analysis at a deep level on some physiological process. I also first learned from him the joys of expeditionary biology, and regret not having made better subsequent use of that lesson. We used to catch terns on Penikese Island, and then drive them in a station wagon to release them on remote little airfields in New Hampshire; the idea was to see if they took the correct initial headings toward home. Later Don thought it was a fine idea to economize on birds, and we hatched a scheme to get several initial headings out of the same one. My dubious talents as a light-tackle fisherman were called into service; I still remember playing a spirited pigeon on two-pound monofilament with my spinning rod. The experiment was a disaster for tackle and pigeon alike, and I got a draw at best.

But these ventures provided the side benefit of a lot of personal instruction. I recall a late-night lecture on the cochlear microphonic on Route 28, with a dusty dash for a blackboard. *That* taught me that teaching and research are kingdoms with poorly mapped boundaries. Most of all, I learned from Don a respect for diversity, for the instructive power of the odd mechanism or the chance solution—in short, the lesson that evolutionary opportunism provides the investigator with a great natural laboratory.

I finished my doctoral degree with Don, and got a generous assist from George Wald, who was willing to abet an effort on vision even though it was not in his line, and who was as conscious as anyone of the advantages of comparative studies. But, fortunate as I was in that sponsorship, I must nonetheless confess that the larger part of my graduate education at Harvard came at the hands of a remarkable group of fellow graduate students and young faculty (two categories which, as those familiar with Harvard know, mean the same thing). Tom Eisner, Ed Wilson, Roger Milkman, Tim Goldsmith, Ric Miller, Bill van der Kloot, Bill Drury, and many others formed an overlapping group whose members all shared, to a greater or lesser degree, an interest in the behavior of whole animals.

On this same theme, I would acknowledge special debts, incurred during a period of life that most people believe *follows* training! But, having entered on an academic career just before the general availability of the postdoctoral fellowship, I, like many of my friends, received on-the-job training as a teacher and an investigator. I was unusually lucky to fall (there is no more appropriate verb) into a first-rate small zoology department at Syracuse in 1956. I was protected and encouraged by a neurobiologist chairman, Vern Wulff, and helped by sympathetic colleagues like George Holz. I was introduced to neurophysiology by some senior colleagues elsewhere who believed that science is a shared venture and greeted new recruits as converts rather than competitors. In different ways, Kees Wiersma and Ted Bullock did just that.

## A Zoological Perspective

This brief personal history may explain some of the attitudes I will reveal below. The tradition in which I was raised is very separate from that of physiology proper, with its wellsprings in medicine. What does this zoological perspective leave one with, in a roomful of physiologists? There is a temptation to joke: When the physiologist begins, "In the snake...," the zoologist is the spoilsport who interrupts with "What kind?" Respect for diversity is part of the tradition. Another part is the use of an exploitative experimental strategy that seeks situations in which the system under study appears cornered by natural selection—under great pressure to do all it can. Water balance in the desert, temperature control in the arctic, respiration on a mountaintop—each offers opportunities for uncovering mechanisms that are covert under more "normal" circumstances.

The conviction that a given design problem, recurring as it is likely to in different selectional contexts, will result in different solutions is part of the comparative zoologist's basic belief system. It makes us a little skeptical of the odd fact that the great debates in physiology are apt to be couched in universals: Is transmission across THE synapse chemical? Is inward current across THE nerve membrane carried by sodium ions? Such issues are often decided inconclusively, because diversity of mechanisms is simply a fact of life. Inevitably there are some yeses and some nos; it is doubtful if uniformity of physiological mechanism occurs at any level above the genetic code. Yet it is clear that there must be some convergence of *process* in all of the evolutionary solutions to a given problem.

This focus on process can save the comparative zoologist from being condemned to the role of iconoclast. It is no fun at all to run around reminding people that there are probably exceptions to each exciting new finding. Indeed, if we made that our exclusive business, the discoverers could be forgiven for according us the treatment traditionally reserved for bringers of bad news. I believe, however, that the comparative approach can offer more than just a suspicious posture in the presence of generality.

One area in which it may have something to contribute is in *designing the preparation*, as it were, as well as the experiment. I have the prejudice that advances in physiology owe much more to the developers of new experimental material than is usually supposed. We all know how much membrane biophysics owes to Professor Young; if he had not found the giant axon, someone would have had to invent it. Fewer, I suspect, would recognize the preparation-founding contributions of David Keilin to muscle biochemistry, of J. S. Alexandrowicz to receptor physiology, and of Madame Arvanitaki to the analysis of networks of identified nerve cells. Yet each of these people, in very different scientific arenas, found just the right circumstance in the right organism to ask a whole set of new experimental questions. The proper test of the importance of such discoveries is their infectivity. The flight muscles of bees, the stretch receptors of crayfish, and the abdominal ganglion of *Aplysia* have become standard experimental subjects in a number of first-rate laboratories.

The developers of new preparations frequently do not have the last, or even the most important, word about their discoveries. They resemble pioneering plants on new ground: well-adapted for the colonizing role, they ultimately give way to an association of more stable, settled forms. Yet they are essential. What are their characteristics? A certain restlessness I suspect; but more than that, a fascination with the richness of the evolutionary process and a strong faith in the power of the resulting relationships. These are prominent benchmarks in the zoological tradition, but somewhat less visible in the medical one.

The history of such innovation is often fascinating. It tells, not infrequently, of the reluctance of most scientists to move off familiar ground so as to better their angle of attack on the problem that interests them. An illustration is to be found in the history of *Aplysia*, the tectibranch mollusk that has taught us so much about the organization of neuronal networks. The preparation was first used by Madame Arvanitaki in her laboratory at the Oceanographic Institute of Monaco. The first experiments were done just before World War II; a more extended series of papers (e.g., Arvanitaki and Chalazonitis, 1949) began appearing just after the war, describing the identity, spontaneous activity, photosensitivity, and synaptic connections of *Aplysia* neurons. It soon became clear that the preparation had unique advantages. The process of contagion began in the late 1950s when Ladislav Tauc took up the preparation in Fessard's laboratory in Paris. I can remember noting this fact with a distinct feeling of pleasure, since Tauc had been carrying on a dangerous flirtation with crayfish at the time.

Tauc produced a brilliant series of descriptions of the relationship between cell morphology and patterns of synaptic activation in the giant cells and in some of the smaller neurons as well (e.g., Tauc, 1962). The infection leaped the Atlantic in 1960, when Arvanitaki and Chalazonitis showed Felix Strumwasser the preparation at Woods Hole, and soon afterward Eric Kandel went to spend a year in Paris with Tauc. Then the rush was on, and by the late 1960s an uncounted number of laboratories were filled with Aplysiasts. So effective was the recruitment that it spawned a brisk cottage industry of *Aplysia* collection on the West Coast. Only edibility could have made sea hares a hotter market item.

At this point the main interest in *Aplysia*, largely under Kandel's influence, had been pointed toward the connectivity patterns of neuronal clusters in the abdominal ganglion (see Kandel, 1969). These efforts had proven unusually successful, with one exception: the networks had not yet been tied to behavioral action. They were, in a sense, all dressed up with no place to go.

Now, no one realized this problem more clearly than Kandel, who did two things about it. First, he moved away from an exclusive preoccupation with the abdominal ganglion and began with his group to study the brain and buccal ganglion systems that control the apparatus of feeding; and second, he demonstrated outputs to visceral organs from some of the abdominal ganglion networks.

An entirely different response was given by others. These, and they include the more zoologically inclined, examined the situation and decided to select new preparations using behavior as the selection criterion. A first thrust in this direction is represented by the work of Willows (1968) on *Tritonia*. Later there was a coherent move by several laboratories toward mollusks with predacious habits and a more stereotyped and interesting feeding behavior: the work on *Pleurobranchaea* by Davis and his coworkers (Davis, Siegler, and Mpitsos, 1973) and on *Helisoma* by Kater's group (Kater and Rowell, 1973) are good examples. The logic of the move to these new preparations for the specific purpose of associating neural circuits with behavior was sound, and like all developments of this kind they were initiated by a small fraction of the investigating population. The characteristics of the exploring group, as far as I can tell, included a primary interest in the behavior itself and, it must be admitted, a minimal investment in the "standard" preparation.

We have had only one adventure of this sort in my laboratory. It was exciting to us, though it led to nothing as widespread as the things I have been discussing. The story is worth relating mainly because it suggests the major roles played by preparedness and luck in finding the right system. Since graduate student days I have had an interest in primary transductions and network organization in the visual system. I turned from the vertebrate retina almost as soon as I could, after finishing my thesis, and started working on some simple neural photoreceptors in invertebrates that promised to reveal something about primary processes. In a new sensory system in a lamellibranch mollusk, which I discovered in Woods Hole one summer, I found evidence that primary photoreceptors can exhibit primary photoinhibition (Kennedy, 1960)—a proposal first actually made by Hartline and subsequently confirmed by many others (Land, 1968; Gorman and McReynolds, 1969; Mpitsos, 1973). This made me anxious, when I first moved to Stanford, to find a better molluskan light receptor—one in which intracellular recording was possible and a limited number of cells were involved. A graduate student, John Barth, later developed enthusiasm for the search, and we decided that the place to prospect was in a diverse group of gastropods with eyes not borne on tentacles. The reasoning was fairly straightforward: gastropods as a group apparently went in for large neurons early in their evolution, and they have them in all parts of the nervous system; other molluskan groups do not. Although the tentacular eyes of snails and their relatives seem mostly to consist of a large number of small cells, the marine nudibranch mollusks have smallish internal eyes located right over the brain. Dozens of species were available; it seems, in retrospect, as though John must have tried nearly all of them before finding a suitable preparation in the small aeolid nudibranch called *Hermissenda*. Its eye contains only five cells, each large enough to penetrate with microelectrodes (Figure 3.1). Soon after John Barth had

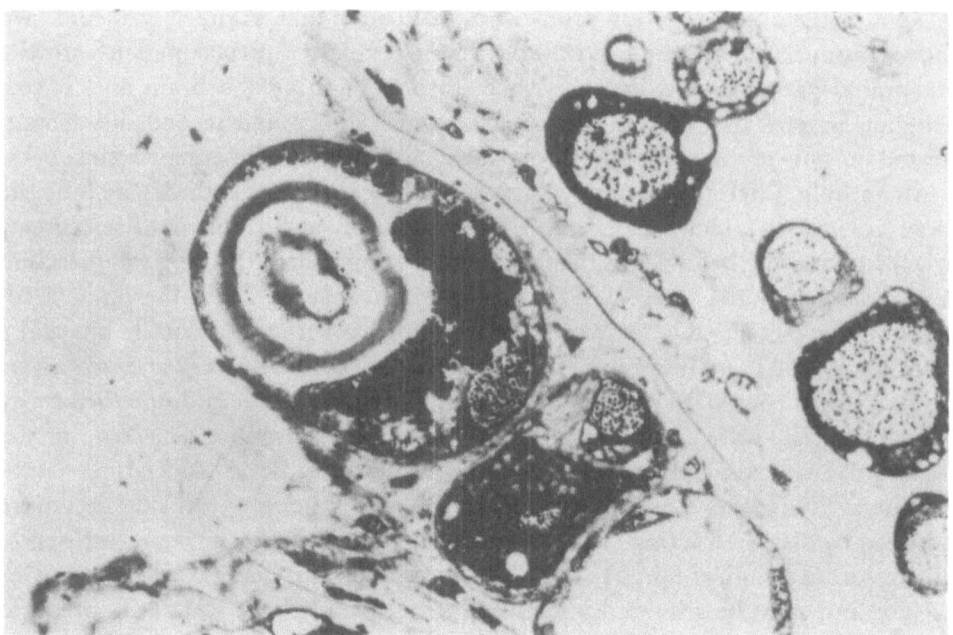

Figure 3.1   Eye of the nudibranch mollusk *Hermissenda*. The lens (upper structure) measures about 50 μm across; below it a ring of screening pigment and the nucleus of one of the five photoreceptor neurons are visible. Photograph of a 1 μm epon-embedded section prepared by the late John Barth.

completed a brief description of the situation, he died in a tragic automobile accident. Michael Dennis later took up the work with *Hermissenda* and investigated the visual cells in detail. His experiments showed that one of the processes we hoped to find there, primary inhibition, is missing. Instead there is a network of lateral inhibitory synapses that is the smallest one known, and very amenable to analyses of visual integration in a simplified system (Dennis, 1967). Since other groups have found the preparation useful for further studies, I am not much bothered by the fact that we came away with something other than what we went in to get.

A second outcome of the comparative approach is the *deliberate analysis of convergence*. We will only be able to mount an attack on higher functions in the nervous system if we can find some generalities. At what level do they occur? One approach is to inspect a variety of systems, each independently evolved to perform the equivalent output function, to see if all of the solutions have common properties. If they do, then those properties are likely to be found in all other systems, and we may be able to approach new ones with some sensible experimental priorities.

We have barely begun to assemble a list of neural networks that have common circuit properties. One entry is the common design of sensory networks that deal with discontinuities in the spatial distribution of stimulus energy. In a number of such systems receptors or higher-order elements are connected by lateral inhibitory synapses, which function to enhance the contrast level. The arrangement is so common in independently evolved systems dealing with different modalities that experiments on a novel preparation can proceed economically on the assumption that it will be found.

A principle of equivalent breadth, though of more recent origin, is that within a given system of parallel elements, neurons of increasing size show increasing thresholds to natural input and become more phasic. As far as I know, this general notion was first adumbrated by Bullock in 1953. It was shown by careful experiment to hold for mammalian motoneurons by Henneman and his colleagues (Henneman, Somjen, and Carpenter, 1965), and it was subsequently extended in intracellular studies on identified crustacean motoneurons in our laboratory by W. J. Davis (1971). This principle probably has somewhat less explanatory power than has been accorded it by its most enthusiastic admirers: size is not the *only* threshold-determining parameter among motoneurons, nor are size differences by themselves always sufficient to produce the differences in threshold with which they are correlated. Nonetheless, the size principle is a generality with an impressively broad application to systems with independent evolutionary histories.

In systems for the control of load-moving skeletal muscle, a connection principle is emerging that seems to show the same kind of evolutionary convergence. Most muscles that operate skeletal joints in higher animals possess parallel muscle-receptor organs. These are excited when the muscles are stretched passively; in turn they activate motoneurons returning to these same muscles and so mediate resistance reflexes that tend to stabilize their length despite imposed changes. Many such receptors also receive efferent innervation. That to mammalian muscle spindles, first defined experimentally by Hunt and Kuffler (1951), excites the main sensory endings and thus produces a reafferent discharge when the spindle is not unloaded by a greater contraction in the extrafusal, working muscle fibers. Various views of the functional utility of this arrangement were presented in the 1950s; some emphasized the role efferent discharge might play in maintaining the sensitivity of the stretch reflex during contraction of the main muscle, while others proposed that the spindle efferents might "lead" in initiating voluntary movements (Granit, 1955).

It now appears that elements of each view are correct. The efferent innervation of the muscle spindle is critically involved in voluntary movements; but activation of the motor pathways to the spindle and to the working muscle via the large motoneurons appears to be simultaneous, with discharge to the spindle efferents often predominating by virtue of the lower thresholds of these smaller motoneurons (for a clear early statement of this relationship see Hunt and Perl, 1960). This coactivation of the two kinds of motoneurons produces a situation in which the spindle afferents exhibit a discharge maximum during the contraction of the muscles in which the spindles are contained, a fact that seems paradoxical since they are being unloaded. But this arrangement provides a sensitive measure of the load opposing the contraction, since additional load will have the effect of slowing the shortening of both muscles and adding to series tension in the spindle. The accelerated spindle discharge then feeds additional excitation to the large motoneurons, using the stretch-reflex pathway. The participation of such circuits in load compensation has been directly demonstrated for the mammalian respiratory system by Corda, Eklund, and von Euler (1965).

A circuit controlling the contraction of abdominal extensor muscles in crayfish has almost identical formal properties (Figure 3.2). The receptor muscle is in parallel with the functional extensors, and passive stretch excites the receptor neuron. It connects monosynaptically with motoneurons that innervate only the working muscle, just as in the stretch reflex of mammals. Control circuits in the central nervous system activate motoneurons that innervate both the receptor muscle and the working muscle; although the peripheral pattern of connection is different, it has exactly the same result as in mammals: there is afferent return from the muscle receptor organ during extension. These features were worked out in collaboration with Howard Fields, who demonstrated most of the connections (Fields, 1966). Later we showed that the central pathways for load-compensated (or, in current terms, "servo-assisted") movements (those involving the coactivation of receptor muscle and working muscle) were independent of those producing activation of only the working muscles (Fields, Evoy, and Kennedy, 1967). Recently Sokolove (1973) has obtained more direct evidence for load-compensating ability.

Now, it seems quite remarkable to me that a single engineering problem has produced two entirely independent solutions that resemble one another so exactly. It is interesting to note where the similarities begin and end. The resemblances at

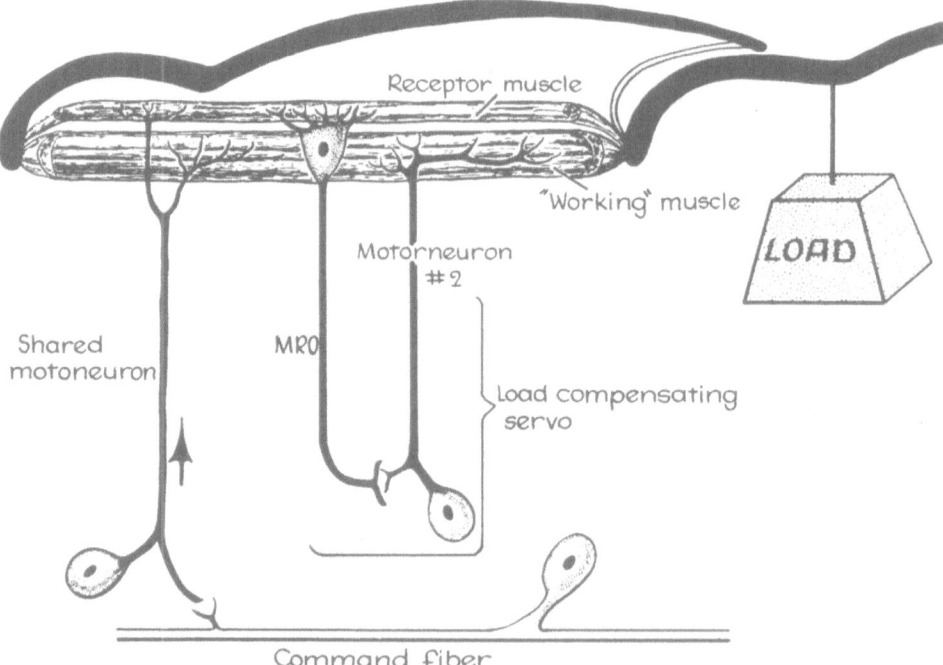

Figure 3.2  Diagram of the connections linking the muscle receptor organ of the crayfish to identified motor neurons in the central nervous sytem. The amount of reflex excitation sent around the load-compensating servo loop is proportional to the load placed upon receptor and "working" muscles during a commanded coactivation.

the cellular level are not good at all: spindles look nothing like crayfish stretch receptors, the method of achieving coactivation is quite different, and so on. Working with these disparate starting materials and with special constraints in each case, evolution has nonetheless produced a *formally* identical structure, which is to say that the same input produces similar outputs. This is not surprising since natural selection acts upon outputs and not on internal structure; but it says something about the levels at which we should look for similarities in organisms. That is what I mean by commonness of *process*. Its implication is an article of faith for me: selection will shape behavioral controls in the same way if the output objective is similar. The corollary is that we are free to pick the most advantageous preparation in which to pose a question, because the results will be transferable if we correctly interpret the similarities of process.

## Special Situations

Comparative studies need not, of course, always end by focusing on similarities of organization. Sometimes they lead instead to singular solutions which, by virtue of their uniqueness, provide special opportunities for the experimenter. The electron micrograph in Figure 3.3 is a cross section of part of a large motor axon in the crayfish. As you can see, it contains all of the ultrastructural features analyzed in much earlier electron micrographs of nerve axons by Frank Schmitt and his colleagues: neurotubules and neurofilaments are there, along with mitochon-

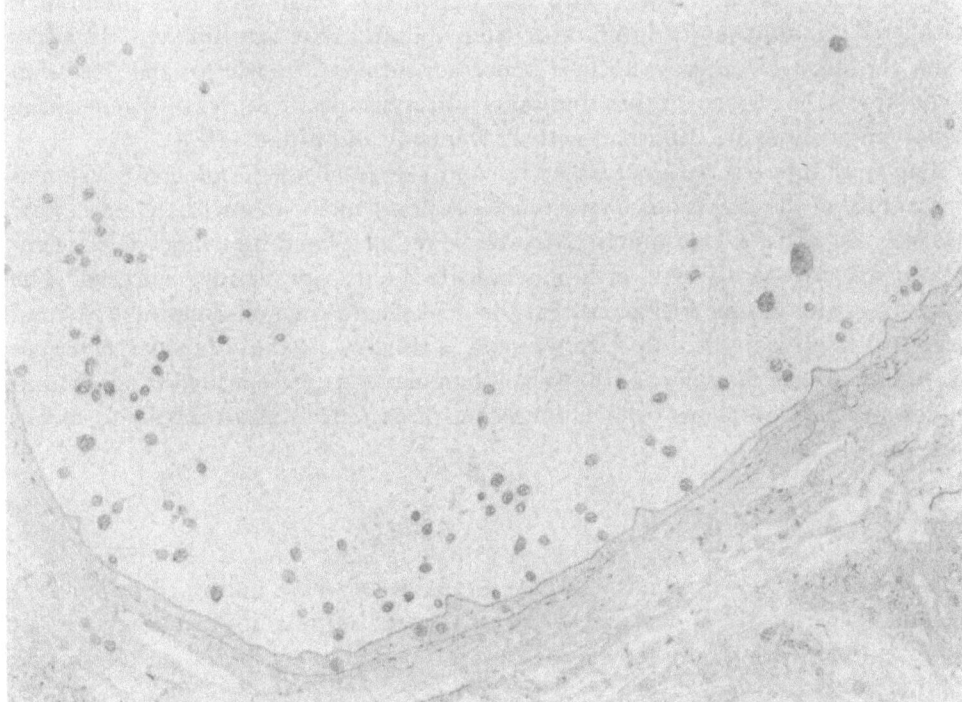

Figure 3.3  Electron micrograph of a motor axon in the crayfish claw that had been surgically separated from its cell body in the central nervous system 37 days earlier. The diameter of the axon is approximately 15 $\mu$m.

dria, a normal looking membrane, and the like. In fact, there is absolutely nothing here worth making a fuss over, except that this axon had been separated from its cell body and nucleus more than a month earlier.

Isolated axon segments prepared, like this one, by sectioning the nerve proximally and removing a large segment to prevent regeneration, can also retain their ability to conduct impulses and to release normal amounts of transmitter. We first showed that the motor axon innervating the opener muscle in the crayfish claw, even three months after amputation from the soma, will produce facilitated excitatory junctional potentials even on the 100,000th stimulus (Hoy, Bittner, and Kennedy, 1967). George Bittner, who played a major role in these early experiments, has now pushed the survival time to nearly one year in work in his laboratory at the University of Texas.

Why should such pieces of nerve retain this competence long after a vertebrate axon would have degenerated? We believe it is part of a reconnection strategy well-suited to the special needs of the organism. Where peripheral nerves take a superficial course in appendages vulnerable to injury, there must be a requirement for economical repair in case of interruption. The surviving distal segment of an efferent neuron affords its outgrowing central counterpart an opportunity to reestablish functional relations with 60,000 peripheral connection sites (Bittner and Kennedy, 1970) without the considerable expense of growing new terminals for each one. Both the cost of new synthesis and the possibility of making incorrect terminations are avoided if the regenerating central process simply fuses with the distal segment. In agreement with this hypothesis we showed that function is recovered simultaneously in co-innervated muscles that are different distances from the point of outgrowth. This is not conclusive evidence for the "refusion hypothesis," but we now think that other ultrastructural features of regenerating distal motor axons are consistent with it (Kennedy and Bittner, 1974).

But more interesting questions lie beyond our reach at the moment. What is the nature of the provisions for metabolic support in such enucleated processes? Do they depend for transmitter biosynthesis on long-lived messenger RNA templates? Or are there sources of supply outside the neuron, in adjacent cells? The only suggestive datum we have is that the glial sheaths surrounding these isolated segments show considerable enlargement, especially in the adaxonal layer, as though they were unusally active. This system may thus provide a good opportunity for studying problems of metabolic control as well as those involving reconnection specificity.

## Behavior as Adaptation

A third area in which a concentration upon the adaptive nature of behavior is helpful is the matter of deciding what inputs and outputs are meaningful to an animal—which we need to know if we are to unravel connectivity patterns in the central nervous system.

It may be a truism to say that the evolutionary process governs behavior, just as it does other dimensions of the phenotype, but the neurophysiologist ignores it at his peril. A knowledge of the importance of real stimuli in the life of an animal can be an effective predictor of the kinds of neural networks that function as sensory filters in its central nervous system. This should hardly be a necessary reminder at MIT, where the imaginative use of "real-world" stimuli, not to say dioramas, made its debut in visual physiology. I was enormously impressed by the success of this approach, and it has been widely copied: the techniques that disclosed bug perceivers have since revealed many other higher-order neurons that qualify as "feature detectors." A good discussion of the issue is to be found in *Auditory Processing of Biologically Significant Sounds* (Worden and Galambos, 1972). Though the concept has had to survive some heavy weather, it is not readily assailable when applied to input configurations that have directly demonstrable survival value. That is possible, for example, where the systems are concerned with species recognition in courtship—a circumstance that allows a direct assessment of the coding significance of particular stimulus properties on behavior as well as on neural responses.

These requirements are met in the acoustical courtship behavior of crickets (Bentley and Hoy, 1972). In this system the technique of hybridization has made it possible to test genetically controlled features of the output pattern as well (Bentley, 1971). Female crickets have neurons that respond to calls of the male, and it has recently been shown that hybrid females orient to the songs of sibling hybrid males in preference to those of either parental species (Hoy and Paul, 1973). By thus linking the genetic control of input selectivity to that of pattern generation, this result suggests a genetic attack on the question of how input connections accomplish recognition—the logical testing ground for notions of feature detection.

On the output side the question is a little different. What are the neural elements that control behavior? In particular, how large are the controlling ensembles? The discovery that single, identified neurons release coordinated behavior patterns (Kennedy, Evoy, and Hanawalt, 1966; Willows, 1968), and the related evidence that sensory feedback modulates but does not phasically structure the motor output, contribute to a new view of the evolution of behavioral routines. Changes in input connections to such command neurons could, in principle, account for basic phenomena that have been of much concern to comparative ethologists, including ritualization.

In conclusion, there is in our invitation to this symposium language that can be construed as a license to talk rather personally about the style or circumstances in which we like to do science. I haven't tried a lot of ways, so there may be a little rationalization in what follows. Nevertheless, two aspects of my professional circumstance have always struck me as especially valuable.

The first is the opportunity to have worked for most of my productive scientific life in a special kind of department. Much as I value the close association of other neurophysiologists, I value even more the opportunity to interact with different kinds of biologists—developmentalists, students of evolutionary theory, molecular biologists, and so on. A biology department of modest size that must cover the whole waterfront makes such interaction almost inevitable. But at Stanford I have had the extra advantage of a remarkable group of colleagues who deliberately seek that kind of relationship and value the results it generates.

The second is a modus vivendi that somehow came into being in our laboratory. It was never announced; indeed I never planned it in any deliberate, formal way. But it came about that each graduate student and postdoctoral fellow in the group usually picked his or her own research program, even if the choice involved some fairly painful fumbling around; and that the direction and rate of progress were determined more by personal appetite than by schedule. There is no denying that this leads to some disorganization, but I think the benefits in two areas outweigh the costs. First, a broader diversity of work goes on at any moment; second, there is greater sense of independence and responsibility for one's own work on the part of each individual.

The best argument for this loose arrangement has to do with the students themselves: I think it makes them broader, more curious about more things, and more able to do for themselves later on. It comes at some cost to the proprietor, since it requires a relatively more scattered investment of time. But this particular sacrifice is not a difficult one to make, and the returns are rich because the variety of projects itself guarantees an interchange of suggestions and, in many cases, a shifting flow of collaborations. Though I cannot be certain of it, I think this system generates novel ideas at a faster rate than one in which the group pursues a defined problem. I hope that is true, because to me diversity is as exciting in the laboratory as it is in nature.

## References

Arvanitaki, A., and Chalizonitis, N. (1949): Prototypes d'interactions neuroniques et transmissions synaptiques. Données bioélectriques de préparations cellulaires. *Arch. Sci. Physiol.* 3:547–566.

Bentley, D. R. (1971): Genetic control of an insect neuronal network. *Science* 174:1139–1141.

Bentley, D. R., and Hoy, R. R. (1972): Genetic control of the neuronal network generating cricket (*Telegryllus gryllus*) song patterns. *Anim. Behav.* 20:478–492.

Bittner, G. D., and Kennedy, D. (1970): Quantitative aspects of transmitter release. *J. Cell Biol.* 47:585–592.

Bullock, T. H. (1953): Comparative aspects of some biological transducers. *Fed. Proc.* 12:666–672.

Corda, M., Eklund, G., and Euler, C. v. (1965): External intercostal and phrenic $\alpha$ motor responses to changes in respiratory load. *Acta Physiol. Scand.* 63:391–400.

Davis, W. J. (1971): Functional significance of motoneuron size and soma position in swimmeret system of the lobster. *J. Neurophysiol.* 34:274–288.

Davis, W. J., Siegler, M. V. S., and Mpitsos, G. J. (1973): Distributed neuronal oscillators and efference copy in the feeding system of *Pleurobranchaea*. *J. Neurophysiol.* 36:258–274.

Dennis, M. J. (1967): Electrophysiology of the visual system in a nudibranch mollusc. *J. Neurophysiol.* 30:1439–1465.

Fields, H. L. (1966): Proprioceptive control of posture in the crayfish abdomen. *J. Exp. Biol.* 44:455–468.

Fields, H. L., Evoy, W. H., and Kennedy, D. (1967): Reflex role played by efferent control of an invertebrate stretch receptor. *J. Neurophysiol.* 30:859–874.

Gorman, A. L. F., and McReynolds, J. S. (1969): Hyperpolarizing and depolarizing receptor potentials in the scallop eye. *Science* 165:309–310.

Granit, R. (1955): *Receptors and Sensory Perception.* New Haven: Yale University Press.

Henneman, E., Somjen, G., and Carpenter, D. O. (1965): Functional significance of cell size in spinal motoneurons. *J. Neurophysiol.* 28:560–580.

Hoy, R. R., Bittner, G. D., and Kennedy, D. (1967): Regeneration in crustacean motoneurons: Evidence for axonal fusion. *Science* 156:251–252.

Hoy, R. R., and Paul, R. C. (1973): Genetic control of song specificity in crickets. *Science* 180:82–83.

Hunt, C. C., and Kuffler, S. W. (1951): Stretch receptor discharges during muscle contraction. *J. Physiol.* 113:298–315.

Hunt, C. C., and Perl, E. R. (1960): Spinal reflex mechanisms concerned with skeletal muscle. *Physiol. Rev.* 40:538–579.

Kandel, E. R. (1969): The organization of subpopulations in the abdominal ganglion of *Aplysia.* *In: The Interneuron* (Proceedings of a Conference Sponsored by the Brain Research Institute, U. C. L. A., September 1967). Brazier, M. A. B., ed. Berkeley and Los Angeles: University of California Press, pp. 71–111.

Kater, S. B., and Rowell, C. H. F. (1973): Integration of sensory and centrally programmed components in generation of cyclical feeding activity of *Helisoma trivolvis.* *J. Neurophysiol.* 36:142–155.

Kennedy, D. (1960): Neural photoreception in a lamellibranch mollusc. *J. Gen. Physiol.* 44:277–299.

Kennedy, D., and Bittner, G. D. (1974): Ultrastructural features of regeneration in crayfish motor axons. *Z. Zellforsch. Mikrosk. Anat.* (in press).

Kennedy, D., Evoy, W. H., and Hanawalt, J. T. (1966): Release of coordinated behavior in crayfish by single central neurons. *Science* 154:917–919.

Land, M. F. (1968): Functional aspects of the optical and retinal organization of the mollusc eye. *In: Invertebrate Receptors* (Proceedings of Symposium of the Zoological Society of London, 30–31 May 1967). Newall, G. E., ed. New York: Academic Press, pp. 75–96.

Mpitsos, G. J. (1973): Physiology of vision in the mollusk *Lima scabra.* *J. Neurophysiol.* 36:371–383.

Sokolove, P. G. (1973): Crayfish stretch receptor and motor unit behavior during abdominal extensions. *J. Comp. Physiol.* 84:251–266.

Strumwasser, F. (1971): The cellular basis of behavior in *Aplysia.* *J. Psychiatr. Res.* 8:237–257.

Tauc, L. (1962): Site of origin and propagation of spike in the giant neuron of *Aplysia*. *J. Gen. Physiol.* 45:1077–1097.

Willows, A. O. D. (1968): Behavioral acts elicited by stimulation of single identifiable nerve cells. *In: Physiological and Biochemical Aspects of Nervous Integration*. Carlson, F. D., ed. Englewood Cliffs, N. J.: Prentice-Hall, pp. 217–243.

Worden, F. G., and Galambos, R., eds. (1972). Auditory processing of biologically significant sounds. *Neurosci. Res. Program Bull.* 10:1–119.

Gerald M. Edelman (b. 1929, New York, New York) received the Eli Lilly Award in 1965 and the Nobel Prize in 1972 for contributions to the field of immunology. His research has focused on such topics as the structure of antibody molecules, the control of cell growth, and the architecture of the cell surface. Dr. Edelman is currently Vincent Astor Distinguished Professor at Rockefeller University.

# 4

# Molecular Recognition in the Immune and Nervous Systems

## Gerald M. Edelman

My first encounter with the Neurosciences Research Program left me pleased but a bit puzzled. In the fall of 1964 I was invited to come to Boston to tell the members of this group about the latest ideas in immunology, and I still vividly remember my impressions. In a small meeting room I was surrounded by a number of scientists of large reputation in a variety of disciplines, whose attitude was one of joyous receptivity and open questioning. It did not seem to be a very professional group, however, for the tone of the meeting contrasted strongly with the tightly scheduled and somewhat "held-in" atmosphere of the usual specialist societies.

Within the year I was to learn more of the NRP and of the infectiously curious and insatiable man who was at its head. In that first meeting he ranged over many fields, trying to integrate, to probe, and to define things without pretense. My first impression was of strength but also of some vagueness. Later experiences taught me that this slight blur was the price paid by Frank Schmitt for a very profound gift: that of pointing the nose of curiosity in the directions of greatest intellectual importance and interest. Given this man and this company, and above all the feeling of intellectual freedom they engendered, I felt privileged to be asked to join them as an Associate early in 1965.

At the time I could understand that they might want to have an immunologist in their midst if only for general reasons, for an immunologist was no more removed than a mathematician from the central interests of the neurosciences, and there were several mathematicians around. But it was only with the passing of time that I began to see the specific as well as the general relationships between the immune and nervous systems. It might be of some value to talk of those relationships here, and, in the tradition set by Frank Schmitt, to see where they might lead.

### Recognition Systems, Molecular and Otherwise

The difficulties of comparisons of biological systems can be covered by a metaphor in an easy and shallow fashion. At this level of "analysis" it is not difficult to see that both the brain and the immune system are recognition systems. Both can recognize and therefore distinguish positively among different objects in a set (in the one case via sensory signals, in the other via molecular complementarity between the shapes of antigens and the combining sites of antibodies). By positive recognition I mean that they do not merely exclude an object by subjecting it to a match with a fixed pattern, but rather that they can name or tag an object uniquely. This is a much more powerful kind of recognition than the exclusive one embodied, say, in the construction of a combination lock. Furthermore, both systems have the capacity to store a recognition event ("memory" and "immunological memory") as well as the capacity to forget (Edelman, 1967; Jerne, 1967; Nossal, 1967).

Is this metaphor useful? In a specific sense I believe it is not, for the detailed components of the nervous system are obviously very different from those of the immune system. But in the heuristic and exemplary sense I believe it is worth some discussion, for the immune system is the best understood molecular recognition system, and the principles embodied in such a system during evolution may turn out to be general ones.

The main principle is that the immune response is selective: all of the information required to recognize various foreign antigens as well as self-antigens is programmed in the system before the first encounter with these antigens. An encounter with a particular antigen serves to elicit or select responses by those antibody-producing cells that happen to have antibodies with sites complementary to the shape of that antigen (Figure 4.1). There is no information transfer from antigen to antibody during synthesis of the antibody molecule; that is, the system is not instructive.

It so happens that the immune system operates by a particular form of selection, known as clonal selection. The response to a selection event is the stimulation of the lymphocyte bearing an antibody on its surface to mature and divide, producing a clone of progeny cells each capable of synthesizing more antibodies of the same kind. The production of daughter cells that do not exhaust themselves in the division process leaves an imprint on the system that can serve as a "memory" for a second encounter. Should such an encounter take place the response is greater

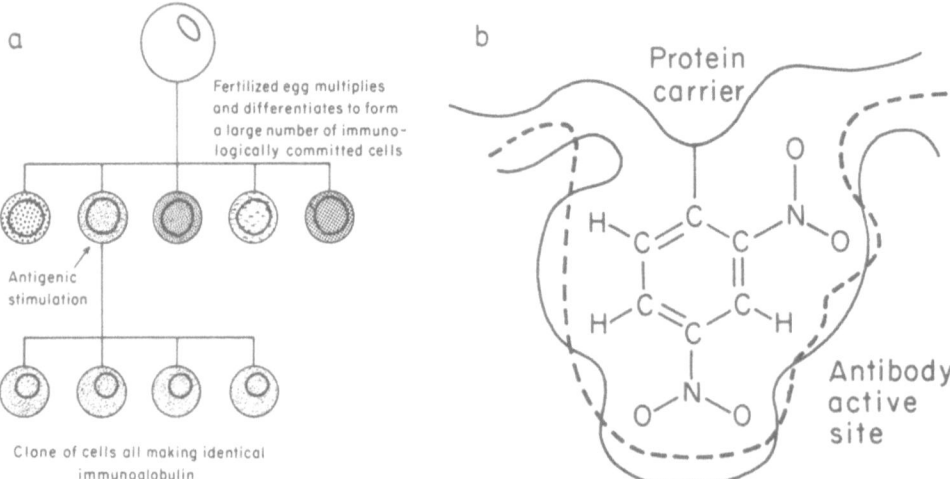

Figure 4.1  (a) Diagram illustrating the fundamental idea of clonal selection after development of a repertoire of antigen-binding cells carrying different antibody molecules on their surfaces. Encounter by an antigen with the appropriately complementary surface antibody molecule results in cell maturation, mitosis, and increased production of antibodies of the same type by the daughter cells. (b) Diagram illustrating the principle of molecular complementarity by means of which an antibody molecule can bind an antigen of a given shape.

and swifter because there are more daughter cells, but their response obeys the same principles as the original stimulation.

Although it is very unlikely that the nervous system operates in this particular fashion, it may be true that principles of selection are of fundamental importance for its functioning. For this reason it is worth asking what, in general, are the minimal requirements of a selective system. First, it is important to have a large enough repertoire of diverse recognition units, and this requirement is an absolute one. In the case of the immune system, we now know that the repertoire is provided by a large collection of antibody molecules, each with a different combining site. The basis of the diversity rests in differences in the amino-acid sequences of the light and heavy chains of the antibody molecule, each of which has a "variable" and a "constant" segment (Figure 4.2).

Although we do not know the genetic origins of this diversity, we do know that some of it is determined during evolution and that it is expressed very rapidly during ontogeny. As shown schematically in Figure 4.3, a selective system must generate sufficient diversity above a certain threshold in order to function at all. Above a certain amount, however, more diversity would provide only greater redundancy or degeneracy to the system. Perhaps the most important characteristic of such systems is this degeneracy: there are *several* more-or-less effective ways in which a recognition can take place. This is an obligatory property of selective-recognition systems, and it is expressed in the immune system by the existence of a heterogeneous collection of antibodies to even a single well-defined antigen.

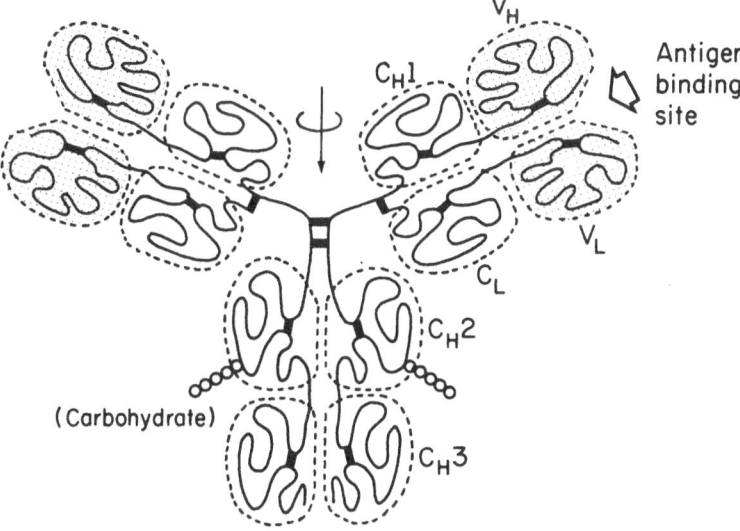

Figure 4.2 A model of the structure of a human IgG molecule. The variable regions of heavy and light chains ($V_H$ and $V_L$), the constant region of the light chain ($C_L$), and the homology regions in the constant region of the heavy chain ($C_H1$, $C_H2$, and $C_H3$) are thought to fold into compact domains (delineated by dotted lines), but the exact conformation of the polypeptide chains has not been determined. The vertical arrow represents the twofold rotation axis through the two disulfide bonds linking the heavy chains. A single interchain disulfide bond is present in each domain. Carbohydrate prosthetic groups are attached to the $C_H2$ regions.

The second requirement of a selective-recognition system is a means to ensure reasonably frequent encounters between the members of the repertoire (e.g., the lymphocytes carrying the antibodies on their surface) and the object to be recognized (e.g., the antigen). In the immune system this trapping is carried out by a complex recirculation of various types of cells, by restrictions of anatomy, and by cooperation between two systems of cells. One type, the T cell or thymus-derived lymphocyte, can somehow present the antigen it traps via its receptors to the B cells or bone-marrow-derived lymphocytes that eventually secrete humoral antibodies.

The third requirement for selective recognition is, like the first requirement, absolute: it is the ability to respond to an encounter by an amplified response. In the immune system the amplification takes place by an increase in cellular protein synthesis among the antigen-binding cells as well as by an increase in the number of cells producing antibodies after clonal expansion. To produce a sufficiently great number of cells and specific antibody molecules constituting the response, the gain of this amplification must be fairly high.

One of the most important features of this amplification is that it is not only a threshold event, it also provides enhancement of specificity. Only cells with antibodies of high binding affinity appear able to be triggered by an antigen (Figure 4.4). Inasmuch as antibodies of higher affinity are in general more specific, this explains how specificity becomes a dynamic property of the system and not just a static property of antibodies in the degenerate repertoire. If this thresholding for specificity did not occur, perhaps 1 in 100 cells would produce antibodies upon stimulus, and the system would show relatively low specificity. In most immune systems of higher animals, however, only 1 in $10^5$ cells actually responds, despite the fact that 1 in $10^2$ cells binds that antigen specifically. One might say that the system behaves as an amplifier with a high-pass filter on its input and that such a filter acts as an affinity or "high-free-energy" filter.

How do the ideas of selection, repertoire, degeneracy, amplification, and affinity

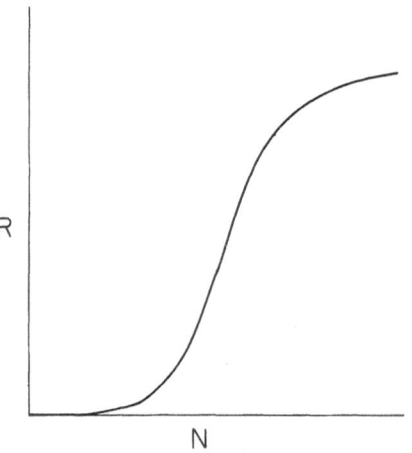

Figure 4.3 Hypothetical diagram illustrating the relationship between the number of different binding sites (N) on antibody molecules and the likelihood (R) of recognition of a variety of antigens.

filtering apply to the brain? Obviously, in the absence of much more detailed knowledge they are not of much direct use. Nevertheless, the immune example may be valuable both in relation to the general nature of the central nervous system and also to certain specific cellular details. I should like to discuss these briefly and also to consider how immunology may be used as a tool to explore certain molecular aspects of specificity in the nervous system.

### The CNS as a Selective System

The early hopes that there would be direct coding for memory into macro-molecules always seemed to me to be naive, mainly because, even in the vastly simpler immune system, additional cellular and dynamic characteristics and properties are required for specific storage of a recognition event. I therefore do not think that it surprises immunologists that the direct-coding hypotheses have received little or no support (Quarton, Melnechuk, and Schmitt, 1967; Schmitt, 1970). Does this mean that molecules are not key mediators of memory? Obviously not, but merely that both complex molecular *and* cellular structures must be involved.

One of the first questions to ask about such structures is whether they operate by selective principles. This is equivalent to asking what the repertoire is, what the trapping mechanism is, and what the amplification mechanism is. As already discussed by Jerne (1967), it is not at all a vain hypothesis to consider the brain as a selective system. The problem is, of course, to define the unit of selection. Is it the neuron, the individual synapse, or a larger multicellular structure? Whatever it turns out to be, it is interesting to consider that the spatially distributive aspects of

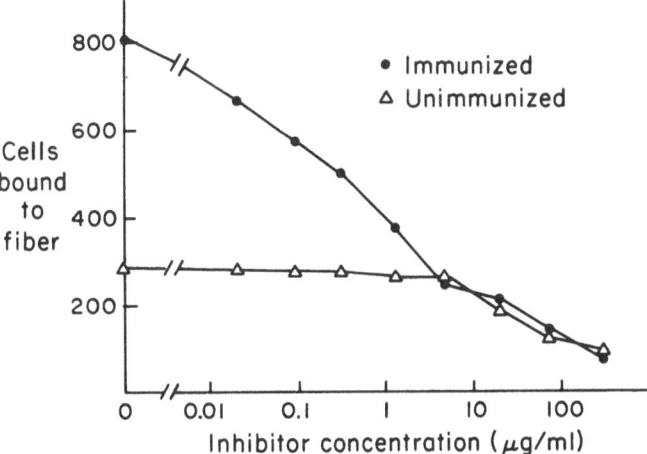

Figure 4.4   Inhibition by soluble dinitrophenyl bovine serum albumin of spleen-cell binding to dinitrophenyl-derivatized fibers. Cell numbers are the fiber edge counts for a 2.5-cm fiber segment, and each point represents a separate binding experiment in the presence of the indicated amount of inhibitor. Note that only a small proportion of the nonimmune-cell binding is inhibited at low concentrations of inhibitor, whereas over 60% of immune-cell binding is prevented by concentrations of less than 4 $\mu$g/ml. Above 4 $\mu$g/ml, the two cell populations behave essentially the same. This indicates that immunization results in a selective increase in the number of cells with a high affinity for the soluble inhibitor.

long-term memory within the brain may be related to an obligate requirement for degeneracy: it may be a necessary consequence of the structure of the system that *many different* recognition units, not necessarily structurally isomorphous in the strict sense, may serve to process a single report.

Assuming that the CNS is selective, I would speculate that the unit of selection in the nervous system is a network consisting of at least three cells, that the synapse is the obvious locus for the amplifier acting via the transmitters and the protein molecules that modulate their behavior, and that the clue to the mechanism of memory lies in the specific molecules and dynamics of the cell membrane, just as the clue to the triggering of lymphocytes rests in their cell membrane. If we accept these speculations for now, what suggestions for research can we take from the immunological example?

## Cell Surface Specificity and Dynamics

The lymphocyte sits poised with its immunoglobulin receptors at its surface, waiting for an antigen to define them as antibodies. It must not be imagined, however, that this poised state is a static one. The lymphocyte receptors are diffusely distributed (about $5 \times 10^4$ per cell) and anchored (in a fashion not understood), but they are still free to diffuse in the plane of the membrane. They are a jostling crowd, capable of being aggregated into patches by multivalent antigens, and, in their movement, they are more-or-less independent of other receptors and surface molecules. What does the binding of an antigen by such receptors do to signal that a cell should mature and divide?

We do not yet understand the nature of this signal, but in the course of explaining it much has been learned of the behavior of surface receptors in general, and it is likely that this knowledge will bear directly upon the neuron. Indeed, if its properties are to be understood, the neuron needs to be mapped by the same immunological methods as has the lymphocyte, which is so far the most extensively mapped eukaryotic cell. For example, besides its immunoglobulin molecules, the lymphocyte has $\theta$ markers (shared with the nervous system), TL antigens, and various histocompatibility antigens.

One may ask whether the distribution and properties of such surface receptors are modulated by structures within the cell, and vice versa. There is now evidence to suggest that the motion of these receptors is modulated by cross-linkage of other receptors and by colchicine-binding proteins in the cell. The addition of antibody to immunoglobulin receptors results in a kind of nucleation event that cross-links the receptors into patches. The actively metabolizing cell can gather these patches into caps at one pole, presumably as a result of cell and membrane motion (Edelman, Yahara, and Wang, 1973). We have recently observed that the specific cross-linkage of glycoprotein receptors on lymphocytes by plant proteins called lectins results in immobilization of cell-surface receptors (Figure 4.5). This inhibition depends upon the valence of the lectin, and also upon structures that are responsive to drugs known to bind to microtubules, including colchicine and the Vinca alkaloids. The addition of these drugs to the lymphocyte reverses the inhibition of re-

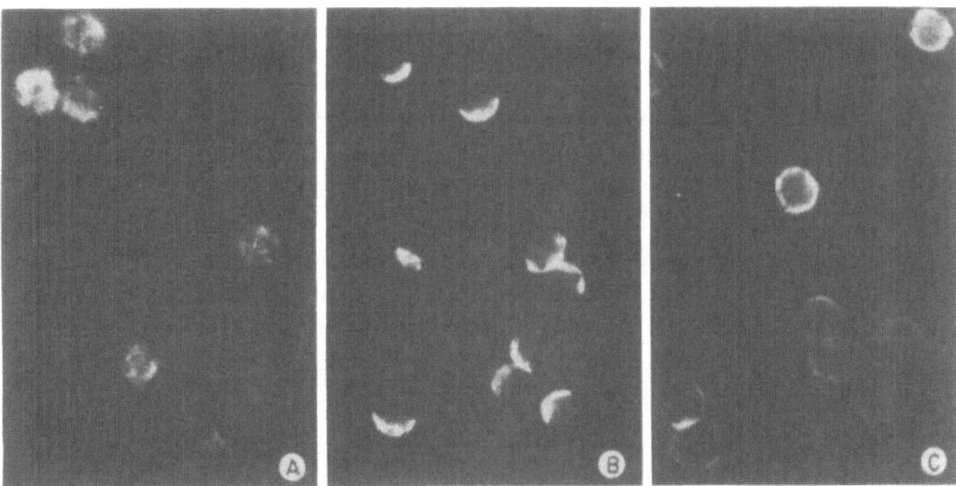

Figure 4.5   (A) Labeling patterns of cells with fluorescein-labeled anti-immunoglobulin (80μg/ ml) plus NaN₃ (10mM) at 21° C for 30 min showing patches. (B) Cells incubated with fluorescein-labeled anti-immunoglobulin (80μg/ml) at 20°C for 30 min showing caps. (C) Cells incubated with fluorescein-labeled anti-immunoglobulin (80μg/ml) plus Con A showing diffuse patterns.

ceptor mobility (Edelman, Yahara, and Wang, 1973) and also alters the capacity of the cell to take on certain shapes (Rutishauser, Yahara, and Edelman, 1974). On the basis of these observations we have postulated the existence of an assembly of colchicine-binding proteins (CBP) beneath the cell membrane, responsible for modulating the anchorage and mobility of cell receptors.

What is the relevance of these observations to specificity and selection in the nervous system? One of the first conclusions to force itself on our attention is the fact that certain receptors in the nervous system must remain at certain locations, while others must be free to move, particularly in a system that depends so strictly upon geometry in structure and in function. Clustering of receptors leads not only to topographical specificity, but also to avidity effects, in which the capacity to bind a given ligand (polyvalent drug, transmitter) is enhanced because there are multiple receptor sites in a single place. Moreover, the neuron is particularly rich in microtubular and filamentous structures, and, in analogy to the lymphocyte, it would be interesting to see whether such structures modulate receptor distribution. A classical instance for such an exploration is denervation sensitivity. Could it be that the so-called trophic effects of neurons have to do with an intracellular system for modulating the distribution of surface receptors, and that this system is capable of being triggered by cell-cell interaction?

Before we can know the answer to this question, we must certainly carry out more detailed analyses of specific neuronal proteins at the cell surface and elsewhere. Here the use of specific antiserums should be a powerful tool for both the chemistry and the cytochemistry. I shall cite here a single example from my own laboratory, both because it is pertinent and because it is intimately tied to the development of similar interests by other NRP Associates.

## The Isolation by Immunological Assay of a Specific Nervous System Protein, Antigen α

In the 1960s, motivated by a variety of interests, several researchers were looking for proteins that were specific for the nervous system. The first to be found was the protein S-100, isolated by Moore and his colleagues and shown to be a glial protein by Levine and his colleagues, largely as a result of the promptings of Frank Schmitt. We had been chatting about these things for some time at the NRP, and my student Gudrun Bennett and I decided to use an immunological approach (Bennett and Edelman, 1968). We prepared an antiserum to rat-brain extracts in rabbits and showed that aqueous extracts of rat brain, spinal cord, and sciatic nerve contained an antigen that was not detected in extracts of other rat tissues (Figure 4.6).

This protein, which we designated antigen α, had a high mobility in zone electrophoresis at pH 7.4, and, like the S-100 protein, it was acidic. After purification of antigen α, we found that it consisted of a single type of subunit with a molecular weight of 39,000. The subunits formed dimers linked by disulfide bonds, but they could also form higher aggregates. Antigen α is antigenically and compositionally unrelated to S-100, is neuronal rather than glial in origin, and has recently been shown (Bennett, 1974) to be antigenically identical to protein 14–3–2, isolated by Moore from beef brain. Cytochemical studies are under way to determine its location in the cell.

The existence of such nerve-specific proteins suggests that they have a unique function. It is interesting that, in our experience so far, none of the *soluble* proteins of the brain are microheterogeneous in the same way as immunoglobulins, and therefore there is no suggestion of a repertoire of binding sites within proteins of

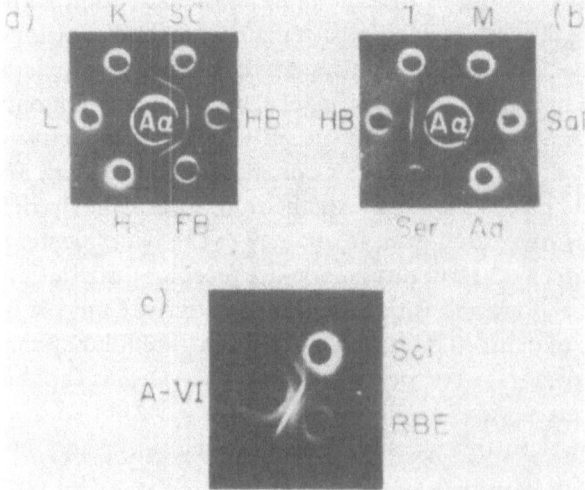

Figure 4.6   Immunodiffusion showing organ specificity of antigen α. All tissues are from rats: FB, forebrain; HB, hindbrain; SC, spinal cord; K, kidney; L, liver; H, heart; T, testes; M, skeletal muscle; Sal, salivary gland; Ad, adrenal gland; Ser, serum; Sci, sciatic nerve. The center wells in a and b contain antiserum to antigen α (Aα); the well in c contains antiserum VI, prepared against an electrophoretic fraction rich in antigen α.

the same family. It is conceivable, however, that such a repertoire may be formed from a great number of *different* proteins. In the course of our isolation procedures we observed at least 100 different proteins in the soluble brain fractions, which in fact constitute only 30% of the total protein.

Metabolic alterations, such as those induced by unilateral cortical spreading depression, appear to affect all of these proteins rather than one or another uniquely. Bennett and I were able to show that such events have large-scale effects on metabolic pools but not on the synthesis of individual proteins such as antigen $a$ (Bennett and Edelman, 1969). It seems likely, therefore, that drugs affecting protein synthesis in the nervous system usually do not have a single brain protein as their target, although many more proteins must be isolated and analyzed in terms of structure and cellular function before we can be sure of this conclusion.

The continuation of this program of the characterization of brain proteins should be specifically linked to immunochemical exploration of the neuronal and glial surface, both in the developing and adult animal. Further progress should enable us to do for the neuron what has been done for the lymphocyte, and we can look forward to some illuminating surprises similar to those already encountered in immunology.

### The Scientist as Impresario and the Invisible College

I cannot close this informal account of research and speculation without a comment upon the role of personality both in scientific research and in the coalescence of knowledge. After my years in the NRP I am no longer puzzled by the organization, but I am just as pleased as I was in my initial encounter. I now understand that the main source of my puzzlement was a failure to understand the role of such an organization as contrasted with that of the more traditional professional coteries. Frank Schmitt has helped to bring together a most unusual combination of talented scientists in an organization that frees them of the prejudices and preconceived ideas that impose themselves on any guild.

For this reason it is my conviction that Frank will be celebrated, not only for his important contributions to biophysics and protein chemistry, but also as one of the most original and generous of the scientist-impresarios of our time. Lest this seem invidious, I should hasten to say that he has persuaded me that a key ingredient in scientific research, as important as the imagination of the solitary worker, is the association of sympathetic and curious minds, versed in a variety of specialized areas, who because of their curiosity and also because of their mutual confidence are free to ask basic and even naive questions. For perhaps adventitious reasons, such a group forms an invisible college of a kind that brings elements to the analysis of a problem that cannot be provided by specialist groups.

The NRP has been such an invisible college, and its fruits have been many: some of the best small meetings in the world (the Work Sessions), the monumental volumes resulting from the ISPs, and, above all, the feeling of intellectual community among scientists from many disciplines. All of us who have had the privilege of association with the enterprise have had our hunger for generality (which I believe is the sharpest of all scientific hungers) satisfied in a meaningful and not just a vacuous or trivial fashion. The credit for the invention of this particular social mode of neuroscience must go to F.O. Schmitt, who stands as its main if not "onlie begettor": generous, energetic, hungering after knowledge, aware of both tradition and the interaction of individual talents, the authentic scientist as impresario.

## References

Bennett, G. S. (1974): Immunologic and electrophoretic identity between nervous system-specific protein antigen $\alpha$ and 14–3–2, *Brain Res.* 68: 365–369.

Bennett, G. S., and Edelman, G. M. (1968): Isolation of an acidic protein from rat brain. *J. Biol. Chem.* 243: 6234–6241.

Bennett, G. S., and Edelman, G. M. (1969): Amino acid incorporation into rat brain proteins during spreading cortical depression. *Science* 163: 393–395.

Edelman, G. M. (1967): Antibody structure and diversity: Implication for theories of antibody synthesis. *In:* Quarton et al. (1967), pp. 188–200.

Edelman, G.M., Yahara, I., and Wang, J. L. (1973): Receptor mobility and receptorcytoplasmic interactions in lymphocytes. *Proc. Natl. Acad. Sci. USA* 70: 1442–1446.

Jerne, N. K. (1967): Antibodies and learning: Selection versus instruction. *In:* Quarton et al. (1967), pp. 200–208.

Nossal, G. J. V. (1967): The biology of the immune response. *In:* Quarton et al. (1967), pp. 183–187.

Quarton, G., Melnechuk, T., and Schmitt, F. O., eds. (1967): *The Neurosciences: A Study Program.* New York: Rockefeller University Press.

Rutishauser, U., Yahara, I., and Edelman, G. M. (1974): Morphology, motility and surface behavior of lymphocytes bound to nylon fibers. *Proc. Natl. Acad. Sci. USA* 71: 1149–1153.

Schmitt, F. O., editor-in-chief. (1970): *The Neurosciences: Second Study Program.* New York: Rockefeller University Press.

Paul A. Weiss (b. 1898, Vienna, Austria), professor emeritus at Rockefeller University, has had a distinguished career as teacher, researcher, and scientific statesman. His research in developmental, cellular, and neural biology included the experimental analysis and theoretical interpretation of growth control, differentiation, cell behavior, regeneration, wound healing, and the coordination of nerve centers.

# 5
# Neural Specificity: Fifty Years of Vagaries

## Paul A. Weiss

### Preamble

The sampling in the present book, the stream of progress along the "paths of discovery in neurosciences," is a fitting tribute to Frank Schmitt's 70th birthday. It will serve as a reminder of how many notable tributaries he has added to that stream by his spirit, vision, and hardy work. But, from a broader perspective, no date can be more than arbitrary for taking stock of the progress made up to that point by a process such as the advancement of human knowledge. In the same sense discovery is more than sheer disclosure, as in spotting a wellspring in a mountain meadow and then letting it seep into the ground: it calls for grooving a bed that will allow the new source to become confluent with, or even divert, the older course of the mainstream. In our present context, discovery then ceases to be an isolated event; it appears as just a conspicuous marker for a major contribution to the continuous stream of man's growth of knowledge. And because knowledge cannot be equated with a sheer pile of amassed items of information (Weiss, 1971a: Chapter 9), its growth can not be measured in terms of bulk production, like mining ore without going on to process it, but can be assessed only by its success in furthering rational understanding of nature. Without entailing such progress, discovery remains stale. Therefore, in rating "paths of discovery," we must not just look at the act; we must trail each path, if possible, from its source through its sequential effects.

Other chapters deal with discoveries from which have issued major straight roads to a wider vista and deeper insight into the nervous system. To balance that score I shall discuss one of my own discoveries, whose outgrowths seem not to have been pursued with equal vigor and direction: the fact that each muscle has a specific "personal" identity that is instrumental in its correct communication with the nerve centers.

### Fundamentals of "Myotypic" Response

This discovery just "happened" to me, exactly fifty years ago, when I was studying the morphodynamics of regeneration of amphibian limbs. To test possible influences of their location on the body, I tried transplanting fully functional limbs to foreign sites. Not only did the transplants heal in, but, wholly unexpectedly, they also resumed motor functions of such curious regularity that they immediately diverted some of my attention from the original purpose of the experiments. And so, in 1923, I could give the first report on the basic principle governing the mode of function, which became later known as "homologous" or "myotypic" response (Weiss, 1922, 1923). My studies had revealed the existence of a highly specific form of neuromuscular communication such that a spinal nerve center can call each given muscle in its domain into action by its "name" as it were; and the signaled muscle unfailingly responds correctly and blindly, with no regard to the resulting functional appropriateness or inappropriateness to the body (Weiss, 1928, 1936b, 1941, 1950a, b, 1971b: Chapters 4 and 5).

There are, at a minimum, forty muscles in the limb of a salamander, newt, or frog, and each one of those muscles seems to have its distinctive private code, matched by a corresponding activator in the central system. Similarly the spinal system can discriminate a sensory message from the periphery in terms of the name of the sender, regardless of the latter's "local sign." To sketch the evidence for these conclusions I shall have to condense the 1500 pages of print I have devoted to their critical description by a factor of 100:1. This should be borne in mind when I use imagery.

Just how do the motor centers in the spinal cord manage to deal *coordinately* with the specific sets of muscles under their control? For a conventional answer we might use the analogy of a piano: the keyboard represents the columns of motor neurons; the strings, the individual muscles; and the hammers, the synapses. For the moment we ignore the pianist, who strikes the keys in ordered combinations to make the strings yield chords and melodies instead of noise. Just ponder if harmonic sounds could ever emerge from such an instrument unless there were a standard grid of point-to-point connections between each given key and its corresponding string; that is, unless the pianist could know which key to strike to yield a given sound.

Yet this premise became questionable when I discovered that the strict correspondence between a central key and its peripheral effector persisted even if the pattern of their connections had been arbitrarily disarrayed. In those days of the emergence of wireless broadcasting, it struck me at once that the neurons might, by analogy, serve simply as a medium over which coded signals activate a correspondingly tuned receiver by physical *resonance*. This hypothetical notion was not intended to "explain," but rather to point to, an underlying principle: each muscle has a unique property that allows the nervous system to communicate with it selectively without a prearranged set of private lines. However, whatever the explanation turns out to be, the crux of the discovery lies in the evidence for a sharp qualitative diversity among the individual muscles, paired with a matching diversity of either units or emission codes in the respective nerve centers. This basic fact had previously been neither suspected nor postulated. Muscles were just muscles, red or white, slow or fast, but without further individualization.

It is crucial to stress this *factual* side of the phenomenon, and not the various speculations offered to explain it, including my own. As far as *explanation* is concerned, I would assert that the matter has rested just about where it stood when it first came to light, even though some more data have in the meantime been amassed. It makes me question whether we are not too shortsighted in groping for the solution exclusively within the confines of orthodox neurophysiological experience, and should not rather let the phenomenon lead us to expand our range of vision. Today I can do no better than cite Adrian's conclusion upon describing the phenomenon in *The Mechanism of Nervous Action* more than forty years ago: "The neurons have a fairly simple mechanism when we treat them as individuals. Their behavior in the mass may be quite another story, but this is for future work to decide" (Adrian, 1932). Unfortunately, it is still undecided. Some twenty years later Alexander Forbes said, in a similar vein, that "Dr. Weiss has certainly given us something that we cannot brush aside casually, and we must find a way to explain it, presumably through some property of specificity, such as he suggested, but for the life of me I cannot visualize just how it works. However, I hope some day somebody can" (Weiss, 1950b: p. 21).

Thus far nobody has. There have been certain vague generalities and circumlocutions, but also a few serious efforts—considering the time lapse of half a century, pitifully few—to explain the facts in terms of the existing range of neurobiological knowledge and to try new experiments to support one or another conjecture. The trouble is that almost all such trials have been focused on one facet of the phenomenon only, or on one testing method only, and they have thus lacked conclusiveness when viewed within a comprehensive framework of *all* the features of the phenomenon already known. Since this trend can in large part be traced to insufficient familiarity with the bulk of pertinent data available, I feel prompted to give a critical summary of the hard core of relevant facts in their pristine simplicity, for to ignore any of them is bound to lead to misconceptions.

## Transplants as Monitors

I shall select from the experimental material just a few pregnant examples, all referring to amphibians. In essence the basic experiment consists of "tapping" the line of communication between the spinal cord and a leg by inserting into the line system a supernumerary leg as a live listening device. Such a transplant can be fastened in arbitrary positions and orientations so that, on recovering motility, its movements will be useless or even adverse to the locomotion of the animal. For a source of regenerative nerve supply, I severed a minor fraction of the nerve plexus to the neighboring normal limb. Upon recovery the transplants moved with unfailing consistency as follows: every muscle in the extra leg contracted whenever, and only whenever, its namesake—the synonymous muscle in the normal leg—contracted, both contractions being of corresponding strengths. The latter fact was revealed by the congruity of the angles of the respective synonymous joints, even in cases of great inequalities of size (up to 1:10) between the transplant and the neighboring normal limb.[1]

To prove the involvement in this of each individual muscle as a separate entity, rather than as a member of a functional group such as "flexors," "extensors," etc., let us consider first the absurd movement of a right limb inserted next to a

1. The motions of the animals and their limbs were recorded in motion pictures, sequences of frames being used for detailed analysis and illustration. A set of titled sample strips has been assembled in a coherent demonstration film, copies of which are available on request.

left limb, in which case the musculature of the graft is the mirror image of that of the original limb (Figure 5.1; see Weiss, 1937a). According to the myotypic rule the simultaneous motions of the two limbs should then be the exact reverse of each other, and so they are (Figure 5.2). When I crowded two or three grafts into one location, the precise unison among the synonymous muscles in each triplet or quadruplet of legs became even more striking.

Even more instructive was the opposite procedure of removing the native fore-limbs completely and replacing them by other forelimbs of opposite laterality, all muscles thus coming to lie in inverse positions relative to the body and to be reinnervated from the normal brachial limb plexus. The actual movements of such grafts during the locomotion of the animal were always true mirror images of what the movements of the absent limbs would have been at those instants (Weiss, 1937d). As a result, whenever the hind limbs moved the body forward, the reversed

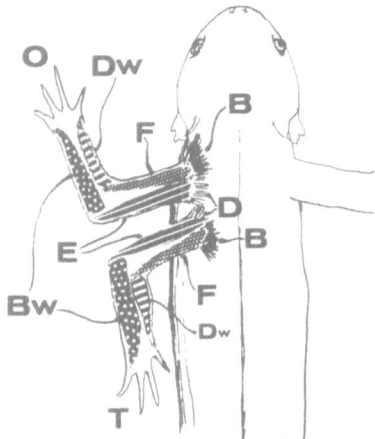

Figure 5.1   Diagram of a supernumerary right limb (T) transplanted behind the normal left limb (O) of a host, showing the positions of six sample pairs of muscles of the same name in both: B, shoulder abductor; D, shoulder adductor; F, elbow flexor; E, elbow extensor; Bw, wrist abductor; Dw, wrist adductor.

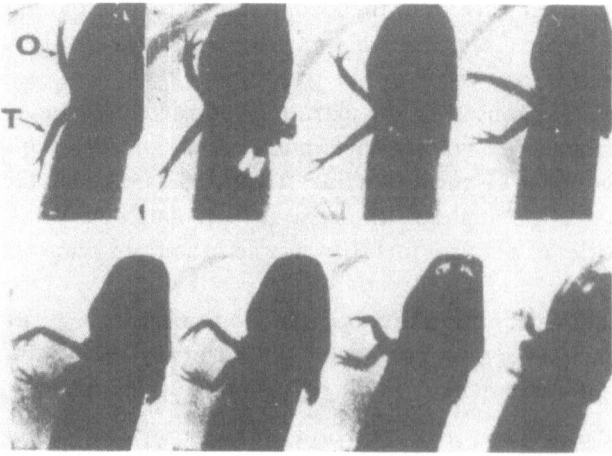

Figure 5.2   Eight stages of a single stroke of an indigenous left forelimb (O) and transplanted right forelimb (T) in swimming.

forelimbs shoved it backward, making it shuttle back and forth on the same spot for a lifetime. In other words every single muscle of the substituted limb kept operating blindly in the stereotyped sequence programmed centrally for its namesake in a normal limb.

Let these examples stand as evidence for the general conclusion drawn from a much wider array of experiments. According to it there exists in each half of the spinal cord a finite repertoire of rigorously defined and immutable patterns of excitatory processes, called "coordination," which through specific means, whether attuned signals or separate "private" lines, call forth unique sequences of muscular responses, regardless of the anatomical disposition and the functional serviceability to the body of the responding muscles. This is the gist of "myotypic response." Let me translate it into some extremely schematized diagrams.

Figure 5.3 shows a sequence of four muscles selected from the forty engaged in a forelimb during ambulation: an elbow flexor, F; a shoulder abductor, B; an elbow extensor, E; and a shoulder adductor, D. We could compare the parts assigned to individual muscles in such time patterns of coordination to the assignments for different instruments in an orchestral score (Figure 5.3, top), leaving the composition and nature of that central score wide open. Prior to the discovery of myotypic correspondence between center and periphery one could legitimately presume that the activity circle F→B→E→D→F (flexor, abductor, extensor, adductor) was carried out by the respective muscles because of their rigorously stereotyped grid of nerve connections with the appropriate corresponding emission stations, as indicated by the arrows in Figure 5.4a. Now we must examine this premise in a new light. Let me pretend for the moment, pending the detailed evidence which I shall adduce presently, that a transplanted supernumerary muscle, in order to operate always in precise synchronism with its namesake according to the myo-

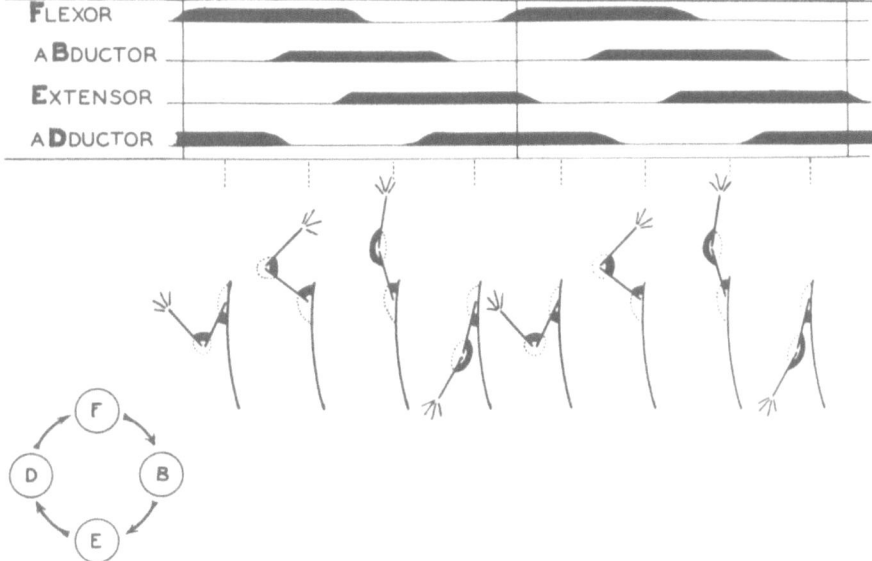

Figure 5.3   A one-and-a-half cycle sequence of left forelimb movement in ambulation, indicating the activity pattern of four muscles, labeled as in Figure 5.1.

typic rule, need not share the latter's original nerve connections. This is schematized in Figure 5.4b. A muscle B is randomly inserted in a place where it is reinnervated by some severed nerve fibers that formerly supplied the indigenous muscle D. Since the intruder B is henceforth actuated in unison with the conventional normal B, rather than with D, the original premise would have to be modified by letting the central wiring diagram be superseded by the net of arrows shown in Figure 5.4b. If we have one or two additional transplants B around, as described above, the rewiring problem would become rather perplexing, and even more so if we expand it from our single sample muscle B to the forty-times-larger total array of selectively responding muscles of a limb.

This then is the concrete situation to which any effort to "explain" the myotypic principle must address itself. A number of such attempts have been made, trying to get by conservatively with recourse to conventional terms. They have all been refuted by experimental evidence. Let me briefly state a few that, despite refutation, have lingered on.

## Unsupported Interpretations

It has been suggested that afferent sensory impulses might somehow reflexly link the motor centers of synonymous muscles together. Aside from its vagueness, this notion is contradicted by two facts: first, the *patterns* of motor coordination of amphibians can develop quite normally despite the complete absence of sensory input to the cord (produced by early deafferentation) (Weiss, 1936a); and second, consequently, transplanted limbs also operate myotypically regardless of whether sensory innervation is present or has been radically eliminated by prior removal of the dorsal-root ganglia (Weiss, 1937c). On the other hand, grafts that do acquire sensory connections render a further service by revealing that myotypic specificity

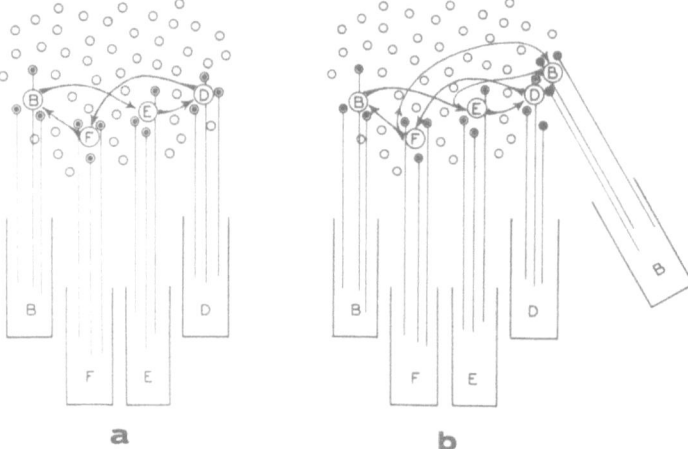

a                                                b

Figure 5.4   (a) A theoretical path of excitation that would yield the cyclic muscular sequence of Figure 5.3, on the assumption of standard stereotyped point-to-point nerve-fiber connections between spinal motor cells and muscles. (b) The complex distortion of this putatively preestablished path of excitation brought about by a duplicate muscle (B) of a supernumerary grafted limb, arbitrarily inserted into the native forelimb nerve plexus.

is a feature not solely of the efferent sector of central muscular communication, but of the afferent sector as well.

This was ascertained by watching the proprioceptive ("Eigen") reflexes, in which stretch applied to a given muscle is followed by a spinal reflex contraction of the very same muscle, as is commonly tested clinically in man by tapping the tendon over the kneecap. In the clumsy salamanders this can be demonstrated only crudely for the separate sets of flexors and extensors of the wrist. But observations and cinematographic records of a frog (which is a more highly differentiated animal) with two supernumerary limbs proved that the phenomenon holds rigorously for single *individual* muscles as such (Verzár and Weiss, 1930). For example, lifting the tip of one of the digits of the triplet of hands dorsally, thus stretching the muscle on the plantar side of that terminal finger joint, immediately yields, not only an isometric reflex contraction of the stretched muscle itself, but also strictly localized isotonic contractions in precisely its namesakes in the two other hands, which, not having been touched at all, could register the response by a crooking of the corresponding joint. Conversely, forcing one of the fingers to bend toward the palm, stretching its dorsal muscles, promptly registers as a dorsal straightening of the corresponding fingers on the other two. Since these correct myotypic responses can be evoked unfailingly with the same degree of selective discrimination regardless of whether the afferent stimulation is set off in a muscle of the native limb or in one of its doubles, it is clear that the spinal centers must somehow have become enabled to identify by "name" the individual sender of the message. Further comments on sensory specificity will follow later.

Since this and much related evidence for myotypic specificity as a distinctive property of each *individual* muscle has been amply described, published, illustrated, and projected in motion pictures in many lands, it seems odd still to encounter efforts to oversimplify the phenomenon beyond recognition by blurring or glossing over its basic features, either from insufficient familiarity with them or from failure to take all of them into account. For instance, the fact that each separate muscle acquires during ontogeny its own identifying and discriminative response code has at times been watered down by the pretension that there is just a *general* proclivity for functional groups to work together, such as flexors with flexors, extensors with extensors, and so forth. To be factually correct, though, one would also then have to differentiate between abductors and adductors, which incidentally are alternately engaged simultaneously with either flexors or extensors (see Figure 5.3), then further subspecify the individual joints involved, and so on all the way down to the localized involvement of a single finger joint—thus ending up by having trimmed down a superficial generalization to the factual experience one could have acknowledged in the first place. And those who are enticed to take that circuitous route, but stop before reaching its conclusion, risk being sidetracked into blind alleys.

Another major recurrent proposition has been that, having recognized some forty different preprogrammed specificities in the limb musculature, we might as well concede to the motor cells in the cord a like set of native prespecializations, provided that each motor neuron can manage to connect selectively with its matching muscle type, or, as it has been loosely expressed, "to find its proper muscle." Leaving aside the massive evidence against such a rigorously discriminative faculty of motor neurons, let me just rule it out for the case at hand.

First, in cases in which I had chosen deliberately for the regenerative reinnervation of the whole-limb transplant a small branch to a far distal muscle group in the normal limb, *all* muscles of the graft functioned myotypically. Figure 5.5, for instance, shows a graft I had provided with a nerve branch that had formerly supplied only flexor muscles for the wrist and hand of the neighboring native limb. Yet in the transplant, it reinnervated all the muscles, retracing the old deserted nerve tracts of that limb (note its bifurcation at the arrow into branches to the extensor and flexor sides, respectively), resulting in the restoration of function of the entire limb musculature, coordinated myotypically with the neighboring donor.

Second, when a limb is transplanted to the flank (Figure 5.6), its muscles do acquire reinnervation from intercostal nerves beyond the brachial segments of the cord and contract vigorously; yet their contractions are en masse, without coordination. This means that intercostal neurons can make transmissive connections with limb muscles, but the orderly central pattern for limb coordination is lacking outside the limb region of the cord. Both of these examples document the fact that a regenerating motor neuron cannot be claimed to be predestined from its source for a given muscle and thereby either be somehow guided to its singular destination or, if it arrives at an incorrect one, be barred from entry. This does not,

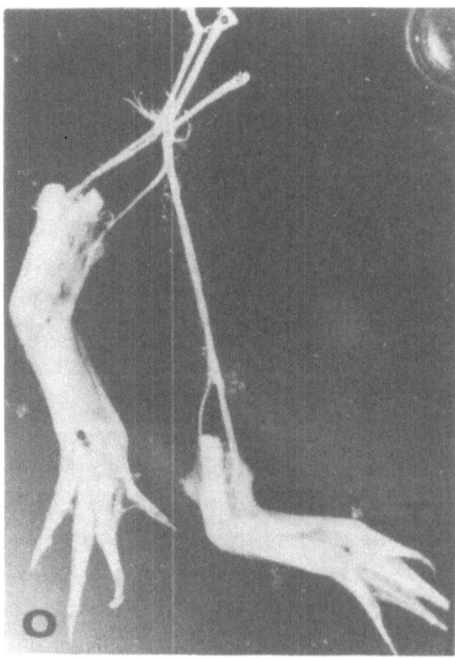

Figure 5.5   Innervation of a grafted limb (T) by a branch of the forearm innervation of the local normal limb (O), severed below the elbow of O and inserted into T at its shoulder girdle, six months after the operation. (The upper parts of both limbs above midhumerus have been removed to expose the nerve pattern. The broken line connects the positions of the cut end of the deflected nerve branch before and after its insertion in T.)

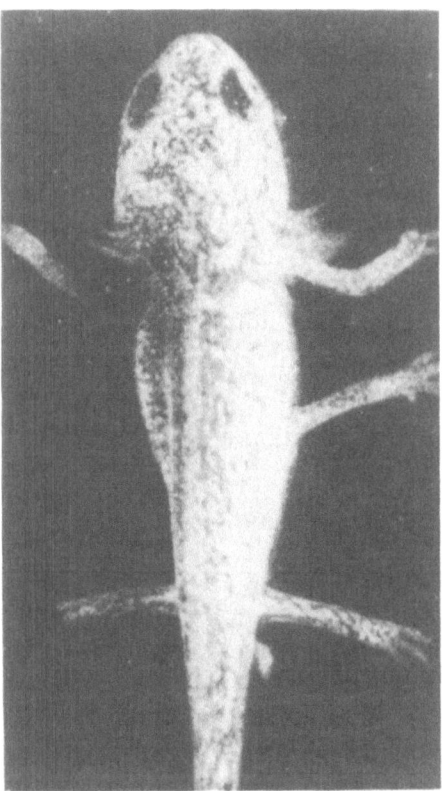

Figure 5.6   Limb transplant in midtrunk innervated by thoracic nerves.

however, exclude all degrees of selectivity in the outgrowth of nerves (see below);
it simply excludes the high degree on which myotypic response is predicated.

The difference between our earlier examples of myotypic coordination and the
massive discoordination in the last example is highly significant and calls for comment. As I have stressed repeatedly in the past, the spinal centers for given peripheral areas are strictly circumscribed to the corresponding monosegmental or
plurisegmental cord districts, and myotypic correspondence holds only within a
given domain. In amphibians the domain for forelimb-specific coordination is
restricted to spinal segments 3, 4, and 5, which act as a single unit and whose
motor neurons can therefore be used promiscuously, as has been done in our
experiments. On the other hand, reports on *eye-muscle* translocations seem to intimate that eye-muscle centers, especially the oculomotor and the trochlear ones,
each represent a separate domain (Sperry and Arora, 1965). This could explain
some discrepancies between the results of eye coordination versus limb coordination upon reinnervation from nerves other than those of their respective regular
central domains.

Another question arising from the last example is, of course, why the limb
muscles respond at all to activations for intercostal muscles. Obviously the answer
is deeply relevant to the whole concept of neuromuscular specificity. All I can
contribute is the observation that the same "unspecific" mass response precedes,

though only for a brief transition period, the strictly myotypic function of limb muscles innervated from the proper limb segments of the cord. I shall return later to this point.

These examples, which could be greatly multiplied, have only illustrated why spinal motor neurons must not be assumed to be immutably prespecified for "their" preassigned muscles, and no others. I shall now show why, even if they were so preassigned, they would have zero probability of ever reaching "their" muscular match in our experiments. The evidence comes from a closer look at the numerical details of the regenerative reconnections from cord to transplant. Sprouts from a severed axon grow out along essentially fortuitous routes, branching on the way. Branches with access to degenerated fiber bundles are guided by them to whatever stations lie at the end (Weiss and Taylor, 1944; Weiss and Hoag, 1946). However, whether the nerve stumps are left in a transplant or are extirpated makes no difference in the results. Any fraction of the contingent of fibers in the limb plexus can by peripheral branching take over the full quota of innervation of the transplant, as indicated in Figure 5.7. Also note the im-

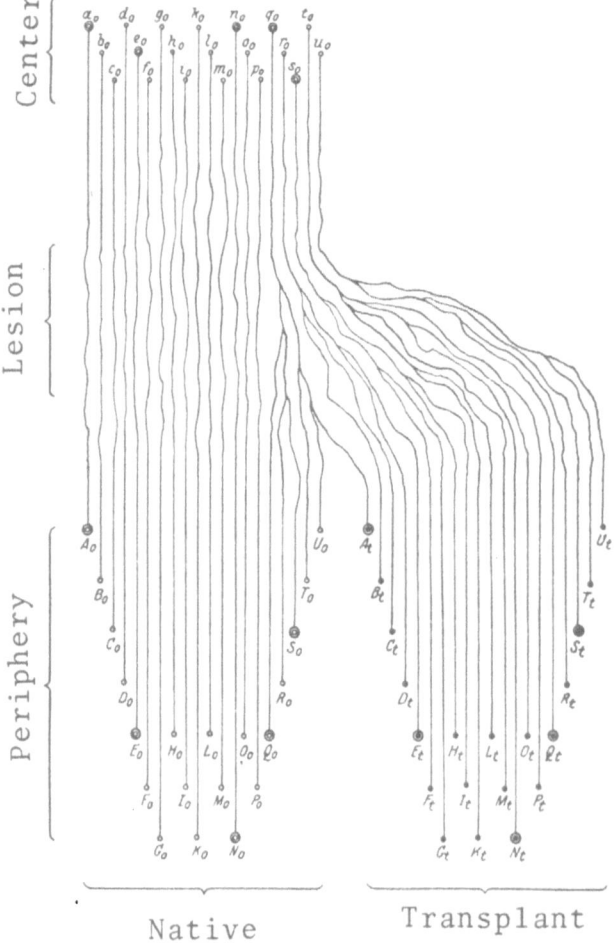

Figure 5.7   Schematized model of the regenerative innervation of the full set of muscles in a transplanted limb and the few muscles of the indigenous limb that were denervated in the operation.

plication that, contrary to expectations, addition of a graft does not entail an increase in the neuron population that leaves the cord segment involved in the accessory load of peripheral innervation. This is one of the salient features that have been overlooked in so many speculations.

## Sources and Distributions of Innervation Quantified

Here are the facts (Weiss, 1937b). Let me return to the grafts with inverse symmetry shown in Figures 5.1 and 5.2. Their main new nerve source (e.g., Figure 5.8) is the severed fifth root (save for a small branch from nerve 4). A tally of myelinated fibers in the right and left brachial plexuses of this case six months after reinnervation showed similar numbers on both sides (Figure 5.9), with the left plexus, which supplies two limbs, containing no more than the right (control) one. More broadly, in nine cases counts of myelinated neurons in the proximal spinal nerve trunks, supplying from one to three transplanted extra legs in addition

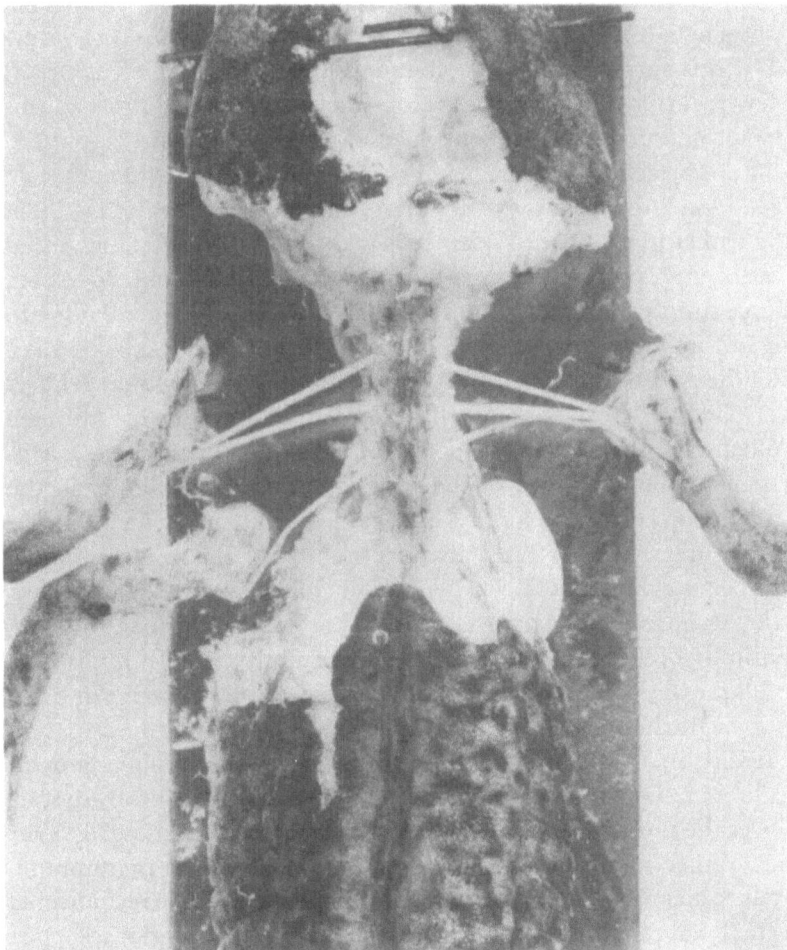

Figure 5.8   Dissection of the brachial nerve plexus (segments 3, 4, 5), with the original left limb (O) and the transplanted right limb (T) on the left side and the control limb on the right side, three months after operation. The transplant is almost entirely reinnervated by the fifth nerve root, with only a thin n. pectoralis contribution from the fourth.

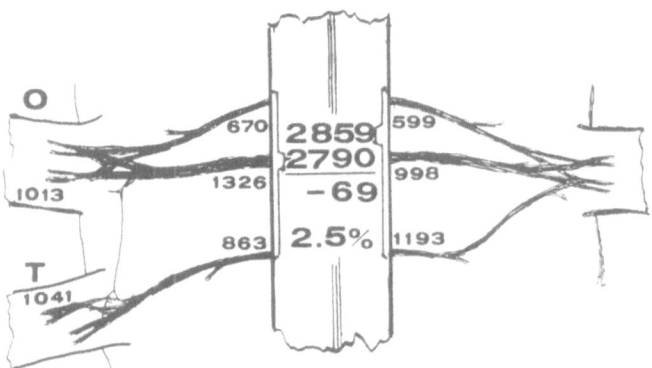

Figure 5.9  A tally of myelinated nerve fibers from the left and right roots of the plexus of Figure 5.8 and within the two limbs, O and T, supplied by the left plexus.

to the original limb, showed an average 7.5% surplus of fibers on the overloaded compared with the single-legged (control) side. In proportion to the muscle over-load of 100–300%, this surplus cannot be considered relevant. (It might be at-tributable to a few recurrent fiber branches from the lesion.) In contrast to this central pool of motor neurons, the complements of fibers *inside* the multiple limbs, which had to share that common and unenlarged central source, were close to normal when faced with the heavy extra load of muscles imposed upon them (confirmed in the triplet frog set; see Weiss, 1931); that is, a transplant acquired about as many fibers in proportion to its size as the nearby original limb possessed, most of whose nerve supply had been left intact. This outcome is evidently due to ample branching of the fibers growing from the severed nerve stems.

As an example, the graph in Figure 5.10 gives the number of myelinated fibers counted at corresponding cross sections through the upper arms of thirteen normal limbs (filled circles) and their associated transplants (open circles). Dif-ferences within pairs could be traced to the wider margin of variability in the sizes of the transplants. Accordingly, for a valid assessment of the peripheral nerve supply, the *density* of fibers, rather than absolute numbers, must be compared. Plotting densities (i.e., the number of fibers per unit of cross section) as double bars, shaded lightly for the transplant, darker for its normal control partner, one rec-ognizes the quantitative correspondence in the nerve supply of both limbs of a pair, indicating full saturation of the musculature with nerve connections in both. We must bypass some further interesting details, such as the fact that the apparent excess of fiber density in the transplants is attributable to their slight growth retardation by suboptimal vascularization; or that the mass of a tissue exercises a marked control over the number of fibers admitted to entry (Litwiller, 1938). Our present concern is, rather, what bearing our fiber census has on the presumption that any doublet or triplet of synonymous muscles might have become reconnected with precisely its matching sender amidst the original forty central stations.

Let us again look at the facts. The number of myelinated fibers entering a limb base varies between eight hundred and twelve hundred. According to Agduhr (1933), about 60% of them are sensory, which leaves about four hundred motor

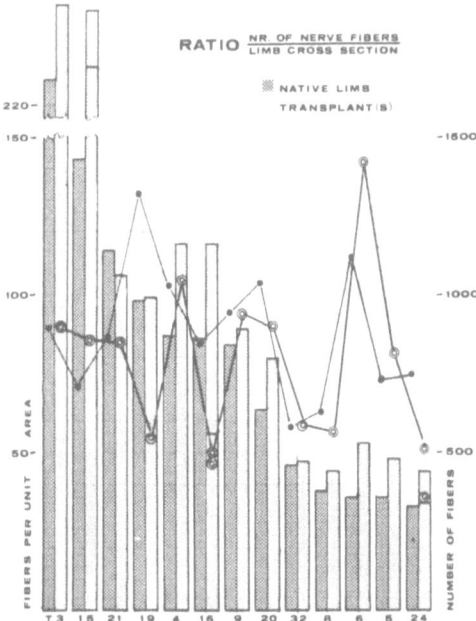

Figure 5.10 *Solid lines* connect myelinated-nerve-fiber counts in eleven pairs and two triplets (T16 and T24) of native and conjoint grafted limbs, marked by filled and open circles, respectively; there are marked differences in numbers in some cases (e.g., T19, T16). *Double bars* represent the ratios of fiber numbers over cross-sectional areas (size index) of native (dark) and grafted (lighter) limbs; there is a much closer correspondence here.

fibers to supply all forty muscles—about ten fibers per muscle. Returning to our specific sample (Figure 5.9), we find the fifth nerve, the major source for the graft, emerging from the cord with about 340 motor neurons, half of them going to the trunk. So only about 170 pass on into the limb. These have then branched further to give the graft its full fiber complement. Thus, on an average, no more than four neurons from the spinal source are available for each muscle, and these are subsequently dispersed by branching and errant courses. Now, faced with these realities, could one ever see a reasonable chance for any one of four prespecified motor-cord cells, let alone every one of forty, to have "found" its preordained matched ending? Let those who think so either prove it or forget it.

Further disproof comes from our admittedly primitive and scanty *electric* tests of connections, which can be carried out in specimens with nerve stems shared by both members of a pair (Weiss, 1937b: pp. 504–509). Never did a single stimulus to an anastomosing branch yield contractions of precisely *synonymous* muscle groups in both limbs. In some such cases, by cutting all nerve trunks and stimulating only the peripheral anastomoses, some so-called "axon reflexes" between two muscle groups could be obtained, both of them again disparate.

In sum, the wide array of facts thus far available would seem to dispose of the notion that the myotypic specificity of communication between centers and *effectors* can be explained in terms of microstructural improbabilities. And this conclusion extends to the *receptor* field as well.

## Selectivity in Afferent Somatic and Intracentral Communication

I have already said that each proprioceptive input to the cord is identified centrally, according to the name of the muscle from which it came, by being answered
myotypically (Verzár and Weiss, 1930). Later experiments confirmed this sensory
phenomenon for the cornea and the skin. Touch to the cornea in metamorphosed
amphibians entails a prompt localized lid-closure reflex of the eye. Stimulation of
the cornea of a supernumerary eye, transplanted in the place of the ear and innervated by a trigeminal nerve branch that had formerly elicited flexion of the
neck, yielded instead closure of the ipsilateral normal eye in front (Weiss, 1942).
Later Miner (1951) exchanged back and belly skin in frogs by 180° rotation. It
was known that normally a stimulus applied to back or belly would yield wiping
reflexes appropriately oriented to the dorsal or ventral side, respectively, executed
in the former case by the hind leg, and in the latter case by the forelimb. (Figure
5.11a). Stimulation of back skin (darkly shaded), translocated to the belly (light),
and vice versa, was reported to elicit a wiping reflex directed to the spot and
executed by the limb appropriate for the original rather than the new location of
the graft, even though its nerve supply had come from local sensory fibers at the
new site (Figure 5.11b). Thus, here too, the sensory input seems to have been
identified centrally by the specific signal of its sender, that is, its back-skin-specific
or belly-skin-specific character, rather than by the topographic sign of its nerve
supply.

True, repetitions of these last experiments have yielded mixed results and interpretations. But it seems that these ambiguities have arisen from the failure of
various investigators to take cognizance of the wide variation in behavior and fate
after transplantation of the two different component layers of the skin: the epi-

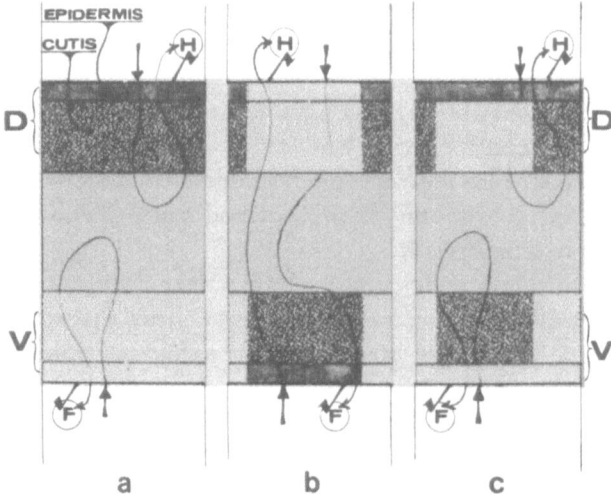

Figure 5.11 Diagrams of characters of dorsal (D, darkly shaded) and ventral (V, light) frog
skin: a, in intact animal; b and c, after exchange of dorsal and ventral skin areas either by 180°
rotation of a belt of skin or by reciprocal grafting of skin patches. In b both cutis and epidermis of
the graft have persisted, while in c local epidermis has grown over the cutis of the graft. Lines
symbolize the reflex sequences between sites of stimulation (arrows) and wiping responses by hind-
limb (H, inverse check) or forelimb (F, check), respectively.

dermis, housing the specific cutaneous receptors, and the cutis, containing or lacking the pigment granules that mark skin for the onlooker as "back" or "belly" skin, respectively (Figure 5.11a). Cutis (dermis) is more apt to heal in firmly, whereas epidermis more readily sloughs off, either entirely or in spots, to be replaced by the peripatetic cells of the normal *local* epidermis surrounding the graft (Figure 5.11c). Therefore the color of the grafted "skin" cannot be taken as a criterion of whether an afferent message from there has originated in transplanted foreign epidermis or in its replacement by cells of the local type. Once more, pending a more detailed critical accounting of such underlying facts, general conclusions drawn from them must remain in suspense.

Taken all together, the sensory experiences reported in the preceding section have in common with the experiments in muscle transplantations that the centers have to "know" what specific receptor or effector lies at the end of a given line. And since the grafting experiments have ruled out prescience in the laying of lines from spinal cells to their terminals, the centers must evidently get that knowledge by information passed on to them from the receptor and effector organs themselves. Just how this is accomplished remains a matter of conjecture.

However, given the current state of uncertainty and equivocation, it would seem indicated not to jump at broad generalizations from a limited set of data, experimental procedures, assaying techniques, forms of animal, types of organ, and so forth, by lumping together results obtained under circumstances that are not strictly comparable. In the present context this is pertinent to comparisons between the sensory-central rapport just illustrated, in which a sensory impulse *ascends* from a peripheral receptor and is properly distinguished by the central cell body, and the series of experiments inaugurated by Sperry (1943) that demonstrated correct restoration of the topical projection of *efferent* excitations from a central organ—the retina—upon another central organ that acts as effector—the optic tectum—after dislocations or lesions of the retina, combined with transection of the optic nerve.

Very briefly, there seems to be consensus about the fact that the topographical maps of stimulated points on the retina and responding points on the tectum retain their mutual congruency despite the various experimental manipulations of the retina and the temporary interruption of fiber connections (Gaze, 1970). This has been widely ascribed to (1) a critical diversity of specificities in the mosaic of retinal ganglion cells as sources of the optic-nerve fibers, (2) a correspondingly specified set of matching tectal neurons, and (3) an enigmatic faculty of the regenerating optic fibers to "find" their matching tectal mates.

Leaving aside a notable variance among the various degrees of sharpness and of discriminative specificity and precision ascribed to the matching, in essence points (1) and (2) are fully consonant with the general and fundamental thesis I have derived from the myotypic studies: namely, that neural communication operates through a *principle of mutually attuned specificities*. The matter of point (3), however, seems to me still wide open, largely because too many crucial questions of plain fact have not yet been asked, let alone answered. To narrow down the problem let me start by discounting a priori objections to proposal (3).

Developmental prespecification of central neurons is, just as that of muscles, a fact well documented—for example, by the distinctions between adrenergic and cholinergic neuron types, or the specialized single pair of Mauthner's cells in fish and amphibian brains. Early observations by Hamburger (1929), later expanded by Taylor (1944), revealed the intrinsic constitutional differences between sensory and motor neurons, which reach their destinations by separate routes, proving pairwise distinctions between both the respective neuron types and their matching guide routes. My concept of "selective fasciculation" of nerves (Weiss, 1941b), according to which younger axons tend to apply themselves in their outgrowth to older, already established ones of the same specific type, likewise affirms the existence of prematched correspondences. Perhaps the broadest biological corroboration might be adduced from my analysis of the self-sorting of cells in mixed populations according to their type specificities (Weiss, 1958), which I referred hypothetically to mutual molecular recognitions between the contacting cell surfaces (Weiss, 1968: Chapters 11, 19, 21).

There is no counterpart in my own experience to corroborate the report that the severed cerebral stump of the optic nerve might retain some mosaic of specific guidelines for regenerating purportedly specified optic fibers (Sperry, 1963), but we may let it pass for lack of experimental evidence to the contrary. The difficult questions arise at the retinal end of the optic nerve and at the arrival of the fibers in the tectum.

At the retinal end, proposition (3) pretends tacitly that a visual point stimulus to the eye has separate well-defined radial connections leading from the visual receptor cells straight to the source of the optic nerve fibers in the ganglionic layer. This is wrong, for it ignores the fact that there are interspersed between the optic and the ganglionic layers interlayers of amacrine and bipolar cells. These cells, by their horizontal network of processes connecting with the other layers and thereby entailing both scattering and reintegration of impulse conduction, rule out the notion that there is a direct anatomical point-to-point projection from a given visual element to a given optic-ganglion cell. Thus the question of how a given pattern of *visual* point stimuli can register in rather correct replication on the screen of *ganglion* cells in the first place precedes the question of how in the next step a given *ganglionic* cell group can pass on the respective fraction of the received impulse "image" (alternatively conceived of in terms of "field" or of "mosaic" states) to sets of *tectal* neurons in topographically corresponding configurations. Once we better understand the preservation of the integrity of the visual image in its passage across the retina, we might perhaps get a clue as to how the further rectified projection from the ganglionic to the tectal layer proceeds without having to postulate that each regenerating fiber has a homing instinct.

The mode of the tectal reunion also calls for further comment. In spite of profitable starts, the reinnervation of the tectum has not yet received a microscopic and quantitative analysis as detailed as that made for the myotypic studies; and this makes it difficult to interpret the meaning of such inferences as that a fiber "finds" its proper connection. Even if we concede that each arriving neuron actually carries a specific mark prefitted for a matching mark on a tectal cell, like key to lock, does the fiber aim at that mark beeline-fashion, or does it roam blindly till it happens to meet its match and get trapped? Given all the evidence from the study of nerve outgrowth over long distances in development and regeneration, the first alternative can be dismissed decisively. For the second there is at least some parallel in the fact that in normal development, when motor and sensory fiber populations pervade peripheral tissues, each type is accepted for synaptic connection only by the correlated tissue type. The lamination of the tectum would permit a profuse interfacial spreading and extensive arborization of the fiber tips; but since the latter are at the limit of microscopic resolution, it would be technically difficult to test the point. So this question also remains unanswered. At least, contrary to muscle reinnervation, here the whole fiber contingent arrives from the chiasma in a single assemblage, free to spread, instead of in widely separated, discrete bundles no longer fit to anastomose and "search."

**Fixation of Specificity?**

I have raised these few elementary questions merely as examples of the dearth of the factual information that would be indispensible for the formulation of conclusive explanations for any one of the various scattered experiments and observations, let alone for a comprehensive unified theory. To reinforce this argument further let me add just a few more random scraps to the heap of uncertainty. All positive conclusions on selectivity in central-peripheral communication have been derived from animals short of full maturity. The epigenetic emergence of differential specificities is a time-limited process, seemingly tapering off fast or slowly toward a terminal steady state of fixation. When, and at what rate the process proceeds varies so greatly among groups of animals, developmental stages, types of organs and tissues, criteria used, and so on, that it would seem futile to do more than accept the fact. In this sense the "age" of a neuron and its nerve connections, for instance, by no means bears the simple relation to the age of the animal that it was once assigned, when mature neurons were rated as "nongrowing" cells because they mostly stop dividing early in life. But with the discovery of axonal flow (see Weiss and Hiscoe, 1948) and the attending demonstration of the rapid rate of continuous reproductive renewal of the macromolecular (e.g., protein) population of the nerve-cell body and its processes (about once to several times a week) (Weiss, 1970), this old aspect of fixity has to give way to the realization of a high degree of potential impermanence and adaptive flexibility in the constitution of neurons and, consequently, their interrelations.

To study this new aspect is, again, a matter for the future. But it already places the specificity problem into a new light by establishing a realistic, indeed molecular, foundation for the phenomena of despecification and respecification of neurons. If "specification" signifies the gradual emergence of central-peripheral and central-central response relations of specific mutual correspondence, then "despecification" must be postulated whenever and wherever that correspondence has become less discriminative (more "unspecific"). This applies regardless of whether one looks at it as a waxing promiscuousness in regenerative connectivity or, as I do, as a waning sharpness of the power to distinguish coded messages. Now, lack of specificity in the response of transplants is a well-attested fact. For instance, the muscles of transplanted *limbs* innervated by *intercostal* nerves (see above), as well as of deplanted limbs innervated from isolated and randomly *disarrayed* pools of spinal neurons (Weiss, 1941c, 1950c), respond massively to the discharges coming to them from these alien neurons, which are decidedly not specified for limb muscles. These data evidently dispose of the speculative alternative of selective connectivity.

Before discounting that alternative altogether, however, one might wish to salvage it by assuming that the vast majority of regenerating motor fibers are not muscle-specific, and hence can establish transmissive connections with any muscle, but are gradually displaced at their terminals by collaterals from the sparse minority of "correct" fibers that might happen to be among them. Although the high improbability of this presumption becomes clear from the actual numerical census of the regeneration process I have given above, one might still advance in its favor the proven fact that there are imperfections in the myotypic function of transplanted limbs during the first week or so of signs of beginning reinnervation. But it seems more plausible to ascribe this transitory phenomenon either to the simple fact that not all muscles receive their reinnervation at exactly the same time, or to a true process of "despecification" of the regenerating branches, followed by "respecification" in accordance with the encountered muscle.

There are still many loose ends to the questions on the specification of neurons, such as: when and how it occurs; at what rate it proceeds; how long it remains reversible prior to becoming fixed; whether all neuronal types ever become totally single-tracked or whether some varieties (e.g., the "internuncial gray") retain adaptability; how those variables differ among different animal groups; and so forth. Too many unknowns, as much as generalized verbiage, still block concrete answers. To cite just one example, myotypic correspondence can be changed readily in larval and early transmetamorphic *amphibians*; yet the specificity of *motor* fibers in *mammals* seems to be firmly fixed in prenatal stages (Sperry, 1941), and nerves of the mammalian *vegetative* system, crossed postnatally to foreign terminations, can still become respecified to yield fully functional results there, tending, however, to be replaced later competitively by fibers from the original uncrossed source (Guth and Bernstein, 1961). Such gradual "returns to normalcy," also noted after crossing eye nerves in fish (Marotte and Mark, 1970), while definitely disproving any *absolute* exclusion of "incorrect" unmatched central-peripheral connections, would still seem to intimate some competitive *superseding* of usurping neurons by fibers of the standard origin. Thus there is certainly a call for more detailed study of the steps in which the process occurs. Since a supernumerary "monitoring" organ was present in neither of these experiments and the crossed nerves came from separate central "domains" (see above), a direct comparison with the results of our myotypic series would carry some risk. Vaguely one might even think that some trait of "selective fasciculation" (Weiss, 1941b) is somehow involved, but this is another instance where speculation had better defer to factual information yet to come. On the whole it would seem to me rash to try to reduce the conclusions from a still incoherent smattering of data to a common denominator.

My attitude during this half century of vagaries that has elapsed since my discovery of the myotypic principle has been the one stated in my introductory remarks: to lay primary stress on the *factual* substance of the principle and the *logical* conclusions to be drawn from it, rather than on the various *theoretical* interpretations that try either to squeeze them into conformance with existing concepts or to remodel the latter to accommodate the novel insights.

This applies equally to my own doctrine—by no means a dogma—of neural "modulation." Therefore, in order to practice what I preach, I shall conclude the relevant factual part of my paper at this point with the contrite confession that, notwithstanding varied efforts from many sides, the discovery of the myotypic principle is still in a very infantile state of maturation. It is in need of far more unambitious factual elaboration before it can "explain," returning to my metaphor of the piano, not only how a keyboard of central neurons selectively activates each separate muscle without a rigidly preset and stereotyped grid of strings, but also the far more enigmatic problem of the "piano player." How does "he" strike the keys in such appropriate combinations, sequences, and timings as to yield chords and melodies instead of noise? Is "he" manipulated, puppet-like, by another variable net of strings or perhaps by a stencil-like machine or, as a last resort (perhaps the most likely one), by the orderly, patterned dynamics of the collective activity of entities of neuronal cell groups and connecting networks and glial environment in interaction? To approach answers is the challenge of the future.

My own reward for half a century of concern about these problems I have expressed in this ditty (Weiss, 1969):

Something we find with intention
Commonly is called invention.
If the goal was practical,
Then, of course, it's tactical
To exploit its fruits for money,
Much as bees milk plants for honey.
But when I, of mind more humble,
Just observing nature, stumble
Upon a *discovery*,
What it holds as prize for me
Is the thrill to have detected
Something wholly unexpected.

## Postscript

For those who care to know my personal opinions about potential "explanations" of neural specificity, I might add the following brief commentary.

In early ontogenetic stages neurons multiply and differentiate;[2] that is, neurons

2. For a comprehensive analytical review of the processes of neurogenesis, see Weiss (1971b: Chapter 4.2).

become progressively diversified or, as we may now call it, "specified," like cell strains in any other proliferating tissue, such as muscle and epidermis. Yet the stoppage of the proliferative phase of a cell need not denote the dead end of its diversification. Many cellular modifications lumped under the name of "modulations" occur postmitotically (Weiss, 1968; Chapters 6, 8). I therefore see no reason to deny the same potential to the "developed" neuron, the offspring of a line of neuroblasts.

In trying to give neuronal modulation a more concrete meaning, I assume that each individual muscle (or cornea or skin sector, etc.), in its embryonic differentiation, develops a protein component typical of it. On contact with an innervating axonal tip of less specific constitution, this component then induces in the tip, through an antigen-antibody-like interaction, a reciprocally conforming subspecification of its neuronal protein, which, by an ascending chain reaction, gradually extends to the cell body and tags the entire surface of that neuron correspondingly. A neuron thus becomes the representative, carrying the identifying "name," of its terminal effector or sender. As such it either (1) accepts centrally synaptic connections only with other neurons of identical surface molecules, or (2) accepts such connections with "unspecified" neurons to which it passes on its own acquired specificity, or (3) is physiologically receptive, regardless of connectivity, solely to those central excitatory dynamic patterns of coded configurations with which its own specific constitution is in resonance. In view of the incessant renewal of protein in the perikaryon (Weiss, 1970), this scheme is, of course, predicated on the older specified form of the respective protein acting as a template for the subspecification of the newly formed ones.

This whole hypothesis, which explains most of the facets of the myotypic phenomenon, still remains short of conclusiveness as long as it treats the axon as an indivisible all-or-none entity. It cannot then account for the independent functioning of *different* muscles connected with branches of *common* fibers (see the note on "axon reflexes" on p. 89). One would have to "miniaturize" the scheme to the subcellular level by conceding to a single neuron protein chains of *multiple* specificities, monotonic between ending and branching point but combined in the common stem. This would, of course, presuppose that impulse conduction is somehow predicated on the uninterrupted uniformity in the specificity of the molecular chains involved in the electrochemically propagated disturbance; this would allow a common effect track to be used from center to branching point, each branch then admitting only its properly attuned component for single-tracked propagation to its terminal. As one can readily see, the whole mechanism would hinge on a principle of *resonance between macromolecules of identical "conformation,"* transmission resulting from the coded pattern of emission of one unit being "recognized," and hence selectively responded to, by a conforming one, and so on down the line. In terms of the analogy of acoustic resonance, one would compare the process to the selective resonance of a given string of a harp to the sound of the appropriate wavelength, whereas the wind blowing through would set all the strings in motion. One wonders whether the artificial electrical, chemical, or

mechanical stimulation of a nerve might not introduce an analogous feature of
noise into the system in contrast to the "adequate," specifically coded, messages
from the body's own devices.

In its bearing on neural specificity, this scheme may look like idle phantasy. But
it looks much less so in a broader cell-biological perspective, based on my own
explorations on specificity in growth control (Weiss, 1971b: Chapter 3.7) and
intercellular relations (Weiss, 1968: Section 4). Indeed, the antigen-antibody
concept of complementariness among macromolecules as a template-antitemplate
patterning device for selective intercellular communication, recognition, and
adaptation has proved so widely valid as a model that the idea of testing its appli-
cability to problems of neurobiological concern does not seem too farfetched.
Pending such tests, the field for competitive ideas lies wide open, with the caveat, of
course, that they should be not incompatible or irreconcilable with any of the well-
documented facts already at our disposal. At any event, as Adrian said forty years
ago, "this is for future work to decide."

**Addendum**

While this paper was in press, an interesting set of articles on "Development and
regeneration in the nervous system" appeared in the *British Medical Bulletin* (vol.
30, no. 2, May 1974), containing a four-page summary (pp. 122–125) on "Selective innervation of muscle" by R. F. Mark. In trying to salvage the idea that given
individual motor neurons are specifically preprogrammed for their predestined
muscles, Mark replaces the discredited contention of an anatomically rigid *primary* point-to-point connectivity by a *secondary* competitive "functional" exclusion
of the "improper" connections. In order to preclude further confusion of the facts
on which any future explanation of the myotypic principle will have to rest,
and without entering into any disputation, I shall provide the reader here simply with a correction of a few factually questionable assertions in the article.

The statement (p. 122) that "electromyography has shown that the synchrony
of movement of individual muscles is not as precise as Weiss claimed," is evidently
due to a total unfamiliarity with the extensive, detailed motion-picture analysis of
which I have given a few samples above. In fact, the latest reference to my
results cited by the author in his bibliography is my unillustrated report of 1936,
thus missing the elaborately illustrated and documented 190 pages of description
of the factual details in 1937 (a,b,c,d) and the earlier, likewise cinematographic,
study by Verzár and Weiss (1930). The further statement (p. 122) that "animals
that can recover completely from nerve repair, such as teleost fish and salamanders, have muscle cells [*sic!*] which may receive more than one ending from different motoneurones," is factually incorrect if the "*may* receive" is then tacitly
expanded to "*do* receive," as it is in the speculative interpretations by the author
diagrammed in his Figure 4. No more than about 2–4% of muscle fibers (not
"cells"), already innervated by one neuron, ever accept an additional neuronal
connection. For the most decisive experimental evidence in mammals (rats), see

Weiss and Hoag (1946); but since the author confines his conclusions to lower vertebrates, the most extensive experimental confirmation for frogs is found in the 89-page dissertation by Fort (University of Chicago, 1940). My evidence (see above) that a forelimb is activated by a three-segmental spinal sector as a single unit clearly refutes such statements as (p. 124): "The pattern of segmental inner-vation of the limb is reconstituted by nerve regeneration, even when nerve roots are cut and deliberately placed in the wrong part of the limb [sic]."

These few comments have been meant only to put the reader on guard. How-ever, I might point out that at the end of May 1974 I arranged a personal clarify-ing session with the author, which, if it had taken place prior to the publication of his article, might have rectified such factual discrepancies as those singled out in this addendum.

## References

Adrian, E. D. (1932): *The Mechanism of Nervous Action. Electrical Studies of the Neurone*. Philadelphia: University of Pennsylvania Press.

Agduhr, E. (1933): Vergleich der Neuritenanzahl in den Wurzeln der Spinalnerven bei Kröte, Maus, Hund und Mensch. *Z. Anat. Entwicklungsgesch.* 102:194–210.

Gaze, R. M. (1970): *The Formation of Nerve Connections. A Consideration of Neural Specificity Modula-tion and Comparable Phenomena*. New York: Academic Press.

Guth, L., and Bernstein, J. J. (1961): Selectivity in the re-establishment of synapses in the superior cervical sympathetic ganglion of the cat. *Exp. Neurol.* 4:59–69.

Hamburger, V. (1929): Experimentelle Beiträge zur Entwicklungsphysiologie der Nervenbahnen in der Froschextremität. *Wilhelm Roux Arch.* 119:47–99.

Litwiller, R. (1938): Quantitative studies on nerve regeneration in amphibia. II. Factors con-trolling nerve regeneration in regenerating limbs. *J. Exp. Zool.* 79:377–397.

Marotte, L. R., and Mark, R. F. (1970): The mechanism of selective reinnervation of fish eye muscle. I. Evidence from muscle function during recovery. *Brain Res.* 19:41–51.

Miner, N. (1951): Cutaneous localization following 180° rotation of skin grafts. *Anat. Rec.* 109: 326–327 (abstract).

Sperry, R. W. (1941): The effect of crossing nerves to antagonistic muscles in the hind limb of the rat. *J. Comp. Neurol.* 75: 1–19.

Sperry, R. W. (1943): Visuomotor coordination in the newt *(Triturus viridescens)* after regeneration of the optic nerve. *J. Comp. Neurol.* 79:33–55.

Sperry, R.W. (1963): Chemoaffinity in the orderly growth of nerve fiber patterns and con-nections. *Proc. Natl. Acad. Sci. USA* 50:703–710.

Sperry, R.W., and Arora, H. L. (1965): Selectivity in regeneration of the oculomotor nerve in the cichlid fish, *Astronotus ocellatus. J. Embryol. Exp. Morphol.* 14:307–317.

Taylor, A. C. (1944): Selectivity of nerve fibers from the dorsal and ventral roots in the development of the frog limb. *J. Exp. Zool.* 96:159–185.

Verzár, F., and Weiss, P. A. (1930): Untersuchungen über das Phänomen der identischen Bewegungsfunktion mehrfacher benachbarter Extremitäten. Zugleich: Direkte Vorführung von Eigenreflexen. *Pfluegers Arch.* 223:671–684.

Weiss, P. A. (1922): Die Funktion Transplantierter Amphibienextremitäten. *Anz. Akad. Wiss. Wien* 59:199–201.

Weiss, P. A. (1923): Die Funktion transplantierter Amphibienextremitäten. II. Kompensatorische Reflexe. III. Histologische Untersuchungen über die Nervenversorgung der Transplante. IV. Theorie: Die Erfolgsorgane als Resonatorensystem. *Anz. Akad. Wiss. Wien* 60:57–61.

Weiss, P. A. (1928): Erregungsspezifität und Erregungsresonanz. Grundzüge einer Theorie der motorischen Nerventätigkeit auf Grund spezifischer Zuordnung ("Abstimmung") zwischen zentraler und peripherer Erregungsform (Nach experimentellen Ergebnissen). *Ergeb. Biol.* 3:1–152.

Weiss, P. A. (1931): Die Nervenversorgung der überzähligen Extremitäten an dem von Verzár und Weiss in Bd. 223 dieser Zeitschrift beschriebenen hypermelen Frosch. *Pfluegers Arch.* 228:486–497.

Weiss, P. A. (1936a): A study of motor coördination and tonus in deafferented limbs of amphibia. *Am. J. Physiol.* 115:461–475.

Weiss, P. A. (1936b): Selectivity controlling the central-peripheral relations in the nervous system. *Biol. Rev.* 11:494–531.

Weiss, P. A. (1937a): Further experimental investigations on the phenomenon of homologous response in transplanted amphibian limbs. I. Functional observations. *J. Comp. Neurol.* 66:181–209.

Weiss, P. A. (1937b): Further experimental investigations on the phenomenon of homologous response in transplanted amphibian limbs. II. Nerve regeneration and the innervation of transplanted limbs. *J. Comp. Neurol.* 66:481–535.

Weiss, P. A. (1937c): Further experimental investigations on the phenomenon of homologous response in transplanted amphibian limbs. III. Homologous response in the absence of sensory innervation. *J. Comp. Neurol.* 66:537–548.

Weiss, P. A. (1937d): Further experimental investigations on the phenomenon of homologous response in transplanted amphibian limbs. IV. Reverse locomotion after the interchange of right and left limbs. *J. Comp. Neurol.* 67: 269–315.

Weiss, P. A. (1941a): Self-differentiation of the basic patterns of coordination. *Comp. Psychol. Monogr.* 174:1–96.

Weiss, P. A. (1941b): Nerve patterns: The mechanics of nerve growth. *Growth* 5(Suppl.):163–203.

Weiss, P. A. (1941c): Autonomous versus reflexogenous activity of the central nervous system. *Proc. Am. Philos. Soc.* 84: 53–64.

Weiss, P. A. (1942): Lid-closure reflex from eyes transplanted to atypical locations in *Triturus torosus*. Evidence of a peripheral origin of sensory specificity. *J. Comp. Neurol.* 77:131–169.

Weiss, P. A. (1950a): Experimental analysis of coordination by the disarrangement of central-peripheral relations. *Symp. Soc. Exp. Biol.* 4:92–111.

Weiss, P. A (1950b): Central versus peripheral factors in the development of coordination. *Proc. Assoc. Res. Nerv. Ment. Dis.* 30:3–23.

Weiss, P. A. (1950c): The deplantation of fragments of nervous system in amphibians. I. Central reorganization and the formation of nerves. *J. Exp. Zool.* 113:397–461.

Weiss, P. A. (1958): Cell contact. *Int. Rev. Cytol.* 7:391–423.

Weiss, P. A. (1968): *Dynamics of Development: Experiments and Inferences.* New York: Academic Press.

Weiss, P. A. (1969): "Panta' rhei"—and so flow our nerves. *Am. Sci.* 57:287–305.

Weiss, P. A. (1970): Neuronal dynamics and neuroplasmic flow. *In: The Neurosciences: Second Study Program,* Schmitt, F. O., editor-in-chief. New York: Rockefeller University Press, pp. 840–850.

Weiss, P. A. (1971a): *Within the Gates of Science and Beyond: Science in Its Cultural Commitments.* New York: Hafner.

Weiss, P. A. (1971b): *Bio-Medical Excursions: A Biologist's Probings into Medicine.* New York: Hafner.

Weiss, P. A., and Hiscoe, H. B. (1948): Experiments on the mechanism of nerve growth. *J. Exp. Zool.* 107:315–395.

Weiss, P. A., and Hoag, A. (1946): Competitive reinnervation of rat muscles by their own and foreign nerves. *J. Neurophysiol.* 9:413–418.

Weiss, P. A., and Taylor, A. C. (1944): Further experimental evidence against "neurotropism" in nerve regeneration. *J. Exp. Zool.* 95:233–257.

# Nerve Cells and Brain Circuits

John Szentágothai (b. 1912, Budapest, Hungary) is professor and head of the anatomy department at the Semmelweis University Medical School in Budapest. His extensive neuroanatomical researches have dealt with such topics as the structural bases of inhibition, the mapping of cerebellar pathways and the neuronal machinery of the cerebellar cortex, and the anatomical basis of nervous control of endocrine functions.

# 6

# From the Last Skirmishes around the Neuron Theory to the Functional Anatomy of Neuron Networks

John Szentágothai

## Reticularism or the Neuron Concept

When I was "awakening" on the scene of neurobiology in the early thirties, the neuron concept was once again under heavy barrage from the "reticularists," then led by Jan Boeke at Utrecht and by Philipp Stöhr, Jr., at Bonn. My first impressions about this nineteenth-century type of scientific *Streitschrift* (debate) derived from the hopeless struggle of my revered teacher in anatomy, Michael von Lenhossék, to refute the claims of Boeke (1926) about the existence of a "periterminal network" in the motor endplate, which would allegedly prove the continuity between neural and muscle substance. Von Lenhossék had too critical a mind not to realize that it was impossible then to beat Boeke with his own weapons, the sole use of neurofibrillar impregnation methods: so he refrained from publishing anything about the results of his last researches. However, I became "imprinted" with a fundamental distrust of the "histological picture" as a source of information about things that were not obvious in routine preparations and not visible with the mediocre resolving power of available optic systems.

A beginner like myself was, of course, not supposed to meddle in such high-level affairs, so I was commissioned by von Lenhossék to study the connections of the geniculate ganglion. I immediately became fascinated with the possibilities offered by experimental (Wallerian) degeneration and by cell and fiber counts (Schimert, 1936b).[1] Very soon, however, my interests shifted to another field: the structure of the vegetative nerve terminals, which at the time was the main battlefield of the "reticularists." The apparently reticular structures labeled *Grundplexus* (ground plexus) by Boeke (1933a,b) and *Terminalreticulum* by Stöhr (1935) were considered as clear evidence that neurofibrillar structures could invade other cells, mainly Schwann cells, and (according to Stöhr) also other tissues. The histological pictures furnished by various modifications of the Bielschowsky method were indeed deceptive, and I might have succumbed to the beauty of the seemingly reticular structures and the bombastic rhetoric of the chief reticularists had it not been for the advice of my immediate laboratory chief, P. von Mihálik, a somewhat strange personality but a man of exceptional clarity of mind and an unusual knowledge and sound criticism of the cytological and histological literature.

Having had some experience with experimental secondary degeneration, I soon concluded that of the two principal options—one, that the silver-stained filamentous elements were neurofibrils, as claimed by the "reticularists," or, two,

1. In publications until 1940 I used my original family name, Schimert.

that they were axons, as assumed by the "neuronists"—option one was not disproved by degeneration occurring after the removal of ganglia supplying the terminal plexus (secondary degeneration in vegetative terminals had been shown in 1934 by Lawrentjew and several of his coworkers). However, option two would and could be proven exclusively correct if it could be shown that the filamentous elements underwent partial degeneration in the same Schwann-cell process after partial transection of the nerves feeding into the network.

The experiments showed very clearly that some impregnated filaments underwent degeneration, while others in the same Schwann cell did not (Figure 6.1A; see Schimert, 1936a, 1937, 1938a). The conclusion was that these elements were nerve fibers (axons) from various sources, with up to ten or more axons embedded

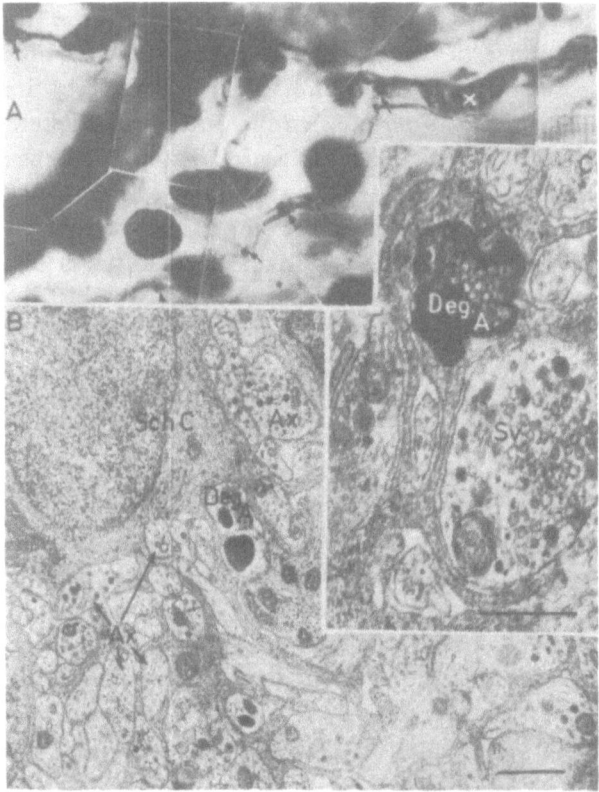

Figure 6.1    A comparison between the degeneration evidence, available on the light-microscopic level in the mid-thirties, and as seen today under the electron microscope. (A) A partly degenerated "ground plexus" in the myocardium of the cat three days after removal of one stellate ganglion. The degeneration fragments are clearly visible as black dots (arrows) together with normal axons in the same Schwann-cell process (x marks the Schwann-cell nucleus). From Schimert (1937). (B) The honeycomb structure of a Schwann cell (Sch. C.) with numerous small axons (Ax) embedded. One axon is in the early so-called cytolysomal stage of degeneration. (C) The same area under somewhat higher power, where besides normal axons (one with varicose thickening showing accumulation of synaptic vesicles = Sv), a fiber fragment is visible in later stage of degeneration (Deg. A). From an intestinal terminal plexus and the material of Ungváry and Léránth. The evidence, hence, was as good almost forty years ago as it is today; however, practically nobody cared for it.

in the same Schwann-cell process. Needless to say, my humble efforts did not at all impress the "reticularist establishment" of the time, beyond some remarks about the "childish attempts of some nincompoop beginner" (see Stöhr, 1939: pp. 603–604). It is fair, though, to mention that the situation was this bad only in continental European histology; in American and Soviet neurohistology, reticularist views exercised less influence. Nevertheless, I was the only man who stood up in defence of the neuron theory when the Anatomische Gesellschaft, at the 1937 Congress at Königsberg, sought to "bury the neuron theory" as announced by Boeke in his characteristic flowery speech.

It is not exactly gratifying to find that now, when this issue has been clarified beyond any doubt through electron microscopy (Figure 6.1B and C), hardly anybody remembers that exactly the same conclusions were drawn forty years ago on the basis of indirect evidence, and that this evidence was as good forty years ago as it is today. However, it seems to be the general fate of circumstantial evidence that, when more direct evidence turns up (irrespective of whether it is really more "direct" or only apparently so), it is immediately superseded and soon forgotten.

The neuron theory may be now due for certain revisions. The observation of presynaptic dendrites and somata, mainly in Golgi type-2 cells, brings into question the general validity of the principle of "histodynamic polarity"; tight gap junctions, fused membranes with possible loci of specific permeability between different neurons, etc., may call for certain reassessments in the traditional concept. However, it is unlikely that such a revision will invalidate the basic principles of the theory. The antineuronist claims raised in the classical period of the debate have all been shown to be nonrealistic—many even as irrational. If the theory has to be revised, as I am sure it will eventually, the observations for this will come from an entirely different level of investigations and understanding.

### Identification of Synapses by Experimental Degeneration

Around 1937 my interests shifted to the possibility of using axonal secondary degeneration to trace pathways directly to their synaptic contacts.

I was certainly not the first to attempt this, but I had perhaps more luck than most other investigators. Most earlier workers (Hoff, 1932; Hoff and Hoff, 1934; Foerster, Gagel, and Sheehan, 1933) used the Cajal impregnation techniques, but their results were not really convincing (see criticism by Phalen and Davenport, 1937) because the Cajal procedures do not generally stain degeneration fragments of axons; neither did the Bodian and other protein silver stains, although they were otherwise superior to the classical procedures. The only author who could observe unequivocal terminal degeneration was Miskolczy (1931), who looked for the degeneration of the mossy fibers in the cerebellar cortex, which could be well stained in various stages of degeneration with the Cajal methods. In retrospect I do not doubt that Ramón y Cajal would have been the first to exploit this technique in the central nervous system if it were not for the unfortunate circumstance that his silver stain gave unsatisfactory results with degenerating terminal elements.

A successful staining procedure for degenerating terminal axons and synapses actually had been developed much earlier by Rasdolsky (1923, 1925), a neuro-pathologist from Leningrad, and his pictures and indeed his results were convincing. However, his technique—a modified Alzheimer stain—was extremely cumbersome and involved, and after some unsuccessful attempts (probably due to some difference in the dyes used) I gave the venture up, especially since I was satisfied with the results of the Bielschowsky silver stains (Schimert, 1938b, 1939; Szentágothai-Schimert, 1941). Strangely, when I met Rasdolsky for a few moments many years later in Leningrad, during a memorial meeting at the graveside of I.P. Pavlov on the occasion of the 1949 Pavlov Centennial, he was not even aware of any development in this field. Then an aging man and apparently in failing health, he had some difficulty recalling his own studies when I told him with enthusiasm that his work 25 years earlier had begun to bear fruit. To be exact, my optimism was still not too well founded at that time and was shared by few neuroscientists, because this was long before the eventual "breakthrough" of the "suppressive silver stains" of Nauta and Gygax (1954).

The early history of the axonal and synaptic degeneration method is little known and rarely if ever appropriately cited.[2] Nobody could deny that the real breakthrough in the application of experimental axonal degeneration came with the introduction by Nauta and Gygax (1954) of silver stains that make it possible to suppress the staining of normal elements, and their recent modification by Fink and Heimer (1967) (and an array of more recent formulas) for more delicate terminal fibers. However, if criticism is permitted by an old hand at this game, I would have wished that students in this field had not forgotten about the original Bielschowsky stains—especially for identification of the real synaptic terminals—in their enthusiasm over the easy results gained with the modern suppressive stains. The consequence of this neglect was much uncertainty and discussion about whether the fragments corresponded to synaptic terminals or to preterminal fibers. Some of my first drawings, published in 1938 and reproduced here in Figure 6.2, might indicate what I mean. Although it was difficult to find degenerated terminals among the hundreds of thousands of normals, the few found could be clearly identified as real synaptic contacts.

I did not think it fortunate that, as a result of a modification introduced by Glees (1946), whose work otherwise had considerable merits in the early phase of the degeneration studies, the expression "bouton degeneration technique" became popular for some years before the introduction of the Nauta technique. This expression was based on the appearance of large, heavily stained ringlike structures in the regions of axonal degeneration. We know now from electron microscopy that these are rings of neurofilaments that undoubtedly occur in the so-called filamentous stage of degeneration. But they are not confined to terminals, and they are not characteristic of all neuron types. However, all these minor details have now become irrelevant, since degeneration of synaptic terminals can be better identified under the electron microscope, and degeneration techniques soon

2. However, I cannot, of course, complain about neglect in the citation of my early papers if a more recent author finds himself in disagreement with one of my minor statements or interpretations.

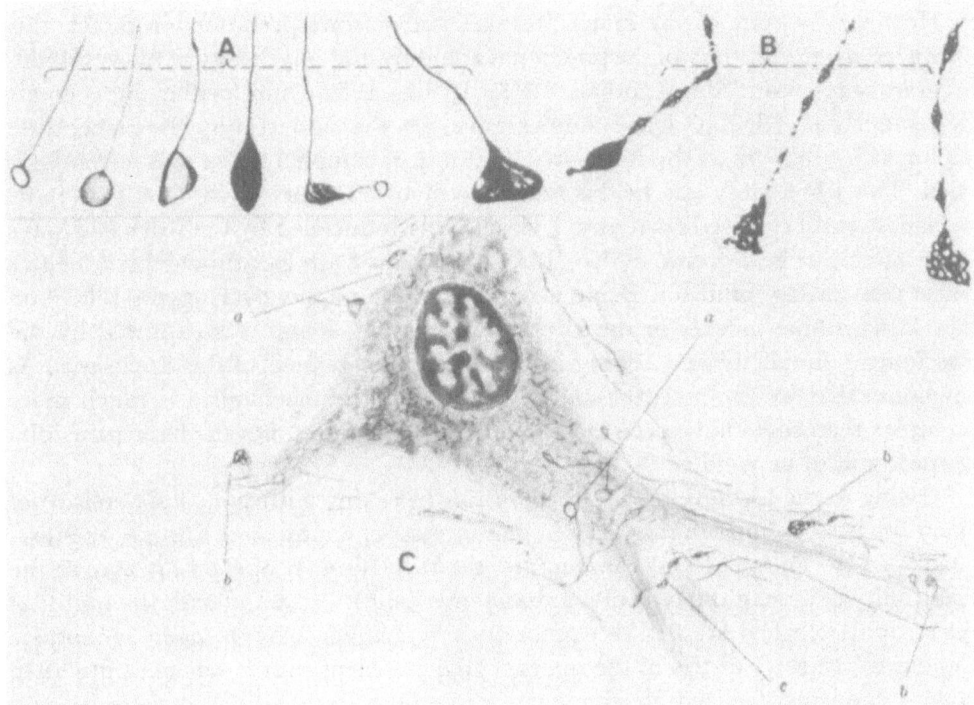

Figure 6.2   Drawings of synaptic degeneration in the spinal cord as revealed by the ordinary Bielschowsky stain. (A) Normal terminals showing a wide variation of size and staining but a clearly visible thin preterminal fiber. (B) A degenerated terminal with the characteristic "foamy" appearance (its equivalent is well recognizable now under the electron microscope), and the clear fragmentation of the presynaptic axon. (C) The relation to a nerve cell body of normal boutons (a), degenerated terminals (b), and preterminal fiber (c). From Schimert (1938b).

will be superseded by the new radioautographic methods using uptake by the cell bodies of tritiated amino acids (Lasek et al., 1968; Cowan et al., 1972).

## Stereotaxic Techniques and the Functional Anatomy of Reflex Arcs

Luckily, P. von Mihálik had earlier called my attention to the possibilities offered by the stereotaxic technique of Horsley and Clarke (1908), which after a delay of more than 20–25 years had begun to be routinely used by S. W. Ranson and his group in Chicago. Starting in the early forties with my own homemade apparatus (a polar-coordinate apparatus for cats, another for rats built of old microscope stages, and a special instrument for the spinal cord; see Szentágothai, 1951), I began exploring the sites of motoneurons supplying individual muscles. This was not exactly a fortunate choice, as my methods for exploring the somatotopic localization of various muscles by electrical stimulation (Szentágothai, 1942, 1948a, 1949) and also by axonal degeneration after minor lesions had been placed into various sites of the nuclei were fraught with many sources of error. I made more blunders in this work than I care to think of, even though Jan Jansen, Sr., tried to caution me in his gentle way—I am afraid with little success—against overrating the capacities of my technique.

However, in spite of the many mistakes and misinterpretations in detail, this work led me to the study of the functional anatomy and physiology of the vestibulo-ocular reflex arc (Szentágothai, 1943, 1950a, 1952) and ocular movements (Szentágothai, 1950b). These studies gave me the opportunity to concentrate upon such problems as the functional anatomy of complex reflex arcs and inhibition. This whole story and field is much too involved to be retold here, even in its broadest outlines. Needless to say, I was heavily influenced by the work and ways of reasoning of R. Lorente de Nó (1933), to whom I am greatly indebted for this most stimulating influence. Being able to read the papers by Högyes (1880)[3] on the labyrinthine reflexes in the Hungarian original, I also was captured by the fascinating simplicity and beauty in the basic arrangement of this apparatus. As demonstrated by the recent study of Ito (1973), the mechanism is much more complex than could have been imagined in the early fifties, but the basic principles appear still to be valid.

Owing to the low level in the early forties (speaking with some euphemism) of neurophysiology in Hungary, and in fact in most of continental Europe, the textbook of J. F. Fulton (1938) came to me as a revelation. It opened my eyes to the possibility of joining the type of neuroanatomy that I was doing with the results of electrophysiological studies. I cannot now remember why I made stereotaxic lesions in the early forties in the mesencephalic nucleus, but it was not until 1946 that I suddenly realized its significance for the study of the stretch reflex arc. Everything went very smoothly from then on. The first muscle spindle in the masseter muscle I looked at had a clearly degenerated anulospiral terminal, and nice terminal degeneration was found in the mesencephalic nucleus of the trigeminal nerve of the same preparation.

My paper on the anatomical basis of the monosynaptic reflex arc (1948b) was hailed and publicized by J. F. Fulton in the kindest fashion, and I became convinced that this was the course neuroanatomy had to take. Indeed, the search for parallelisms and eventually a dialogue between neuroanatomy and neurophysiology proved to be rewarding, particularly in Clarke's column (Szentágothai and Albert, 1955) and in the spinal motor nucleus (Szentágothai, 1958, 1961, 1967b). The latter studies came about under the influence of a fruitful correspondence beginning in 1954—and later developing into a friendship and collaboration "à distance"—with Sir John Eccles. His book *The Neurophysiological Basis of Mind* (1953), and subsequent studies on two elementary mechanisms of inhibition (Eccles, Fatt, and Koketsu, 1954; Eccles et al., 1954), prompted me to try my hand at the synaptic basis of inhibition. Although such attempts at the light-microscopic level were premature, some of their results were essential in preparing the way for the functional anatomical approach to neuron nets. Neurophysiology being far in the lead, I soon incurred the kind criticism that said: "If Sir John makes some claim of a new physiological mechanism, this guy (J.Sz.) soon finds an

---

3. It was a most unfortunate circumstance that Högyes published these studies only in Hungarian and, additionally, in a very heavy and cumbersome style of expression. In the pre-Sherringtonian era it was nearly impossible even for the reader familiar with Hungarian to grasp the fundamental importance of these studies. When the papers were eventually translated into German in 1911, unfortunately without the necessary editorial adaptation to the neurophysiology of the time, they understandably exercised little influence.

appropriate anatomical basis." This was true for the time being, but the time was soon to come when anatomy would take its turn in the forefront. However, first the story of my hypothalamic adventure has to be told.

## Hypothalamic Adventures

Although I had tried to exploit secondary degeneration in the hypothalamus as early as 1943–44, especially in the supra-optico-hypophyseal pathway, which I knew from the classical descriptions and the studies of S. W. Ranson's group at Northwestern University, the results were marginal. I had some notion of the neurosecretory concept of the Scharrers (1937, 1940), and I had the vague idea that it might be rewarding to see how a neurosecretory pathway degenerates. Otherwise my ignorance of hypothalamo-hypophyseal relationships and endocrine mechanisms could not have been more complete. However, when I took over the chair of anatomy at Pécs University in early 1946, I was joined by a young postdoctoral fellow, N.S. Halmi (now at the University of Iowa), who had a remarkable knowledge of the then-emerging literature of this field and a clear idea of what he wanted to do with my stereotaxic instruments. Indeed, the first experiments with hypothalamic (tuberal) lesions produced spectacular changes in the endocrine organs. Although Dr. Halmi left soon afterwards for the United States, where his remarkable studies in this field and in pituitary cytology became well known, some of my young coworkers were already infected with the hypothalamic bug, and, with some reluctance, I had to go along with them.

Due to my ignorance, our progress was very slow and circuitous in the beginning. Foreign travel was out of the question for us in the late forties and early fifties; but gradually my coworkers, Drs. Flerkó, Mess and Halász, developed their own experimental models, some of them quite ingenious and now well known by neuroendocrinologists. Looking back on our early naive exploits, I have the feeling that isolation for some limited time, when a group is left largely to its own intellectual resources, may be quite beneficial. My task on this team was the study of the general neuroanatomy and circulation of the hypothalamo-hypophyseal complex. Since axonal and synaptic degeneration gave very meager results with the methods then available—even with the Nauta stains—I soon had to turn to the Golgi methods.

The Golgi pictures had fascinated me even as a high-school student when, having a student microscope and a hobby-book on microtechniques (but no microtome), it was quite logical to play around with Golgi methods in which the thick sections could be cut with a razor blade. Later, however, as an undergraduate and then research assistant in the anatomy department, I succumbed to the general opinion of the time that nothing new could be expected from Golgi's method after the studies of Ramón y Cajal and other classical authors. The observations on the hypothalamus introduced me to the problems of local connectivity in neuron networks and gave me the feeling that the Golgi method still had much to offer for the understanding of synaptic architecture. This impression was soon strengthened by the remarkable papers on Golgi architecture of the Scheibels (1954, 1955, 1958) and of Fox and Barnard (1957). Still, my observations on the small neurons in the mediobasal hypothalamus, and the identification, in 1962, of the parvicellular tuberoinfundibular neurosecretory system and its terminals in the surface zone of the median eminence, which could be seen so clearly in the Golgi stains, did not impress most neuroendocrinologists (some even thought the whole story ridiculous) until the terminals could be seen under the electron miscroscope. So, eventually, this enterprise turned out quite well, and various editions of the monograph *Hypothalamic Control of the Anterior Pituitary*, prepared jointly with Drs. Flerkó, Mess, and Halász, were received favorably. Looking back now on this period, I am quite conscious of many naive, romantic features in the book especially in the first edition, for which I have to assume sole responsibility. Drs. Flerkó and Halász were of more realistic disposition and tried to dissuade me from various unorthodox notions. Even so our efforts to explain the results of hypothalamic interferences by various types of hypothalamic feedback (negative, positive, short-circuit, purely neural, etc.) made a modest contribution to the understanding of these complex mechanisms.

## The Cerebellar Cortex and the Modular Concept of Neuron Nets

Reverting to the main course of my exploits, I must now tell my side of the cerebellar story. The whole thing started innocently enough with the identification of the climbing fibers as the terminations of the olivocerebellar pathway (Szentágothai and Rajkovits,1959).[4] Due to the bad staining qualities of degenerated fibers in the molecular layer of the cerebellum, the evidence was somewhat meager—as has been pointed out with some relish by critical authors up to this day. However, I still suffer from the misconception that it is the interpretation that counts and not whether the finding is obvious and beautifully demonstrated, or is barely recognizable. I have told the long and romantic story of the climbing fibers elsewhere (Eccles, Ito, and Szentágothai, 1967: pp. 37–38), so I need not repeat it here. This finding gave considerable leverage to the application of microelectrophysiological methods to the cerebellar cortex by Sir John Eccles and his associates.

My present approach to the functional anatomy of neuron nets began when I was trying to convey to my undergraduate medical students, in the usual course on neuroanatomy, an idea about how excitation may spread in a piece of cerebellar cortex. Since specific inhibitory interneurons were not yet known, I speculated at the blackboard[5] that excitation might spread along the longitudinal axis of the folium via the parallel fibers, thought then to be about 3 mm in length, and across the folium via the basket axons (i.e., about 1 mm to both sides of a simultaneously excited beam of parallel fibers). When the concept of specific inhibitory interneurons gradually became established and generalized in the minds of neuroscientists in the late fifties, I suddenly realized that this speculation on the cerebellar cortex would become more elegant if the basket cells were en-

4. Actually I had made the first observations of this in 1937 and had mentioned in my discussion with Boeke (Schimert, 1937–1938) that, according to degeneration findings, climbing fibers originated from the inferior olive.

5. I ought perhaps to confess that, apart from refreshing my memory on the Latin nomenclature in anatomy and looking up new developments in fields other than my own, usually in *Scientific American,* I do not like to prepare lectures.

dowed with inhibitory properties. (Perhaps my readers will not be unduly shocked if I confess that the elegance and beauty of a concept is much more attractive to me than the dreary details of fact.)[6] Eventually I gave a formal lecture on this at our Academy at the beginning of 1963. In Figure 6.3 the pictorial representation of the concept (published only in the Hungarian report of this lecture) is reproduced. The figure gives an overall view of a field of excited Purkinje cells in the form of an elongated mountain (corresponding to the longitudinal course of the parallel fibers), with the valleys on either side representing the regions where the Purkinje cells are inhibited by the basket cells. I will not give the details of the concept because they have been reproduced on several occasions and publicized by Sir John (see Szentágothai, 1963: Figure 4; 1965: Figure 19; Eccles, Ito, and Szentágothai, 1967: Chapter 12, Figure 115).

The rationale of this approach was that, knowing the anatomy of any neuron network, one might make arbitrary assumptions about the excitatory or inhibitory nature of the various neuron types constituting the network and then try to understand how such a network might function under the assumed circumstances. This concept of a "neuronal machine" in the cerebellar cortex worked very nicely as long as local stimulation was applied to the cerebellar cortex, probably because local stimulation was essentially a repetition of the original anatomical speculation. Difficulties were soon encountered when researchers in this field began to apply more physiological conditions, and the recent studies of Eccles and coworkers (1971) and Llinás, Precht, and Clarke (1971) have shown clearly that closely neighboring Purkinje cells may behave in very different ways in the same situation, something not foreseen in the original, simplified concept.

My own peace of mind (or "tao" in the sense of Lao-Tse) in this matter had been disturbed before by the uneasy feeling that any speculation on the possible functions of neural nets would be on extremely shaky ground unless exact numerical and metrical relations of the neuron network were known. We started, therefore, a major enterprise in stereological analysis of the cerebellar cortex. The results (Palkovits, Magyar, and Szentágothai, 1971a,b,c, 1972) were shocking at first, because they showed that all numerical and many metrical conclusions obtained from earlier naive counts and measurements were grossly off the mark. On second thought, however, the accumulated data, and especially the total numerical transfer functions of the cerebellar neuron chain, make some sense and offer distinct possibilities for realistic modeling of the network functions by digital computer. For this purpose, each of the sequential links of the neuron chain have only to be compressed into a two-dimensional matrix (Szentágothai and Arbib, 1974: Figure 19) with the correct positions and the relative densities of the several elements. Any arbitrary input can then be investigated by the computer as it is successively processed through the stacked two-dimensional matrices (Pellionisz

6. Perhaps I should have become a mathematician, as in fact was suggested by some of my teachers in late high school. However, a strange "disability" in both reading and arithmetic in the early grades left me with a lasting shock. Fortunately, this disability was not considered a pathological condition in my time, as it sometimes is now, so that I was able to drag myself (or was dragged) through the lower grades. Somehow my disability disappeared; I certainly did not do much to overcome it.

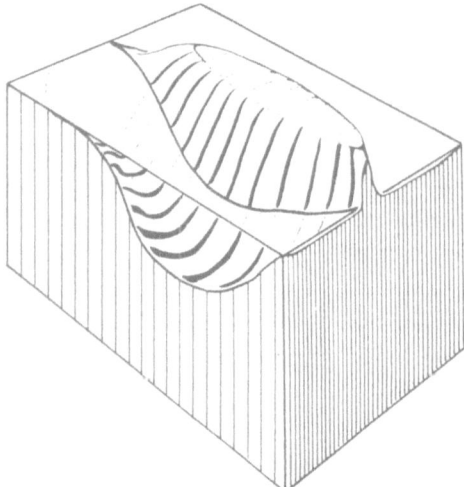

Figure 6.3   Pictorial representation of the effect on Purkinje-cell activity of a bundle of fibers imagined to run parallel to the longitudinal axis of the brick-shaped piece of molecular layer illustrated. The field of excited Purkinje cells is represented as a "mountain ridge" standing out from the plane, which symbolizes the resting state. In the foreground, the depressed state of activity—due to the action of the basket cells—is represented as a depression of the ground running parallel to the ridge.

and Szentágothai, 1973, 1974). This modeling leads to most interesting predictions that I cannot go into here. It might suffice to mention that closely neighboring Purkinje cells behave in very different ways to the same input, exactly as found by the physiologists.

This example does not by any means show that the type of reasoning about neural networks that led originally to the neuronal-machine concept is useless or basically false. On the contrary, it is the only way to look at neuron networks from the viewpoint of functional anatomy. One simply has to be more cautious in drawing far-reaching functional conclusions from the apparent overall features of the network. For example, the apparent features of the neuron network of the cerebellar cortex are a longitudinal spread of excitation and a transverse spread of inhibition. But if all the conditions are considered—something only the computer can do because of the large number of elements and variables—it emerges that any focal excitation reaching the granular layer by the mossy fibers can be conveyed to, and brought into interaction with, neighboring foci in the longitudinal and the transversal axes of the folium as well.

My anatomical speculations did not remain confined to the cerebellar cortex. I tried to generalize this way of looking at neuron nets into a general concept of "higher integrative units of the nervous system" (Szentágothai, 1967a). By this term I meant the minimum amount of neural tissue that would be able to perform the essential part of the information processing for which that piece of tissue is built. Now, after six years, these attempts look shockingly naive to me and might appear even more so to others. I cannot say more in their defense than that, however crude these models might be, they still offered stepping-stones over which some progress could be made. Consider, for example, the "iterative process" by which the basic circuit model of the neocortex could gradually be approached (Szentágothai, 1969, 1970, 1971, 1972a,b, 1973a,b). I do not for a moment believe that even the most recent of my circuit models (still unpublished) can make any claim to be "correct," but it is a way to raise questions and focus assumptions that can be tested by neurohistology and eventually by physiology.

By a fortunate convergence, the important concept, first raised by the Scheibels (1958), of a quasi-"modular" structure of the nervous tissue will soon be joining this general line of neuroanatomical reasoning. Considering the spinal segmental apparatus as a sequence of transversely oriented modules of neuropil (Scheibel and Scheibel, 1968; Szentágothai and Réthelyi, 1973)—something like stacked "chips" or microintegrated circuits in modern electronics technology—might prove eventually to be an important structural framework for the better understanding of nervous system design (see Szentágothai and Arbib, 1974). Similar modules of structure can be recognized everywhere in the neural centers. The cerebellar cortex itself is the best example of a modular design on the cellular level (Szentágothai and Arbib, 1974: Figure 42A). Here the separation of the arborization spaces of the Purkinje cells automatically brings about a regularity of space arrangement that results in a whole chain of regularities and in the final "quasi-crystalline" structure of the network. Similar regularities can be detected in seemingly irregular entanglements of dendritic and axonal arborizations, where the repetition at regular intervals of arborizations that are in themselves irregular leads to interference (or moiré) patterns, as I have tried to show for the lateral geniculate body (Szentágothai, 1972b, 1973b). There is a continuous line of these structures, from elementary modules up to large organoid modules of neural structure such as the cortical barrels of Woolsey and Van der Loos (1970). We may still be at the very beginning of the path to the understanding of neural nets, but I am convinced that this will be one of the most fruitful routes neuroanatomy can take.

It would be impossible, for obvious reasons, to tell here the detailed story of the developments in synaptology that went parallel with the quest for the understanding of neuronal nets. In the last fifteen years these developments were intimately tied up with the discovery of ultrastructural architectonics in synaptic systems through the use of the electron microscope. My own attitude and role here were determined by my fascination with the essence and the possible functional significance of the design and by my lack of interest in, and patience for, detail. I do not want to find excuses for the many mistakes, misinterpretations, or premature conclusions that I made and might have avoided with a more restrained attitude. Nor do I think that excuses are necessary, because a more imaginative approach—even though burdened with a broad margin of error—sometimes does more for progress than the "correct" inductive path to discovery. To be exact, both are needed. I might perhaps return to the example of the glomerular synapses in the lateral geniculate body, which I predicted purely on the basis of light-microscopic studies in a lecture at Atlantic City in April 1961 (Figure 6.4; the original diagram was in color so this figure does not do full justice to the original). I had not seen until then a single electron micrograph of this region. This stereoscopic view was very naive indeed by our present standards, but it does

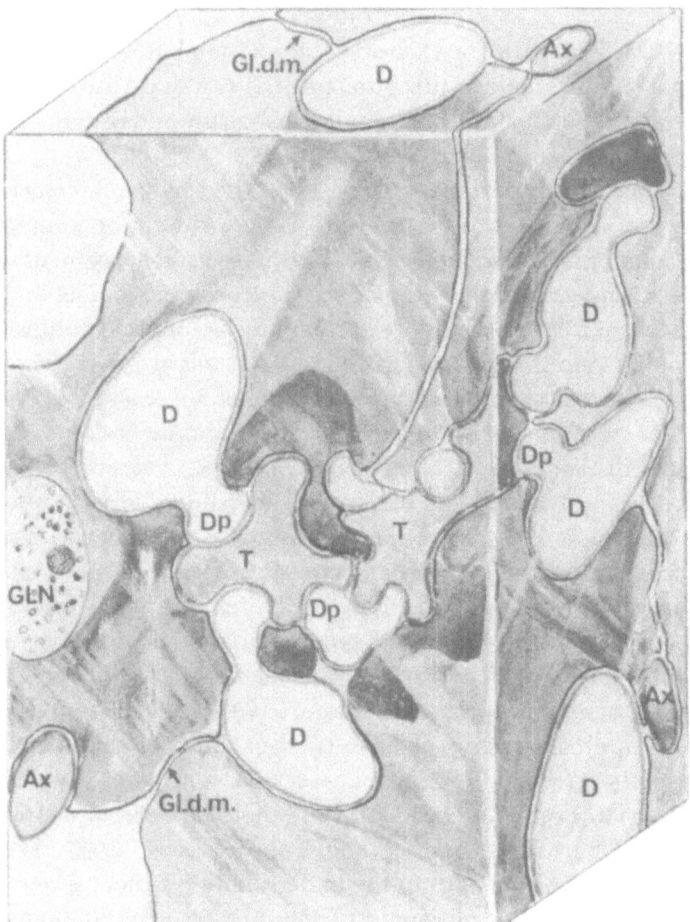

Figure 6.4 Stereoscopic "fantasy view" of a "glomerular" synaptic arrangement in the lateral geniculate nucleus as predicted on the basis of light-microscopic studies, showing the dendrites (D) of the relay cells with their protrusions (Dp), and the convergence of different types of axon (Ax) terminals (T). In spite of its many naive features, much of this view has been substantiated by subsequent electron-microscopic studies. The representation of the glia (GLN=glial-cell nucleus) was (unconsciously) adapted from the periphery; however, the occurrence of double glial membranes (Gl.d.m.) at sites where territories of adjacent cells (more correctly of processes) meet was predicted.

demonstrate that certain essentials of some very complex reality can be predicted quite well on the basis of extremely incomplete information.[7]

7. It is interesting, and quite characteristic of how the mind works under such circumstances, that the structure of the glia is represented very much in the fashion then known for the peripheric nerves (Schwann cells and mesaxons). One is reminded of the attempt by Edgar Allan Poe in his essay "The philosophy of composition" to explain in retrospect the creation of his poem "The Raven" as a process of conscious design. I am advocating here the reverse, and one might well argue which is the true mechanism. However, I share with Poe the emphasis on "undercurrents of meaning," even in scientific reasoning.

**Outlook**

My basic stand, for the time being, is a firm conviction that the real breakthrough that will bridge the still widening gap between the "behavioral" (whatever this includes, from simple conditioned reflexes, through memory and learning, to complex built-in behavior patterns and psychology and even up to the phenomenon of consciousness) and the "cellular" (including anatomy, physiology, biochemistry, molecular neurobiology, etc.) hinges on an understanding of the neuron networks. Whatever the molecular mechanisms of the elementary memory trace (most probably something in the material-producing machinery of the neuron), and whatever still unknown molecular or even charge-transfer or cooperative trickery lies behind the mechanism of "plasticity" in synaptic transmission (transduction, etc.), I believe that nature would not have brought forth the neuron network if the basic riddle and/or solution of behavior lay somewhere else. The immense degree and complexity of internal connectivity in the higher integrative centers, hypothalamus, and the limbic structures, striatum, and especially the cerebral cortex, gives me the strong feeling that single (or groups of) neurons may drag into action neighboring ones (or groups) not immediately connected with the chain of neurons to which the input is delivered. This concept can be imagined pictorially in very crude analogy to "flow patterns" of fluids (Bénard patterns or Zhabotinsky patterns; see Katchalsky, Rowland, and Blumenthal, 1974). Here I must give my tribute to the late Aharon Katchalsky who, in the last scientific meeting that he attended, raised this issue as a case analogous to cooperative phenomena. In this instance, again, we are treading in almost complete darkness, but, nevertheless, some of us may have the feeling that it is here that the solution of the ultimate riddle, consciousness, will be found. The tasks we are facing are well beyond the means and understanding as well as the technical apparatus of the biologist. What we need is the cooperation of theoreticians of various branches. Therefore, I am full of gratitude to the man—in honor of whom this symposium has been called together—who has founded in the Neurosciences Research Program a place where, and an apparatus with whose help, the students of these various fields can meet.

Trying to evaluate, as a function of my own personality, my own humble role in the development of the neurosciences in the last few decades, I am made painfully aware of my inadequacies. However, I have enough sense of humor and resignation to visualize myself as a romantic whose life—in spite of its outward smoothness and largely undeserved success—has been an uninterrupted sequence of quixotic ideas and enterprises, full of premature attempts and mistakes—some of which might perhaps be excused by the limitations in my circumstances (in equipment and technical resources)—more appropriate sometimes for one of the classic films of Charlie Chaplin than for the biography of a scientist. Or, on a more metaphorical level, I have to accept for my own fate the words of Saint Paul in the Epistle to the Romans (9:16): "It is not to those who will, neither to those who run, but to the mercy of God."

# References

Boeke, J. (1926): Die Beziehungen der Nervenfasern zu den Bindegewebselementen und Tast-zellen. Das periterminale Netzwerk der motorischen und sesibeln Nervenendigungen, seine morphologische und physiologische Bedeutung, Entwicklung und Regeneration. *Z. Mikrosk. Anat. Forsch.* 4:448–509.

Boeke, J. (1933a): Innervationsstudien. III. Die Nervenversorgung des M. ciliaris und des M. sphincter iridis bei Säugern und Vögeln. Ein Beispiel plexiformer Innervation der Muskelfasern. *Z. Mikrosk. Anat. Forsch.* 33:233–275.

Boeke, J. (1933b): Innervationsstudien. IV. Die efferente Gefässinnervation und der sympathische Plexus im Bindegewebe. *Z. Mikrosk. Anat. Forsch.* 33:276–328.

Boeke, J. (1939): Innervationsstudien. X. Sympathischer Grundplexus und Bindegewebsstruktur-en (Reticulinfasern des Bindegewebes und des Sarkolemmas). *Z. Mikrosk. Anat. Forsch.* 46:488–519.

Cowan, W. M., Gottlieb, D. I., Hendrickson, A. E., Price, J. L., and Woolsey, T. A. (1972): The autoradiographic demonstration of axonal connections in the central nervous system. *Brain Res.* 37:21–51.

Eccles, J. C. (1953): *The Neurophysiological Basis of Mind. The Principles of Neurophysiology.* Oxford: Clarendon Press.

Eccles, J. C., Faber, D. S., Murphy, J. T., Sabah, N. H., and Táboříková, H. (1971): Investiga-tions on integration of mossy fiber inputs to Purkinje cells in the anterior lobe. *Exp. Brain Res.* 13:54–77.

Eccles, J. C., Fatt, P., and Koketsu, K. (1954): Cholinergic and inhibitory synapses in a pathway from motor-axon collaterals to motoneurones. *J. Physiol.* 126:524–562.

Eccles, J. C., Fatt, P., Landgren, S., and Winsbury, G. J. (1954): Spinal cord potentials generated by volleys in the large muscle afferents. *J. Physiol.* 125:590–606.

Eccles, J. C., Ito, M., and Szentágothai, J. (1967): *The Cerebellum as a Neuronal Machine.* New York: Springer-Verlag.

Fink, R. P., and Heimer, L. (1967): Two methods for selective silver impregnation of degenerating axons and their synaptic endings in the central nervous system. *Brain Res.* 4:369–374.

Foerster, O., Gagel, O., and Sheehan, D. (1933): Veränderungen an den Endösen im Rücken-mark des Affen nach Hinterwurzeldurchschneidung. *Z. Anat. Entwicklungsgesch.* 101:553–565.

Fox, C. A., and Barnard, J. W. (1957): A quantitative study of the Purkinje cell dendritic branch-lets and their relationship to afferent fibres. *J. Anat.* 91:299–313.

Fulton, J. F. (1938): *Physiology of the Nervous System.* New York: Oxford University Press.

Glees, P. (1946): Terminal degeneration within the central nervous system as studied by a new silver method. *J. Neuropathol. Exp. Neurol.* 5:54–59.

Hoff, E. C. (1932): Central nerve terminals in the mammalian spinal cord and their examination by experimental degeneration. *Proc. R. Soc. B.* 111:175–188.

Hoff, E. C., and Hoff, H. E. (1934): Spinal terminations of the projection fibres from the motor cortex of primates. *Brain* 57:454–474.

Högyes, A. (1880): Az akaratlan együttjáró (associált) szemmozgások idegmechanizmusáról [On the nervous mechanism of the involuntary associated movement of the eyes]. *Orv. Hetil.* 23:17–29.

Horsley, V., and Clarke, R. H. (1908): The structure and functions of the cerebellum examined by a new method. *Brain* 31:45–124.

Ito, M. (1973): The vestibulo-cerebellar relationships: Vestibulo-ocular reflex arc and flocculus. *In: Handbook of Sensory Physiology.* Vol. 6. *Vestibular System.* Kornhuber, H. H., ed. New York: Springer-Verlag.

Katchalsky, A. K., Rowland, V., and Blumenthal, R. (1974): Dynamic patterns of brain cell assemblies. *Neurosci. Res. Program Bull.* 12:1–187.

Lasek, R., Joseph, B. S., and Whitlock, D. G. (1968): Evaluation of a radioautographic neuro-anatomical tracing method. *Brain Res.* 8:319–336.

Lawrentjew, B. J. (1934): Experimentell-morphologische Studien über den feineren Bau des autonomen Nervensystems. IV. Weitere Untersuchungen über die Degeneration und Regeneration der Synapsen. *Z. Mikrosk. Anat. Forsch.* 35:71–118.

Llinás, R., Precht, W., and Clarke, M. (1971): Cerebellar Purkinje cell responses to physiological stimulation of the vestibular system in the frog. *Exp. Brain Res.* 13:408–431.

Lorente de Nó, R. (1933): Vestibulo-ocular reflex arc. *Arch. Neurol. Psychiatr.* 30:245–291.

Miskolczy, D. (1931): Über die Endigungsweise der spinocerebellaren Bahnen. *Z. Anat. Entwicklungsgesch.* 96:537–542.

Nauta, W. J. H., and Gygax, P. A. (1954): Silver impregnation of degenerating axons in the central nervous system: A modified technique. *Stain Technol.* 29:91–93.

Palkovits, M., Magyar, P., and Szentágothai, J. (1971a): Quantitative histological analysis of the cerebellar cortex in the cat. I. Number and arrangement in space of the Purkinje cells. *Brain Res.* 32:1–13.

Palkovits, M., Magyar, P., and Szentágothai, J. (1971b): Quantitative histological analysis of the cerebellar cortex in the cat. II. Cell numbers and densities in the granular layer. *Brain Res.* 32:15–30.

Palkovits, M., Magyar, P., and Szentágothai, J. (1971c): Quantitative histological analysis of the cerebellar cortex in the cat. III. Structural organization of the molecular layer. *Brain Res.* 34:1–18.

Palkovits, M., Magyar, P., and Szentágothai, J. (1972): Quantitative histological analysis of the cerebellar cortex in the cat. IV. Mossy fiber-Purkinje cell numerical transfer. *Brain Res.* 45:15–29.

Pellionisz, A., and Szentágothai, J. (1973): Dynamic single unit simulation of a realistic cerebellar network model. *Brain Res.* 49:83–99.

Pellionisz, A., and Szentágothai, J. (1974): Dynamic single unit simulation of a realistic cerebellar network model. II. Purkinje cell activity within the basic circuit and modified by inhibitory systems. *Brain Res.* 68:19–40.

Phalen, G. S., and Davenport, H. A. (1937): Pericellular endbulbs in the central nervous system of vertebrates. *J. Comp. Neurol.* 68:67–81.

Rasdolsky, J. (1923): Über die Endigung der extraspinalen Bewegungssysteme im Rückenmark. *Z. Gesamte Neurol. Psychiatr.* 86:361–374.

Rasdolsky, J. (1925): Beiträge zur Architektur der grauen Substanz des Rückenmarks (Unter Benutzung einer neuen Methode der Färbung der Nervenfasernkollateralen). *Virchows Arch. (Pathol. Anat.)* 257:356–363.

Scharrer, E., and Scharrer, B. (1937): Über Drüsen-Nervenzellen und neurosekretorische Organe bei Wirbellosen und Wirbeltieren. *Biol. Rev.* 12:185–216.

Scharrer, E., and Scharrer, B. (1940): Secretory cells within the hypothalamus. *Res. Publ. Assoc. Res. Nerv. Ment. Dis.* 20:170–194.

Scheibel, M. E., and Scheibel, A. B. (1954): Observations on the intracortical relations of the climbing fibers of the cerebellum. A Golgi study. *J. Comp. Neurol.* 101:733–763.

Scheibel, M. E., and Scheibel, A. B. (1955): The inferior olive. A Golgi study. *J. Comp. Neurol.* 102:77–131.

Scheibel, M. E., and Scheibel, A. B. (1958): Structural substrates for integrative patterns in the brain stem reticular core. *In: Reticular Formation of the Brain.* Jasper, H. H., Proctor, L. D., Knighton, R. S., Noshay, W. C., and Costello, R. T., eds. Boston: Little, Brown, pp. 31–55.

Scheibel, M. E., and Scheibel, A. B. (1968): Terminal axonal patterns in cat spinal cord. II. The dorsal horn. *Brain Res.* 9:32–58.

Schimert, J. (1936a): Untersuchungen über den Ursprung und die Endausbreitung der Nerven der Iris. *Z. Zellforsch. Mikrosk. Anat.* 25:247–258.

Schimert, J. (1936b): Der Nervus intermedius und das Ganglion geniculi nervi facialis. *Z. Mikrosk. Anat. Forsch.* 39:35–44.

Schimert, J. (1937): Die Nervenversorgung des Myokards. *Z. Zellforsch. Mikrosk. Anat.* 27:246–266.

Schimert, J. (1937–1938): Discussion of paper by J. Boeke, Dritte Wissenschaftliche Sitzung. *Verh. Anat. Ges.* 45–46:158–159.

Schimert, J. (1938a): Die Endigungsweise des Tractus vestibulospinalis. *Z. Anat. Entwicklungsgesch.* 108:761–767.

Schimert, J. (1938b): Die "Syncytielle Natur" des vegetativen Nervensystems. *Z. Mikrosk. Anat. Forsch.* 44:85–118.

Schimert, J. (1939): Das Verhalten der Hinterwurzelkollateralen im Rückenmark. *Z. Anat. Entwicklungsgesch.* 109:665–687.

Stöhr, P., Jr. (1935): Beobachtungen und Bemerkungen über die Endausbreitung des vegetativen Nervensystems. *Z. Anat. Entwicklungsgesch.* 104:133–158.

Stöhr, P., Jr. (1939): Über "Nebenzellen" und deren Innervation in Ganglien des vegetativen Nervensystems, zugleich ein Beitrag zur Synapsenfrage. *Z. Zellforsch. Mikrosk. Anat.* 26:569–612.

Szentágothai, J. (1942): Die innere Gliederung des Oculomotoriuskernes. *Arch. Psychiatr.* 115:127–135.

Szentágothai, J. (1943): Die zentrale Innervation der Augenbewegungen. *Arch. Psychiatr.* 116:721–760.

Szentágothai, J. (1948a): The representation of facial and scalp muscles in the facial nucleus. *J. Comp. Neurol.* 88:207–220.

Szentágothai, J. (1948b): Anatomical considerations of monosynaptic reflex arcs. *J. Neurophysiol.* 11:445–454.

Szentágothai, J. (1949): Functional representation in the motor trigeminal nucleus. *J. Comp Neurol.* 90:111–120.

Szentágothai, J. (1950a): The elementary vestibulo-ocular reflex arc. *J. Neurophysiol.* 13:395–407.

Szentágothai, J. (1950b): Recherches expérimentales sur les vois oculogyres. *Sem. Hop. Paris* 26:2989–2995.

Szentágothai, J. (1951): Short propriospinal neurons and intrinsic connections of the spinal gray matter. *Acta Morphol. Acad. Sci. Hung.* 1:81–94.

Szentágothai, J. (1952): *Die Rolle der einzelnen Labyrinthrezeptoren bei der Orientation von Augen und Kopf im Raume.* Budapest: Akadémiai Kiadó.

Szentágothai, J. (1958): The anatomical basis of synaptic transmission of excitation and inhibition in motoneurons. *Acta Morphol. Acad. Sci. Hung.* 8:287–309.

Szentágothai, J. (1961): Anatomical aspects of inhibitory pathways and synapses. In: *Nervous Inhibition* (Proceedings of the Second Friday Harbor Symposium). Florey, E., ed. Oxford: Pergamon Press, pp. 32–46.

Szentágothai, J. (1963): Ujabb adatok a synapsis funkcionális anatómiájához [New data on the functional anatomy of synapses]. *Magy. Tudom. Akad. Biol. Orv. Tudom. Osztal. Közl.* 6:217–227.

Szentágothai, J. (1965): The use of degeneration methods in the investigation of short neuronal connexions. *Prog. Brain Res.* 14:1–32.

Szentágothai, J. (1967a): The anatomy of complex integrative units in the nervous system. In: *Recent Developments of Neurobiology in Hungary.* Vol. I. *Results in Neuroanatomy, Neurochemistry, Neuropharmacology and Neurophysiology.* Lissák, K., ed. Budapest: Akadémiai Kiadó, pp. 9–45.

Szentágothai, J. (1967b): Synaptic architecture of the spinal motoneuron pool. *Electroencephalogr. Clin. Neurophysiol.* Suppl. 25:4–19.

Szentágothai, J. (1969): Architecture of the cerebral cortex. In: *Basic Mechanisms of the Epilepsies.* Jasper, H. H., Ward, A. A., Jr., and Pope, A., eds. Boston: Little, Brown, pp. 13–28.

Szentágothai, J. (1970): Les circuits neuronaux de l'écorce cérébrale. *Bull. Acad. R. Med. Belg.* 10:475–492.

Szentágothai, J. (1971): Some geometrical aspects of the neocortical neuropil. *Acta Biol. Acad. Sci. Hung.* 22:107–124.

Szentágothai, J. (1972a): The basic neuronal circuit of the neocortex. *In: Synchronization of EEG Activity in Epilepsies* (Symposium Organized by the Austrian Academy of Sciences, Vienna, September 12–13, 1971). Petsche, H., and Brazier, M. A. B., eds. Vienna: Springer-Verlag, pp. 9–24.

Szentágothai, J. (1972b): Lateral geniculate body structure and eye movement. *Bibl. Ophthalmol.* 82:178–188.

Szentágothai, J. (1973a): Neuronal and synaptic architecture of the lateral geniculate nucleus. *In: Handbook of Sensory Physiology.* Vol. VII/3B. *Central Processing of Visual Information.* Jung, R., ed. Berlin: Springer-Verlag, pp. 141–176.

Szentágothai, J. (1973b): Synaptology of the visual cortex. *In: Handbook of Sensory Physiology.* Vol. VII/3B. *Central Processing of Visual Information.* Jung, R., ed. Berlin: Springer-Verlag, pp. 269–324.

Szentágothai-Schimert, J. (1941): Die Endigungsweise der absteigenden Rückenmarksbahnen. *Z. Anat. Entwicklungsgesch.* 111:322–330.

Szentágothai, J., and Albert, Á. (1955): The synaptology of Clarke's column. *Acta Morphol. Acad. Sci. Hung.* 5:43–51.

Szentágothai, J., and Arbib, M. A. (1974): Conceptual models of neural organization. *Neurosci. Res. Program Bull.* 12:305–510.

Szentágothai, J., Flerkó, B., Mess, B., and Halász, B. (1962): *Hypothalamic Control of the Anterior Pituitary. An Experimental-Morphological Study.* Budapest: Akadémiai Kiadó (2nd ed., 1965; 3rd ed., 1968).

Szentágothai, J., and Rajkovits, K. (1959): Über den Ursprung der Kletterfasern des Kleinhirns. *Z. Anat. Entwicklungsgesch.* 121:130–141.

Szentágothai, J., and Réthelyi, M. (1973): Cyto- and neuropil architecture of the spinal cord. *In: Developments in Electromyography and Clinical Neurophysiology.* Vol. 3. Desmedt, J. E., ed. Basel: S. Karger, pp. 20–37.

Woolsey, T. A., and Van der Loos, H. (1970): The structural organization of layer IV in the somatosensory region (S I) of mouse cerebral cortex. The description of a cortical field composed of discrete cytoarchitectonic units. *Brain Res.* 17:205–242.

Alf Brodal (b. 1910, Oslo, Norway) is professor of anatomy at the University of Oslo. His neuroanatomical researches have dealt particularly with the structural-functional relationships between various parts of the brain. Dr. Brodal has received the Bárány Medal (for vestibular research) and the Jahre Prize (for outstanding medical research in the Nordic countries).

# 7

# The "Wiring Patterns" of the Brain: Neuroanatomical Experiences and Their Implications for General Views of the Organization of the Brain

## Alf Brodal

According to the suggestions of the organizing committee, contributors to this symposium are free to talk about whatever they prefer. I have chosen to use this rare opportunity to present some personal points of view on the place of neuroanatomy within the neurological sciences and on the question of how morphological research may contribute to our understanding of the brain and its organization.

Let me first make clear a few points. I will use the term neuroanatomy or neuromorphology as a common denominator for all kinds of research on the structure of the nervous system, from gross comparative-anatomical observations to the study of cell organelles and membranes at the macromolecular level. My personal research covers, of course, only a small sector of this extensive field. It has chiefly been devoted to attempts to disentangle, experimentally, selected parts of the vast and complicated network of fiber connections in the brain, to determine what one may call its "wiring patterns." The complex and yet extremely orderly arrangement of this pattern is indeed fantastic and has never ceased to fascinate me. After more than thirty years of occupation with this subject, I am more convinced than ever that a knowledge of the structure of the brain in its minutes details is a prerequisite for meaningful interpretations of observations in all other fields of the neurosciences. Some of you may not share this conviction. In the following I will try to explain how it has matured in my mind, and to give some reasons for my view. Presumably it is a consequence of inborn personal characteristics as well as of experiences and external influences.

### Why I Chose Anatomy

I have a strong suspicion that the budding scientist's choice of a particular field of research is often not as haphazard as it may appear. Without, perhaps, reflecting much about the matter, unconsciously we have a feeling of where we have our strengths and weaknesses, of what appeals to us and what does not.

As for myself, I believe that the choice of anatomy as my field of research was partly a consequence of a preference for the concrete aspects of life, a deep-rooted skepticism against nonverifiable claims and postulates. Airy speculations have never offered much fascination. Even the most intensive training would never have made me become a philosopher! The fact that in anatomy you can directly see the things you are studying has always been a great satisfaction to me. Of course, here as elsewhere, you may be biased in one direction or another when you make your observations, and you cannot avoid the possibility that your judgments may get a subjective tinge.

However, the barren recording of structural data has never carried any appeal to me. From the very beginning the study of structure has been meaningful only insofar as it contributes in some way to an understanding of function. My preference for the functional aspects of structural features, for "functional anatomy," was considerably strengthened during my almost three years' work in clinical neurology (under the late Professor G.H. Monrad-Krohn) and psychiatry (under the late Professor R. Vogt). The derangements of function we see in the human being when parts of the central nervous system are destroyed constantly present us with questions concerning possible structural-functional relationships. I found this clinical study so fascinating that for a time I was in doubt as to whether I should choose clinical neurology or neuroanatomy as my field of research.

Looking back, I can say with assurance that I have never regretted staying in anatomical research. It has, I believe, suited my mind better than any other research field. Perhaps it was this feeling of satisfaction that gave me strength to resist the advice of many senior colleagues when in 1940 I was awarded a Rockefeller Fellowship for one year's study abroad: they recommended that I turn to neurophysiology, which was then at the beginning of its expansion in our time.

**The First Steps and the Search for a Method**

The directing of my early interest in morphology more particularly to neuroanatomy was due, at least in part, to the fascinating teaching I received from the late Professor K. E. Schreiner, head of the Anatomical Institute in Oslo until he retired in 1945. When I finished my preclinical training in 1932, I took a part-time position in the Anatomical Institute. My teacher, Jan Jansen, at that time prosector (associate professor), suggested that I should do some research in neuroanatomy, a field into which Jansen had been introduced by C. J. Herrick. My clinical neurological orientation at that time made me consider the anatomical basis of cerebral apoplexia as an interesting subject for research. To my great disappointment, however, Jansen strongly advised me against attempting a study of this kind. I realized that he had very good reasons for his attitude. Jansen suggested that I undertake an experimental-anatomical study of the olivocerebellar projection. A few human cases, described by Henschen (1907), Holmes and Stewart (1908), and some other authors, had produced fairly convincing evidence that in man there is a clear topographical relation between the inferior olive and the cerebellar cortex, an olivocerebellar localization.

The plan for my work was to make localized lesions in the cerebellum in cats and to map the distribution of retrograde cellular changes, which would presumably appear in the olive, in the course of a few days. To my astonishment it was not possible to find a single paper in the literature up to that time (1936) dealing with the subject in animals. By and by the reason for this became clear to me: the task was far from being as simple and straightforward as we had imagined.

We decided then that I should undertake a systematic study of the changes in the olive following cerebellar lesions and of their temporal course. The meager result of this was that after survival periods of several weeks I found some cell loss in the contralateral olive. However, it was not possible to make a valid distinction between affected areas and normal ones. Even with great variation in the survival times of the experimental animals, I never found convincing cell changes in the olive. Most likely other people before me had also had no success with this problem, but had not published their negative results. I remember well my feeling of frustration when the time-consuming microscopic scrutiny of large numbers of serial sections through the cerebellum and olive of some thirty rabbits, fifty mice, and twenty cats did not give any indication of how I could solve my problem.

During this period I had gone through much of the literature on changes in the central nervous system following lesions. I was particularly fascinated by a report of Gudden from 1870. Having made lesions in the brains of newborn animals, he found after about six weeks a massive cell loss in certain other parts (for example, in the thalamus following lesions of the cerebral cortex). Later on von Monakow (1882) had shown that some of these changes must be retrograde. Gudden's method, however, had some serious disadvantages. The operative mortality was very high, and, more important, when the animals lived for several weeks the shrinkage of the affected parts caused a distortion of the normal topography of the region in question, which made it difficult to identify particular structures. It occurred to me that the shrinkage and glial proliferation might be avoided if the survival period was reduced to just the time necessary for the affected cells to disappear. Furthermore, could it be that the cell loss which Gudden described might develop even if the animals were some days old instead of quite newborn? And finally, if the cell loss was due to a retrograde infection, could it be that the cellular changes preceding cell death in young animals might be more clearly identifiable than corresponding changes in adult animals?

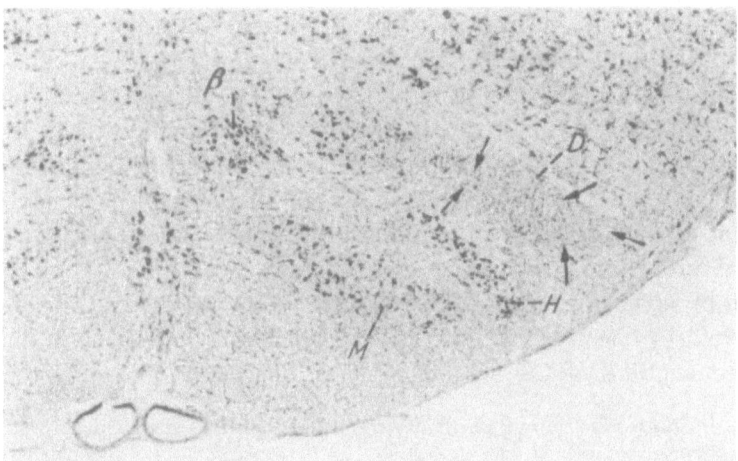

Figure 7.1 *Retrograde cell loss in very young animals (modified Gudden method) may be used to determine the origin of fibers.* A photomicrograph from a thionine-stained section through the inferior olive of a kitten. At the age of ten days a circumscribed lesion was made in the cerebellum (anterior lobe). The animal was killed five days later. In the area outlined by arrows in the contralateral dorsal accessory olive (D) practically all nerve cells have disappeared and there is a marked glial proliferation. The retrograde cell loss shows that this part of the olivary complex projects to the destroyed part of the anterior lobe: β, nucleus β; H, principal olive; M, medial accessory olive. Magnification: × 25. From Brodal (1972a).

With considerable enthusiasm I set about to test these questions. Cerebellar lesions were made in a great number of mice, rabbits, and kittens, aged from one to about fourteen days, and survival times were varied from a few hours to three weeks. To my great satisfaction it turned out that the "newborn" type of reaction of the nerve cell persisted for eight to ten days in the olive. Furthermore, the cell loss occurred extremely rapidly. As early as four to five days after a cerebellar lesion, practically all cells had disappeared in circumscribed parts of the contralateral olive and there was a marked localized glial reaction (Figure 7.1). Finally, the morphological changes in the olivary cells before they disintegrated resembled the classical picture of the *primäre Reizung* or retrograde cellular changes described by Nissl. Thus, by using quite young but not necessarily newborn kittens, and by killing them after a few days, I had a method (which I termed "the modified Gudden method") that made it possible to map the olivocerebellar localization with considerable precision (Brodal, 1940a).[1]

**Further Paths**

This experimental study of the olivocerebellar projection was submitted as my doctoral thesis in 1940 (Brodal, 1940b). I have dealt with it at some length because this, my first serious work in neuroanatomy, has certainly been influential in determining the line of my further research. In the first place, I became intrigued by the cerebellum, a subject to which a fair part of my later research work has been devoted. Second, this study directed my attention to the problem of topographical localization in the nervous system. Furthermore, it taught me emphatically the necessity of using reliable methods in research.

When in the following years, during the Second World War, Norway was scientifically isolated from most countries, Jansen and I worked together on other cerebellar connections. With the Marchi method we undertook a study of the cerebellar corticonuclear projection in the rabbit, cat, and monkey, and described for it a pattern of localization and a longitudinal zonal arrangement (Jansen and Brodal, 1940, 1942). As a continuation of my studies of the olive we used the same material for a study of the pontocerebellar projection. I also looked at the projections to the cerebellum from other brainstem nuclei.

The development of silver-impregnation methods for demonstrating degenerating fibers about 1950 (by Glees, Nauta, and others) opened a new epoch in the study of the fiber connections of the nervous system. The axons, not only the

1. The map worked out in this study relates a particular cerebellar lobule to one particular circumscribed part of the olivary complex. Studies undertaken in 1974, using horseradish peroxidase as a marker for retrograde protein transport in nerve cells, indicate that the pattern is far more complex than was originally supposed. Following injection in a small part of a single cerebellar lobule, labeled cells are found in localized patches in up to four different places in the olivary complex. A tentative explanation of the different results of the two methods is that each olivary neuron supplies several cerebellar subdivisions by way of collaterals, as has been concluded from neurophysiological studies. Since these collaterals are presumably not yet developed in almost-newborn animals, while the main fiber has already reached its destination, only the main line of the olivocerebellar projection is revealed by the modified Gudden method.

sheaths of myelinated fibers, could be traced, and the site of termination of transected fibers could be determined with considerable precision. With many of these methods even degenerating terminal boutons appear to be impregnated. The original belief that they could give information concerning the presence of synaptic contacts was, however, not tenable. This became clear when experimental electron-microscopic studies were later undertaken.

Nevertheless, used critically, the silver-impregnation methods have provided a wealth of information about the fiber connections of the central nervous system and the principles of their organization. In our laboratory a considerable part of the studies devoted to tracing fiber connections since the early 1950s has been done with one or more of these methods, in part combined with the modified Gudden method for studies of the origin of connections. Parallel experimental electron-microscopic investigations have also been made to some extent.

It has meant very much to me in these studies to have had the advantage and the pleasure of a close collaboration with young colleagues and with a number of research fellows from many countries. I am greatly indebted to them in many ways, as scientists and friends.

## A Central Theme: Topographical Localization in the Nervous System and Its Relation to Function

It has been said—I do not remember by whom—that in every scientist's work there is a central theme to which he always returns. I believe this is true, and I have little doubt that in my case the theme is: To what extent is there a topographical localization, especially a somatotopical localization, in the organization of the central nervous system?

It is a challenge to the function-oriented anatomist whenever physiological or clinical evidence for topographical relations in the brain cannot be explained by reference to known anatomical data. Challenges of this kind have induced us to study in some detail several of the fiber connections of the cerebellum and the brain stem. When the presence of a somatotopical localization in the cerebellum was demonstrated physiologically in the early 1940s (Adrian, 1943; Snider and Stowell, 1942, 1944; and others), little was known of the anatomical organization of those pathways that might mediate the cerebellum's somatotopical relations with the spinal cord and the cerebral cortex. It has been a source of great satisfaction that we have succeeded in showing that this functional pattern does indeed have a morphological substrate (for example, in some of the spinocerebellar pathways, in the projection of the cerebellum onto the red nucleus and the nucleus of Deiters, of these two nuclei onto the spinal cord, and of the cerebral cortex onto the red nucleus, the inferior olive, and the pons; for references see Brodal, 1969). Numerous other instances have been revealed by other researchers (for example, with regard to motoneurons, dorsal columns, thalamus, and basal ganglia). In other instances our anatomical observations of somatotopical patterns have later been confirmed physiologically.

I would like to illustrate this by the results of recent studies on the projection from the cerebral cortex onto the pontine nuclei, performed by Per Brodal (1968a, b, 1972). While a clear somatotopical relation had been demonstrated physiologically between certain cerebral cortical areas and some of the cerebellar lobules, anatomical studies had disclosed only a very crude pattern of localization in the corticopontine projection. By utilizing minute cortical lesions it has, however, been possible to show that there is indeed a precise topographical anatomical arrangement in the corticopontine projection, the first link in the quantitatively most important cerebrocerebellar pathway.

Figure 7.2 shows on the left a photomicrograph of a section from the pontine nuclei in the cat. The section was treated with the silver-impregnation method of Fink and Heimer for demonstrating degenerating nerve fibers and terminals. The animal was killed five days after a small, purely cortical lesion was made in the striate area (P. Brodal, 1972). As seen from the figure, the pontine area containing black particles, which represent degenerating fine fibers and presumably some terminals, is very circumscribed. On the right is another photomicrograph from a case with a large lesion of the sensorimotor cortex, which of course results in a more extensive area of degeneration (P. Brodal, 1968a). It is significant that the transition between the area showing degeneration and that free from degeneration

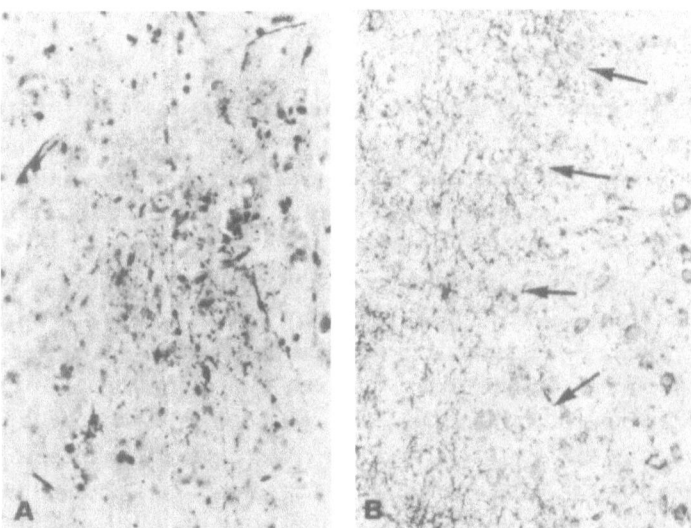

Figure 7.2 *Silver-impregnation methods may be used to determine localized sites of termination of fibers.* (A) A photomicrograph from the pontine nuclei of a cat, killed five days after a small, purely cortical lesion had been made in the ipsilateral striate area. Degenerating fine fibers and, in part, terminals appear as black particles in the section (impregnated with the stain of Fink and Heimer) and indicate the area of termination of fibers from the cortical lesion. Note the spatially restricted distribution of degeneration. Magnification: x 320. From P. Brodal (1972). (B) A photomicrograph from the pontine nuclei of a cat, killed five days after a lesion of almost the entire primary motor and sensorimotor cortex. The distribution of degeneration (to the left) is more extensive than that following the small lesion in the case shown in A. Note, however, the sharp transition (arrows) between the terminal field of the degenerating fibers and the normal areas to the right. The Nauta staining method was used. Magnification: × 150. From P. Brodal (1968a).

is remarkably sharp. By combining the results of many cases with small, differently placed cortical lesions, it has been possible to map several components of the corticopontine fiber connections in the cat.

Figure 7.3 shows a diagram of the projections to the pons from the cortical sensorimotor areas (MsI, SmI and SmII) in the cat (P. Brodal, 1968b). Each of these areas sends its fibers to two (or three in the case of SmII) longitudinal cell columns in the pons. Their positions are shown below in the transverse sections through the pons. In each column there is a somatotopical pattern (with a somewhat different orientation). Thus the corticopontine projection is very precisely organized. Let me add that, in view of this, one might postulate that there must be a correspondingly precise arrangement in the second link of the corticocerebellar projection, the pontocerebellar. Otherwise it is difficult to see that the corticopontine localization can have any functional meaning. I have no doubt that it will be possible in the future to demonstrate a pontocerebellar localization, using either the recently developed methods for tracing fibers following injection of tritium-labeled proteins or the retrograde axonal transport of horseradish peroxidase.

## A Precise and Specific Organization to the Minutest Details

The rather common occurrence of patterns of somatotopical-anatomical relations is only one piece of evidence for precision in the structural organization of the nervous system. Additional evidence comes from studies of other types of con-

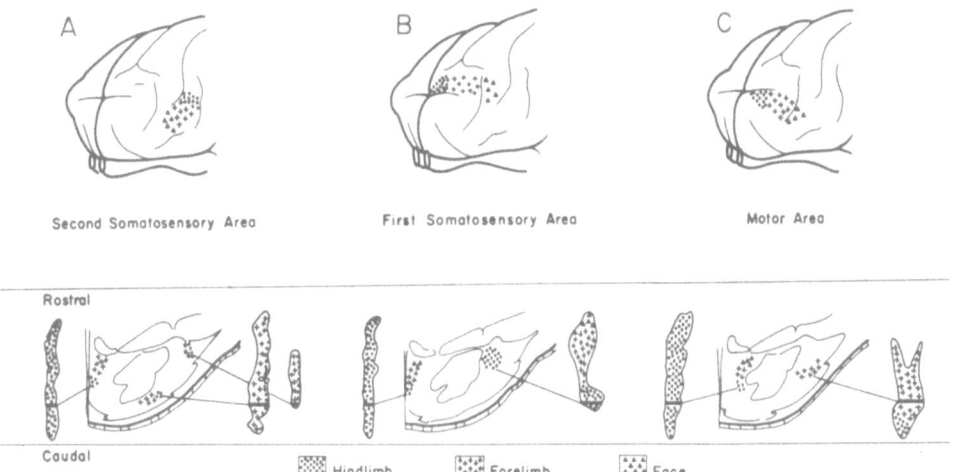

Figure 7.3 *Examples of precise somatotopical fiber connections.* The projections from sensorimotor cortical regions (MsI, SmI, and SmII) onto the pontine nuclei in the cat are somatotopically organized. The longitudinal cell columns in the pons projected upon by the three cortical areas are represented on each side of a transverse section through the lower part of the pons. Note the different orientations of the somatotopical patterns. The first and second somatosensory areas have a medial column in common; otherwise the three cortical areas project to different pontine areas. From P. Brodal (1968b).

nections. Figure 7.4 shows a diagram of the sites of termination in the vestibular nuclei of fibers from the vestibulocerebellum (Angaut and Brodal, 1967). The fibers from its three regions (the flocculus, nodulus, and uvula) have their particular principal sites of ending in the nuclei, even if they overlap in certain regions. This illustrates a principle that we have found to be valid for a number of nuclei: when a nucleus receives fibers from several other sources, the various afferent contingents do not intermingle diffusely and without order. On the contrary, each contingent has its preferential area of supply. We have shown this to be the case, for example, for the arrangement of other types of afferents to the vestibular nuclei and for the reticular nuclei of the brainstem.

The examples I have mentioned illustrate a general experience we have had: in a number of fiber systems it is possible to get a rather precise indication of the sites of termination of afferent fibers from a particular destroyed part of the brain.

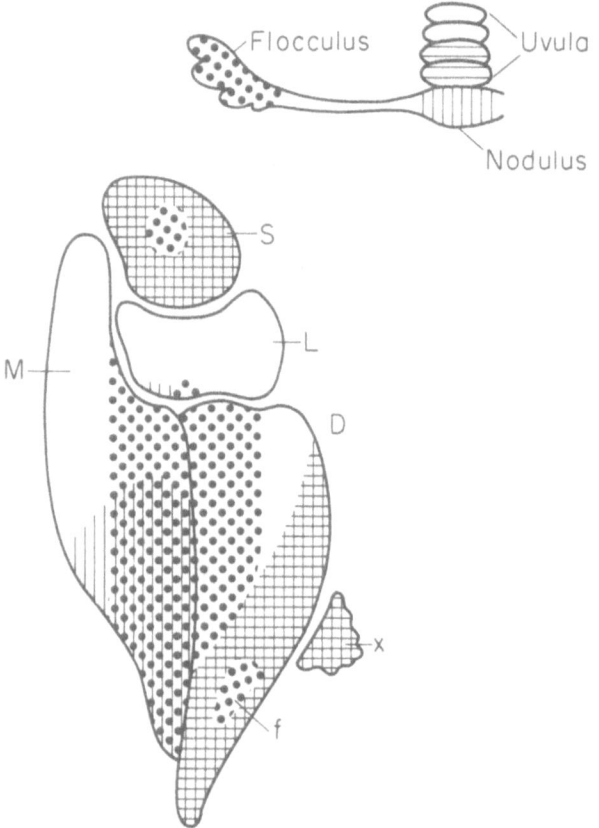

Figure 7.4  *Afferent fibers from different, although functionally closely related, parts may each have their particular and restricted areas of distribution within a nuclear complex.* The diagram shows how fibers from each of the three main parts of the vestibulocerebellum in the cat (above) are distributed to particular regions within the vestibular nuclear complex, shown schematically below as seen in a projection in the horizontal plane. Some nuclear regions are supplied only by one of the afferent contingents, others by two, and still others are free: D, L, M, S, descending, lateral, medial, and superior vestibular nucleus, respectively; x and f, particular small cell groups within the nuclear complex. From Walberg (1972), redrawn from Angaut and Brodal (1967).

If there are fairly precise borders to the receiving areas, one can demonstrate this. It should be emphasized that, in order to bring out topographical relations of this kind, it is necessary to have small lesions and to use serial or approximately serial sections. Furthermore, the plane of sectioning of the specimen is important if you are searching for definite borders. The geometrical arborization patterns of the fibers in question are also relevant, as I have discussed elsewhere (Brodal, 1972b). If regard is given to these factors, I feel that anatomical findings like those described here are reliable evidence of precise anatomical topographical relationships.[2]

It may be objected that if you study Golgi sections you get an entirely different impression. You may find dendrites that extend for some distance outside the area determined as the terminal site, perhaps even into a neighboring nucleus. Fine axonal ramifications of the afferent fibers may likewise extend beyond the alleged terminal site. While these circumstances induce features of complexity, they do not by themselves disprove the presence of localization. It should be recalled that in Golgi sections you see only scattered elements, so that quantitative estimates cannot be made. What the anatomical tracing with our present methods can give us is only the *principles* underlying the *patterns* of the topographical relations. However, there are good reasons to believe that the functional topographical relations may be even sharper than it is possible to show anatomically. The detailed anatomical arrangements of cells and cell processes, such as internuncial cells and collaterals, and their physiological properties, may contribute to this, for example, by way of lateral or surround inhibition. Thus, while we could only demonstrate the general pattern of localization in the projection of the anterior lobe onto the nucleus of Deiters (see Brodal, Pompeiano, and Walberg, 1962; Walberg, 1972), it has been shown physiologically that a cell in Deiters's nucleus may be influenced from a single folium of the anterior lobe but not from the immediately adjoining folia (Pompeiano and Cotti, 1959). On the other hand, the presence of a localization within a projection may be missed in physiological studies if the afferent input to the nucleus under study is derived from a source that is a seat of convergence of impulses from different regions (for example, from different parts of the body; see Künzle, 1973, on the lateral reticular nucleus).

There is little doubt that, as a rule, the morphological details that can be shown in studies of architectonics and fiber connections are of functional relevance. It has been highly rewarding, therefore, to see that the results of modern neurophysiological investigations agree very well with the anatomical data (for example, with regard to the vestibular nuclei) which we have studied in some detail (for partic-

2. Experimental electron-microscopic studies (or tracing of fibers with tritium-labeled proteins) usually confirm that the area interpreted as a site of termination of a fiber system with silver methods is its true ending place. It has recently been shown, however, that with these methods the terminal site is occasionally found further "distally," because the silver-impregnation methods do not visualize the final "preterminal" and terminal branches, presumably because they are of too fine caliber. (See, for example, Sterling, 1973, on afferents to the superior colliculus, and Grofová and Rinvik, 1975, on the ending of the cerebellofugal fibers in the thalamus.) This urges caution against accepting findings obtained with silver-impregnation methods as final in all instances.

ulars see Brodal and Pompeiano, 1972). Even a very small cell group, which we originally distinguished on a cytoarchitectonic basis and called group z (Brodal and Pompeiano, 1957), consisting of only a few hundred cells, has recently been shown to be a particular unit physiologically as well as anatomically (Landgren and Silfvenius, 1971; Grant, Boivie, and Silfvenius, 1973).

The more we study the brain, the more evidence we find for the view that there is an extremely specific organization in its structure. Each small unit seems to have its particular morphological characteristics. This view receives support from other fields of light-microscopic research. Thus the reaction of nerve cells to transection of their axons varies widely among nuclei and among animal species. The recent renaissance in the use of the Golgi method has given further examples. When one turns to electron microscopy, there appears to be almost no end to the structural differences between minor regions with regard to synaptic arrangements, morphology of synapses, patterns of degeneration of cells, fibers, and boutons, and so on. Chemical studies tell the same story, for example, of chemical differences between layers and areas in the cerebral cortex. It appears as if every nucleus or cell group has, so to speak, its own "personal characteristics."

## The Overwhelming Complexity in Interconnections between Minor Units

The structural complexity of the nervous system is overwhelming, as is emphatically demonstrated by any consideration of the interconnections or lines of communication between its various parts. When you trace these connections, you are struck over and over again by their profusion. New ones are still being discovered! Even if some of these are quantitatively rather modest, we are scarcely entitled to disregard them when we attempt to understand the brain. If they were functionless, they would probably have fallen victim to disuse atrophy. Some features in the arrangement of this multitude of connections appear to be general. Thus there are a large number of reciprocal connections between any two structures, producing feedback pathways that often pass through one or more intercalated stations (a fact not always considered in functional studies). Furthermore, it seems to be the rule rather than the exception that a given unit, or complex of units, has at its disposal more than one pathway by which it can influence another particular unit or area. These parallel pathways differ in their structural organization, in the stations intercalated along them, and in other respects. As an example, Figure 7.5 shows a simple diagram of the main cerebrocerebellar pathways. Their differences with regard to sites of origin, numbers of fibers, and terminal areas in the cerebellum are not shown, but it is seen that they pass through different intercalated nuclei (the reticular nuclei, pontine nuclei, etc.). It is particularly important to realize that these intercalated nuclei differ with regard to their noncortical afferents, their feedback connections with the cerebellum, and their intrinsic organization (for some data see Brodal, 1972a). We can scarcely avoid the conclusion that these pathways play functionally different roles in the cerebrocerebellar cooperation.

The cerebrocerebellar pathways are only one example of a pattern of multiple,

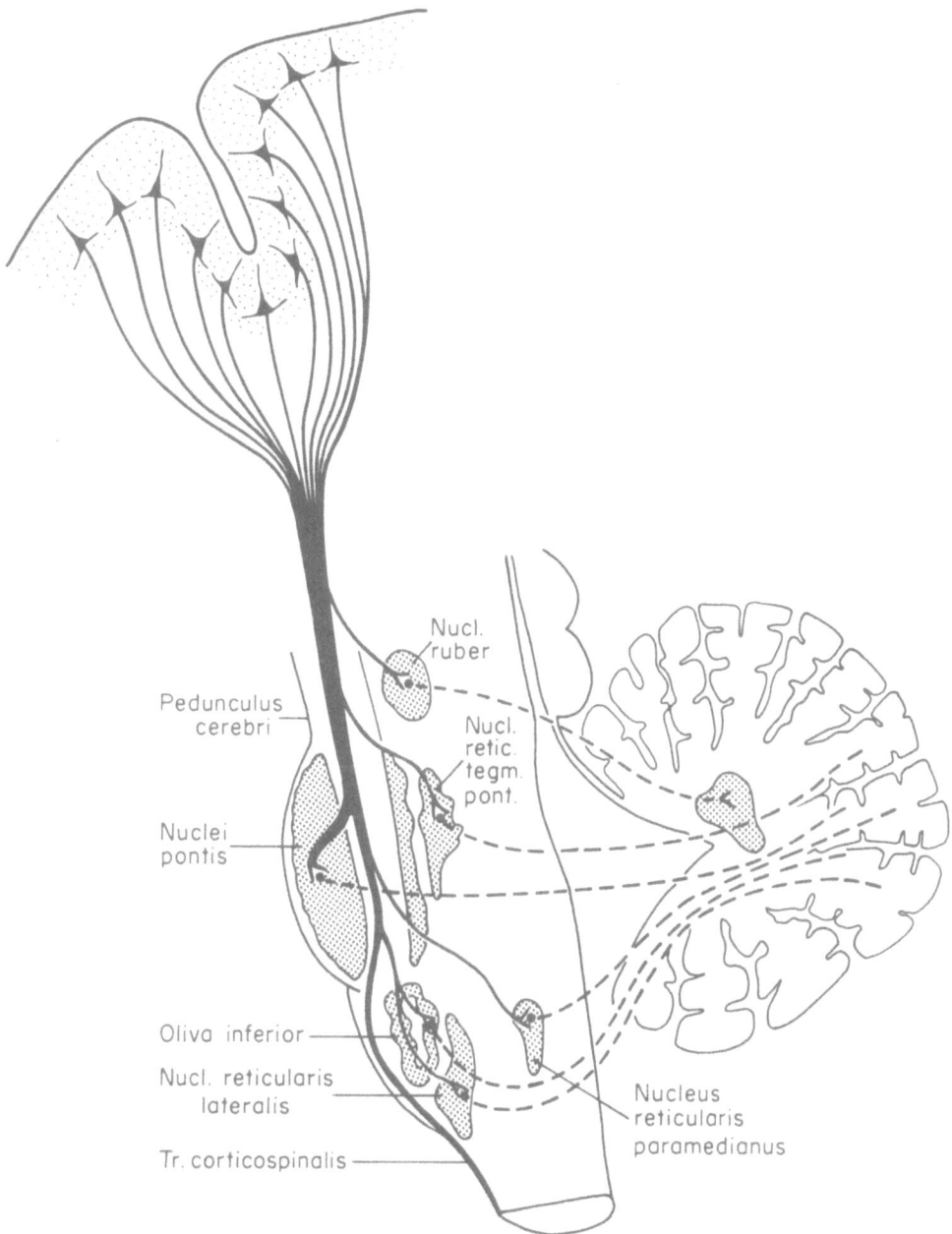

Figure 7.5  *A region of the brain may be connected with another by several routes.* The figure shows as an example six neuronal pathways that lead from the first motor area of the cerebral cortex to the cerebellum. Some of the intercalated nuclei receive fibers from more cortical areas than MsI. They differ with regard to the patterns of their projections onto the cerebellum, their afferent connections, and in other respects, and they must be assumed to play different functional roles. From Brodal (1972a).

more-or-less parallel connections between two brain regions. There are other patterns. The organization of the corticopontine projection (see Figure 7.3) illustrates a pattern of interconnections that may turn out to be more common than hitherto realized. A particular unit may have projections to more than one small part of a larger complex (for example, the pontine nuclei in the case of the first sensorimotor region). We have likewise found that the first motor area (MsI) sends fibers to three different and separate parts of the inferior olive (Sousa-Pinto and Brodal, 1969), and that the crus II and the paramedian lobule of the cerebellum both have their particular projection fields in the nucleus interpositus anterior, interpositus posterior, and in the lateral nucleus (Brodal and Courville, 1973; Courville, Diakiw, and Brodal, 1973). Other examples may be quoted. The more we learn about the fiber connections, the more complex seems to be the pattern of their organization. But in this apparent chaos of "wires" there is certainly a very precise order, an order down to the minutest details.

Recent experimental electron-microscopic studies lend further support to this view. I would like to mention as a pertinent example some recent studies performed in our department by Grofová and Rinvik (1975). The afferents to the ventral lateral nucleus of the thalamus from the cerebellum, the pallidum, and the motor cortex differ with regard to their synaptic relations, such as types of terminal boutons and contacts with the somata or dendrites of cells. A single afferent fiber may contact relay cells as well as internuncial cells, even with the same terminal bouton, and the dendrites of these cells may be linked by dendrodendritic synapses.

**A General View**

In the general view of the nervous system that every neurobiologist forms in his mind, his own personal experiences will loom heavily. If I am to integrate the data that I have discussed into a general concept, I am forced to the following conclusion (which is by no means original): we may consider the brain as consisting of a multitude of small units, each with its particular morphological (and presumably functional) features. These units collaborate by way of an immensely rich, complicated, and differentiated network of connections, which are very precisely and specifically organized. The anatomical possibilities for (more-or-less direct) cooperation between various parts of the brain must be almost unlimited!

This view of the organization of the brain is neither purely localistic nor purely holistic, but is, I believe, a fruitful combination of both. It is, of course, very general, you may say vague, and as such it does not lead us very far in our attempt to understand the functions of particular parts of the brain. However, it has important consequences for our approach to the study of the brain.

**The Importance of Neuroanatomy for Research in Other Fields of the Neurosciences**

Thus, in all types of functional studies, not least in microphysiology, it will be essential in the future that particular attention be directed to the exact determination of sites of recording or stimulation. It may not be at all irrelevant if one's electrode is in the dorsal or ventral, rostral or caudal part of a nucleus! Histological control is necessary; one cannot, for example, rely on stereotactic coordinates in an atlas. Such technical difficulties can usually be overcome if sufficient care is taken. In neurochemical studies the finer geography of the brain must likewise be considered. A statement that "half of the cells in the reticular formation of the pons and rostral medulla are influenced by noradrenaline," does not tell us much, since this part of the brain is made up of a great number of different "units."

In experimental physiology one will always have to study a particular part, function, or "system" in isolation. One deals with only a fraction of what actually goes on in the brain (for example, on natural stimulation of receptors, or on electrical stimulation of a nucleus), and one sees only a fragment of a very complex picture. When attempts are made to formulate a hypothesis about the function of the particular nucleus or fiber system under study, consideration of relevant data on the possible morphological basis is essential. If, for example, one disregards other anatomically verified possible routes for transmission than those one has in mind, the hypotheses set forth will most probably be misleading. Take as an example the cerebellar climbing fibers. The view that all these fibers come from the inferior olive has been doubted by anatomists for years and appears now to be disproved (see O'Leary et al., 1970; Murphy, O'Leary, and Cornblath, 1973). There is also physiological evidence against it (see Brodal, 1972a, for some references). Furthermore, the olive also appears to give rise to mossy fibers. It is, therefore, scarcely permissible to interpret a climbing-fiber response in the cerebellar cortex as evidence that the impulse pathway concerned has a last relay in the inferior olive and to assume that a mossy-fiber response has not. Another example: in physiological studies of the cerebrocerebellar relations one has to take into account the fact that the impulses reaching the cerebellum from the cerebrum may be transmitted along several more-or-less parallel, but differently organized, routes, as shown in Figure 7.5. We may safely assume that each of them has its particular task in cerebrocerebellar cooperation.

Much still remains to be done in the exploration of the functional properties of the various links in this multichannel pathway. And how do they work together? The situation is complicated by the fact that (unfortunately!) only a few of the cortical fibers to the relay nuclei can be collaterals of the cortical fibers passing to the spinal cord (which in man make up only about 1 million of some 21 million fibers in the cerebral peduncle; see Tomasch, 1969).

In attempts to explain brain functions on the basis of an animal's behavior following brain lesions or stimulation of a particular region or nucleus, knowledge of the relevant structural features is likewise necessary for proper conclusions. The derangements of its interplay with several other nuclei or regions, demonstrated by the presence of fiber connections of a very detailed character, may be more decisive for the observed changes in behavior than are the effects of the lesion or stimulation of the target structure.

In a corresponding way, attempts to construct working models of the brain on the basis of incomplete knowledge of structure are likely to be misleading. Even if one restricts oneself to analysis of the working of a particular nucleus, such as the inferior olive, there is a great risk of overlooking some of the many anatomical possibilities of its being directly or indirectly influenced by, and influencing, other nuclei or cell groups. And what about the temporary changes in chemical, ionic, and hormonal conditions that may influence the physiological properties of cells, nerve fibers, and synaptic function? I am convinced that in functional studies of the brain it will become increasingly essential to take the morphological basis into consideration. It is encouraging to notice that in contemporary research there appears to be a growing tendency to realize this.

Even if there are exceptions, it is generally true that differences in structure reflect functional differences. The overwhelming degree of specificity of structural units, their combination into units of a higher order, and their multiple and differently organized interconnections are tangible evidence of a functional complexity that we can scarcely imagine, particularly when we take into account recent electron-microscopic morphological data. We must assume that the many interconnections present are morphological evidence that an extremely complex interaction is possible among almost all parts in any brain function. Such mutual influences may be subtle and are, therefore, often not recognized in physiological or clinical studies. However, they can not be disregarded when we are attempting to understand the brain properly. As a single illustration, let me mention the observation (Brodal, 1973) that, following a small embolic lesion in the right hemisphere in a right-handed person, there may be disturbances in the subject's handwriting, appearing as evidence of incoordination. For optimal function we certainly need the entire intact brain!

## Brain Research in the Future: A Credo

The anatomical data cannot be disregarded if one wants to formulate tenable views on the working of the brain! The morphological complexity of the brain confronts neurophysiologists with tremendous tasks. One may indeed ask: Do we have sufficiently subtle methods? Another question also arises: Will it be necessary to revise the classical neuron doctrine? (See Shepherd, 1972.)

Faced with the tremendous complexity in the structure and function of the brain, we have to realize that our knowledge is very modest. (What about the role of the many unmyelinated fibers, the glia-neuron relations?) We still have a very long way to go before we will understand much of it.

I strongly believe that in our attempts to study the brain it is a first imperative that we increase the precision in our observations as far as possible, if they are to have lasting value and to form a solid and reliable basis for future research. In addition to accurate observations we need, of course, thoughts, reflections, and hypotheses. Hypotheses are certainly important and necessary tools in research. I agree entirely with my former teacher, Sir Wilfred Le Gros Clark, when he says that "every scientist is entitled to offer an interpretation of his observations even if it can only be a provisional one, without running the risk of being discredited as a scientist because subsequent investigations replace it by an alternative inter-pretation" (Clark, 1955–56, p. 17). The problem with hypotheses is that it is all too often, and too easily, forgotten that they are indeed provisional. Once set forth, they have a remarkable tendency to longevity, to become fixed and to persist in the conceptual world of scientists and the public. All too often hypotheses are taken as "truths," especially by young scientists. In this way they may hamper and delay progress in research instead of furthering it.

Even in science we too often meet the kind of magic, antithetical thinking to which the human mind all too easily takes resort: What is not positive is negative, what is not white is black! Probably related psychologically to this and to the ten-dency to accept fascinating hypotheses as truths, is our naive inclination to believe that if you find a name for a phenomenon, you understand it! Foggy concepts such as "the extrapyramidal system" and, even worse, "the limbic system" have given rise to a wealth of uncritical and sterile speculations. This tendency of the human mind is aptly ridiculed by Goethe when he says (in one of my favorite quota-tions):

Denn eben wo Begriffe fehlen,
da stellt ein Wort zur rechten Zeit sich ein.
Mit Worten lässt sich trefflich streiten,
mit Worten ein System bereiten.[3]

These words may be a useful reminder, not only for politicians, but for scientists as well.

As you will perceive, I take a skeptical attitude toward many present views, explanations, and hypotheses offered for various problems in the neurosciences. However, even if I doubt that man will ever be able to understand his own brain, I do not think we should discontinue our attempts to do so. There are good the-oretical and practical reasons (concerning daignosis and therapy) to go on. But we will not progress further by pretending that we really understand much. As to what one might call one of the neurosciences' final goals, to understand the mind-brain relations, I belong to those who believe that this problem will forever remain insoluble! This attitude is in part a consequence of my lifelong preoccupa-tion with the structure of the brain, our best constructed and most complex organ. In part my view may be a result of personal traits that make me an inveterate agnostic. We have to accept that there are things we do not know, in science as in life. We must leave the question open! But I think it is good this way. The recogni-tion of our ignorance is our most effective incentive to do research. How dull life would be if we knew everything! No doubt the brain, in its organization and func-tion, harbors secrets that will keep neuroscientists busy for generations. Let us go on with the fascinating job of unraveling these secrets, for the sake of our own pleasure and for the benefit of mankind!

3. "For just where fails the comprehension,/A word steps promptly in as deputy./With words 't is excellent disputing;/ Systems to word 't is easy suiting. (J.W. Goethe, Faust, Part 1, Scene 4. B. Taylor, trans. London: Straham & Co., 1871.)

## References

Adrian, E.D. (1943): Afferent areas in the cerebellum connected with the limbs. *Brain* 66:289–315.

Angaut, P., and Brodal, A. (1967): The projection of the "vestibulo-cerebellum" onto the vestibular nuclei in the cat. *Arch. Ital. Biol.* 105:441–479.

Brodal, A. (1940a): Modification of Gudden method for study of cerebral localization. *Arch. Neurol. Psychiatr.* 43:46–58.

Brodal, A. (1940b): Experimentelle Untersuchungen über die olivo-cerebellare Lokalisation. *Z. Gesamte Neurol. Psychiatr.* 169:1–153.

Brodal, A. (1969): *Neurological Anatomy in Relation to Clinical Medicine.* 2nd ed. New York: Oxford University Press.

Brodal, A. (1972a): Cerebrocerebellar pathways. Anatomical data and some functional implications. *Acta Neurol. Scand.* Suppl. 51:153–195.

Brodal, A. (1972b): Some features in the anatomical organization of the vestibular nuclear complex in the cat. *In: Basic Aspects of Central Vestibular Mechanisms.* Brodal, A., and Pompeiano, O., eds. Amsterdam: Elsevier, pp. 31–53.

Brodal, A. (1973): Self-observations and neuro-anatomical considerations after a stroke. *Brain* 96:675–694.

Brodal, A., and Courville, J. (1973): Cerebellar corticonuclear projection in the cat. Crus II. An experimental study with silver methods. *Brain Res.* 50:1–23.

Brodal, A., and Pompeiano, O. (1957): The vestibular nuclei in the cat. *J. Anat.* 91:438–454.

Brodal, A., and Pompeiano, O., eds. (1972): *Basic Aspects of Central Vestibular Mechanisms. Prog. Brain Res.* 37:3–656 (Amsterdam: Elsevier).

Brodal, A., Pompeiano, O., and Walberg, F. (1962): *The Vestibular Nuclei and Their Connections, Anatomy and Functional Correlations* (The Henderson Trust Lectures). Edinburgh: Oliver and Boyd.

Brodal, P. (1968a): The corticopontine projection in the cat. I. Demonstration of a somatotopically organized projection from the primary sensorimotor cortex. *Exp. Brain Res.* 5:210–234.

Brodal, P. (1968b): The corticopontine projection in the cat. II. Demonstration of a somatotopically organized projection from the second somatosensory cortex. *Arch. Ital. Biol.* 106:310–332.

Brodal, P. (1972): The corticopontine projection from the visual cortex in the cat. I. The total projection and the projection from area 17. *Brain Res.* 39:297–317.

Clark, W. Le Gros (1955–1956): Hypothesis and speculation in scientific research. *In: Lectures on the Scientific Basis of Medicine.* Vol. 5. London: Athlone Press, pp. 1–18.

Courville, J., Diakiw, N., and Brodal, A. (1973): Cerebellar corticonuclear projection in the cat. The paramedian lobule. An experimental study with silver methods. *Brain Res.* 50:25–45.

Grant, G., Boivie, J., and Silfvenius, H. (1973): Course and termination of fibres from the nucleus z of the medulla oblongata. An experimental light microscopical study in the cat. *Brain Res.* 55:55–70.

Grofová, I., and Rinvik, E. (1975): Light and electron microscopical studies of the normal structure and main afferent connections to the nucleus ventralis lateralis thalami of the cat. (Proceedings of 6th Symposium of International Society for Research in Stereoencephalotomy, Tokyo, 1973.) *Confin. Neurol.* (in press). See also *Anat. Embryol.* (1974) 146:57–93, 95–111, 113–132.

Gudden, B. (1870): Experimentaluntersuchungen über das peripherische und centrale Nervensystem. *Arch. Psychiatr. Nervenkr.* 2:693–723.

Henschen, F., Jr. (1907): Seröse Zyste und partieller Defekt des Kleinhirns. *Z. Klin. Med.* 63:115–152.

Holmes, G., and Stewart, T.G. (1908): On the connection of the inferior olives with the cerebellum in man. *Brain* 31:125–137.

Jansen, J., and Brodal, A. (1940): Experimental studies on the intrinsic fibers of the cerebellum. II: The cortico-nuclear projection. *J. Comp. Neurol.* 73:267–321.

Jansen, J., and Brodal, A. (1942): Experimental studies on the intrinsic fibers of the cerebellum: The corticonuclear projection in the rabbit and in the monkey (*Macacus rhesus*). *Nor. Vid. Akad. Oslo, Avh. I, Mat.-Naturv. Kl.* No. 3:1–50.

Künzle, H. (1973): The topographic organization of spinal afferents to the lateral reticular nucleus of the cat. *J. Comp. Neurol.* 149:103–116.

Landgren, S., and Silfvenius, H. (1971): Nucleus z, the medullary relay in the projection path to the cerebral cortex of group I muscle afferents from the cat's hind limb. *J. Physiol.* 218:551–571.

Monakow, C.v. (1882): Weitere Mitteilungen über durch Exstirpation circumscripter Hirnrindenregionen bedingte Entwicklungshemmungen des Kaninchengehirns. *Arch. Psychiatr. Nervenkr.* 12:535–549.

Murphy, M.G., O'Leary, J.L., and Cornblath, D. (1973): Axoplasmic flow in cerebellar mossy and climbing fibers. *Arch. Neurol.* 28:118–123.

O'Leary, J.L., Dunsker, S.B., Smith, J.M., Inukai, J., and O'Leary, M. (1970): Termination of the olivocerebellar system in the cat. *Arch. Neurol.* 22:193–206.

Pompeiano, O., and Cotti, E. (1959): Analisi microelettrodica delle proiezioni cerebello-deitersiane. *Arch. Sci. Biol. (Bologna)* 43:57–101.

Shepherd, G.M. (1972): The neuron doctrine: A revision of functional concepts. *Yale J. Biol. Med.* 45:584–599.

Snider, R.S., and Stowell, A. (1942): Evidence of a representation of tactile sensibility in the cerebellum of the cat. *Fed. Proc.* 1:82–83.

Snider, R.S., and Stowell, A. (1944): Receiving area of the tactile, auditory and visual systems in the cerebellum. *J. Neurophysiol.* 7:331–358.

Sousa-Pinto, A., and Brodal, A. (1969): Demonstration of a somatotopical pattern in the cortico-olivary projection in the cat. An experimental-anatomical study. *Exp. Brain Res.* 8:364–386.

Sterling, P. (1973): Quantitative mapping with the electron microscope: Retinal terminals in the superior colliculus. *Brain Res.* 54:347–354.

Tomasch, J. (1969): The numerical capacity of the human cortico-ponto-cerebellar system. *Brain Res.* 13:476–484.

Walberg, F. (1972): Cerebellovestibular relations: Anatomy. *In: Basic Aspects of Central Vestibular Mechanisms.* Brodal, A., and Pompeiano, O., eds. Amsterdam: Elsevier, pp. 361–376.

# Membrane Excitability and Synaptic Transmission

Kenneth S. Cole (b. 1900, Ithaca, New York) has combined research, teaching, and work as a government official. His research interests have included the measurement and analysis of nerve-membrane capacity and conductance, the electrical analysis of nerve impulses, and the theory of membrane ionic conductances. He is a recipient of the U.S. National Medal of Science. Dr. Cole is currently professor of biophysics at the University of California, Berkeley, and research biophysicist at the Laboratory of Biophysics, NINCDS, NIH.

# 8
# Neuromembranes: Paths of Ions

## Kenneth S. Cole

It has seemed to me that probably the simplest, and certainly the most basic, activity underlying the functions of the nervous system is that of communication, and that the development of fast, reliable, and unlimited communication by axons is one of its most superb achievements. Whatever the evolutionary process by which axons appeared, certainly the crucial step was the creation of an electrical control of cell-membrane permeability to ions—now usually sodium and potassium—even in the absence of metabolism. Rather than speculating on how this marvelous machine came about, I will only sketch something of what I know about how we found out what it does and our efforts to find out how it does them. Nor will I speculate on the dynamics of my part along the path—which is only a path in retrospect. My choices have always been of only the next step or two, and in these I can only claim an extraordinary amount of very good luck. Even luckier has been the tender, loving care I have been given along the way by more persons than I can count, and I regret that I have seldom appreciated it adequately. Of course, I've sometimes gone far astray and also I've been battered around a bit. But I survived, and besides, you can't be unlucky all the time either.

### From Physics into Membranes

I suppose that my addiction to the paths of ions began when I first found out something about membranes, half a century ago. I had been interested in electrical engineering, influenced particularly by the grandeur of the Niagara plants and the Panama Canal locks; but as a physics major, I was encouraged to try research. A year at Schenectady was the turning point: a challenging high-vacuum gas-analysis problem; contacts with most of the General Electric Research Laboratory; spankings by Irving Langmuir and W. R. Whitney, the superb director, among others; and above all my start as a protégé of F. K. Richtmyer with his lectures on "Modern Physics." When a sign appeared on the Rockefeller Hall bulletin board reading, "Wanted—Two Biophysicists at the Cleveland Clinic," Richtmyer said, "Darned if I know what a biophysicist is—but I'll tell you something I think is biophysics." After telling me of a day with W. J. V. Osterhout at Woods Hole and his explanation of the electric-current flow through a sea weed, he concluded, "I think he's right and it looks like darned good fun."

I managed to spend the summer in Hugo Fricke's laboratory, and gained a tremendous respect for his elegant combination of theory and experiment, as well as a considerable acquaintance with medicine and surgery and the futility of a few experiments of my own. To complete the circle, I found that G. W. Crile had made mammalian tissue measurements following Osterhout's work and had hired Fricke to explain them! Just the spring before, in 1923, Fricke had published his demonstration of the molecular thickness of intact red-blood-cell membranes. He interpreted his measurement of almost one microfarad/cm² for the electrical capacity as a 33 Å layer of hydrocarbon separating impermeable ions. This was only shortly before Gorter and Grendel gave their lipid-extraction and spreading experiments as the basis for the biomolecular membrane—which, incidentally, Fricke never accepted.

After another year of graduate work and a summer at Woods Hole, measuring the heat production of sea-urchin eggs, I was thoroughly oriented toward biology because I saw so many more interesting and useful things that I thought I could do. So, as I was finishing my physics degree, I applied for a fellowship to follow Fricke and measure the membrane capacity of sea-urchin eggs, which would be so much easier to study than mammalian red cells because they are so large and beautifully spherical. But my application was quickly rejected, first by physics and then by biology, only to be granted after Richtmyer and Osterhout persuaded the sporting board that it should bet on such a long shot. The experiments were clever but disastrous; the paper still reminds me to be generous to brash youngsters, and it was ten years before the job was properly done. However, the analyses and calculations made in the course of my struggles are of continuing value. To understand and improve on Fricke, I discovered the Maxwell equation and also the "circular-arc" expression of the electrical behavior, which has found its way into engineering, chemistry, and physics—in part because it is a useful meeting point for the reality of experiment and the inadequacy of theory.

Over the decades the 1 $\mu$F/cm² has become so firmly established as a universal membrane characteristic by so many electrical measurements on so many living cells—and a few tissues—as to demand special explanations for apparent exceptions. Then, when the electron microscope finally produced such vivid evidence as the railroad tracks and a similar generality, one might have thought that the basic membrane structure was quite certain. Yet there was considerable disbelief and numerous attacks until just a few years ago, when apparently all possible objections were satisfactorily answered and the lipid bilayer was generally accepted.

Also within the past decade, many laboratories with new techniques, instruments, and analyses far too numerous to mention have added tremendously to this fundamental structure, mostly under the rubric of molecular biology. Although the detailed membrane pictures are changing rapidly, considerable similarity is emerging. One very striking generality is the failure to include more than guesses about the machinery for ions to go into, through, and out of the membrane.

A friendly biologist asked me at Woods Hole why I wasn't working on nerves, which were slim, long, and carried messages like the cables physicists were supposed to know something about. My curiosity was aroused when he confessed ignorance of such elementary things as conductivity of the core and the capacity and leakage of the insulation. So I tried for a fellowship with A. V. Hill—but he wouldn't have me unless I'd work on heat again. Then I was thoroughly excited when Osterhout suggested I go with Peter Debye—whom I had met when he talked about electrolyte theory at Cornell. But he wouldn't have me either—until after a timely visit from Richtmyer.

The year with Debye was wonderful, and I cherish the memories of him and of Leipzig. I had rather hoped to work on dielectric loss—which is still in limbo—but my first assignment was to read Lewis and Randall's *Thermo-dynamics* in German to improve two of my weak spots. Then I was led on through what has come to be called Nernst-Planck theory. After we had outlined one of the myriad problems I couldn't solve, I asked, "What has this to do with membranes?" Debye got up from the conference sofa and stomped, muttering, to the window. After my puzzled, "I'm sorry, I didn't hear what you said," he turned and glared at me. "I said, 'You and your damned membranes.' " Then we both laughed and the phrase became the password at our many and happy meetings during the rest of his life. When, as usual, I was stuck on a problem, he'd ask how I had simplified it, and might make a suggestion. Finally he got exasperated—"If you can't solve a problem, make approximations until you do get an answer and then go back and find how bad they are." This has been very useful in experiment as well as theory, despite such experiences as when I got an electrodiffusion explanation for our inductive reactance in a squid axon membrane and then went back to my Leipzig notebooks and found that I had assumed myself out of that answer!

It was indeed a great privilege to work as closely as I did with Debye, and my admiration, respect, and friendship have been far more valuable than the small paper I wrote. On the other hand, Debye was disappointed and frustrated that we hadn't come up with something immediately useful. I think, and hope, that was the basis for his letter to Richtmyer: "I've enjoyed having Cole here—but please don't send me more like him." He gave me many and helpful suggestions over the years as we did more and puzzling experiments. I wish he were still around!

In the middle of that year and completely out of the blue, H. B. Williams offered me an assistant professorship of physiology at Columbia College of Physicians and Surgeons. I barely knew Williams by name as the author of two papers on string-galvanometer theory and design, but it was a good offer and in spite of my misgivings and of communication difficulties between Athens, Leipzig, and New York, I accepted. Only much later did I find that Williams, a cardiologist who had majored in mathematics and been in charge of sound ranging during World War I, had fought and pleaded for years until he had gotten such an absurd thing as a post for a physicist in a medical center. And then he had asked Richtmyer for a recommendation!

The years at Columbia-Presbyterian were the best I have had. It was rough going at times for such a stranger in such a strange land. But Williams was sympathetic, understanding, and helpful as I struggled and slowly established myself, with some assistance from his many and highly diversified friends—all the way from the hospital and school to Bell Labs.

## The Squid Axon

Just as there had been a long debate over even the existence of a membrane, there were then widely differing opinions as to its permeability to ions. A moving ion is an electrical current, and no membrane currents were to be found—except in giant single plant cells—partly because the ions would rather go around the cells than through them. So when John Young showed H. J. Curtis and me his squid giant axon, it seemed worth a try—using *Nitella* as a winter warm-up. They both had respectable membrane capacities, which changed little, if at all, during an impulse. Both showed increased ion permeabilities—measured as electrical conductances—during the impulse. This was an enormous increase over the very small resting conductance for *Nitella* and turned out to be a fortyfold increase for the axon, but only after A. L. Hodgkin and I successfully measured the resting conductance as about one millimho per square centimeter. This was about $10^{13}$ ions/cm$^2$/sec for a mV, but for a 100 Å membrane it was only about $10^{-8}$ of the conductivity of sea water. So ions could go through the membrane, and much faster during an impulse, and only a small part of the membrane seemed to be involved. These conclusions have come to be considered as a starting point for much of the membrane work since 1938, perhaps in part because we have had such photogenic results as those shown in Figure 8.1.

Steady currents changed the membrane conductance without affecting the capacity, in much the same way as a passing impulse. This showed that both were thoroughly indecent violators of Ohm's Law, requiring very special rules and procedures to cope with their considerable nonlinearity.

Completely incidentally, Hodgkin and I did make a discovery in the classical style. For only the good reason of curiosity, we measured an axon at alternating-current frequencies to 10 Hz—far below what we needed—and found the impossible. The normal capacity representing the membrane decreased, disappeared, and turned into an inductance of $\frac{1}{5}$ henry/cm$^2$, which is usually made by many thousands of turns of wire wound on a sizeable core of iron. To be sure some oscillations had suggested such a component before—only to be dismissed as absurd—but the direct measurement was not to be denied. Several such anomalous reactances were well known, but none would fit until the new Bell Labs thermistor, because of its nonlinearity, did the same thing.

I looked upon the squid axon with considerable respect as a delicate and fragile structure. So in spite of all the empty space inside, it seemed quite rash to insult the axon by pushing an electrode in for a couple of centimeters. But it was quite tolerant and gave rectification ratios up to a hundred to one, while our potential measurements suggested that the resting membrane was mostly permeable to potassium, as Bernstein had suggested early in the century. Then, too, in support of this idea, current flowed outward across the membrane from the high potassium interior more easily than inward, and the time it took to get more potassium into the membrane could account for the incredible inductive reactance of the membrane. It thus seemed likely that Bernstein's other postulate might be true, namely,

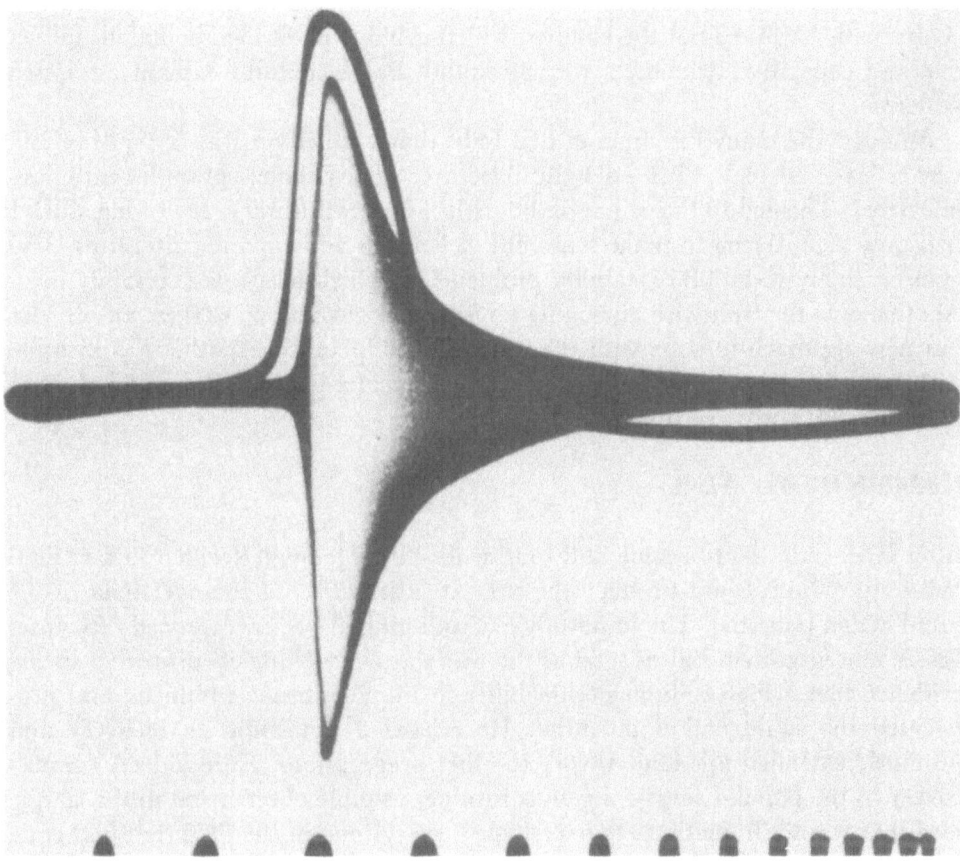

Figure 8.1  Oscilloscope records of action potential, line, and membrane-conductance increase from the squid giant axon during passage of an impulse. The time marks are at 1 msec intervals. From Cole (1968).

that a small increase in potential made the membrane permeable to all ions—but how?

Up to this point almost everything was quite straightforward and rather obvious. After Fricke had found the membrane, the squid axon made it possible to find an ion permeability with its increase in activity and to blame it all on potassium. But then came the shocker. Just before World War II hit, it was found that during an impulse the membrane potential doesn't just rise to near zero, but most emphatically goes well beyond. Several others, including me, had recorded the same thing before but had blamed it—if at all—on various artifacts. But this overshoot was a blatant violation of acceptable dogma, which had to wait out the war for the right time, the right atmosphere, and the right man to be understood.

I made primitive attempts to explain the new experimental results, but it was not until I took my only sabbatical, on a fellowship at Princeton, that I could get out of the medical atmosphere and go to work on them. Debye had suggested Mott's new semiconductor theory as an analogue for a potassium membrane, which led to the now-classical Goldman equation between potential and ion concentrations. I worked out the kinetics, which could explain the anomalous inductive and capacitive reactances we had found. But I couldn't explain an action potential.

Amongst the many resources of Bell Labs that I called on was L. A. MacColl, who referred me to V. Bush's delightful lecture, "The engineer grapples with nonlinearity." This led to many happy hours in Fine Hall library, reviewing work I had only seen, trying to make sense out of Russian and Japanese literature, and gaining an appreciation of stability problems (of which nerve was certainly one). Another was the iron-wire analogue, for which I was able to suggest an obvious but new approach to cope with the negative conductance. Another gratification was that I turned S. Lefschetz onto the topological path he followed for the rest of his life. But I was overdue for war work.

## Hodgkin-Huxley Axon

After RAF radar had become dull routine for him, Hodgkin decided that sodium and only sodium could produce the prewar observation of the overshoot of the squid action potential. The importance of sodium had been recognized fifty years before and forgotten. But as soon as the war was over, Hodgkin presented strong evidence that a high sodium permeability in the membrane produced and propagated the squid action potential. He relegated potassium to recovery and promptly extended this ionic theory to other preparations. Here indeed was discovery in the popular sense—a genius turning a simple observation into a simple idea that was so dramatic as to overshadow everything in the field that had gone before and to completely dominate everything thereafter.

At about the same time, on this side of the Atlantic, the axon was invaded by a massive long electrode and surrounded by another, with a guard at each end. Thus impulse propagation was eliminated, and the membrane potential and current density were uniform in the central measuring region. Many responses of the membrane potential below and above the threshold for the all-or-none action potential were measured directly for the first time using one of the war's technical spin-offs: electronic control of membrane current. But the membrane was still a disappointing and disgustingly unstable device, and my years of reading and of experience on nonlinearity, control, and stability converged to urge a changing of the electronics to control the membrane potential. Not only was there no threshold, but the now all-too-familiar patterns of early inward currents changing to steady outward currents were enough to account qualitatively for threshold, spike height, and propagation in an unmolested axon.

While I was involved with national defense work, Hodgkin and Huxley had, by 1952, and with almost incredible speed, thoroughly proven Hodgkin's ionic hypothesis and described it in quantitative form. They had (1) adopted the potential-control concept and, with improved techniques, christened it the "voltage clamp," (2) separated the current into early transient sodium and slowly appearing steady-state potassium components, (3) reduced the resulting sodium and potassium conductances to a family of mathematical equations, and

(4) demonstrated that these equations could produce many normal axon phenomena—among them, subthreshold, threshold, action potential, and speed of propagation.

In essence, they described the membrane behavior in terms of fixed parameters and those depending upon membrane potential (Figure 8.2). The fixed parameters were: membrane capacity; equilibrium electromotive forces for sodium, $E_{Na}$, and potassium, $E_K$, given by their concentration ratios across the membrane (Nernst potentials), and a third, leakage emf of unknown origin; and maximum conductances for sodium, potassium, and the unknown ions as found experimentally. The instantaneous conductances and their rates of change were given in mathematical form in terms of three dimensionless factors that depend upon, and only upon, the membrane potential. These are best described in terms of the usual voltage-clamp current patterns—the "m" factor turning on the inward sodium current in fractions of a msec, the "h" turning it off in a few msec, and the "n" turning the outward potassium current on in msec. Then, to complete the formulation and to give a basis for standardization of all results to 6.3 °C, they showed that the rate constants all increase about threefold for a 10 °C temperature increase.

The Hodgkin-Huxley achievement burst upon an almost unsuspecting audience at the 1952 Cold Spring Harbor Symposium, where Hodgkin presented it with his usual, careful, convincing enthusiasm and, as I remember, F. O. Schmitt was chairman. The immediate reactions ranged from enthusiastic acceptance through utter confusion to the dismay, dislike, and disbelief that continued almost unabated for a decade. The believers started promptly to try to understand the gospel, to

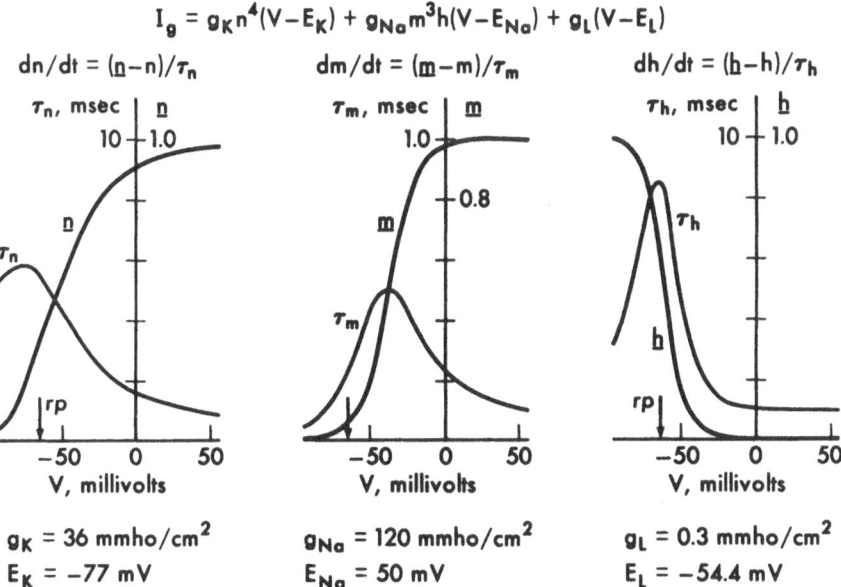

$$I_g = g_K n^4 (V - E_K) + g_{Na} m^3 h (V - E_{Na}) + g_L (V - E_L)$$

$dn/dt = (\underline{n} - n)/\tau_n$      $dm/dt = (\underline{m} - m)/\tau_m$      $dh/dt = (\underline{h} - h)/\tau_h$

$g_K = 36$ mmho/cm$^2$        $g_{Na} = 120$ mmho/cm$^2$        $g_L = 0.3$ mmho/cm$^2$

$E_K = -77$ mV              $E_{Na} = 50$ mV                $E_L = -54.4$ mV

Figure 8.2  Representation of the Hodgkin-Huxley equations for a normal squid axon. The membrane ionic current $I_g$ is given at the top in terms of the membrane potential V, the constants shown at bottom, and the parameters m, n, and h. The steady-state values and time constants in the differential equations are shown graphically as functions of the membrane potential; rp is the potential at rest. From Cole (1968).

interpret, improve, and extend it, and to use it. The heretics began to look for sudden-death arguments and experiments. Certainly one of the first to appreciate and put the concepts to work was J. C. Eccles—leading the way into such complicated things as synapses and cell bodies.

Just as Hodgkin and Huxley had improved upon my original work with an independent, internal potential electrode, J. W. Moore and I got closer to the membrane and got rid of the resting potential as a floating zero by using a reference microelectrode. But as everyone came to use more potent axons, strange things began to appear. When these turned out to be a not-so-simple failure of our potential control to control the potential, it became obvious that we had been lucky on both sides of the Atlantic that our first axons had not really been in very good condition. As more and more investigators have come to apply the voltage-clamp concept, a wide variety of cell and electrode combinations have been tried —some superb, some not so good. The input and output electronics have advanced on the heels of semiconductor developments until now not only are experiments electronically programmed and controlled, but the results are also recorded and analyzed by computers.

I know no good reason why *Nitella* should be excitable and propagate an action potential. So for years I was prepared to find that such an exotic thing as a squid axon might be an anomaly in the animal world. But B. Frankenhaeuser brought that expectation to a spectacular end by capitalizing on medullated axon structure to show that the nodes of frog and toad follow the same general pattern as squid axons. And now voltage clamping has been expanding so fast in so many directions and for such a variety of ions that I seem to be hopelessly behind the times. I had really expected it to reach its limits before now and just fade away. Even if it had, I would still be surprised and very pleased that such a simple idea could be so useful.

Huxley had to do the early computations of the equations with a desk calculator after the Cambridge computer failed him. So I felt lucky when the Bureau of Standards was looking for biological problems to show the potential of its new Standards Eastern Automatic Computer (SEAC). I was a firm believer in the "all-or-nothing" law, and so I had criticized K. F. Bonhoeffer's exothermal-reaction analogy—which involved me in a hassle with M. Delbrück. Even after SEAC gave a threshold for the Hodgkin-Huxley equations, Bonhoeffer was not entirely convinced, Huxley thought it very peculiar behavior, and R. FitzHugh just didn't believe it. Several years later I found consistent peculiar wiggles in some other SEAC computations. FitzHugh and H. A. Antosiewicz ran these down and found that we had made an absurd mistake in programming and that the membrane response was graded over a stimulus strength of one part in $10^8$ at $6.3\,°C$, which was obviously academic. It was not until 1969 that R. Guttman found the response to increase at a maximum rate of only $5\%$ of the stimulus at $35\,°C$, while F. Bezanilla showed the Hodgkin-Huxley computations to give a similar obvious and significant failure of the "all-or-nothing" law for membrane excitation above about $20\,°C$. It was certainly a "discovery"—at least for me—after traveling a long and sometimes bitter path—that the "all-or-nothing" law applied only to the excitation of a propagating impulse, again as given by the Hodgkin-Huxley equations. Nonetheless, it may have at least dampened some of the cooperative, phase-transition explanations that were so much in vogue!

For at least a decade it was singularly difficult for many to have a real appreciation of the voltage-clamp results or the Hodgkin-Huxley work on them. I have often tried to explain that, within wide limits, they summarized in one neat, tidy little package the many thousands of experiments done previous to 1952, and most subsequent ones. On the one hand, with the computers almost everyone has available, the equations can give and seem to have given better-than-approximate answers to about every conceivable experiment. On the other hand, why should a theorist waste his time trying to explain the mechanism of a single experiment when the results of such a multitude of individual experiments are collected in this one challenging summary? Certainly one of the best such encounters I had was with Aharon Katchalsky several winters ago at Berkeley. At his request we spent one whole, uninterrupted week going into Hodgkin-Huxley in depth as well as latitude, longitude, and some amusing trivia. Finally: "Do you mean to tell me *that* is *all* they did? I thought they had a theory of membrane permeability!" And I: "Sure that's all; what do you want for one li'l, ole Nobel Prize?"

If, in my enthusiasm, I've given the impression that the Hodgkin-Huxley equations are perfect descriptions of the squid membrane in all respects, I've been misleading to say the least. Allowing for the fact that most experiments have, for some time, usually been done with rather better axons than those reported on in 1949 and 1952, the equations are not only amazingly good over the range of experimental conditions they were designed to fit, but often also unreasonably good far beyond the limits of the original data. Yet the exceptions can be interesting. As increasing membrane decrepitude required larger steady, hyperpolarizing current to maintain a satisfactory reference potential, we often polarized our current electrode enough to fill the axon with bubbles. So we turned to relatively short prepulses and let the axon and electrode rest most of the time. And for no reason that I remember, we tried various hyperpolarizations. The test pulses showed no effect on the sodium current but a considerable slowing for potassium. Then, at an extreme of 212 mV, hyperpolarizations gave simple delays that were far more than Hodgkin-Huxley could produce. My suggestion that the potassium was given by the 25th—instead of 4th—power of "m" wasn't recognized for tongue-in-cheek, as more serious attempts have been made to explain the effect.

The Hodgkin-Huxley equations appeared twenty-one years ago as the result of prodigious effort and ingenuity. With so many recent increases of power, speed, and accuracy in experiment, analysis, and formulation, perhaps they should not have to wait longer for an adequate theoretical replacement before they are given a face-lifting. I would hope that their beauty and essentially simple elegance need not be lost in the operation.

But of theories there has been a rapidly gathering avalanche. And having been so happy with my prewar potassium success—in spite of inconsistencies—I had to try to include sodium. The troubles seemed so immediate, obvious, and serious that they didn't seem worth increasingly valuable space in the literature. But as even more naive attempts to deal with the membrane as a nonaqueous electrolyte appeared, I announced my intention of publicly repudiating my quarter-century favorite. A friend said, "That I want to see"—so his journal published it. And Hodgkin consoled me, "Perhaps it will still work for potassium." But the approach is still being used with sophistication and variations. It is quite impossible for me to summarize or even classify all of the serious attempts to account for the facts of the sodium and potassium ion permeabilities of the squid axon membrane. Perhaps in a few years this will be a worthwhile project on the paths of discovery, but I would not like it to show how much poor comment and advice, or even evidence, I have given.

I certainly have been too optimistic in my hopes for an understanding of how ions get through membranes. In 1967 we were visited as a sideshow at Woods Hole by H. H. Humphrey and entourage as he was taking up his statutory duties as chairman of the Ocean Resources Board. In a thirteen-minute tour I spieled the exhibits our group had prepared: squid swimming in the tank; stellar nerves in an open mantle and dissected out; a clean axoned under a microscope and another connected to an oscilloscope. I was discouraged to get no comeback on my ploys about pharmacy—he was dead serious and only asked me to confirm his rephrasing of what I had said. Finally, before the oscilloscope, I explained that the potential trace rises as a few sodium ions flow in to push the impulse ahead and then potassiums slide out to ease the potential down again, ready for the next impulse; that since 1952 we had known how many of these ions cross the membrane and when—but that we had no idea how or why they got across; that I'd been predicting for some time that in twenty years we ought to know how these things happen and, if I was right, it would probably be before the November 1972 election. "Well, I sure hope you do and you work at it real hard—we may need just that." My revised prediction has not been for publication.

About then I wrote a short summary of the more interesting approaches and put a long, laborious book to bed. It was a good time, as membranes were just coming into fashion. The literature has been increasing at a fantastic rate with new techniques, results, and interpretations that were barely budding half-a-dozen years ago. I cannot even recount the major attempts and achievements of either theory or experiment as they have grown, and I will limit myself here to a brief outline of some current points of view on ion paths through axon membranes. Although these are mostly rather direct consequences of experiment and are quite widely accepted, they are not yet necessarily correct.

## Theories, Models, and Pictures

Energy considerations have made it very unlikely that ions are at all uniformly distributed in the hydrocarbon core, and there must be a special mechanism for ion flow. Many lines of evidence agree in concluding that, although this core is somewhat asymmetric, the inner and outer surface structures are vastly different. Just as several times before, an intense interest in carrier molecules has come to the conclusion that they are relatively unimportant in neuromembranes.

The tetraethylammonium ion (TEA) was quite spectacular in producing the long-lasting action potentials characteristic of heart muscle. But it has become a standard internal reagent for specifically blocking the squid axon machinery carrying potassium and the ions that can more or less replace it. Tetrodotoxin seemed incredible when I first heard of it on the Stanford campus, and its specificity for externally blocking sodium and substituents is still remarkable—although saxitoxin is more reversible. In proposing TTX as an abbreviation, I was thinking TeTrodo toXin; but everyone uses ToXin, so saxitoxin is STX instead of the SXX I prefer. Since TEA and TTX are almost without cross-effects, the separation of potassium and sodium seemed quite certain—except that L. J. Mullins wanted to see both full-blown currents at the same time instead of replacing each other. Internal pronase has done just that and now makes it possible for us to work with steady sodium currents. Many of us have thought of the Hodgkin-Huxley parameters "m," "n," and "h" as perhaps convenient, but arbitrary, empirical variables and have tried to split and combine them in various ways, so it is something of a shock to have to recognize them as separate functional entities. Here is yet another, but long-delayed, plus for Hodgkin and Huxley. I just can't dismiss it as good luck; it seems too well deserved.

It has become quite usual to think and talk of potassium and sodium channels through the axon membrane—with little, if any, direct evidence for or against such a neat concept. But it is very hard not to believe that the recent work on artificial lipid bilayers is a really good model for a functioning axon membrane. Mueller and Rudin reinvented the Langmuir and Waugh technique, and by adding a bacterial protein they made a brain lipid bilayer conducting and excitable! I then persuaded R. C. Bean to try very small amounts of the protein. So he sent delightful records of current increasing, and sometimes decreasing, in steps of the order of $10^{10}$ ohms. Subsequent work at Bethesda on the simpler activated bilayers of oxidized cholesterol showed that all the channels have the same resistance and act independently, opening and closing with Poisson statistics at rates depending exponentially on membrane potential. These conclusions for a few channels could be explained by a voltage-independent barrier, and when extended to between 100 and 1000 channels they gave potential-conductance relations very similar to those of Hodgkin and Huxley for the squid axon membrane. Of course, it was delightful to have conductance-relaxation plots turn out to be roughly circular with potential as the parameter. But it also turned out that the relative permeabilities of the channels to five cations were proportional to their mobilities in free aqueous solution, so these bilayer channels do not have the high selectivity of the axon membrane channels—if there be such.

In 1939 I tried to stabilize the albumin–amyl acetate bilayers of Dean and Gatty, which we had found remarkably similar to living membranes except for a somewhat high conductance. I found the conductance and capacity to increase, at intervals of from minutes to hours, in less than a millisecond by a small step. Although we discussed it at length, neither Langmuir nor I mentioned channel formation as an explanation, and I think it only barely possible that I could have seen a conductance increase of $10^{-10}$ mho!

Returning from the fascinating, cooperative, and compelling artificial bilayers, we are strongly tempted to apply the results to axon membranes, even though they have not yet been persuaded to show us individual channels opening and closing. The facts that the voltage dependence of the axon conductances and the time constants are so similar to those of multichannel bilayers, which are in turn derived statistically from single-channel data, are very strong presumptive evidence. Although white noise must be generated by any electrical conductor at equilibrium, an assembly of fluctuating conducting elements can give additional noise-spectra components.

I was very disgusted in the 1930s to find that cells and tissues—such as marine eggs and muscle—produced so much more noise than their equivalent circuits that I had to go to increased amplification and filtering to get sufficient bridge sensitivity at low frequencies. I suggested the problem to several students, and I'm glad they were bright enough to pass it up—the first definitive measurements were made, most appropriately in the Netherlands, on single nodes in 1965. A major component, which was attributed to potassium, had a power spectrum inversely proportional to the frequency—as did lobster and squid axon membranes. This was most discouraging because this so-called l/f noise, recognized before 1930, has become an increasingly severe limitation on the use of semiconductors and is without explanation in any of the many nonliving systems in which it appears. However, recent advances of instrumentation and analysis in this already highly sophisticated field have exposed a potassium spectrum for squid and node membranes of the kind expected from fluctuating channels. The mean open time agrees with the Hodgkin-Huxley potassium time constant $\tau_n$ and its variations with temperature and membrane potential, but the number and conductance of individual channels are not yet available. If this component does indeed account for all of the potassium current, the troublesome l/f noise would then seem to be linked to the almost equally mysterious and apparently deleterious membrane leakage current.

The first and apparently only statistical evidence to be interpreted in terms of axon sodium channels came from voltage fluctuations of a node during current flow; it gave conducting units of about $10^{-10}$ mho. Similarly, fluctuations of the opening of motor endplates by acetylcholine and of photoreceptors by light have also given units of the same orders of magnitude. We hope for more and more direct evidence of sodium-channel kinetics.

Probably the most startling and dramatic of recent experiments are the titrations of sodium sites by TTX. These sites seemed incredibly sparse, with an average density of only about seven per $\mu$m for three axon membranes as compared to a phospholipid density of about 1400 per $\mu$m; and the technique must be admired for its simplicity, directness, and ingenuity. It is the kind of idea we had come to expect from Trevor Shaw and will miss so much. These figures again give a channel conductance of about $10^{-10}$ mho.

The onslaught of $10^{10}$ ohm channels makes some consequences of the figure seem worth thinking about. This corresponds to a maximum flow of some $10^8$ ions per sec or $100/\mu$sec. If, as seems most probable, electrostatic repulsion will not permit more than one ion in a channel at a time, then the velocity would be half a meter per second through a 50 Å long channel, and the mobility would be about $10^{-3}$ cm²/volt/sec. This is amazingly close to the mobility in water, but the probable coincidence should not reduce efforts to describe the membrane transport mechanism.

The potassium channels are rather less fun because there is no reasonably reliable count of them. It is generally expected that they are more numerous than the sodium channels, and may have adsorption sites along the way to give a somewhat lower mobility—but certainly nowhere near my 1940 electrodiffusion calculation of $2 \times 10^{-8}$ cm²/volt/sec.

Hodgkin and Huxley suggested the movement of six unit charges to open a conducting path and so explain the rapid increase of steady-state conductance with depolarization. With increasing power and sensitivity, experiments in the near absence of ion conductances have only recently shown charge displacements that increase with depolarization to reach a maximum of about $10^{-8}$ coulombs/cm². The figures lead to densities of gates to be opened, by such concerted action of six charges, that are only several times larger than the TTX channel counts. I have no specific picture of the possible mechanism.

It seems almost certain that the fast (usually sodium) and the slow (usually potassium) conductance mechanisms are separate, independent, and specialized structures bridging the membrane, each with its own selector (Figure 8.3). For sodium, part of this is probably at the outside opening, where the guanidinium groups of TTX and STX can fit neatly, while a properly placed oxygen will sort $3 \times 5$ Å ions according to whether or not they hydrogen-bond it. Similar considerations are expected to determine the permeability series $Li = Na > K > Rb > Cs$. Inner openings that can be blocked by TEA will also admit potassium, which can only proceed after dehydration. Here the sequence to be accounted for is $K > Rb > Na > Cs$.

For quite a few years I was often asked about the relationship between the impedance measurements I was doing and electrokinetic phenomena. I usually explained, patiently I hope, that the ionic strength of most intercellular media, and particularly sea water, was so high that double-layer effects were quite negligible. But I was completely unprepared when a discrepancy between resting and action potentials in axons perfused with solutions of low ionic strength was explained by internal surface charge—reducing the membrane potential below the measured potential by the amount of the double-layer potential. The effects of external calcium on the Hodgkin-Huxley equations have been similarly explained by charge densities of up to 120–200 Å² per charge near the channels. Analyses on red-cell ghosts and mitochondrial membranes saturate for H, Ca, Mg, K, and Na at about 200 Å² per charge. But there are as yet no satisfactory direct measurements on either the inner or the outer squid membrane surfaces.

It was no great surprise to find a squid axon membrane resistance of $10^3$ ohm-cm², if only because it gave a specific resistance in the same general range as those queer semiconductors. After several other cell membranes came in the same magnitude, *Nitella* did look strangely high with $2–5 \times 10^5$ ohm-cm², and now a red

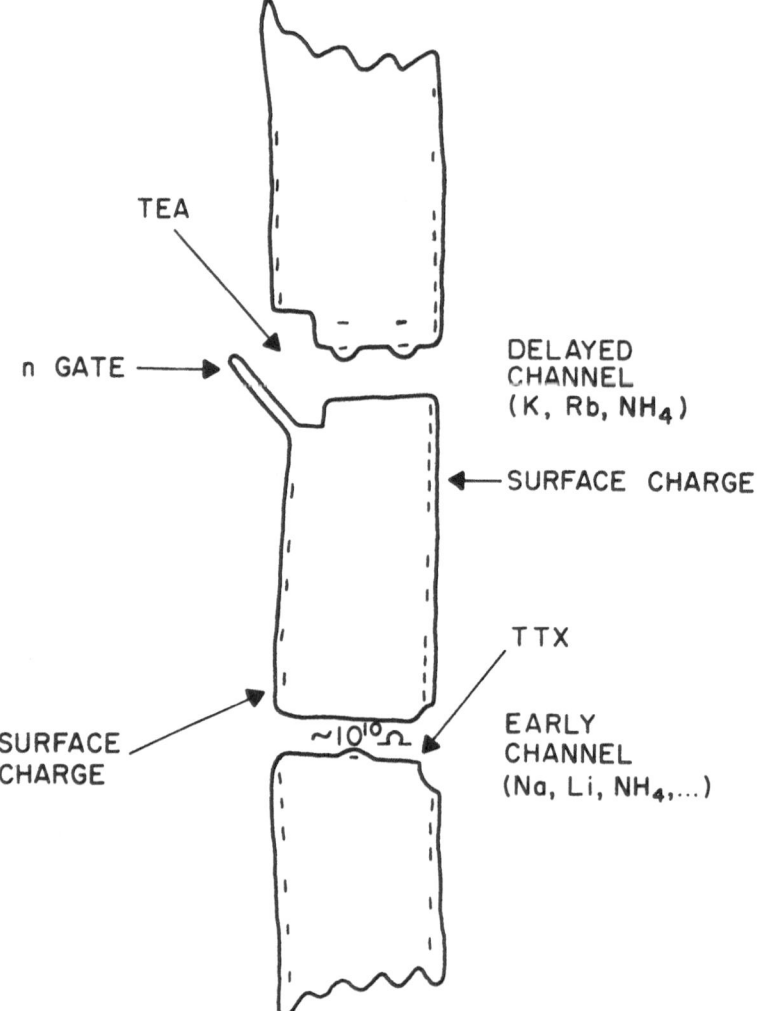

TEA

n GATE ———→

DELAYED
CHANNEL
(K, Rb, NH$_4$)

←—SURFACE CHARGE

TTX

SURFACE
CHARGE

~$10^{10}\,\Omega$

EARLY
CHANNEL
(Na, Li, NH$_4$,...)

Figure 8.3   A schematic representation of the paths of ions in neuromembranes, inside left. From Ehrenstein and Lecar (1972).

cell at 10 ohm-cm$^2$ is an extreme in the other direction, but not so much different from the 25 ohm-cm$^2$ for an active squid membrane. These were all facts, interesting and useful, but not significant or provoking because then—as now—there was no strong theory for the paths of ions through membranes. But I frankly did not believe the evidence that the two membranes between two salivary-gland cells in contact had only an unmeasurably low resistance—a small fraction of an ohm-cm$^2$—instead of something like the 2000 ohm-cm$^2$ that would be right and proper. The measurements and the cable analysis seemed entirely adequate, but my prejudice was so deep-seated that I insisted upon measurements of potentials across internal membrane pairs during current flow with sucrose outside before I would be convinced. And as if that wasn't outrageous enough, sponge cells that had never been near each other before could be made to form such junctions reversibly under proper conditions and calcium concentrations. Membranes are thus getting to be more and more remarkable, but also these "electrotonic"—or, as I prefer, "ionic"—junctions of various kinds are found to provide fast communication between axons where the usual chemical synapses are much too slow. I, for one, have no idea how they get that way.

The increasingly detailed information on the bilayer structure with its lateral fluidity and asymmetries, with proteins embedded from both sides, is gaining a rather general acceptance. There are also proteins crossing the red-cell membrane, but how they might provide the necessary ion paths is not clear. Two interesting hints are that three hemoglobin molecules can combine to provide a 10 Å aqueous channel and that two-to-one mixtures of two proteins have just been found that produce a sodium-specific conductance in an artificial bilayer.

Even as there was an accelerating buildup to the superb breakthrough of 1952, facts and concepts have been developing most rapidly over the past few years. We have been able to establish that there are, almost certainly, separate sodium and potassium channels, and that the three Hodgkin-Huxley parameters for them are separable. The topography of the channel mouths seems quite definite. Wonderful progress has been made about the nature of the sorting mechanism between most favored channels. It seems highly probable that the individual sodium channels have a resistance of about $10^{10}$ ohms and that the potassium channels may be somewhat higher but not vastly different. It is not yet certain that all channels are available all of the time to flicker on and off, as in the artificial bilayers, or that there are gates for each that are operated by the membrane potential. The present estimates are that the ions go through the membrane in almost nothing flat. However, it seems only fair to claim that, in spite of our uncertainties about channel structure, the arrangements and mechanisms of the various controls, and other problems, we are more than creeping up on our goal. But I think that we would be less than candid if we did not admit that we do not have a solid answer for a most fundamental problem of the neurosciences: What are the paths of ions through membranes?

## References

Cole, K.S. (1968): *Membranes, Ions and Impulses.* Berkeley: University of California Press.

Ehrenstein, G., and Lecar, H. (1972): The mechanism of signal transmission in nerve axons. *Annu. Rev. Biophys. Bioeng.* 1:347–368.

Hodgkin, A. L. (1951): The ionic basis of electrical activity in nerve and muscle. *Biol. Rev.* 26:339–409.

Hodgkin, A. L. (1964): *The Conduction of the Nervous Impulse.* Springfield, Ill.: C. C Thomas.

Hodgkin, A. L., and Huxley, A. F. (1952): A quantitative description of membrane current and its application to conduction and excitation in nerve. *J. Physiol.* 117:500–544.

Keynes, R. D. (1972): Excitable membranes. *Nature* 239:29–32.

Sir John C. Eccles (b. 1903, Melbourne, Australia) shared the 1963 Nobel Prize with Alan Hodgkin and Andrew Huxley for their study of the transmission of nerve impulses along a nerve fiber. In the past decade he has pursued the problem of communication at higher levels in the vertebrate nervous system, particularly in the cerebellum, and has dealt extensively with philosophical problems deriving from brain science, particularly the nature of the experiencing self. Sir John is currently Distinguished Professor of Physiology and Biophysics at the State University of New York at Buffalo.

# 9

## Under the Spell of the Synapse

John C. Eccles

### Introduction

My scientific life began when, as a 17- to 18-year-old medical student in Melbourne, I became enthused by the brain-mind problem, in particular as it related to my own experienced self-consciousness. I read avidly all the philosophical and psychological texts available, and came to the conclusion that there was a plethora of dogmatism superimposed on a most inadequate scientific base. At that time it seemed to me that the specialized connections between nerve cells, the synapses, contained the clue not only to the subtlety of the nervous reactions, but even to the problems of brain and mind. I had come under the spell of the synapse!

Thanks to the magnificent work of Ramón y Cajal in the latter part of the nineteenth century, it was known that the brain is composed of thousands of millions of independent nerve cells that communicate with each other by entering into very close contacts, which Sherrington called synapses. In the early decades of this century it was recognized that each nerve cell has over the surface of its body and branching dendrites a large number of these synapses that are made by branches of other nerve cells. These synapses are of two kinds, excitatory and inhibitory. When a nerve cell is sufficiently bombarded by impulses to its excitatory synapses, it is activated to fire brief messages of its own—the so-called nerve impulses—along its own output line or axon. Activation of inhibitory synapses has the opposite effect, tending to prevent discharges. Each nerve cell has a single axonal branch, and these axons form the communication pathways or nerve fibres in the central nervous system and also in peripheral nerves.

Important principles enunciated by Sherrington relate to the branching of a nerve fibre so that it diverges to give synapses to many nerve cells; complementarily, there is convergence on each nerve cell of the axonal branches from many nerve cells. Essentially the operations of the brain can be thought of as occurring at two levels: the transmission of brief all-or-nothing messages or impulses along the communicating lines or nerve fibres; and the transmission across the synapses made by these nerve fibres where they enter into relationship with other nerve cells in the brain and spinal cord, or peripherally on muscles.

### Sherrington as My Master

In the early decades of this century Sherrington was the world's greatest neurophysiologist. So I was very fortunate indeed when, after medical graduation in Melbourne, I went in 1925 to Oxford as a Rhodes Scholar to study under him in the Final Honour School of Natural Science. This move was a turning point in my life because it gave me the opportunity not only of listening to Sherrington and having discussions with him, but of actually working with him during three years (1928–1930) of continual collaboration. I learned from him his unique insights into the mode of operation of the nervous system, his many skills in investigation, and also, very importantly, the cultural outlook of a scholar. In addition I had the advantage during my earlier years at Oxford of working with my very good friends: Ragnar Granit, who later received a Nobel Prize for his work on vision, though it could just as appropriately have been given for his later work on motor control; John Fulton, who soon left Oxford to accept the Sterling professorship at Yale, where he founded one of the great neurophysiological schools of the world; and Derek Denny-Brown, who later at Harvard built up a school of neurology distinguished by its integral relationship with neuroanatomy and neurophysiology.

My greatest indebtedness was to Denny-Brown, who had had two years' research experience at Oxford before I started in research in late 1927. Not only did I profit in countless ways from his criticisms of our experiments and interpretations, but I also inherited from him an experimental procedure with analytical power, the antidromic-impulse technique. I can still vividly recall my delight when he told me in 1927 that he had just been investigating the effects produced in a motoneurone by firing an impulse into it from its axon, i.e., antidromically. I was later to exploit this technique in many researches, the first being in collaboration with Sherrington.

I was particularly happy with the use of the antidromic-impulse technique in studying with H. E. Hoff the regular rhythmic discharge of single motoneurones. We discovered that, as in the heart, there was a compensatory pause in the rhythm, but it was not fully compensatory. The relationship of the abbreviation of the cycle time to the cycle time subsequent to the antidromic invasion was so regular that I was able to produce a mathematical theory that very satisfactorily accounted for this antidromic action on rhythmically discharging motoneurones (Eccles and Hoff, 1932). I still possess a long critical letter by Rushton in 1932 of my half-or-nothing hypothesis! Actually, it is now recognized that several factors are concerned; but the original observations still stand as a model of an early analytical attack on the single neurone, and we were at least correct in assuming that the antidromic impulse acted by destroying an enduring excitatory state set up by synaptic action.

I must confess to several failures to recognize the significance of observations I have made. The most remarkable was the failure in 1930 of Sherrington and myself to appreciate the significance of our discovery that the motor fibres to a muscle come in two sizes. Histograms showed the complete separation of the large fibres (the fibres responsible for the muscle contraction) and much smaller fibres that we thought were motor fibres that in the initial outgrowth failed in the capture of muscle fibres. We completely missed their true function, namely the motor innervation of muscle spindles, which Sherrington had postulated several years earlier in his 1924 Linacre Lecture, and which we mistakenly thought was effected by collaterals of the large motor fibres. It was not until 1945 that Granit, with his pupil Leksell (1945), discovered the correct role of the small or gamma fibres as the motor supply to muscle spindles. In 1952 I tried to tell Sherrington of this remarkable discovery and of our failure in 1930, but he had lost all interest in such problems, being in his last years under the spell of his philosophical interests on the nature of man!

## The Controversy of Chemical versus Electrical Synaptic Transmission

The later years in England (1933–1937) were for me remarkable in that I was engaged in a controversy about the nature of the fast-synaptic-transmission processes in sympathetic ganglia and neuromuscular synapses. Sir Henry Dale and his colleagues, Marthe Vogt, Gaddum, Feldberg, and Brown, had built up very extensive experimental evidence in support of the postulated chemical transmission across synapses. According to his hypothesis (Dale, 1937), when the nerve impulse reaches a synaptic junction, it causes the liberation of a minute quantity of acetylcholine. This substance in turn acts across the synaptic junction, so effecting synaptic transmission. At that time I was advocating a widely held alternative hypothesis, that synaptic transmission is effected by the action currents of the nerve impulse exciting across the synapse by a direct electrical stimulation.

In England there were many good occasions for academic disputation of these rival hypotheses at the meetings of the Physiological Society, and a final one in a symposium at the Royal Society in 1937 just before I left England for Australia. I learned there the value of scientific disputation—that it provides a great incentive to perfect one's experimental work and also to examine it more critically. (Of course the critical appraisal is even more searchingly applied to the experiments of one's opponents!) In 1936 I wrote a rather long review on the whole field of synaptic-transmission processes for the *Ergebnisse der Physiologie*. It was my first effort at an intensive survey of a wide field, and it cost me much labor. But at least it clarified many points, and still is valuable as an historical survey of a controversy. My position was not that chemical transmission did not occur, but that it was a later slow phase of the transmission, the early fast phase being electrical. I was impressed by the extremely fast times of action of synaptic transmission, the "on" and "off" times being often less than 1 msec.

## Neuromuscular Transmission

From 1937 to 1966 I lived and worked in my own native country of Australia, and in New Zealand. However, I had the good fortune to have in Sydney for some years, until the end of 1943, the collaboration of Katz and Kuffler in a little research institute at Sydney Hospital. There, under conditions of very severe academic isolation, we managed to pioneer some concepts in neuromuscular transmission. Of special interest was the discovery, independently made by Göpfert and Schaefer (1938) in Germany, Feng (1940) in China, and ourselves in Australia (Eccles and O'Connor, 1938; Eccles, Katz and Kuffler, 1941), that, when a nerve impulse reaches a neuromuscular junction, it does not directly set up a muscle impulse. The initial action is a local reduction of the electrical charge across the surface membrane of the muscle fibre. If this potential change, the so-called endplate potential, is large enough, it generates the discharge of an impulse along the muscle fibre.

During this Sydney phase of my scientific life I was particularly indebted to Katz, because I was able to learn from him much fundamental biophysics that stood me in good stead in subsequent years. However, it was Kuffler who did the finest experimental work in the laboratory, giving evidence even then of that elegant physiological experimentation by which he has been able to investigate many a problem so that the outcome is a unique and beautiful solution. At that time it already seemed from our pharmacological investigations on neuromuscular transmission that the electrical-transmission hypothesis was severely threatened! Nevertheless I continued to believe, in a way that seems incredible in retrospect, that the fast phase was electrical.

**The Scientific Method**

During the first two of the eight years (1944–1951) I spent in Dunedin, New Zealand, I had the good fortune to be associated with the eminent philosopher of science Karl Popper. I learned from him what for me is the essence of scientific investigation—how to be speculative and imaginative in the creation of hypotheses, and then to challenge them with the utmost rigor, both by utilizing all existing knowledge and by mounting the most searching experimental attacks. In fact I learned from him even to rejoice in the refutation of a cherished hypothesis, because that, too, is a scientific achievement and because much has been learned by the refutation.

Through my association with Popper I experienced a great liberation in escaping from the rigid conventions that are generally held with respect to scientific research. Until 1944 I held the following conventional ideas about the nature of research: First, that hypotheses grow out of the careful and methodical collection of experimental data. (This is the inductive idea of science that we attribute to Bacon and Mill.) Second, that the excellence of a scientist can be judged by the reliability of his developed hypotheses, which, no doubt, need elaboration as more data accumulate, but which, it is hoped, stand as a firm and secure foundation for further conceptual development. Finally, and this is the important point, that it is in the highest degree regrettable and a sign of failure if a scientist espouses an hypothesis that is falsified by new data so that it has to be scrapped altogether. When one is liberated from these restrictive dogmas, scientific investigation becomes an exciting adventure opening up new visions; and this attitude has, I think, been reflected in my own scientific life since that time.

My prolonged isolation in the Antipodes was relieved for three weeks early in 1946 by my first visit to the United States. In those immediate postwar weeks the journey from New Zealand to New York was certainly an unpredictable and exciting experience, but that is another story! The occasion was a meeting of the New York Academy of Sciences organized by David Nachmansohn, and I was

most grateful to him for the opportunity of meeting so many of the leading neurophysiologists of America and also the international visitors from Europe. On that
occasion I developed still further my story of electrical synaptic transmission
(Eccles, 1946). I had been encouraged by Karl Popper to make my hypothesis as
precise as possible, so that it would call for experimental attack and falsification.
It turned out that it was I who was to succeed in this falsification by the discoveries
that came towards the end of my stay in New Zealand.

## The Excitatory Synaptic Mechanism

In New Zealand I had returned to working with the central nervous system. I
studied the monosynaptic reflex in the spinal cord, following work that had been
so well developed by a brilliant investigator, David Lloyd (1946), who had been
my research pupil at Oxford. One example of the monosynaptic reflex that is
known to all is the knee jerk. It was established by Lloyd that the pathway led
from a special type of stretch receptor in muscle through large nerve fibres into the
spinal cord, where these fibres in turn made powerful excitatory synapses on
motoneurones to the muscle from which the stretch-evoked impulses had originated. Thus the monosynaptic reflex is the simplest type of reflex arc, involving
just a single synaptic relay in the central nervous system.

My investigations on the monosynaptic reflex of the spinal cord were in good
agreement with Lloyd's work, but I differed from him and many other investigators in that I believed synaptic facilitation to be a postsynaptic process. The
evidence for synaptic facilitation was given by experiments in which conditioning
impulses in the afferents from a muscle excite the motoneurones too weakly to
cause their discharge but leave behind a residual excitatory state that facilitates
a later testing volley of impulses to evoke a discharge. I attributed this synaptic
facilitation to a residual depolarization of the surface membrane of the postsynaptic cells (i.e., the motoneurones), an effect analogous to the endplate potential of
skeletal muscle. The alternative hypothesis of Lloyd and others was that synaptic
facilitation was due to a prolonged presynaptic depolarization, which would be
restricted to the terminals of the nerve fibres that were making the synaptic contact.

It was in the effort to do a crucial experiment on these facilitation hypotheses
that I came to consider the possibility of intracellular recording from motoneurones. Already there had been the work of Nastuk and Hodgkin (1950) on intracellular recording from muscle, and this had been followed up by the preliminary
report of Fatt and Katz (1950) on a similar study of neuromuscular transmission.
Clearly, here was the opportunity for a crucial experimental test. It seemed
feasible to insert very fine glass microelectrodes with a tip diameter of only $0.5\mu$
into motoneurones without destroying them. If the synaptic facilitation was due to
a depolarization of the postsynaptic membrane, then it would be picked up in a
unique and selective manner by a microelectrode recording from inside the nerve
cell. In this position the electrode would record specifically the postulated electrical depolarization of the surface membrane, just as had been observed by Fatt and

Katz for the endplate potential set up by a nerve impulse on a muscle fibre. On the other hand, no such potential would be recorded if synaptic facilitation was due to depolarization of the presynaptic fibres.

I was fortunate to have in my department in Dunedin an excellent electronic engineer, Mr. J. S. Coombs, who was able to construct a cathode-follower amplifier and all the ancillary equipment, and Dr. L. Brock, who became very skilled in pulling the fine glass microelectrodes that were required. For some years prior to 1951 my colleagues and I in New Zealand (Chandler Brooks, Charles Downman and Wilfrid Rall) had been using rather coarse steel or tungsten electrodes for recording potentials in the spinal cord, even of single motoneurones, so it was not a serious problem to change over to the fine glass microelectrodes required for intracellular work. Furthermore, we had the advantage of long experience with spinal-cord physiology and with all the dissection and fixation equipment required for microelectrode work.

Our initial micromanipulators were very crude, being merely the focusing devices extracted from old microscopes. Nevertheless, right from the start we were successful, though only after great travail. It seemed that our best recordings were always after midnight! The first work in this recording was presented at a meeting of the local society of the Medical School at Dunedin on July 31, 1951. By a strange irony of fate it wasn't my turn to present a paper at the meeting. As a consequence, the paper was merely read by title with not even one slide projected. I was rather piqued by this, remarking in an offhand manner that this work could eventually win a Nobel Prize! This preliminary paper was published in the *Proceedings* of that society (Brock, Coombs and Eccles, 1951), the first publication of intracellular recording from nerve cells in the central nervous system (Figure 9.1).

The immediate outcome of this work was to show that synaptic facilitation could be fully explained by a relatively prolonged postsynaptic depolarization, as we had postulated. It also showed that impulses invaded the whole of the soma of motoneurones, giving spike potentials just as large as had been recorded from muscle fibres or giant nerve fibres. Furthermore, we found that the invasion by an antidromic impulse occurred in two stages, and we correctly attributed this to the axonal region in the first stage and then to a later soma-dendritic invasion. The postsynaptic depolarization was called the excitatory postsynaptic potential (EPSP), and I am happy to say that this terminology has been generally

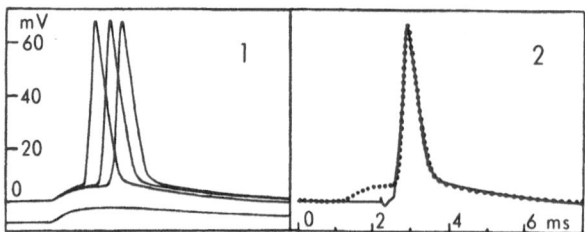

Figure 9.1   Intracellular recording of a motoneurone, activated synaptically in 1, and both synaptically and antidromically in 2. In 1 there is, below, a simple EPSP and, above, EPSPs generating impulse discharges at three different latencies. In 2 there is superposition of a synaptically fired impulse and an antidromically fired impulse. From Brock, Coombs and Eccles (1951).

adopted. An important finding was that when the postsynaptic depolarization or EPSP reached a critical level, it generated the discharge of an impulse from the motoneurone, there being what we called a threshold level of depolarization. A satisfactory explanation was thus provided for all the observations on synaptic facilitation, namely the summation of EPSPs.

## The Inhibitory Synaptic Mechanism

The most fascinating part of these early New Zealand investigations was the recording of the inhibitory postsynaptic potentials as the inverse of the excitatory, being associated with an increased charge on the neuronal surface membrane (which technically is called a hyperpolarization). The occasion of this first recording was especially memorable. It so happened that the experiment was interrupted in the evening because Mr. Coombs's wife was delivered of a baby! After rendering the appropriate obstetrical assistance, Dr. Brock and Mr. Coombs returned in the early hours of the morning to continue our first experimental attempt to determine what intracellular potential changes occurred during synaptic inhibition. This experimental demonstration of the inhibitory postsynaptic potential of motoneurones proved to be quite a crucial test of my electrical hypothesis of synaptic transmission.

Some years earlier Chandler Brooks and I (1947) had developed an electrical hypothesis of synaptic inhibition as a counterpart of the electrical hypothesis for excitatory synapses. Before the experimental test with intracellular recording I had predicted that, on the electrical theory, the intracellular electrode would be positively charged relative to the indifferent ground electrode, while there would be a relative negativity on the chemical theory. So the excitement was quite intense when we had the conditions ready for testing on that fateful night. To my amazement and initial chagrin the recording showed unmistakably an increased negativity; this meant refutation of the electrical-transmission story, a brain-child that I had cherished for so many years! At that early hour of the morning various comments were made that were recorded on a tape, but unfortunately they proved to be virtually indecipherable the next morning. However, the photographic records were beyond all doubt; they still are extant, but the tape has disappeared! I had immediately realized that chemical transmission must obtain for the inhibitory synapses, and therefore most probably also for the excitatory synapses since they resembled so closely the inhibitory in all respects except sign. The inhibitory postsynaptic potential (IPSP) is virtually a mirror image of the EPSP (cf. Figure 9.2 A, B and C, D).

I can vividly remember the gesture of Herbert Gasser a few months later when he looked at our records showing on the same trace the EPSPs and IPSPs, the former being positive, the latter negative (Figure 9.2 E). He recognized immediately that this was the experimental answer to conflicting theories of central excitatory and inhibitory mechanisms. These theories had been the source of debate ever since Sherrington had discovered the existence of an inhibitory mechanism in the central nervous system.

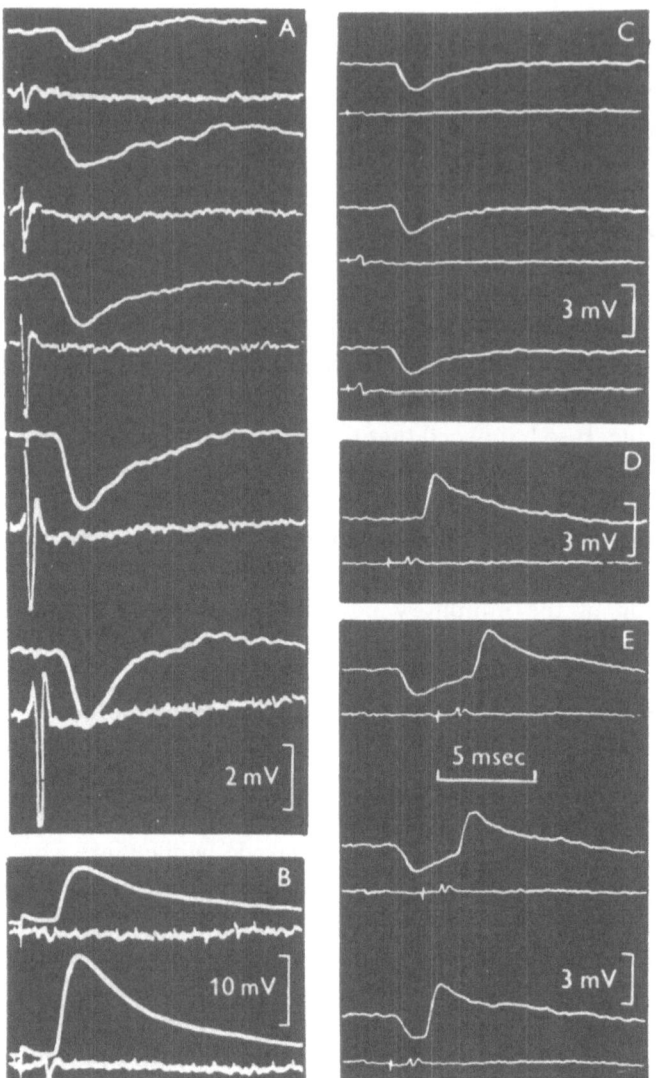

Figure 9.2 (A and C) Intracellular potentials (IPSPs) set up in BSt. (biceps-semitendinosus) neurones by single afferent volleys in quadriceps nerve of increasing size from above downwards. Note the sizes of dorsal-root spikes in A particularly. (B and D) Synaptic potentials (EPSPs) set up by an afferent volley in BSt. nerve in same neurones as A and C, respectively. (E) Potentials set up by quadriceps and BSt. afferent volleys combined at various intervals, the respective controls being seen in first record of C and in D. Note the same time scale for all records, but separate potential scales for each section. From Brock, Coombs and Eccles (1952a).

## Oxford Revisited and the Waynflete Lectures, 1952

That work during the latter part of 1951 was interrupted in December by a long period in England when I gave the Waynflete Lectures at Oxford, and then by the long delays in setting up a scientific laboratory at the Australian National University in Canberra to which I had moved. The Canberra experiments did not start until February 1953. However, on my return to England in January 1952, after nearly fifteen years' absence, there were many good occasions at meetings for giving an account of the intracellular recording from nerve cells and all of its theoretical implications. Needless to say, Sir Henry Dale was delighted at my sudden conversion to chemical transmission, which appeared to him to be of an apocalyptic nature:

Eccles and his team concluded that this positive variation in the motor horn cell could only be due to the release of a chemical agent from the endings of the afferent fibre making synaptic contacts with its surface, and that, if synaptic inhibition was thus chemically transmitted, synaptic excitation was unlikely to be transmitted by an essentially different process, though the transmitter might probably be a different one. By obvious analogy, it was to be supposed that some chemical agent or other would be effective at all central synapses, and that being accepted, Eccles was naturally ready to take cholinergic transmission in the ganglion in his stride. A remarkable conversion indeed! One is reminded, almost inevitably, of Saul on his way to Damascus, when the sudden light shone and the scales fell from his eyes. (Dale, 1954)

I was extremely grateful to my old colleagues at Magdalen College, Oxford, for the invitation to give the Waynflete Lectures because it gave me the challenge and opportunity to present our work on the central nervous system to a distinguished audience, and also to write it up for publication. The scope of the lectures was much broader than the subject of synaptic mechanisms because I was fortunate to be able to give a full account of the magnificent investigations of Hodgkin and Huxley and their colleagues on the nerve impulse and the biophysics of the axonal membrane. This provided a good basis for the presentation of the results of our intracellular recording from nerve cells. Inevitably there were facetious comments on the marriage I essayed between the biophysics of two very different biological systems: giant axons of squids and the neurones of the mammalian central nervous system! I was also fortunate to be able to incorporate a chapter on the remarkable discoveries of Fatt and Katz on neuromuscular transmission.

The lectures were published as *The Neurophysiological Basis of Mind: The Principles of Neurophysiology* by the Clarendon Press in 1953. There was, of course, some adverse criticism of this title, which related especially to the last two chapters and which indicated that through twenty-five years of scientific investigation I had continued to nurture my original philosophical interests. I had, of course, been much encouraged by the remarkable philosophical contribution of Sherrington (1940) in his Gifford Lectures—*Man on His Nature*—and by my two long conversations with him just before he died in 1952. I have returned in the last decade or so to my initial philosophical interests on the nature of the conscious self, but that is another story.

**The Canberra Years**

In the first few years at Canberra I had the good fortune to be associated with Coombs and also with Paul Fatt, who had come from Katz's laboratory in London where they had utilized intracellular recording to elucidate the physiological processes involved in neuromuscular transmission in frogs and crustacea. At this stage of our work we had the great advantage of an electrical stimulating and recording unit (our so-called ESRU), which Mr. Coombs had developed and which was built in 1951 in my old laboratory in New Zealand. At the time this equipment was far superior to anything of the kind in the world, and it was still used by our research team in Chicago until 1968. Now it rests in a sad dismantled form in Buffalo. Even today in Canberra the 1952 copies of this instrument are in use, which is a remarkable tribute to the advanced design it had in 1951! I cherish the ESRU unit because I remember its wonderful service. More than 200 scientific papers have been written from the work that was done with that ESRU, and it operated for over 18,000 hours during the period from 1953 to 1968. Mr. Coombs put it on an hour meter right from its inception!

The other great advantage I had in Canberra was the excellent mechanical equipment designed by Mr. J. G. Winsbury, who was my head technician. This equipment was fabricated in the workshop with its technical staff of 29 that is associated with the School of Medical Research. I venture to suggest that there is no finer workshop in a university medical center. The manipulators and the associated fixation equipment made intracellular recording much more satisfactory and effective. When this equipment came into operation in 1953, we could insert microelectrodes into motoneurones that remained essentially unchanged during several hours of investigation. Mr. Winsbury also designed a microelectrode puller that gave us much better control of the glass pipettes, and that enabled us to manufacture double glass micropipettes that proved to be of great importance during the next stage of work on the biophysics of motoneurones and of the synaptic actions on them.

The work in Canberra with my colleagues Fatt and Coombs, and later David Curtis, continued the investigation of the synaptic processes, using much more refined biophysical methods in the attempt to understand the actual nature of the synaptic events that brought about a depolarization for excitatory synapses and a hyperpolarization for inhibitory synapses.

## Biophysical Studies on Synapses

Perhaps the most remarkable story about this work concerned our efforts to understand the inhibitory synapses. As mentioned above, in Dunedin we had discovered that inhibitory synaptic action produced an increased membrane potential, the IPSP. In one of the early experiments in Canberra, using a rather coarse microelectrode, I noticed to my amazement that the potential changed over quite quickly after the onset of the intracellular recording, so that what had been an inhibitory potential giving hyperpolarization became a depolarizing potential like that produced by excitatory synapses. My colleagues did not believe my excited report. They had not observed it, being otherwise engaged in the experiment. However, a few minutes later we impaled another motoneurone, and I was careful to have them watch the sequence of events which, I am glad to say, repeated themselves just as I had reported before (see Figure 9.3 A–C)! Very quickly we recognized that we had a new discovery, namely that chloride ions diffusing out of the microelectrode (filled as usual with 3M KCl) greatly increase the intracellular concentration of chloride and so cause this inversion of the potential. It is exactly what would be expected if the inhibitory synapses open up ionic gates for chloride ions across the surface membrane of the motoneurone. We realized then that we could inject ions into motoneurones by diffusion out of the microelectrode, provided that the microelectrode was rather larger than normal. The small size of our electrodes in the initial experiments had, of course, prevented this inversion from disturbing and confusing us earlier!

So we studied this inversion by diffusion out of coarse microelectrodes—3 to 5 M Ω resistance—in several experiments (Figure 9.3 A–C). The idea of electrophoretic injection did not come to us until we discovered it by accident. An inadvertent electrical connection onto the microelectrode caused a catastrophic effect on the impaled motoneurone. It was not killed. Instead the inhibitory synapses produced an enormous depolarization. A quick check showed that the current passed through the microelectrode would have caused a large injection of chloride ions out of the microelectrode into the motoneurone. So we had discovered the technique of intracellular ionic injection by accident. It was no longer necessary to use diffusion from the coarse microelectrodes! And we could now also estimate the quantity of ions injected from the coulomb calculation (Figure 9.3 D–F).

I can remember that our initial concept was that chloride was a specific ion for the inhibitory process—an idea that possibly arose from the specificity of the

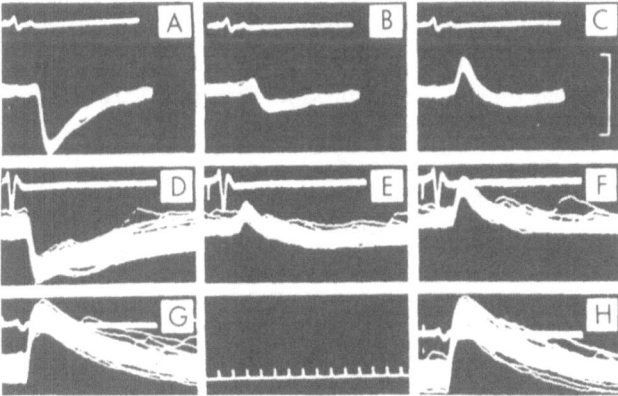

Figure 9.3  Potentials recorded from a motoneurone by an intracellular electrode and generated by a single, direct, inhibitory volley (A to F) or by a single monosynaptic excitatory volley (G, H). Each record is formed by about forty superimposed traces so that random variations are rejected. Upper records show the potential of the inhibitory or excitatory afferent volley as recorded by an electrode in contact with the appropriate dorsal root. A, B, and C reveal the progressive changes in an IPSP attributable to diffusion of Cl⁻ out of a microelectrode filled with 3M KCl, the membrane potential remaining constant at 44 mV. In another experiment an electric current of $10^{-8}$ A flowing for 85 sec through the cell into a microelectrode filled with 3M KCl caused the IPSP to change from D to E, and a further change to F was produced by a current of $2 \times 10^{-8}$ A for 60 sec. Concurrent records G and H show that these combined currents had no significant effect on the EPSP evoked in that motoneurone. The membrane potential was constant at 75 mV. From Coombs, Eccles and Fatt (1953).

ionic mechanisms demonstrated by Hodgkin and his colleagues for nerve impulses. To test this idea we filled microelectrodes with various sodium or potassium salts having several different anions and found to our surprise, and even chagrin, that the anions nitrate, bromide and thiocyanate were just as effective as chloride in causing an inversion of the IPSP. On the other hand, there were a number of anions (sulphate, phosphate, bicarbonate, acetate) that were quite ineffective. When we examined the two sets of anions in the physical tables, we discovered to our delight that the anions causing inversion of the IPSP were all small in the hydrated state, whereas those that failed were all large (Figure 9.4). Hence, on the basis of tests with some nine anions, we postulated that the inhibitory transmitter substance opens up ionic gates or pores in the postsynaptic membrane that allow the free diffusion of anions below a certain critical size in the hydrated state. This hypothesis, of course, demanded experimental testing with other anions in addition to our initial small series. These tests (on 34 species of anions) were performed by my colleagues in Canberra many years later, and revealed only one discrepancy in the critical-size hypothesis (Figure 9.5). This arose from the anomalous behavior of the formate ions, which very effectively pass through the inhibitory ionic gates even though they are slightly larger than three other species of anions that are excluded. This formate anomaly (as we now call it) has been demonstrated with other inhibitory synapses both in vertebrates and invertebrates, but it is still unexplained.

This investigation of critical ionic size and inhibitory transmitter action is a good example of the way in which Popper's concepts of science have helped me to develop hypotheses that have gone far beyond the data and that have consequently challenged experimental testing. Finally, Paul Fatt derived equations for inhibitory synaptic action that were based on the theory of ionic diffusion across membranes proposed some years before by Goldman, and that formed the basis of the mathematical developments of Hodgkin, Katz and Huxley for ionic diffusion through the surface membrane of nerve fibres. It turned out that these equations predicted with surprising fidelity the effects of a wide range of membrane poten-

| Cations | | Anions | |
|---|---|---|---|
| $K^+$ | 1.00 | $Br^-$ | 0.94 |
| | | $Cl^-$ | 0.96 |
| | | $NO_3^-$ | 1.03 |
| | | $SCN^-$ | 1.11 |
| $Na^+$ | 1.47 | $HCO_3^-$ | 1.65 |
| $N(CH_3)_4^+$ | 1.60 | $CH_3CO_2^-$ | 1.80 |
| | | $SO_4^{2-}$ | 1.84 |
| | | $H_2PO_4^-$ | 2.04 |
| | | $HPO_4^{2-}$ | 2.58 |

Figure 9.4   Ion diameters in aqueous solution as derived from limiting-ion conductances and expressed relative to $K^+ = 1.00$. The horizontal broken line gives the division between the small ions that pass readily through the subsynaptic inhibitory membrane and the larger ions that pass with much greater difficulty or not at all. From Eccles (1957).

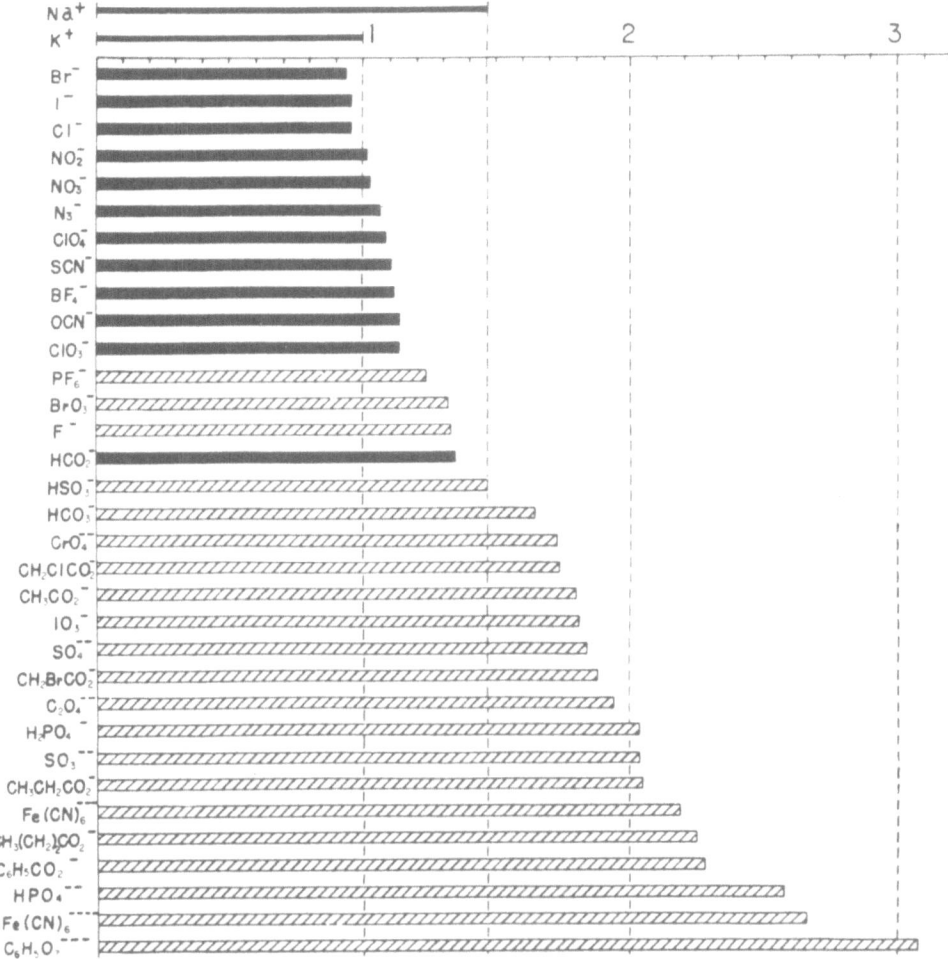

Figure 9.5  Diagrammatic illustration of the correlation between the sizes of ions in aqueous solution and the effects of their injection upon the IPSP. Lengths of bands indicate ion sizes in the aqueous solution as calculated from the limiting conductance in water. The black bands are for anions effective in converting the IPSP into the depolarizing direction, as in Figure 9.3 (A–F), and the hatched bands are for anions not effective. Hydrated sizes of K+ and Na+ are shown above the length scale, the former being taken as the unit for representing the size of other ions. From Eccles (1964b).

tials on the sizes of the IPSPs. Our hypothesis, therefore, was that the inhibitory transmitter opens up ionic gates in the postsynaptic membrane that momentarily allow the free diffusion of anions below a critical size, this ionic movement causing the inhibitory postsynaptic potential. This hypothesis is diagrammatically illustrated in Figure 9.6 for three types of inhibitory synapses. We also included small cations in the ionic inhibitory mechanism, but the evidence had a serious flaw, as was later pointed out by Ito. Nevertheless, investigations with Ito many years later led indirectly to the postulate that small cations such as potassium also pass through the inhibitory ionic gates and contribute to the generation of the IPSP.

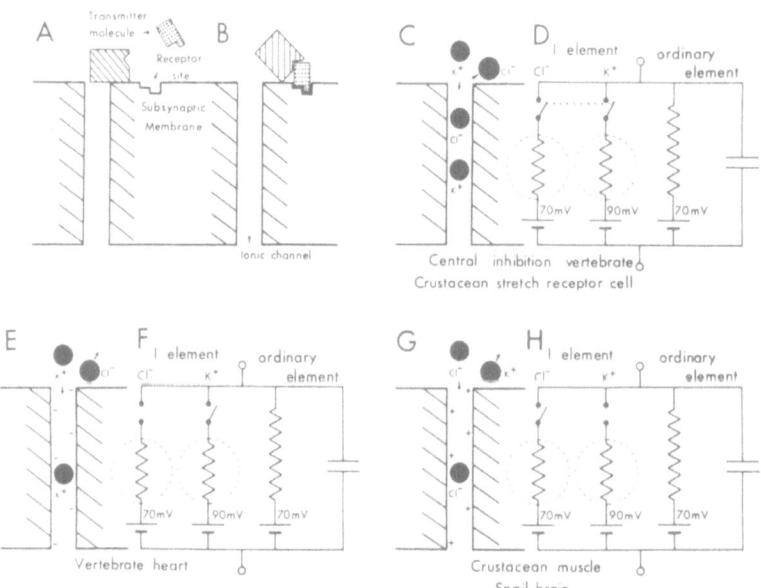

Figure 9.6   Diagrams summarizing the hypotheses relating to the ionic mechanisms employed by a variety of inhibitory synapses in producing IPSPs. (A, B) Schematic representation of the way in which a synaptic transmitter molecule could effect a momentary opening of a pore in the subsynaptic membrane by causing the lifting of a plug. In B the transmitter molecule is shown in close steric relationship both to a receptor site and to the plug, which has been pulled away from the orifice of the pore. As a consequence ions can move freely through pores in the subsynaptic membrane for the duration of the transmitter action on it. (C) A schematic representation of a pore through an activated inhibitory subsynaptic membrane showing the passage of both chloride and potassium ions that is postulated for IPSP production at central inhibitory synapses. (D) A diagram showing the inhibitory element as being composed of potassium and chloride ionic conductances in parallel, each with batteries given by their equilibrium potentials and operated by a ganged switch, closure of which symbolizes activation of the inhibitory subsynaptic membrane. E, F and G, H represent the conditions occurring at inhibitory synapses where there is predominantly potassium ionic conductance, as with the vertebrate heart, or predominantly chloride ionic conductance, as with crustacean muscle or cells in the brain of a snail. It is assumed that the pores are restricted to cation or anion permeability by the fixed charges on their walls, as shown. From Eccles (1964b).

## Neuronal Pathways in the Spinal Cord

In parallel with these investigations on the biophysics of synapses, we were also working in these early years in Canberra on various neuronal pathways in the spinal cord, following up ideas developed by Paul Fatt while he was reading in the library at Canberra from late 1952 through early 1953 before our equipment was working. This enforced experimental idleness proved to be of great value to us all, because Paul came up with two major problems that led to fascinating experimental developments. The first concerned the cells that Renshaw (1946) had described many years before and that he found to be fired repetitively by an antidromic volley in the axons of motoneurones. Renshaw himself thought they could be the cells responsible for the inhibitory effects on motoneurones that he and Lloyd had found to be produced by these antidromic impulses. However, it was Paul Fatt who very clearly defined the problem. He predicted, in accord with Dale's postulate, that the action of the motor-axon collaterals upon Renshaw cells (as we called them) would be cholinergic, acetylcholine being the transmitter, just as at the peripheral neuromuscular synapses made by these same motor axons.

This clear formulation of the problem immediately gave us a clue about how to develop the experimental tests, and in collaboration with Koketsu, who had meanwhile arrived from Japan, we carried out a successful series of experiments (Eccles, Fatt and Koketsu, 1953, 1954). In retrospect this proved to be one of the most elegant experimental investigations that I can remember being associated with. As illustrated in our earliest diagram (Figure 9.7), everything came out as predicted with a full confirmation of all that Lloyd and Renshaw had found, except for the infrequent excitatory action that Renshaw had described. This excitation was later shown by Wilson and Burgess (1962) to be due to a special action of Renshaw cells inhibiting other inhibitory cells of motoneurones. It will be appreciated that the inhibition of a continuous background of inhibition is equivalent to an excitation (technically it is known as a disinhibition). The paper in which we published these experiments is one that I am happy to remember because I feel that it illustrates so well the power of a properly designed experimental attack on a clearly formulated problem.

The other problem that Paul Fatt raised early in 1953 concerned the so-called direct inhibition which Lloyd (1946) had discovered and which was remarkable for the very brief time required for its central action. So short was this time that Lloyd had assumed it to be monosynaptic, just as with the excitation described above. In fact it was produced by impulses in the same large afferent fibres from muscle that are excited by stretch. I too had accepted Lloyd's explanation, even though we had found in our earliest intracellular investigations at New Zealand that the central delay in producing the IPSP was just over 1 msec longer than the central delay for the monosynaptic EPSP.

Paul Fatt questioned our explanation that the long latency was due to slow conduction time in collaterals of the primary afferent fibres in the spinal cord and suggested as an alternative that it was due to the interpolation of an interneurone on the inhibitory pathway. There was just enough time to allow for this synaptic relay. Again, as soon as he raised the problem and had shown its relevance, it was easy for me to design experiments that would test it crucially. The experimental investigations again proved the value of clear formulation of problems, and, in collaboration with Sven Landgren of Sweden, Paul Fatt and I were able to show that the delay of the IPSP was due not to slow conduction but to an interneuronal relay (Eccles, Fatt and Landgren, 1953), and we found cells in the intermediate nucleus of the spinal cord that had properties appropriate for the postulated interneurones (Figure 9.8). It was later shown by Hultborn, Jankowska and Lindström (1971) in Sweden that at least the majority of the inhibitory interneurones lie more ventrally than the interneuronal nucleus we had discovered.

On the basis of these two discoveries of inhibitory cells, namely Renshaw cells and these inhibitory interneurones that were assumed to be on the so-called direct inhibitory pathways, we formulated the much wider generalization that all in-

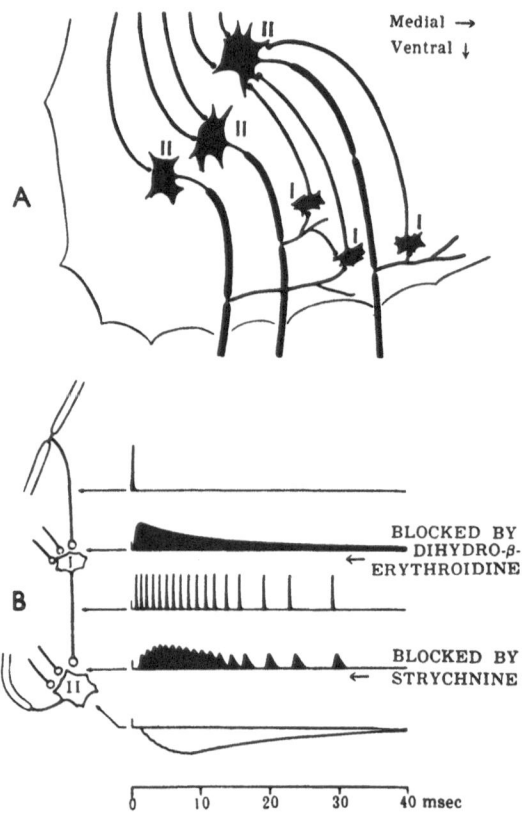

Figure 9.7 (A) Sketch of a proposed neurone system in the ventral horn of the spinal cord. Collaterals are given off by motor axons before they leave the spinal cord, and they are shown making synaptic contact with interneurones (I). The axons of these interneurones proceed dorsally and laterally and make contact with motoneurones (II), which, by this system, are inhibited. Reflexly active afferents are shown descending onto the motoneurones from the dorsal direction. (B) Diagram summarizing the postulated sequence of events from the antidromic impulse in motor axons to the inhibition of motoneurones. All events are plotted on same time scale, and the corresponding histological structures are shown to the left (note indicator arrows). The five events are, from above downwards: (1) impulse in axon collateral; (2) time course of acetylcholine liberated at axon-collateral terminal; (3) repetitive discharge in interneurone; (4) time course of inhibitory transmitter substance liberated at interneuronal terminal; (5) hyperpolarization set up in motoneurone by inhibitory synaptic action. Note additional synapses for convergent pathways both on interneurone and motoneurone. The summation of the synaptic action of several converging interneurones onto a motoneurone is responsible for smoothing the latter part of the motoneurone hyperpolarization. From Eccles, Fatt and Koketsu (1953).

hibitory action in the central nervous system is due to a special class of nerve cell whose sole function is inhibition. Hitherto it had been a general belief that the axons of a nerve cell could excite at some of their synapses and inhibit at others. This erroneous concept was drawn in a classical diagram by Sherrington (1906, Figure 37) and also in a diagram I published in 1952 (Brock, Coombs and Eccles, 1952b). We now developed the hypothesis that the inhibitory interneurone is required in order to change over the transmitter substance that is being manu-

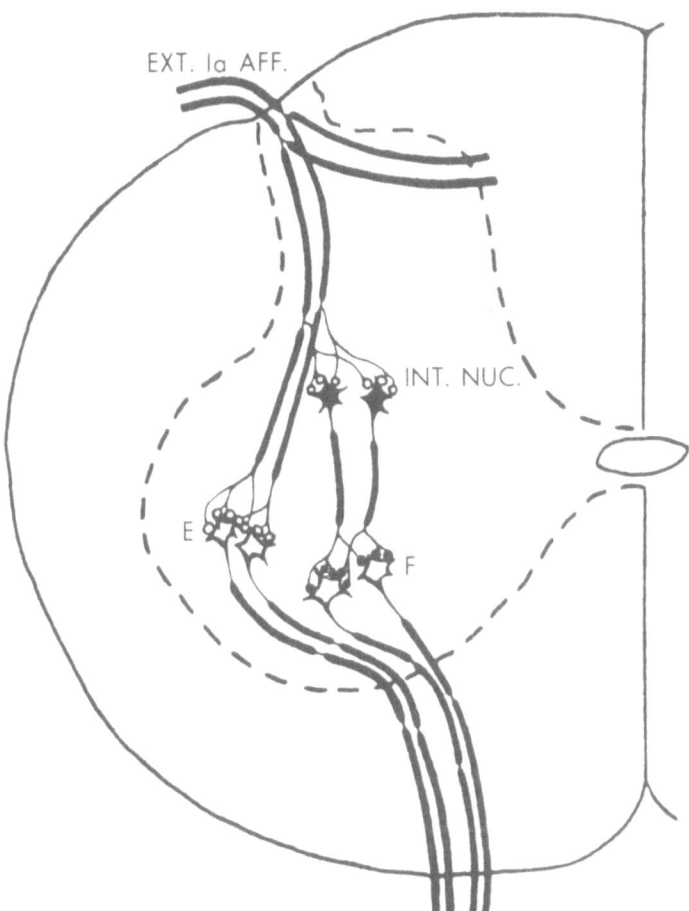

Figure 9.8 Diagram showing the postulated pathways for impulses entering the spinal cord along group Ia afferent fibres from extensor muscles, which give monosynaptic excitation and direct inhibition: INT.NUC., intermediate nucleus; E, extensor motoneurones; F, flexor moto-neurones. Cells shown in solid black, with solid black synaptic knobs, are specifically inhibitory. The other synaptic knobs, shown as open circles, are specifically excitatory. Medullated regions are indicated by thickening along the fibre. A complementary diagram could be drawn for the group Ia afferent fibres from flexor muscles. From Eccles, Fatt and Landgren (1953).

factured from the excitatory to the inhibitory substance. We had some other examples of inhibitory pathways in the spinal cord, and, following the scientific methods advocated by Karl Popper, we formulated a general concept for the cell constitution of the mammalian central nervous system, namely, that it was made up of two classes of neurones—excitatory neurones and inhibitory neurones (Eccles, Fatt and Landgren, 1953, 1956). This postulate is diagrammed in Figure 9.9 and has been rigorously tested by investigators in many parts of the world. On several occasions falsification of this hypothesis has been claimed, but in every case the claim has itself been shown to be unsound. In a recent survey of the central nervous system over thirty separate species of inhibitory neurones have been identi-

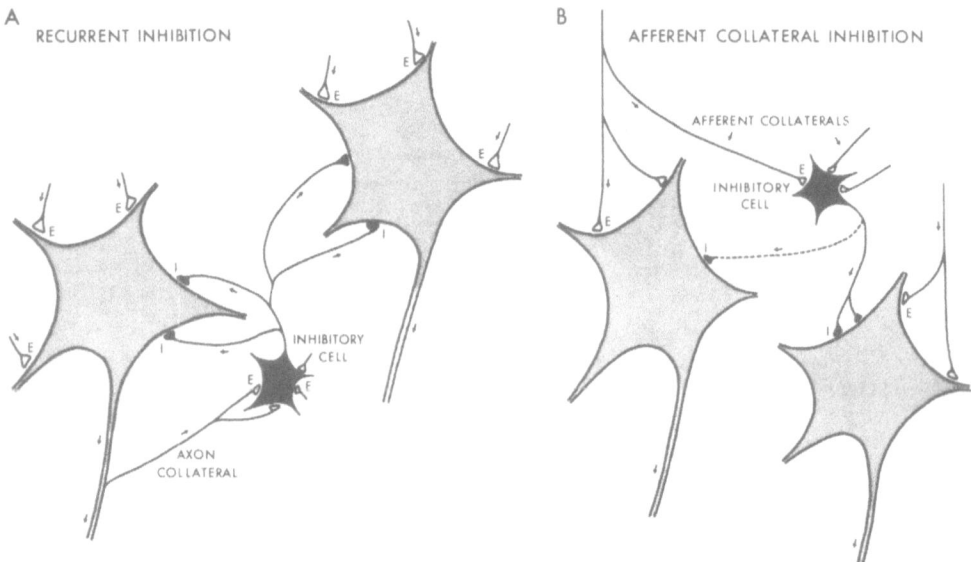

Figure 9.9   Diagrams of the two types of inhibitory pathways (inhibitory cells shown in black), as described in the text.

fied. In every case it appears that these inhibitory neurones have an inhibitory action at all of their synapses. No ambivalent neurones have been discovered in the mammalian central nervous system. However, several exceptions have now been found in invertebrate nervous systems.

## The Herter Lectures, 1955

The occasion to give a comprehensive account of all this work came when I was invited to be the Herter Lecturer for 1955 at Johns Hopkins University. I had been travelling through Europe on a rather extensive tour and eventually arrived in Baltimore in the early autumn to give the lectures. As is my usual custom, I had prepared them only in the form of sketchy outlines, but had a reasonably complete series of slides for illustration. It was planned that three lectures be given in Hird Hall on three successive afternoons. Unfortunately, a severe debilitating illness overtook me during the series, becoming worse each day, so that I had a terrible struggle to survive the last lecture. My friends at Johns Hopkins realized my predicament, but I had a tight schedule for the rest of my journey, which included going to Los Angeles, Seattle and Vancouver and then returning to Australia by ship. In those days air bookings were hard to secure, and, if I delayed my schedule, I probably would not have caught the ship across the Pacific from Vancouver. In retrospect I think I had a virus pneumonia because the symptoms of acute secondary infection developed later during my stay on the West Coast.

The most embarrassing occasion came at the end of the third lecture when the Herter Lecturer is customarily entertained at a dinner at the Macaulay Street Club. I arrived at the dinner verging on collapse and after lying on the floor for half an hour was eventually taken home without attending the dinner. I tell this story because of the sequel, which was that I determined to make amends by writing a very comprehensive account of the Herter Lectures for publication. I decided that these lectures should appear in a monograph form, rather than in the *Bulletin of Johns Hopkins University* as is usually done. So on my journey back across the Pacific I spent all my time in converting the three lectures into six chapters of a book. Since the ideas were all in my mind and the figures were almost all in the slides that I used for the lectures, I was able to compose one chapter every two days for the final twelve days of the voyage. I arrived in Sydney with a fairly completely written manuscript, which was then quickly converted into typescript after I had checked references and added some special sections. This book, *The Physiology of Nerve Cells* (1957), has now sold about 10,000 copies and has been produced as a paperback, which seems to be the ultimate criterion of success in this age!

### Epilogue

From 1955 onwards the synapses have been beautifully displayed in their essential features by the electron microscopists, and their essential modes of operation are being understood at levels that would have astonished the great pioneers Ramón y Cajal and Sherrington. Of course the solution of every problem has entailed yet more problems for investigation. By making a prodigious effort I was able to review in a book almost all that was known about the physiological properties of synapses up to 1963 (*The Physiology of Synapses*, 1964a). Since that time the literature has expanded so enormously that I have had to decline the invitation to prepare a second edition.

My scientific interests have now moved to the problems of neuronal organization at the higher levels of the nervous system. The spell of the synapse has been broken or, rather, transmuted to the challenge of attempting to discover the principles of neuronal organization in the functioning of the higher levels of the nervous system.

I have come to recognize that I was mistaken in my original belief that an understanding of the synapse would contribute substantially to a solution of the brain-mind problem. Of course, synapses still are of fundamental importance in the operation of the neuronal machinery at the higher levels. This is particularly so in relation to the postulated plastic changes in synapses that are presumed to be essentially concerned in learning. But the functional performance of the higher levels requires postulates of a global kind: neuronal systems forming continuously operating and interacting dynamic loops of immense complexity and variability—the "enchanted loom" imagined by Sherrington.

It is in this context that we have to reconsider the brain-mind problem. This problem becomes acute when the effects of brain lesions, particularly the "split-brain" patients of Sperry, lead to the postulate that only a special part of the brain is in liaison with mind, and then only when it is in the appropriate level of dynamic activity. This "liaison brain" is restricted to the dominant hemisphere and probably even to the speech areas and the associated ideational areas. The great adventure now is to discover the special structural and functional properties of this liaison brain. The time of that discovery will be far into the future because the understanding of any area of the mammalian cerebral cortex is still at a primitive level when measured against its presumed immense complexity. Nevertheless, my abiding dedication to the mind-brain problem has led me to be under a new spell—the spell of the liaison brain!

## References

Brock, L.G., Coombs, J.S., and Eccles, J.C. (1951): Action potentials of motoneurones with intracellular electrode. *Proc. Univ. Otago Med. Sch.* 29:14–15.

Brock, L.G., Coombs, J.S., and Eccles, J.C. (1952a): The recording of potentials from motoneurones with an intracellular electrode. *J. Physiol.* 117:431–460.

Brock, L.G., Coombs, J.S., and Eccles, J.C. (1952b): The nature of the monosynaptic excitatory and inhibitory processes in the spinal cord. *Proc. R. Soc. B.* 140:170–176.

Brooks, C. McC., and Eccles, J.C. (1947): An electrical hypothesis of central inhibition. *Nature* 159:760–764.

Coombs, J.S., Eccles, J.C., and Fatt, P. (1953): The action of the inhibitory synaptic transmitter. *Aust. J. Sci.* 16:1–5.

Coombs, J.S., Eccles, J.C., and Fatt, P. (1955): The specific ionic conductances and the ionic movements across the motoneuronal membrane that produce the inhibitory post-synaptic potential. *J. Physiol.* 130:326–373.

Dale, H. (1937): Transmission of nervous effects by acetylcholine. *Harvey Lect.* 32:229–245.

Dale, H.H. (1954): The beginnings and the prospects of neurohumoral transmission. *Pharm. Rev.* 6:7–13.

Eccles, J.C. (1936): Synaptic and neuromuscular transmission. *Ergeb. Physiol.* 38:339–444.

Eccles, J.C. (1946): An electrical hypothesis of synaptic and neuro-muscular transmission. *Ann. NY Acad. Sci.* 47:429–455.

Eccles, J.C. (1957): *The Physiology of Nerve Cells.* Baltimore: The Johns Hopkins Press.

Eccles, J.C. (1964a): *The Physiology of Synapses.* Berlin: Springer-Verlag.

Eccles, J.C. (1964b): Ionic mechanism of postsynaptic inhibition. *In: Les Prix Nobel en 1963.* Stockholm: P.A. Norstedt and Söner, pp. 261–283.

Eccles, J.C., Fatt, P., and Koketsu, K. (1953): Cholinergic and inhibitory synapses in a central nervous pathway. *Aust. J. Sci.* 16:50–54.

Eccles, J.C., Fatt, P., and Koketsu, K. (1954): Cholinergic and inhibitory synapses in a pathway from motor-axon collaterals to motoneurones. *J. Physiol.* 126:524–562.

Eccles, J.C., Fatt, P., and Landgren, S. (1953): The "direct" inhibitory pathway in the spinal cord. *Aust. J. Sci.* 16:130–134.

Eccles, J.C., Fatt, P., and Landgren, S. (1956): Central pathway for direct inhibitory action of impulses in largest afferent nerve fibres to muscle. *J. Neurophysiol.* 19:75–98.

Eccles, J.C., and Hoff, H.E. (1932): The rhythmic discharge of motoneurones. *Proc. R. Soc. B.* 110:483–514.

Eccles, J.C., Katz, B., and Kuffler, S.W. (1941): Electric potential changes accompanying neuromuscular transmission. *Biol. Symp.* 3:349–370.

Eccles, J.C., and O'Connor, W.J. (1938): Action potentials evoked by indirect stimulation of curarized muscle. *J. Physiol.* 94:9–11P.

Eccles, J.C., and Sherrington, C.S. (1930): Numbers and contraction-values of individual motor-units examined in some muscles of the limb. *Proc. R. Soc. B.* 106:326–357.

Fatt, P., and Katz, B. (1950): Membrane potentials at the motor endplate. *J. Physiol.* 111:46–47P.

Feng, T.P. (1940): Studies on the neuromuscular junction. XVIII. The local potentials around N-M junctions induced by single and multiple volleys. *Chin. J. Physiol.* 15:367–404.

Göpfert, H., and Schaefer, H. (1938): Über den direkt und indirekt erregten Aktionsström und die Funktion der motorischen Endplatte. *Pfluegers Arch. Gesamte Physiol.* 239:597–619.

Hultborn, H., Jankowska, E., and Lindström, S. (1971): Recurrent inhibition of interneurones monosynaptically activated from group Ia afferents. *J. Physiol.* 215:613–636.

Leksell, L. (1945): The action potential and excitatory affects of the small ventral root fibres to skeletal muscle. *Acta Physiol. Scand.* 10 (Suppl. 31):1–84.

Lloyd, D.P.C. (1946): Facilitation and inhibition of spinal motoneurons. *J. Neurophysiol.* 9:421–438.

Nastuk, W.L., and Hodgkin, A.L. (1950): The electrical activity of single muscle fibers. *J. Cell. Comp. Physiol.* 35:39–73.

Renshaw, B. (1946): Central effects of centripetal impulses in axons of spinal ventral roots. *J. Neurophysiol.* 9:191–204.

Sherrington, C. (1906): *The Integrative Action of the Nervous System.* New Haven: Yale University Press.

Sherrington, C. (1924): Problems of muscular receptivity. *Nature* 113:892–894 and 929–932.

Sherrington, C. (1940): *Man on His Nature.* London: Cambridge University Press.

Wilson, V.J., and Burgess, P.R. (1962): Disinhibition in the cat spinal cord. *J. Neurophysiol.* 25:392–404.

Ulf von Euler (b. 1905, Stockholm, Sweden) shared the 1970 Nobel Prize with Julius Axelrod and Bernard Katz for their research on transmitter substances in nerve terminals. Dr. von Euler's work has included the discovery, with J.H. Gaddum, of Substance P, of prostaglandin, and the identification of noradrenaline as the neurotransmitter of the sympathetic nervous system. He is professor of physiology at the Royal Caroline Institute and president of the Board of the Nobel Foundation.

# 10

# Discoveries of Neurotransmitter Agents and Modulators of Neuronal Functions

## Ulf von Euler

At a dinner speech in Atlantic City in 1950, Otto Loewi commented on some facets of his work that had led to the discovery of neurochemical transmission. He mentioned that Sir Walter Fletcher, years after the discovery, reminded him that as far back as 1903 Loewi had pointed out to him the possibility of a neurochemical transmission. As we know, this idea gradually developed, and the ripe fruit fell into Loewi's hands in 1921 after a series of genially simple experiments. At the 1926 International Physiological Congress in Stockholm, Loewi demonstrated his experiments, which were to revolutionize the concepts of nerve transmission. As a medical student I had the opportunity to be present and also to make the personal acquaintance of this great scientist and charming and witty man.

Soon after meeting Loewi, I started to do some simple studies in experimental pharmacology relating to the vascular effects of adrenaline. The field of neurotransmitters and their actions seemed to offer many opportunities for further research, and when, after finishing my medical studies in 1930, I had an opportunity to go abroad with the aid of a Rockefeller Fellowship, I eagerly grasped the opportunity to learn the trade and widen my views. At this time I was already set upon doing research work. This decision was no doubt influenced by the fact that both my parents were scientists, and it probably involved some genetic or constitutional factors.

### London and Dale's Laboratory

Through my teacher and friend Göran Liljestrand's wise judgment I spent six months in H. H. Dale's department in Hampstead, London. The National Institute for Medical Research was an ideal school for a young scientist. A great accumulation of knowledge and experience in pharmacology, physiology, biochemistry, and biophysics was to be found there, and I eagerly studied and absorbed the scientific atmosphere and the British way of viewing problems and handling them. It must not be forgotten that Sweden was still mainly in the sphere of influence of German science and German thinking, which differed in several respects from the British style. I soon learned the meaning of the word "evidence," which was always the uncompromising gatekeeper to the green meadows of recognized results. However, despite its importance and value for sound development in science, the craving for evidence sometimes seemed to bar imagination and outlook. Moreover, apparent "evidence"—based on available facts and accepted theories—could lead in the wrong direction and delay progress, though formally unassailable.

At the time of my arrival in London interest was focused on acetylcholine, following the striking discovery that, properly administered, it could elicit a muscle twitch and trigger activity in a postsynaptic neuron. The British School at Hampstead played a dominating role in this work. In a few years the group around Dale, including Feldberg, Gaddum, Bacq, Brown, Vogt, and MacIntosh, created an edifice of knowledge which was widely accepted and even highly normative for thinking all over the world.

As suitable introductory work for the young fellow from Sweden, I was given the task of looking for a release of acetylcholine from the intestine following vagus stimulation. I am afraid I did not succeed, but one day a slight deviation from the program proved fruitful. An extract of rabbit intestine was tested for acetylcholine on a piece of rabbit duodenum, with the unexpected result that the contraction observed still occurred after a dose of atropine, which blocked the action of acetyl-choline. Since the effect could not be attributed to choline or acetylcholine, nor to histamine or any other known amine, I concluded—somewhat hastily—that a new biologically active factor in the intestinal wall had been discovered. This claim was not readily accepted in the laboratory, but I insisted. Dale then wisely proposed that J. H. Gaddum, then working in the laboratory as his first assistant, should undertake with me a more thorough study of the active factor, later known as Substance P. Its presence in the brain, particularly in the hypothalamus and in peripheral nerves, can perhaps justify my mentioning it in this context, especially since it has been identified as an undecapeptide, isolated and characterized by Chang and Leeman (1970). What functions it has—if any—are still unknown, but several observations suggest that it may serve as a modulator of both central and peripheral nervous functions.

## The Search for Active Compounds: Adrenergic Nerves

The introduction to the use of pharmacological tools that I obtained in Dale's laboratory was invaluable, and my finding of a new biologically active factor almost at the turning of the spade convinced me that there was more to find.

Accordingly, on my return to Stockholm, after also working in the laboratories of I. de Burgh Daly, C. Heymans, and G. Embden, I took up the question of active compounds in various tissues.

Although acetylcholine dominated the scene at the time of my stay in London in 1930, Dale always had a vivid interest in what he later termed the adrenergic system. His discovery of the adrenolytic effect of ergot and his concept of two kinds of vascular receptors sensitive to adrenaline, later known as $\alpha$- and $\beta$-receptors, made such an interest natural. In his 1910–1911 work with Barger, Dale had already pointed out the fact that certain non-N-methylated compounds of the catecholamine series mimic the effects of sympathetic-nerve stimulation better than adrenaline. In a later comment (1953) he states:

Doubtless I ought to have seen that nor-adrenaline might be the main transmitter —that Elliott's theory might be right in principle and faulty only in this detail. If I had had so much insight, I might even then have stimulated my chemical colleagues to look for nor-adrenaline in the body; but they would almost certainly have failed to find it, with the methods which were then available. It is easy, of course, to be wise in the light of facts recently discovered; lacking them I failed to jump to the truth, and I can hardly claim credit for having crawled so near and then stopped short of it. But if I had taken the additional step, even in hypothesis, much trouble might, perhaps, have been saved in after years for my late friend, Walter B. Cannon: For the observed differences between the effects of adrenaline and those of "sympathin", as liberated by stimulation of sympathetic nerves, and especially of the hepatic nerves, which led Cannon and his associates to put forward the elaborate theory of the two complex sympathins, E and I, were practically the same as those between adrenaline and nor-adrenaline, with which I was here concerned. It should be noted, indeed, that one who had collaborated with Professor Cannon at one stage, Prof. Z. M. Bacq, actually suggested in 1934 that "Sympathin E" might be nor-adrenaline (*Ann. Physiol.* 10:480); but, in the absence of direct evidence then, and for many years afterwards, of the natural occurrence of nor-adrenaline in the animal body, this suggestion attracted less attention than it deserved.

Back in Stockholm, I found it natural to continue the attempts to obtain more knowledge about the active compounds in tissue extracts. Further studies on Substance P indicated that it was a polypeptide. The search for good sources of this and other biologically active material led me to the vesicular and prostate glands, both of which were unexpectedly rich in what appeared to be adrenaline. This was, in fact, the gate that led to the discovery of the prostaglandins in 1934–1935. After we had shown that these compounds behave chemically as fatty acids, probably unsaturated, the work was continued by my colleague S. Bergström, who isolated them and found their chemical constitution (Bergström, 1967). Their functional connection with the nervous system was revealed only recently, and we now have good evidence that some of them serve as physiological modulators of sympathetic-nerve activity in several organs, including the heart (Hedqvist, 1969).

I have had the privilege to watch the different steps on this particular path that have been taken by young members of a research group in our laboratory. Through their skill and imagination the actions of the two compounds, adrenergic neurotransmitters and prostaglandins, have been integrated into a meaningful symbiosis.

Observations on the occurrence in relatively large quantities of an adrenalinelike substance in the prostate gland, the vesicular gland, and the ampulla ductus deferentis activated the search for similar compounds in other organs. The spleen proved useful as material, as did the splenic nerves since they appeared to be less contaminated with other biologically active compounds. It was already known, especially from the work of Cannon and Lissák (1939), that sympathetic nerves contain an adrenalinelike compound. A study of the active substance on a number of test preparations gradually revealed certain differences when compared to adrenaline as the standard. Once the suspicion had arisen that the compound was not adrenaline itself, Bacq's 1934 suggestion that Cannon and Rosenblueth's Sympathin E (excitatory) might be noradrenaline provided the line for further attacks on the problem. Evidence, both pharmacological and biochemical, soon accumulated that the substance in the splenic nerves actually was noradrenaline, in spite of the conclusion drawn by Cannon and Lissák as late as 1939 that it was adrenaline. It should also be borne in mind that the neurohumor demonstrated by Loewi in the frog's heart (1921) had actually been correctly identified as adrenaline but it was not known at that time that this was a remarkable exception in that most animals use noradrenaline as transmitter.

In this context it seems appropriate to recall that Peter Holtz, then working in Rostock, Germany, had some years previously discovered dopadecarboxylase. In his further studies he detected the $\beta$-hydroxylated reaction product noradrenaline (which he called "Urosympathin") in urine. In the same paper he reported that this substance is also present in extracts of the cat's adrenal medulla. Owing to publication difficulties in Germany following World War II, this paper, submitted in 1944, was not published until 1947, when our results on nerve and tissue extracts were already published.

If Loewi's conclusions about the nature of the sympathetic transmitter were based on the exceptional conditions in the frog, we had a somewhat analogous experience. Since noradrenaline had been identified in adrenergic nerves, we decided to look for it in the suprarenals. A number of analyses of the rabbit's suprarenals seemed to show unequivocally that only adrenaline was present. It was somewhat embarrassing to learn afterwards that this was also an exception, the rabbit being the only laboratory animal that has practically only adrenaline in its suprarenals.

It would be untrue to state that the claim for noradrenaline as a neurotransmitter went home readily. Only in the early 1950s did this concept become generally accepted.

The development of good methods of purification and bioassay greatly helped in the rapid exploitation of this field. Particularly useful were the adsorption on alumina described by Shaw (1938), and the differential estimation of adrenaline and noradrenaline in a mixture using different test preparations. Ehrlén's techniques (1948) for transforming the catechol compounds into strongly fluorescent and stable trihydroxyindole compounds made it possible to include catecholamine estimation in routine clinical tests.

After it had been established that noradrenaline occurs in sympathetic nerves and disappears after nerve degeneration, the question arose: How is it stored? Mainly through the work of Blaschko and Hillarp, it had been demonstrated that the catecholamines in the chromaffin cells of the adrenal medulla are stored in subcellular particles (Blaschko, Hagen, and Welch, 1955; Hillarp, Lagerstedt, and Nilson, 1953). We had good reasons to believe that the noradrenaline is accumulated in a high concentration in the nerve terminals, which show the peculiar swellings first described by Hillarp in 1946. I suggested to Hillarp in 1956 that the noradrenaline in the nerves might also be stored in subcellular particles, and in a series of experiments we actually showed that this is the case. Electron-microscopic pictures from 1957 showed the particles to be in part osmiophilic, with diameters varying from 300 to 1500 Å. These particles are still being investigated to determine their mechanisms of binding and releasing the transmitter, among other things.

I believe that this brief survey describes in a simple fashion the paths followed in the pursuit of the specific neurohumor in the adrenergic nerves. It encompasses, of course, only a small part of our present knowledge of the adrenergic system and its physiology, biochemistry, and pharmacology. Among the most spectacular branches that have grown from the original stem can be mentioned the biochemistry of inactivation and synthesis, where Axelrod and Udenfriend have yielded fundamental contributions. The Falck-Hillarp histochemical method has made it possible to map the distribution of the adrenergic system in the tissues and to reveal its location in the central nervous system. Dopamine, discovered in mammalian suprarenals and hearts by Goodall (1950) in our laboratory and originally considered merely as a precursor of noradrenaline, became recognized after the work of Blaschko (1942) and of Holtz, Credner, and Kroneberg (1947) as a neurotransmitter in its own right (Hornykiewicz, 1971). The action of indirectly acting amines received its explanation when it was recognized that they liberated noradrenaline (Burn and Rand, 1958).

**Paths for the Future**

We may ask where the path leads to. The great problems, as I see them at present, are the nature of the uptake and binding processes in the storage granules and the mechanism of release from the particles. Some seem confident that they already know the answers to these fundamental problems. However, it is useful to remember that thirty years ago almost everybody was convinced that adrenaline was the adrenergic neurotransmitter.

One might, of course, speculate about the many undisclosed mechanisms by which the nervous system transmits the signals to the effector cells. Although this may be outside the scope of this presentation and lead to no new path of discovery, I am tempted to touch briefly upon a few speculations.

First of all: How did noradrenaline come to be the adrenergic neurotransmitter in so many animal classes? We know that certain amino acids in the evolutionary process found use as transmitter substances in the nervous system, and the distance is not too far to the still relatively simply built catecholamines derived from phenylalanine and tyrosine. What made these derivatives useful presumably has to do with their chemical reactivity and their propensity to bind to receptor macromolecules in such a way that conformational adjustments allow translocation of ions or enhance enzymatic actions, as discovered by Sutherland and Robison (1966). The transmitter role of such related substances as octopamine and various hydroxytryptamines also bears on this assumption. Histamine would be expected to belong to this group, and perhaps also piperidine and pyrrolidine, which are both found in the brain. How do they adapt to each other, the prospective transmitter, and the reacting receptor macromolecule? Which comes first, the receptor or the transmitter? This might perhaps be studied in primitive organisms with a rapid life-turnover. Information obtained from enzyme induction experiments might supply a good lead here.

A second point is: When did the catecholamines first appear as transmitters during evolution? This is difficult to assess because the biochemistry of the extinct animal classes is wholly obscure. We have to consider an evolutionary period of about one billion years, and realize that perhaps ten to twenty million animal species are extinct as against some two million now extant. However, the presence of catecholamines in such primitive organisms as protozoans, mollusks, and protostomians seems to indicate that their appearance is not a recent phenomenon. Their double functions as neurohumors and circulating hormones suggest that they were used first as a tissue hormone and later as a neurotransmitter. The occurrence of catecholamines in protozoans is of interest because of their primitive status. It is even possible that members of this group inherited the catecholamine-producing system from plants, since some plants do produce amines of this kind.

In general it seems that dopamine is the dominant catecholamine in invertebrates, accompanied in some cases by noradrenaline, whereas adrenaline is mostly absent. Dopamine has maintained its role as transmitter but shifted mainly to certain parts of the central nervous system. Even in lower vertebrates, such as fish, catecholamines play an important role.

The functions of catecholamines must have been quite useful since the system is so well developed in primates and man, both in the central and in the peripheral nervous system. In man the catecholamines are indeed a prerequisite for the erect position. Their presence in the brain hemispheres is another indication that these amines have not yet had to be replaced by some still more efficient chemical.

Turning to the storage particles, it is generally accepted that these packages enclosing the vulnerable transmitter are being formed by the Golgi apparatus in the cell soma. Since the particles are found all along the axon, they must be somehow transported to the periphery by the axoplasmic flow (Weiss and Hiscoe, 1948), as demonstrated with various techniques (Dahlström, 1965). The nature of this transport is still obscure. After arrival in the peripheral varicosities, the particles apparently have their payload increased and become ready to release it on demand. The rapid equilibration of their contents with labeled noradrenaline, or even other amines such as adrenaline, strongly suggests a continuous exchange with the ambient tissue fluid either inside or through the axon membrane. We know that uptake of noradrenaline in the particles is greatly facilitated by adenosine triphosphate in the presence of magnesium, and that uptake and release are strongly influenced by a number of compounds known to interact with cell metabolism and the respiratory chain. The conclusion that the particles shed their contents by exocytosis through the axon membrane is based on the finding that other granule constituents besides the amine are also released in blood after nerve stimulation. Apparently, high-molecular-weight proteins can pass through the capillary wall into the bloodstream, and perhaps also through the particle and axon membranes in small amounts. Perhaps the particles can last for a limited number of release-uptake cycles. Here, as in so many other membrane-located processes, the primary goal for future research seems to be disclosure of the ion-transfer mechanisms. The turning off and on of transfer channels in the membranes may be one of the main functions of these chemical tools in general. It might be wise not to preclude their involvement in the ion-transfer events accompanying nerve conduction.

## References

Bacq, Z.M. (1934): La pharmacologie du systeme nerveux autonome, et particulierement du sympathique, d' après la théorie neurohumorale. *Ann. Physiol. Physicochim. Biol.* 10:467–528.

Bergström, S. (1967): Prostaglandins: members of a new hormonal system. *Science* 157:382–391.

Blaschko, H. (1942): The activity of (−)-dopa decarboxylase. *J. Physiol.* 101:337–349.

Blaschko, H., Hagen, P., and Welch, A.D. (1955): Observations on the intracellular granules of the adrenal medulla. *J. Physiol.* 129:27–49.

Burn, J.H., and Rand, M.J. (1958): The action of sympathomimetic amines in animals treated with reserpine. *J. Physiol.* 144:314–336.

Cannon, W.B., and Lissák, K. (1939): Evidence for adrenaline in adrenergic neurones. *Am. J. Physiol.* 125:765–777.

Cannon, W.B., and Rosenblueth, A. (1937): *Autonomic Neuroeffector Systems.* New York: Macmillan.

Chang, M.M., and Leeman, S.E. (1970): Isolation of a sialogogic peptide from bovine hypothalamic tissue and its characterization as substance P. *J. Biol. Chem.* 245:4784–4790.

Dahlström, A. (1965): Observation on the accumulation of noradrenaline in the proximal and distal parts of peripheral adrenergic nerves after compression. *J. Anat.* 99:677–687.

Dale, H.H. (1953): *Adventures in Physiology.* London: Pergamon Press.

Ehrlén, I. (1948): *Farmacevtisk Revy (Stockholm)* 47:242–250.

Falck, B., Hillarp, N.-A., Thieme, G., and Torp, A. (1962): Fluorescence of catechol amines and related compounds condensed with formaldehyde. *J. Histochem. Cytochem.* 10:348–354.

Goodall, McC. (1950): Hydroxytyramine in mammalian heart. *Nature* 166:738.

Hedqvist, P. (1969): Prostaglandin E as a modulator of noradrenaline release from sympathetic nerves. *Acta Physiol. Scand.* 77 (Suppl. 330).

Hillarp, N.-A., Lagerstedt, S., and Nilson, B. (1953): The isolation of a granular fraction from the suprarenal medulla, containing the sympathomimetic catechol amines. *Acta Physiol. Scand.* 29: 251–263.

Holtz, P., Credner, K., and Kroneberg, G. (1947): *Arch. Exp. Path. Pharmak.* 204:228–243.

Hornykiewicz, O. (1971): Dopamine: Its physiology, pharmacology and pathological neurochemistry. *In: Biogenic Amines and Physiologic Membranes in Drug Therapy.* Part B. Biel, J.H., and Abood, L.G., eds. New York: Marcel Dekker, pp. 173–258.

Loewi, O. (1921): *Arch. Gesamte Physiol.* 189:239–242.

Shaw, F.H. (1938): The estimation of adrenaline. *Biochem. J.* 32:19–25.

Sutherland, E.W., and Robison, G.A. (1966): Metabolic effects of catecholamines. *Pharmacol. Rev.* 18:145–161.

Weiss, P., and Hiscoe, H.B. (1948): Experiments on the mechanism of nerve growth. *J. Exp. Zool.* 107:315–395.

# Neurotransmitters and Brain Function

Julius Axelrod (b. 1912, New York, New York) shared the 1970 Nobel Prize with Ulf von Euler and Bernard Katz for their discoveries concerning the mechanisms of storage, release, and inactivation of transmitter substances in nerve terminals. His principal areas of research have been the biochemical mechanisms of drug and hormone action and their metabolism and enzymology, the pineal gland, and the biochemistry of sympathetic nerves. Since 1955 Dr. Axelrod has been chief of the Section of Pharmacology at the Laboratory of Clinical Science, NIMH.

# 11

# Biochemical and Pharmacological Approaches in the Study of Sympathetic Nerves

Julius Axelrod

## First Experience in Research

My introduction to biological research was unplanned and accidental. I received a B.S. degree in biology from The City College of New York in 1933 with the hope of studying medicine. At that time, however, it was virtually impossible for a City College graduate to be accepted by a medical school, and I was no exception. Through a friend I heard of an opening for a "volunteer" laboratory assistant in the Department of Bacteriology at New York University. Though this position paid only $25 a month, it gave me an opportunity to work in a research laboratory assisting K. G. Falk in his studies on ester-hydrolyzing enzymes in tumors.

After two years economic necessity forced me to take a position in a food-testing laboratory, where I worked for ten years. The laboratory work was moderately interesting and, at times, challenging since it required modification of analytical methods. In 1946 the laboratory received a contract from a group of pharmaceutical manufacturers to study why methemoglobinemia developed following ingestion of large quantities of acetanilide-containing analgesic preparations. The president of the laboratory, G.B. Wallace, who had retired as head of the pharmacology department of New York University, suggested that I seek advice from his former associate, B.B. Brodie, who was then at Goldwater Memorial Hospital, New York University Division.

I spent an afternoon discussing the problem with Brodie. He invited me to spend a few weeks in his laboratory to do some exploratory experiments, and I accepted. This was a fateful decision for me. At that time Goldwater Memorial Hospital's laboratories were devoted to malaria research under the leadership of James Shannon, who had gathered around him a remarkable group of young investigators, such as B.B. Brodie, Sidney Udenfriend, Robert Berliner, Robert Bowman, Thomas Kennedy, and John Taggart, to name a few. It was in this stimulating environment that I experienced my first taste of research. In my initial meeting with Brodie I learned that drugs such as acetanilide can be transformed in the body. A possible metabolic product of acetanilide that we considered was aniline, which had already been shown to produce methemoglobinemia. I then set about developing a method sensitive enough to measure aniline in blood and urine. Fortunately my experience in devising analytical methods was of great help. Within two weeks a specific and sensitive assay for aniline was developed, and soon thereafter aniline was detected and measured in blood after the ingestion of acetanilide (Brodie and Axelrod, 1948a,b).

After this brief exposure to research I knew that this was what I wanted to do for the rest of my life. The invitation to remain at Goldwater was extended from weeks to months, and then to years. During this period Brodie and I elucidated the physiological disposition and metabolic pathways of a number of analgesic drugs, including acetanilide, acetophenetidin, aminopyrine, and antipyrine (Brodie and Axelrod, 1948b, 1949, 1950a,b). We also developed specific methods for the measurement of these drugs and their metabolites in blood, tissues, and biological fluid. In these studies we showed that the actions of drugs are determined by the biochemical changes they undergo in the body. Drugs can be metabolized to less active, more active, or toxic metabolites. The identification of the various transformation products of many analgesic drugs led to the introduction of N-acetyl-p-aminophenol as a safer analgesic drug (Brodie and Axelrod, 1948b), clarified the action of several drugs such as acetanilide, acetophenetidin, and dicumarol, and indicated that normal metabolic processes are involved in the metabolism of a variety of foreign compounds. These studies laid down the foundations of what was later to become biochemical pharmacology.

Because I lacked a doctorate degree, I knew that I would have no chance for any real advancement in a hospital attached to an academic institution. But I had neither the inclination nor the money to spend three or four years getting a Ph.D. In 1949 Shannon became head of the National Heart Institute in Bethesda and offered me a position, along with many other members of the Goldwater staff. Brodie, Udenfriend, Berliner, Kennedy, Bowman, and I moved to Bethesda en masse.

## Discovery of the Microsomal Drug-Metabolizing Enzyme

During my first two years at the National Heart Institute I carried out studies in collaboration with Brodie and his staff on the metabolism of analgesics, anticoagulants, adrenergic blocking agents, and caffeine. I eventually became dissatisfied with working in a large team, and I was allowed to work independently. In reading the literature I became intrigued by a group of compounds, the sympathomimetic amines. Work on these compounds had begun in 1899, when Abel isolated the principle in the adrenal medulla that elevates blood pressure and identified it as a catechol phenylethylamine derivative. Shortly thereafter Langley (1901) noted that the injection of extracts of the adrenal medulla mimics the effects produced by the stimulation of sympathetic nerves. Then Elliot (1904) made the brilliant suggestion that an adrenalinelike substance released from sympathetic nerves might be responsible for chemical transmission at synapses. In 1910 Barger and Dale, in an incisive paper, reported that numerous $\beta$-phenylethylamine derivatives simulated the effects of sympathetic-nerve stimulation, not only with varying intensity but with varying precision, and coined the term "sympathomimetic amines." Loewi (1921), in an elegant experiment, clearly demonstrated chemical neurotransmission and proposed adrenaline as the transmitter for sympathetic nerves, but von Euler (1946) clearly identified noradrenaline as the sympathetic-nerve neurotransmitter. Sympathomimetic amines not only have pressor action but some, such as amphetamine, mescaline, and ephedrine, also have unusual behavioral effects. However, very little information concerning the chemical fate and physiological disposition of sympathomimetic amines was available in 1952. Because of my background in drug metabolism, I undertook a study on the biotransformation of phenylethylamine derivatives.

The first compound that I studied was ephedrine, a $\beta$-hydroxylated phenylisopropyl amine. (Ephedrine, the active principle of Ma Huang, an herb that has been used by Chinese physicians for over 5000 years, was introduced into modern medicine by Chen and Schmidt in 1930.) Within a year I succeeded in describing the metabolic pathways for ephedrine (Axelrod, 1953). This compound was found to undergo a variety of biotransformations, including N-demethylation, ring hydroxylation, and conjugation. The next compounds to be studied were methamphetamine and amphetamine (Axelrod, 1954). These drugs were also metabolized by N-demethylation, hydroxylation, and glucuronide conjugation. Considerable species variation in the metabolism of these drugs was also observed.

Amphetamine administered to rabbits disappeared without a trace. This puzzled me, and I decided to look for the enzyme that metabolized amphetamine in the rabbit. I had no experience in enzymology, but Building 3 at the NIH was staffed with superb enzymologists, and after a few days of guidance from Gordon Tomkins, who shared my laboratory, I learned the basic elements of enzymology. I soon found an enzyme in rabbit liver that deaminated amphetamine to form phenylacetone and ammonia. This enzyme was localized in the liver microsomes and required both TPNH and $O_2$ (Axelrod, 1955a). I soon discovered a similar microsomal enzyme that N-demethylated ephedrine to norephedrine and formaldehyde (Axelrod, 1955b). Historically, the deamination of amphetamine was the first of a series of microsomal enzyme systems shown to require TPNH and $O_2$. These enzymes were shown to carry out numerous reactions, including O-, N-, and S-demethylation, hydroxylation, and oxidation of alkyl side chains, and to be crucial in determining the duration of action of drugs (Quinn, Axelrod, and Brodie, 1958).

## The Move to Neuroscience

In 1954 I received an offer from Ed Evarts to head a pharmacology section in his laboratory in the National Institute of Mental Health. Though I had only a meager understanding of the mission of the National Institute of Mental Health, I gladly took the job. My impression of neuroscience then was that it was mainly concerned with neurophysiology and brain anatomy. This to me was a somewhat esoteric field concerned with complicated electronic equipment, and I believed that an investigator had to be an especially gifted experimentalist and theorist to work in it. First, I took a year off to get a Ph.D., which involved passing a rigorous examination. Fortunately, my work on the metabolism of sympathomimetic amines was acceptable as a thesis.

Instead of plunging directly into an unfamiliar field in neurobiology, I thought it would be best to work on a problem in which I had some expertise and which, at the same time, would be appropriate to mental-health research. I thus began work on the metabolism and physiological disposition of LSD (Axelrod et al., 1957), and the enzymes involved in the metabolism of narcotic drugs (Axelrod, 1956a). Bob Bowman was in the process of building a spectrofluorophotometer. He was kind enough to let me use one of his experimental models to develop a very sensitive fluorometric assay for LSD. This made it possible to study the metabolism, tissue distribution, and neurophysiology of LSD.

After completing my Ph.D., I turned my attention to metabolism of narcotic drugs. I was successful in finding an enzyme in liver microsomes that N-demethylated narcotic drugs and that was profoundly influenced during the development of tolerance by narcotic-drug antagonists (Axelrod, 1956b). On the basis of these investigations I proposed a theory of tolerance that stimulated a great deal of critical reaction, mostly negative.

Work on narcotic drugs led to a study of the enzymatic synthesis of glucuronide and the discovery of UDPG dehydrogenase, in collaboration with Jack Strominger and Herman Kalckar (Strominger, Kalckar, and Axelrod, 1954). Soon after, in 1956, Rudi Schmid and I described the enzyme that formed the direct-reading bilirubin, bilirubin glucuronide (Schmid et al., 1957). We then showed that there is a defect in the glucuronide-forming enzyme in a mutant strain of jaundiced (Gunn) rats (Axelrod, Schmid, and Hammaker, 1957). To examine whether such a defect is present in humans with a related disease (Gilbert's) subjects were given N-acetyl-p-aminophenol, which we had previously shown to be almost completely conjugated to a glucuronide metabolite. These subjects showed a marked reduction in the appearance of N-acetyl-p-aminophenol-glucuronide in blood, indicating a biochemical lesion in the glucuronide-forming enzyme.

## Initial Work on Catecholamines

Most of the studies mentioned so far had little to do with the nervous system. Though the NIMH administrators were very supportive of the type of research I was doing, I still had a feeling of guilt about not working on some aspect of the nervous system or behavior. Seymour Kety, who was then chief of the laboratory, gave an intriguing account in a seminar of a finding by Canadian psychiatrists that adrenochrome caused schizophrenialike hallucinations (Hoffer, Osmond, and Smythies, 1954). They had on this basis proposed that schizophrenia is produced by an abnormal metabolism of adrenaline to form adrenochrome.

In searching the literature I was very much surprised to find that hardly anything was known about the metabolism of adrenaline. In view of Hoffer and Osmond's provocative hypothesis about the abnormal metabolism of adrenaline in schizophrenia, I decided that a study of the metabolism of catecholamine compounds would be most appropriate. The dogma current in 1957 held that noradrenaline was metabolized and inactivated by monoamine oxidase. However, Greisemer and coworkers (1953) had found that after complete inhibition of monoamine oxidase the physiological actions of injected adrenaline were still rapidly terminated. This clearly indicated to me that there were enzymes other than monoamine oxidase that metabolized catecholamines. It seemed that a possible route of metabolism might involve oxidation of adrenaline, and I spent a frustrating three months looking for such an oxidative enzyme.

In March 1957 I received the latest *Federation Proceedings*, and an abstract by Armstrong and McMillan engaged my attention. They reported the excretion of 3-methoxy-4-hydroxymandelic acid in patients who had been administered noradrenaline and also in patients with adrenaline-forming tumors (pheochromocytomas). From the structure of this compound it appeared that it could be formed by the O-methylation of noradrenaline followed by deamination. That catecholamines could be O-methylated was an intriguing possibility and one that could be checked experimentally.

That afternoon I incubated adrenaline with a rat liver homogenate, ATP, and methionine (which I thought would be a methyl donor). I was surprised and pleased to see that the adrenaline rapidly disappeared. Omitting ATP or methionine resulted in a negligible disappearance of adrenaline (Axelrod, 1957). This was a decisive experiment and indicated that adrenaline is O-methylated in the presence of a methyl donor, presumably S-adenosylmethionine formed from ATP and methionine. Gabriel de la Haba in the laboratory next to mine generously gave me a little S-adenosylmethionine he had synthesized. Incubating liver homogenates with adrenaline and S-adenosylmethionine also resulted in the rapid metabolism of the catecholamine. The most likely site of methylation would be on the phenolic hydroxyl in a position meta to the side chain to form 3-O-methyl adrenaline. I called Bernard Witkop and asked if he or one of his associates would synthesize the O-methylated metabolite of adrenaline. Three days later Siro Senoh, a visiting scientist in Witkop's laboratory, brought me beautiful crystals of meta-O-methyl adrenaline, which we named metanephrine. The metabolite formed enzymatically by incubating adrenaline and S-adenosylmethionine was identified as metanephrine. The O-methylating enzyme was soon purified, and the enzyme was shown to O-methylate catechols but not monophenols (Axelrod and Tomchick, 1958). Other catecholamines O-methylated by this enzyme, which was named catechol-O-methyltransferase, were noradrenaline, dopamine, and dopa. O-methylated catecholamines, normetanephrine and methoxytyramine, were soon found to be normally occurring in brain. At last I was beginning to work with the nervous system. As a result of the discovery of the O-methylation enzyme the pathway for the metabolism of catecholamines was described (Axelrod, Senoh, and Witkop, 1958).

## Radioactive S-Adenosylmethionine and the Discovery of Several Methyltransferase Enzymes

To increase the sensitivity in the assay of catechol-O-methyltransferase, [14]C-labeled S-adenosylmethionine was synthesized enzymatically from [14]C-methyl methionine and ATP. Thus the enzyme transferred the [14]C-methyl group of S-adenosylmethionine to the meta hydroxyl group of noradrenaline. The radioactive product normetanephrine was then extracted into toluene and measured. Because of its ability to label the O- and N-group potential substrates, the availability of $^{14}$C-S-adenosylmethionine made it possible to find a number of methyltransferase enzymes such as histamine N-methyltransferase (Brown, Tomchick, and Axelrod, 1959), hydroxyindole O-methyltransferase (Axelrod and Weissbach, 1961), phenylethanolamine N-methyltransferase (Axelrod, 1962a), and a nonspecific N-methyltransferase (Axelrod, 1962b). Each of these enzymes has some significance to the neurosciences:

1. Histamine N-methyltransferase is highly localized in the brain.
2. Phenylethanolamine N-methyltransferase is the enzyme that converts noradrenaline to adrenaline. It is highly localized in the adrenal medulla, and its synthesis is regulated by glucocorticoids (Wurtman and Axelrod, 1966).
3. Hydroxyindole O-methyltransferase is uniquely localized in the pineal gland, and its activity is regulated by sympathetic nerves.
4. The nonspecific N-methyltransferase can convert tryptamine, a compound normally present in the brain, to N,N-dimethyltryptamine, a psychotomimetic agent.

**The Search for New Biogenic Amines**

The isolation of the specific methyltransferase enzymes made it possible to exploit them in the search for new biogenic amines in the nervous system and to devise sensitive assays to measure noradrenaline, dopamine (Coyle and Henry, 1973), histamine (Snyder, Baldessarini, and Axelrod, 1966), and serotonin (Saavedra, Brownstein, and Axelrod, 1973). These assays involved the incubation of tissue with a selected N-methyltransferase and $^{14}$C- or $^{3}$H-methyl-S-adenosylmethionine. The endogenous substrate accepts a radioactive methyl group and is then separated by extraction into suitable solvents. Using these assays, we were able to detect and measure, in brain and other tissues, octopamine, (Molinoff, Landsberg, and Axelrod, 1969), tryptamine (Saavedra and Axelrod, 1972), phenylethanolamine (Saavedra and Axelrod, 1973), and phenylethylamine (Saavedra, 1974). These enzyme assays made it possible to measure picogram amounts of serotonin, noradrenaline, dopamine, and histamine in 130 separate nuclei in hypothalamus, limbic system, amygdala, and brainstem (Palkovits, Saavedra, Brownstein, and Axelrod, in preparation). In turn, these sensitive assays open up new avenues of research on the pharmacology and biochemistry of specific nuclei in the brain.

**A New Mechanism for Noradrenaline Inactivation**

The elucidation of the metabolic pathways for catecholamines raised more problems. When both major enzymes in catecholamine metabolism, catechol-O-methyltransferase and monoamine oxidase, were inhibited in vivo, the physiological response to injected adrenaline was still rapidly terminated. This indicated that other mechanisms were involved in the physiological inactivation of this compound. The answer to this question came in an unexpected way. After the metabolic pathways and the various urinary metabolites of catecholamines had been characterized, Seymour Kety initiated a study in 1957 to examine the blood and urine of schizophrenics for abnormal catecholamine metabolites. In order to carry out this study he commissioned New England Nuclear Corporation to synthesize radioactive adrenaline and noradrenaline of high specific activity.

In 1958 the first batch of [3]H-labeled adrenaline of high specific activity arrived. Fortunately the [3]H-adrenaline labeled on position 7 was stable. I thought it would be a good idea to examine the tissue distribution of the [3]H-adrenaline after its injection, and Kety agreed to give me a few milligrams of the precious material. At that time Hans Weil-Malherbe was spending three months in my laboratory, and together we developed specific methods for separating and measuring [3]H-adrenaline in tissues. When we injected [3]H-adrenaline into cats, radioactive adrenaline persisted unchanged in heart, spleen, and salivary and adrenal glands long after its physiological effects had dissipated (Axelrod, Weil-Malherbe, and Tomchick, 1959). This phenomenon occurred in repeated experiments, and it puzzled us for many months. When Gordon Whitby joined my laboratory to do a Ph.D. thesis, I suggested that he develop specific methods for measuring [3]H-noradrenaline and its metabolites in tissues. Again, after administering [3]H-noradrenaline to cats, this catecholamine remained in tissues for many hours (Whitby, Axelrod, and Weil-Malherbe, 1961). It became apparent that [3]H-adrenaline and [3]H-noradrenaline are selectively bound to tissues with a rich sympathetic-nerve innervation (i.e., heart, spleen, salivary gland).

Georg Hertting (who spent two productive years with me) and I thought of the crucial experiment that would establish the site of binding of catecholamines. The superior cervical ganglion of a cat was removed unilaterally, causing the sympathetic nerves in salivary gland and eye muscles to degenerate on one side only (Hertting et al., 1961). We then injected [3]H-noradrenaline, and concentrations of the radioactive catecholamine were examined in structures innervated by the cervical sympathetic nerves. There was an almost complete reduction in the levels of [3]H-noradrenaline in the denervated structures. This indicated unequivocally that sympathetic nerves take up and retain the catecholamine. With this knowledge we were then in a position to label the sympathetic nerves selectively by injecting [3]H-noradrenaline and then to study its fate during and after nerve stimulation. In a series of experiments Hertting and I (1961) showed that [3]H-noradrenaline is taken up in nerves and released on nerve stimulation. As a result of these findings we proposed and established that the noradrenaline is rapidly inactivated by reuptake into the sympathetic nerves.

**The NIH Research Associate Program**

In 1961 Linc Potter joined me as my first Research Associate. The Research Associate Program at the NIH provides an opportunity for recent Ph.D.'s or M.D.'s to spend two years in Bethesda doing full-time research. There are many applicants for this program, and investigators in the intramural research program at the NIH are thus able to get the brightest young men who want postdoctoral training. About a dozen of these research associates have worked in my laboratory, and all have subsequently gone on to independent productive careers in research. These postdocs have created an open atmosphere in the lab, and the free exchange of ideas has made it possible to try novel approaches to problems. Though many experiments have not worked, a few have, with important consequences.

Potter and I turned our attention to finding the intraneural site where noradrenaline is taken up and retained. Since we had demonstrated that radioactive noradrenaline can be taken up by sympathetic nerves, such a subcellular site could be labeled. After injection of $^3$H-noradrenaline in rats, the heart and other sympathetic innervated tissues were removed, homogenized, and the various subcellular fractions separated in a continuous sucrose gradient (Potter and Axelrod, 1962). There was a sharp peak of $^3$H-noradrenaline in the microsomal fraction, which coincided with endogenous catecholamines and dopamine-$\beta$-hydroxylase, the enzyme that converts dopamine to noradrenaline. The noradrenaline-containing particles exerted their physiological pressor response only when they were lysed. This suggested that the catecholamine was present in a bound, physiologically inactive form. We anticipated that radioautography and electron microscopy could be used to visualize the intraneural storage site of radioactive noradrenaline. We injected $^3$H-noradrenaline into rats, and the pineal, because of its extensive sympathetic-nerve innervation, was used for radioautography. The electron microscope revealed a clear localization of photographic grains overlying nerve terminals that contained granulated vesicles of about 500 angstroms (Wolfe et al., 1962).

Later studies with another Research Associate, Dick Weinshilboum, showed that on sympathetic-nerve stimulation noradrenaline is discharged from terminals together with dopamine-$\beta$-hydroxylase by a process of exocytosis (Weinshilboum et al., 1971). Upon depolarization of the nerves, the noradrenaline storage vesicle presumably fuses with the nerve terminal, followed by an extrusion of neurotransmitter, dopamine-$\beta$-hydroxylase, and the soluble contents of the vesicle into the synaptic cleft. These findings led Weinshilboum to postulate that the released dopamine-$\beta$-hydroxylase should appear in the blood. This prediction was soon established (Weinshilboum and Axelrod, 1971), and a means was thus provided to study sympathetic-nerve activity by measuring dopamine-$\beta$-hydroxylase in the blood. Our laboratory and others then found abnormally low levels of dopamine-$\beta$-hydroxylase in familial dysautonomia and Down's syndrome and very high levels of the enzyme in torsion dystonias, neuroblastomas, and hypertension.

## Drugs and the Disposition of Noradrenaline

The uptake and storage of radioactive noradrenaline in sympathetic nerves made it possible to study by relatively simple experiments the action of a variety of drugs on the physiological disposition of $^3$H-noradrenaline. Cocaine was chosen as the first drug to study. It had been postulated that cocaine produces supersensitivity by blocking the inactivation of the neurotransmitter. Because of our previous work, it had become apparent that uptake of noradrenaline into sympathetic nerves was an important mechanism for the rapid inactivation of this catecholamine. Knowledge available about the physiological disposition of $^3$H-noradrenaline made it possible to examine whether cocaine interfered with the uptake of the neurotransmitter. Cats were pretreated with cocaine, followed by an injection of $^3$H-noradrenaline. There was a marked reduction in the levels of the $^3$H-catecholamine in sympathetically innervated tissues and an increase in the plasma level of the $^3$H-noradrenaline (Whitby, Hertting, and Axelrod, 1960). These results showed that cocaine blocks the uptake of noradrenaline into nerves and thus allows larger amounts of the catecholamine in the synaptic cleft to react with the postsynaptic receptor. Using a similar approach, we found that other drugs such as imipramine and sympathomimetic amines (amphetamine) also block the uptake of noradrenaline (Axelrod, Whitby, and Hertting, 1961).

In another series of experiments we administered $^3$H-noradrenaline to label the intraneuronal stores of the transmitter and then examined the effects of drugs in releasing the catecholamine. If a drug releases noradrenaline, the tissue level of the radioactive catecholamine should be lowered. Indeed, we did show that amphetamine, tyramine, and reserpine reduce the level of noradrenaline, while imipramine and other tricyclic antidepressants do not (Axelrod, Hertting, and Potter, 1962). In another experiment using an isolated beating heart whose nerves had been previously labeled with $^3$H-noradrenaline, we demonstrated that the physiological actions of sympathomimetic amines on the heart are mediated by the discharge of noradrenaline from sympathetic nerves (Axelrod et al., 1962).

## First Experiments on the Dynamics of Transmitters

The initial, but rather primitive, experiment to measure the rate of utilization of neurotransmitter was made with radioactive noradrenaline. We injected $^3$H-noradrenaline to label the nerves of the salivary gland (Hertting, Potter, and Axelrod, 1962). The disappearance of radioactive noradrenaline, we believed, reflected the rate of release or turnover of the neurotransmitter. When the nerve impulses to the sympathetic nerves of the salivary gland were interrupted by decentralization of the superior cervical ganglia, the rate of disappearance of $^3$H-noradrenaline was markedly slowed. Similar results were obtained by using ganglion blocking agents. These experiments on the rate of disappearance of $^3$H-noradrenaline were the first attempts to study the dynamics of neurotransmitters. Subsequent work by others has produced more sophisticated techniques for measuring the turnover of neurotransmitters. The observation that noradrenaline in the nerves is in a constant state of flux proved to be an important concept.

## The Effect of Drugs upon Noradrenaline-Containing Nerves in the Brain

Up to 1964 almost all of our work on the biochemistry of the adrenergic nervous system was concerned with the peripheral nerves. Dahlström and Fuxe (1964) demonstrated that the brain has noradrenaline-containing nerves similar to those that are present in the peripheral organs. We (Weil-Malherbe, Axelrod, and Tomchick, 1959) observed that there is a blood-brain barrier to noradrenaline that makes it impossible to study the metabolism, storage, and release of noradrenaline in the brain by administering it through peripheral injection. It was Jacques Glowinski, as a Visiting Scientist in my laboratory, who circumvented this problem. He devised a technique to introduce ³H-noradrenaline directly into the brain by injection into the lateral ventricle. The problem was to establish whether the radioactive noradrenaline injected in this manner mixed with the endogenous catecholamine in the brain. In a series of experiments we showed that ³H-noradrenaline is highly localized in noradrenergic and dopaminergic nerve terminals (Glowinski and Axelrod, 1966). Radioautographic studies demonstrated that the radioactive noradrenaline injected into the lateral ventricle is associated with dense-core vesicles (Aghajanian and Bloom, 1966). The radioactive noradrenaline is also distributed in different brain areas in about the same proportions as the endogenous amine, except in the corpus striatum where it is highly localized. As in the peripheral nervous system, the radioactive neurotransmitter was found to be metabolized in the central nervous system by O-methylation and deamination. Studies by Iversen and Glowinski (1966) also demonstrated that the ³H-noradrenaline could be used to study turnover in the rat brain, and various regions of the brain were found to have different turnover rates. The use of the turnover of noradrenaline in the measurement of nerve activity, rather than the endogenous levels, added another dimension to neurobiology. All of these observations indicated that ³H-noradrenaline could serve as a useful tool in studying the activity of brain adrenergic nerves.

After labeling brain noradrenergic neurons, Glowinski and I (1964) then examined the effect of psychoactive drugs on brain biogenic amines. We found that antidepressant tricyclic drugs block the reuptake of ³H-noradrenaline in noradrenergic-nerve terminals. This, together with the observations that monoamine-oxidase inhibitors have antidepressant action and that reserpine, a biogenic-amine depletor, sometimes causes depression, led to the formulation of the catecholamine hypothesis of depression (Schildkraut, 1965). We also found that amphetamine has many actions on catecholamines in the brain. It blocks uptake and causes release of the transmitter, inhibits monoamine oxidase, and diverts the metabolism of noradrenaline (Glowinski, Axelrod, and Iversen, 1966) from deamination to O-methylation. In addition, amphetamines block uptake into corpus striatum while the tricyclic antidepressants do not. These studies on brain catecholamines were followed by a great deal of research in many laboratories on the relationship between the biochemistry, pharmacology, and behavior of noradrenergic nerves in various brain areas.

## Studies on the Regulation of Catecholamine Biosynthetic Enzyme

The adrenergic nervous system can undergo rapid changes in activity; yet it maintains a constant level in the content of noradrenaline in nerves and of adrenaline in the adrenal medulla. In the past few years the complex mechanisms controlling adrenergic activity have been unraveled. One such mechanism, the induction of catecholamine biosynthetic enzyme, was uncovered in an unexpected manner. Hans Thoenen of Basel asked if it would be possible to spend a sabbatical year in my laboratory. He and Tranzer (1968) had just found that injected 6-hydroxydopamine selectively destroys sympathetic-nerve terminals. I told Thoenen that he would be most welcome to spend a year in our laboratory, especially if he brought some 6-hydroxydopamine with him.

To exploit the remarkably selective properties of this drug we examined the effect of 6-hydroxydopamine on the catecholamine biosynthetic enzymes, tyrosine hydroxylase and dopamine-$\beta$-hydroxylase, in various tissues. The extent to which these enzymes were diminished following destruction of sympathetic nerves would be a measure of their localization in these nerves. As expected, there was a complete disappearance of tyrosine hydroxylase in heart and other sympathetically innervated tissues after chemical denervation with 6-hydroxydopamine. To our surprise, however, tyrosine hydroxylase activity was elevated in the adrenal medulla (Mueller, Thoenen, and Axelrod, 1969). One possible explanation for this phenomenon was the capacity of 6-hydroxydopamine to lower blood pressure and thus cause an increase in sympathoadrenal activity; the increased formation of tyrosine hydroxylase would follow in response to prolonged sympathetic-nerve firing. This supposition was confirmed by an elevation of tyrosine hydroxylase activity in the adrenal gland and in sympathetic cell bodies and nerve terminals following administration of reserpine and the $\alpha$-adrenergic-blocking agent, phenoxybenzamine. Both of these compounds cause a heightened sympathetic-nerve activity, but by different mechanisms. Subsequent work showed that the increased sympathetic nerve activity results in the synthesis of new tyrosine hydroxylase molecules in nerve cell bodies and adrenal medullae. The induced tyrosine hydroxylase in the cell bodies is then transported down the axon to the nerve terminals. The induction of tyrosine hydroxylase was found to be a transsynaptic event. Similar results were obtained with another catecholamine biosynthetic enzyme, dopamine-$\beta$-hydroxylase.

Another regulatory mechanism for adrenaline formation was observed by asking the right questions rather than by serendipity. It had been recognized that the ratio of adrenaline to noradrenaline in the adrenal medulla was dependent upon the extent to which the adrenal cortex envelops the medulla (Coupland, 1965). In those species which have limited cortex the preponderant catecholamine in the medulla was noradrenaline, while in those species in which the adrenal medulla is completely surrounded by a cortex the methylated catecholamine, adrenaline, was preponderant. Dick Wurtman, a Research Associate, suggested an elegant experiment to establish the role of the adrenal cortex in the regulation of adrenaline formation: remove the pituitary and measure the effect on the adrenaline-forming enzyme, phenylethanolamine N-methyltransferase, in the adrenal medulla. The ablation of the pituitary caused an 80% decrease in the adrenaline-forming enzyme in the adrenal gland (Wurtman and Axelrod, 1966). The administration of either ACTH or the potent glucocorticoid dexamethasone increased the adrenaline-forming enzyme in hypophysectomized rats.

## Pineal Research

About fifteen years ago Lerner and coworkers (1958) isolated a potent amphibian skin-lightening agent from bovine pineals and identified it as 5-methoxy N-acetylserotonin (melatonin). The unusual localization of melatonin in the pineal and its chemical structure aroused my curiosity. At that time serotonin was very much implicated in abnormal brain function. The presence of a serotonin nucleus and a methoxy group prompted Herb Weissbach and me to look into the biosynthesis of melatonin. Weissbach and his colleagues had just elucidated the biosynthesis and metabolism of serotonin, and I had described an O-methylating enzyme, catechol-O-methyltransferase.

In the first experiment we examined the capacity of pineal extracts to O-methylate the two possible methyl acceptors, serotonin and N-acetylserotonin. We suspected that S-adenosylmethionine could serve as the methyl donor. To detect the enzymatically formed O-methylated indoleamine it was necessary to use [14]C-methyl-S-adenosylmethionine, which Donald Brown and I had just synthesized. Upon incubation of pineal extracts with the labeled methyl donor and the indoleamine, we found that N-acetylserotonin was easily O-methylated but serotonin was not (Axelrod and Weissbach, 1961). The O-methylated product of N-acetylserotonin was identified as melatonin, and the enzyme, hydroxyindole O-methyltransferase, was characterized. To our surprise we found that the enzyme was present only in the pineal gland of mammals. Soon thereafter Weissbach and I worked out the biosynthetic pathway of melatonin in the pineal as follows: tryptophan $\longrightarrow$ 5-hydroxytryptophan $\longrightarrow$ serotonin $\longrightarrow$ N-acetylserotonin $\longrightarrow$ melatonin.

At that time, 1961, the pineal gland was considered to have little physiological activity because it was calcified. The presence of an unusual enzyme that was highly localized in the pineal made it obvious to us that this organ should be studied further. Just about that time it was found that the pineal contains relatively large amounts of such biogenic amines as serotonin, noradrenaline, and histamine (Giarman and Day, 1958), and that it is heavily innervated by sympathetic-nerve terminals (Kappers, 1960). Thus the potential role of the pineal in neurobiology was beginning to evolve.

It had been found that environmental lighting inhibits the activity of the gonads, and this appeared to be mediated by pineal melatonin. This prompted a study with Dick Wurtman on how light and darkness affect the melatonin-forming enzyme. When rats were exposed to long periods of light, the activity of the melatonin-forming enzyme was suppressed (Wurtman, Axelrod, and Phillips, 1963). These findings raised the question of how information about environmental light reaches the pineal, which lies deep in the skull between the cerebral hemispheres. The most likely possibility was a neural route, and the sympathetic nerves innervating the pineal were a prime candidate. Denervation of sympathetic nerves to the pineal by removal of the superior cervical ganglia abolished the effects of light on the melatonin-forming enzyme (Wurtman, Axelrod, and Fischer, 1964), and with this experiment we realized that the pineal could serve as a useful model to study the interactions between nerves and responding cells.

The discovery of a circadian rhythm in the pineal by Quay (1963) initiated another facet in pineal research. How are biological rhythms generated, and what role does the nervous system play? My Research Associates and I found that the 24-hour rhythm of serotonin is controlled by the sympathetic nerve (Snyder et al., 1965) and that there are diurnal changes in the turnover of noradrenaline (Brownstein and Axelrod, 1974). The greater amounts of noradrenaline released from the sympathetic nerves during the night stimulate a $\beta$-adrenergic receptor on the pineal cell, which in turn induces the enzyme that metabolizes serotonin by N-acetylation (Brownstein, Holz, and Axelrod, 1973). Pineal serotonin thus falls at night when the acetylating enzyme is elevated and rises during the day when the enzyme is low. Diurnal changes in sympathetic-nerve activity appear to be generated by a biological clock arising in the suprachiasmatic nucleus in the hippocampus. Mechanisms generating circadian rhythms are one of the more intriguing problems in biology, and the pineal gland has provided many insights to this problem.

Another important phenomenon in neurobiology is that of denervation supersensitivity. Takeo Deguchi spent the past two years working with me and uncovered some new concepts concerning the development of supersensitivity and subsensitivity. The solution of this problem depended on devising a rapid and sensitive assay for measuring the pineal enzyme serotonin N-acetyltransferase. The enzyme is inducible by stimulation with noradrenaline of the $\beta$-adrenergic receptor on the pineal. Reducing the level of neurotransmitter by denervation, decentralization, or drug treatment results in a marked superinduction of N-acetyltransferase by noradrenaline (Deguchi and Axelrod, 1973). Subjecting the pineal gland to large amounts of catecholamine causes a reduced response in induction of the enzyme by noradrenaline. The problem of change in the response of cells is a relatively neglected area of research, and the pineal provides a productive model for further work.

## Conclusions

In retrospect, I find that productivity in research depends primarily on a proper laboratory environment, one that is open, stimulating, and critical. Ideas, no matter how bizarre, should be considered and discussed (if not too time-consuming) and subjected to experiment. The ability to devise specific, sensitive, and simple methods is crucial. I also have found that it takes almost as much time and effort to work on trivial problems as on important ones. Finally, it is of paramount importance to ask the right questions at the right time—and to have a little luck.

## References

Abel, J.J. (1899): Ueber den blutdruckerregenden Bestandtheil der Nebenniere, das Epinephrin. *Hoppe Seylers Z. Physiol. Chem.* 28:318–362.

Aghajanian, G.K., and Bloom, F.E. (1966): Electron microscopic autoradiography of rat hypothalamus after intraventricular $H^3$-norepinephrine. *Science* 156:402–403.

Armstrong, M.D., and McMillan, A. (1957): Identification of a major urinary metabolite of norepinephrine. *Fed. Proc.* 16:146.

Axelrod, J. (1953): Studies on sympathomimetic amines. I. The biotransformation and physiological disposition of l-ephedrine and l-norephedrine. *J. Pharmacol. Exp. Ther.* 109:62–73.

Axelrod, J. (1954): Studies on sympathomimetic amines. II. The biotransformation and physiological disposition of d-amphetamine, d-p-hydroxyamphetamine and d-methamphetamine. *J. Pharmacol. Exp. Ther.* 110:315–326.

Axelrod, J. (1955a): The enzymatic deamination of amphetamine (Benzedrine). *J. Biol. Chem.* 214:753–763.

Axelrod, J. (1955b): The enzymatic demethylation of ephedrine. *J. Pharmacol. Exp. Ther.* 114:430–438.

Axelrod, J. (1956a): The enzymatic N-demethylation of narcotic drugs. *J. Pharmacol. Exp. Ther.* 117:322–330.

Axelrod, J. (1956b): Possible mechanism of tolerance to narcotic drugs. *Science* 124:263–264.

Axelrod, J. (1957): O-Methylation of epinephrine and other catechols in vitro and in vivo. *Science* 126:400–401.

Axelrod, J. (1962a): Purification and properties of phenylethanolamine-N-methyl transferase. *J. Biol. Chem.* 237:1657–1660.

Axelrod, J. (1962b): The enzymatic N-methylation of serotonin and other amines. *J. Pharmacol. Exp. Ther.* 138:28–33.

Axelrod, J., Brady, R.O., Witkop, B., and Evarts, E.V. (1957): The distribution and metabolism of lysergic acid diethylamide. *Ann. NY Acad. Sci.* 66:435–444.

Axelrod, J., Gordon, E., Hertting, G., Kopin, I.J., and Potter, L.T. (1962): On the mechanism of tachyphylaxis to tyramine in the isolated rat heart. *Brit. J. Pharmacol.* 19:56–63.

Axelrod, J., Hertting, G., and Potter, L. (1962): Effect of drugs on the uptake and release of $H^3$-norepinephrine in the rat heart. *Nature* 194:297.

Axelrod, J., Schmid, R., and Hammaker, L. (1957): A biochemical lesion in congenital, non-obstructive, non-hemolytic jaundice. *Nature* 180:1426–1427.

Axelrod, J., Senoh, S., and Witkop, B. (1958): O-Methylation of catechol amines *in vivo*. *J. Biol. Chem.* 233:697–701.

Axelrod, J., and Tomchick, R. (1958): Enzymatic O-methylation of epinephrine and other catechols. *J. Biol. Chem.* 233:702–705.

Axelrod, J., Weil-Malherbe, H., and Tomchick, R. (1959): The physiological disposition of $H^3$-epinephrine and its metabolite metanephrine. *J. Pharmacol. Exp. Ther.* 127:251–256.

Axelrod, J., and Weissbach, H. (1961): Purification and properties of hydroxyindole-O-methyl transferase. *J. Biol. Chem.* 236:211–213.

Axelrod, J., Whitby, L.G., and Hertting, G. (1961): Effect of psychotropic drugs on the uptake of H³-norepinephrine by tissues. *Science* 133:383–384.

Barger, G., and Dale, H.H. (1910): Chemical structure and sympathomimetic action of amines. *J. Physiol.* 41:19–59.

Brodie, B.B., and Axelrod, J. (1948a): The estimation of acetanilide and its metabolic products, aniline, N-acetyl p-aminophenol and p-aminophenol (free and total conjugated) in biological fluids and tissues. *J. Pharmacol. Exp. Ther.* 94:22–28.

Brodie, B.B., and Axelrod, J. (1948b): The fate of acetanilide in man. *J. Pharmacol. Exp. Ther.* 94:29–38.

Brodie, B.B., and Axelrod, J. (1949): The fate of acetophenetidin (phenacetin) in man and methods for the estimation of acetophenetidin and its metabolites in biological material. *J. Pharmacol. Exp. Ther.* 97:58–67.

Brodie, B.B., and Axelrod, J. (1950a): The fate of aminopyrine (pyramidon) in man and methods for the estimation of aminopyrine and its metabolites in biological material. *J. Pharmacol. Exp. Ther.* 99:171–184.

Brodie, B.B., and Axelrod, J. (1950b): The fate of antipyrine in man. *J. Pharmacol. Exp. Ther.* 98:97–104.

Brown, D.D., Tomchick, R., and Axelrod, J. (1959): The distribution and properties of a histamine-methylating enzyme. *J. Biol. Chem.* 234:2948–2950.

Brownstein, M., and Axelrod, J. (1974): Pineal gland: 24-hour rhythm in norepinephrine turnover. *Science* 184:163–165.

Brownstein, M., Holz, R., and Axelrod, J. (1973): The regulation of pineal serotonin by a *beta* adrenergic receptor. *J. Pharmacol. Exp. Ther.* 186:109–113.

Chen, K.K., and Schmidt, C.F. (1930): Ephedrine and related substances. *Medicine (Baltimore)* 9:1–117.

Coupland, R.E. (1965): *The Natural History of the Chromaffin Cell.* London: Longmans.

Coyle, J.T., and Henry, D. (1973): Catecholamines in fetal and newborn rat brain. *J. Neurochem.* 21:61–67.

Dahlström, A., and Fuxe, K. (1964): Evidence for the existence of monoamine-containing neurons in the central nervous system. *Acta Physiol. Scand.* (Suppl). 232:1–55.

Deguchi, T., and Axelrod, J. (1973): Supersensitivity and subsensitivity of the β-adrenergic receptor in pineal gland regulated by catecholamine transmitter. *Proc. Natl. Acad. Sci. USA* 70:2411–2414.

Elliot, T.R. (1904): The action of adrenaline. *J. Physiol.* 31:xx–xxi.

Euler, U.S. von (1946): A specific sympathomimetic ergone in adrenergic nerve fibers (sympathin) and its relations to adrenaline and nor-adrenaline. *Acta Physiol. Scand.* 12:73–97.

Giarman, N.J., and Day, M. (1958): Presence of biogenic amines in the bovine pineal body. *Biochem. Pharmacol.* 1:235.

Glowinski, J., and Axelrod, J. (1964): Inhibition of uptake of tritiated noradrenaline in the intact rat brain by imipramine and structurally related compounds. *Nature* 204:1318–1319.

Glowinski, J., and Axelrod, J. (1966): Effects of drugs on the disposition of H³-norepinephrine in the rat brain. *Pharmacol. Rev.* 18:775–785.

Glowinski, J., Axelrod, J., and Iversen, L.L. (1966): Regional studies of catecholamines in the rat brain. IV. Effects of drugs on the disposition and metabolism of H³-norepinephrine and H³-dopamine. *J. Pharmacol. Exp. Ther.* 153:30–41.

Greisemer, E.C., Barsky, J., Dragstedt, C.A., Wells, J.A., and Zeller, E.A. (1953): Potentiating effects of iproniazid on the pharmacological action of sympathomimetic amines. *Proc. Soc. Exp. Biol. Med.* 84:699–701.

Hertting, G., and Axelrod, J. (1961): Fate of tritiated noradrenaline at the sympathetic nerve-endings. *Nature* 192:172–173.

Hertting, G., Potter, L.T., and Axelrod, J. (1962): Effect of decentralization and ganglionic blocking agents on the spontaneous release of H³-norepinephrine. *J. Pharmacol. Exp. Ther.* 136:289–292.

Hertting, G., Axelrod, J., Kopin, I.J., and Whitby, L.G. (1961): Lack of uptake of catecholamines after chronic denervation of sympathetic nerves. *Nature.* 189:66.

Hoffer, A., Osmond, H., and Smythies, J.R. (1954): Schizophrenia: A new approach. *J. Ment. Sci.* 100:29–45.

Iversen, L.L., and Glowinski, J. (1966): Regional studies of catecholamines in the rat brain. II. Rate of turnover of catecholamines in various brain regions. *J. Neurochem.* 13:671–682.

Kappers, J.A. (1960): The development, topographic relations and innervation of the epiphysis cerebri in the albino rat. *Z. Zellforsch. Mikrosk. Anat.* 52:163–215.

Langley, J.N. (1901): Observations on the physiological actions of extracts of the supra-renal bodies. *J. Physiol.* 27:237–256.

Lerner, A.B., Case, J.D., Takahashi, Y., Lee, T.H., and Mori, W. (1958): Isolation of melatonin, the pineal gland factor that lightens melanocytes. *J. Am. Chem. Soc.* 80:2587.

Loewi, O. (1921): Über humorale Übertragbarkeit der Herznervenwirkung. *Pfluegers Arch. Gesamte Physiol.* 189:239–242.

Molinoff, P.B., Landsberg, L., and Axelrod, J. (1969): An enzymatic assay for octopamine and other β-hydroxylated phenylethylamines. *J. Pharmacol. Exp. Ther.* 170:253–261.

Mueller, R.A., Thoenen, H., and Axelrod, J. (1969): Increase in tyrosine hydroxylase activity after reserpine administration. *J. Pharmacol. Exp. Ther.* 169:74–79.

Potter, L.T., and Axelrod, J. (1962): Intracellular localization of catecholamines in tissues of the rat. *Nature* 194:581–582.

Quay, W.B. (1963): Circadian rhythm in rat pineal serotonin and its modifications by estrous cycle and photoperiod. *Gen. Comp. Endocrinol.* 3:473–479.

Quinn, G.P., Axelrod, J., and Brodie, B.B. (1958): Species, strain and sex differences in metabolism of hexobarbitone, aminopyrine, antipyrine and aniline. *Biochem. Pharmacol.* 1:152–159.

Saavedra, J.M. (1974): Enzymatic isotopic assay for and presence of $\beta$-phenylethylamine in brain. *J. Neurochem.* 22:211–216.

Saavedra, J.M., and Axelrod, J. (1972): A specific and sensitive enzymatic assay for tryptamine in tissues. *J. Pharmacol. Exp. Ther.* 182:363–369.

Saavedra, J.M., and Axelrod, J. (1973): Demonstration and distribution of phenylethanolamine in brain and other tissues. *Proc. Natl. Acad. Sci. USA* 70:769–772.

Saavedra, J.M., Brownstein, M., and Axelrod, J. (1973): A specific and sensitive enzymatic-isotopic microassay for serotonin in tissues. *J. Pharmacol. Exp. Ther.* 186:508–515.

Schildkraut, J.J. (1965): The catecholamine hypothesis of affective disorders: A review of supporting evidence. *Am. J. Psychiat.* 122:509–522.

Schmid, R., Hammaker, L., Axelrod, J., and Maxwell, E. (1957): The enzymatic formation of bilirubin glucuronide. *Arch. Biochem. Biophys.* 70:285–288.

Snyder, S.H., Baldessarini, R.J., and Axelrod, J. (1966): A sensitive and specific enzymatic isotopic assay for tissue histamine. *J. Pharmacol. Exp. Ther.* 153:544–549.

Snyder, S.H., Zweig, M., Axelrod, J., and Fischer, J.E. (1965): Control of the circadian rhythm in serotonin content of the rat pineal gland. *Proc. Natl. Acad. Sci. USA* 53:301–305.

Strominger, J., Kalckar, H., Axelrod, J., and Maxwell, E. (1954): Enzymatic oxidation of uridine diphosphate glucose to uridine diphosphate glucuronic acid. *J. Am. Chem. Soc.* 76:6411–6412.

Thoenen, H., and Tranzer, J.P. (1968): Chemical sympathectomy by selective destruction of adrenergic nerve endings with 6-hydroxydopamine. *Naunyn-Schmiedebergs Arch. Pharmacol.* 261:271–288.

Weil-Malherbe, H., Axelrod, J., and Tomchick, R. (1959): Blood-brain barrier for adrenaline. *Science* 129:1226–1227.

Weinshilboum, R., and Axelrod, J. (1971): Serum dopamine-beta-hydroxylase activity. *Circ. Res.* 28:307–315.

Weinshilboum, R., Thoa, N.B., Johnson, D.G., Kopin, I.J., and Axelrod, J. (1971): Proportional release of norepinephrine and dopamine-$\beta$-hydroxylase from sympathetic nerves. *Science* 174:1349–1351.

Whitby, L.G., Axelrod, J., and Weil-Malherbe, H. (1961): The fate of $H^3$-norepinephrine in animals. *J. Pharmacol. Exp.* 132:193–201.

Whitby, L.G., Hertting, G., and Axelrod, J. (1960): Effect of cocaine on the disposition of noradrenaline labelled with tritium. *Nature* 187:604–605.

Wolfe, D.E., Potter, L.T., Richardson, K.C., and Axelrod, J. (1962): Localizing tritiated norepinephrine in sympathetic axons by electron microscopic autoradiography. *Science* 138:440–442.

Wurtman, R.J., and Axelrod, J. (1966): Control of enzymatic synthesis of adrenaline in the adrenal medulla by adrenal cortical steroids. *J. Biol. Chem.* 241:2301–2305.

Wurtman, R.J., Axelrod, J., and Fischer, J.E. (1964): Melatonin synthesis in the pineal gland: Effect of light mediated by the sympathetic nervous system. *Science* 143:1328–1330.

Wurtman, R.J., Axelrod, J., and Phillips, L.S. (1963): Melatonin synthesis in the pineal gland: control by light. *Science* 142:1071–1073.

Floyd E. Bloom (b. 1936, Minneapolis, Minnesota) has the distinction of being the youngest Associate of the Neurosciences Research Program. His work has focused on the fields of cellular neuropharmacology and neurocytochemistry. Dr. Bloom is chief of the Laboratory of Neuropharmacology, St. Elizabeths Hospital, Washington, D.C., and acting director of the Division of Special Mental Health Research Programs, NIMH.

# 12

## The Gains in Brain Are Mainly in the Stain

## Floyd E. Bloom

### Introduction

Surely Francis O. Schmitt cannot be seventy years old—his wisdom and perception require at least twice that many years while his actions and acuity are unequaled by men half that age. I consider it a great privilege to act as one of the representatives of the NRP Associates in this symposium honoring our chairman. Since my admission ticket to stroll along the paths of discovery in the neurosciences depended upon the relationship between my age and that of the other Associates, I feel quite unqualified to attempt to provide any historical perspectives on the development of the NRP and will leave that task to others. Even if the myopia of youth and inexperience blurs my personal views, I nevertheless feel competent to express my own esteem for Professor Schmitt based upon the inspiration and perception I have received from him and the NRP.

Poincaré wrote that science is a monument whose construction requires centuries, for which each of us must carry a stone (Poincaré, 1963). If we apply this philosophy to the neurosciences, the critical role Professor Schmitt has played in this constructive task can be seen as that of the intuitive developer who has wielded the power of his personality and his scientific reputation to plant the seeds, witness the growth, and guide the refinement of the ideas and tools to be employed by the subsequent generation of scientists in this evolving discipline. Admittedly, he has not been alone in recognizing the critical importance of advances in the neurosciences, but few others have matched his skill in properly organizing ventures and providing sufficient logistical support to turn a research frontier into a series of well-founded settlements.

Mercifully, insufficient time has passed to determine whether my own contributions to the growth of the monument of the neurosciences will be that of an aspiring bricklayer or merely a minor pebble-pusher. In the hope that the short paths of discovery which my colleagues and I have been allowed to pursue do not turn out to have been blind alleys, I shall reflect briefly on the evolution of our work. In order to go beyond that objective of somewhat dubious value, I shall try to provide support for my belief that the gains in brain are mainly in the stain.[1] Viewed from the bridge that spans my brief fledgling career in the neurosciences, this catchy phrase carries significant insights on a necessary strategy for current

---

1. This phrase was first related to me by Dr. Harvey J. Karten, who enfranchised me to use it for the Third Nicholas Giarman Memorial Lecture at Yale University, April 6, 1973. The originator of this paraphrasing from "My Fair Lady" lyrics, however, was Dr. Stanley Yolles, then director of the NIMH.

research problems in neuropharmacology as well as a possible means of expanding our work into a new approach to the problems of the brain. Viewed from the vantage points of the renowned scientists gathered together for the symposium, these remarks may well seem either facetious or presumptuous, but I assure them that the title is chosen with great seriousness, as I shall endeavor to demonstrate.

**Genesis**

Trying to recall my first step along the paths of the neurosciences was as hard as finding the proper sequence of words with which to describe it. In both cases the successful maneuver seemed to consist of a lot of wheel spinning, keeping busy while waiting for inspiration to clarify a train of thought long enough to illuminate the next several steps. In retrospect, it may seem as though many of my initial attempts at experimentation were simply busy work, but there are threads of thought I can retrace, which began with my days as a medical student at Washington University. My organic chemistry professor at Southern Methodist University, Harold Jeskey, had recommended that I attend Washington University, and, as he correctly predicted, the education I received there has stimulated a lifelong pursuit of scientific knowledge.

Here I was merely following in the footsteps of many graduates of this university, including F. O. Schmitt, who have played important roles in the development of the neurosciences. My research as a student was done with Gordon Schoepfle in the physiology laboratory that had once belonged to Joseph Erlanger, and that still contained the original wet chamber in which Erlanger and Gasser had studied the action potentials of nerves on their earliest oscilloscopes. The thread of interest that weaves my work in Saint Louis into that of the present time is a desire to understand the action of drugs on the nervous system and, in turn, to use the action of drugs as a tool with which to explore nervous tissue. At Washington University I combined the biophysical tools I had learned from Schoepfle with pharmacological interests developed in association with David MacDougal and Oliver Lowrey. These interests were allowed further development at the NIMH, where I was able to work with Nino Salmoiraghi in exploiting the technique of microiontophoresis, which Salmoiraghi had set up within a few months after the method was first applied to the central nervous system by Curtis and Eccles (1958).

In studies on the peripheral nervous system, especially those done in vitro, the access of drugs into the tissue is limited only slightly by the physical chemistry of the drug molecules and the enzymatic and diffusional barriers of the tissue. In the central nervous system it had been almost impossible to do drug studies with physiological methods in vitro, and the application of drugs to brain tissue in vivo had been quite problematic because the time scale of the events one could record from neurons with microelectrodes was so much shorter than the access time of parenterally injected drugs. Furthermore, delivering drugs in a general way, such as topically or intravascularly, could lead to many actions that would indirectly produce an alteration of the activity of the cells one was testing.

With microiontophoresis, however, these problems were greatly simplified since the drug ions were delivered by small electrical currents flowing through the channel of a microelectrode, which was assembled with other drug electrodes into a multiple-electrode array complete with a conventional recording microelectrode. In this way it was possible to deliver test substances into the immediate extracellular space of a neuron and to determine which test substances could alter the activity of the cell. Such activity testing is an essential step in determining which of the chemicals present in the brain can serve as the chemical messengers that nerve cells secrete onto one another at synaptic transmission sites. At that time only a few laboratories in the world were set up to do microiontophoresis, and each of these laboratories concentrated on those transmitter candidates that their own background and experience tended to favor. In our lab, thanks in part to a collaboration with Mimo Costa, who introduced Salmoiraghi and me to the intricate pharmacology and intense spirit of competition that permeated the central monoamines field, we concentrated on catecholamines and serotonin as transmitter candidates.

Although considerable data could be generated with this method of testing, the various labs that used the method came up with conflicting interpretations of which substances were most likely to be considered as "real" transmitter candidates. Part of the reason for the interpretive discrepancies were variations in the finer points of applying the methods, such as how the animals were prepared for the experiment, whether anesthesia or decerebration was used to eliminate the pain of the surgery, and how the control observations required to avoid the acceptance of artifacts (e.g., pH, electrical current, or indirect actions) as data were obtained. These variables were not too difficult to discern, and were widely discussed, if not universally controlled (cf. Salmoiraghi and Bloom, 1964).

However, I was more bothered by a variable for which there was no immediate control: Could the responses we observed in the testing of a class of neurons with a particular substance really be taken to indicate the existence of synaptic receptors for that substance on these neurons? It seemed to me that, unless we could establish that the neurons in question actually received such synapses, observing that the neurons would respond to the substance would mean little in physiologic terms.

The early assumption of iontophoresis—that sensitivity must imply synapses—was denied by the fact that amino acids such as gamma aminobutyrate and glutamate, respectively, inhibited or excited cells in every brain region with equal potency, while the "lesser" substances, such as acetylcholine and the monoamines, influenced only a few cells in each area, sometimes exciting them and at other times inhibiting them. The widespread actions of the amino acids were interpreted to mean that their actions were nonspecific, while the less widespread but relatively low-key actions initially observed with the monoamines were taken to mean they were unimportant. We were eventually able to show that norepinephrine has a reproducible inhibitory action on neurons when the responsive cells are anatomically and physiologically categorized and tests are restricted to a given class (Bloom et al., 1964; Bloom, Costa, and Salmoiraghi, 1964). But it was still impossible to assert that these responses were indicative of a noradrenergic synapse on the responsive cell population, even though a variety of pharmacological tests established a consistent correlation between the drug actions of norepinephrine iontophoretically applied to olfactory-bulb mitral cells and the same drug actions on the synaptically elicited recurrent antidromic inhibition of mitral cells (Salmoiraghi, Bloom, and Costa, 1964).

Although our results were partially disappointing in that none of the drugs—even the best antagonists of the mitral-cell responses to norepinephrine—could completely block the recurrent inhibition, the major stumbling block in my mind was our inability to demonstrate that the relatively small amounts of norepinephrine in the olfactory bulb were present in axons that terminated on mitral cells. It seemed to me that in order to get around this impasse, some histochemical method was needed to stain the synapses for their transmitter content and to relate these staining patterns at the electron-microscopic level to the responses of the cells to microiontophoretic testing.

At that time the idea that "the gain in brain" could lie "in a stain" was beginning to ripen in brain research. Dahlström and Fuxe (1964) had applied the fluorescence-histochemical method of Falck and Hillarp to the central nervous system and described at the light-microscopic level the distribution of the monoamine cell bodies and their axon terminals in a general way (Fuxe, 1965). The answer to my problem, in interpreting iontophoretic responses by determining the content of the terminals on the cells I wished to test, seemingly hinged on being able to extend such cytochemical methods to the electron-microscopic level.

In the mid-1960s electron-microscopic cytochemistry was focused mainly on the localization of enzymes rather than on molecules as small as monoamines. However, the situation was not as hopeless as it seemed since methods had been worked out that positively "stained" norepinephrine granules within adrenal medullary cells (Wood and Barrnett, 1964). Furthermore, it had been demonstrated that the most likely source of transmitter storage within nerve terminals were the synaptic vesicles (see de Robertis, 1966) and that the vesicles inside of peripheral sympathetic terminals could be "stained" by certain methods of fixation (Grillo and Palay, 1962; Richardson, 1962, 1963, 1964, 1966). The sympathetic nerve terminals could be further demonstrated by autoradiography after they had been exposed to small amounts of radioactive norepinephrine (Wolfe et al., 1962). I was, therefore, eager to take my primitive skills in the methods of electron microscopy (which I had picked up from a short training period with Keith Richardson at NINDS) to Yale with Russ Barrnett and attempt to develop methods of electron-microscopic cytochemistry of central synaptic transmitters.

The results of the studies that were initiated at Yale, and that were extended and refined through collaboration with George Aghajanian, Luke Van Orden, and Nick Giarman, can best be judged in the light of my recent reviews on the subject (Bloom, 1972a, 1973) and the extent to which the application of these methods (see below) has fulfilled the purpose for which they were pursued. To test the function of central noradrenergic synapses in a meaningful way we had to determine which neurons received such connections, and here the gains were almost exclusively in the stain.

## Evolution

### Neurohistochemistry with optical microscopes

Our present understanding of the location of norepinephrine-containing synapses depends primarily on the "stain" procedure developed by Falck and Hillarp (see Dahlström and Fuxe, 1964) in which freeze-dried brain is exposed to paraformaldehyde vapors. The formaldehyde couples with the monoamines of the tissue in situ and yields a highly fluorescent molecule with an emission peak characteristic of the monoamine. When this method was applied to the central nervous system, it was demonstrated that the norepinephrine neurons are concentrated into several discrete nuclear masses within the pons, medulla, and mesencephalon (Dahlström and Fuxe, 1964; Ungerstedt, 1971). Furthermore, it now appears from this method that the norepinephrine-containing nerve terminals within the cortical areas of the cerebrum, cerebellum, and hippocampus are derived from only one of these nuclei, the locus coeruleus (Olson and Fuxe, 1971; Ungerstedt, 1971; Hoffer et al., 1973; Pickel, Krebs, and Bloom, 1973).

The fibers that fluoresce for norepinephrine exhibit repeated enlargements, descriptively termed "varicosities," which are presumed to correlate with synaptic terminals and, inferentially, with sites of transmitter release. However, because of the limited resolution of the light microscope relative to the intricate network of neuronal cell processes in the brain, electron-microscopic studies are required to demonstrate which neurons receive synaptic contact from the norepinephrine-containing neurons.

### Neurohistochemistry with electron microscopes

Unfortunately, no single electron-microscopic histochemical procedure has as yet achieved the consistency and selectivity of "localization-by-staining" that is required to analyze synaptic transmission. Fixation of the tissue with permanganate salts (see Richardson, 1966; Hökfelt, 1968; Bloom, 1973) offers the most direct approach in that the primary fixation provides a selective "staining" for norepinephrine: synaptic vesicles within certain nerve terminals become electron-dense, revealing the major morphological details exhibited by the norepinephrine-containing axons of the sympathetic nerve system, namely, small granular vesicles (see Bloom, 1972a, 1973). However, this method had only been successfully applied to those brain regions, such as the hypothalamus, with the very highest concentration of norepinephrine, and the physiologic studies we wished to pursue seemed more likely to succeed in areas with substantially less tissue content. Therefore, alternatives and supplements to the permanganate-fixation staining approach were necessary. For our purposes, in the analysis of norepinephrine synaptic functions, we found two methods most useful. One was the autoradiographic location of nerve terminals that can accumulate $H^3$-norepinephrine (Glowinski and Iversen, 1966) after tracer doses have been injected into the cerebrospinal fluid bathing the surface of the cerebellum (Figure 12.1). The second method was based on the acute degeneration that follows the accumulation of the selective neurotoxin 6-hydroxydopamine (Malmfors and Thoenen, 1971; Bloom et al., 1969; Bloom, 1971).

### Localization of noradrenergic synapses in cerebellar cortex

The first area in which we attempted to correlate the electron-microscopic identification of norepinephrine nerve terminals with an electrophysiological analysis of the effects of norepinephrine applied by microiontophoresis was the cerebellar cortex (Hoffer, Siggins, and Bloom, 1969; Siggins, Hoffer, and Bloom, 1969). We chose this area, not because it was in any known way representative of the actions of central norepinephrine neurons, but because my colleagues, Barry Hoffer and George Siggins, felt that the cerebellar circuitry described by Eccles, Ito, Llinás, and Szentágothai (see Eccles, Ito, and Szentágothai, 1967) would make it easier for us to discriminate the physiological properties on which norepinephrine was able to produce an effect. When the light- and electron-microscopic methods for norepinephrine were applied to the cerebellar cortex, we saw that the norepinephrine-containing synaptic terminals made synaptic contact with the only output cell of the cerebellar cortex, the Purkinje cell (Figure 12.2). This correlated well with the responses observed in the initial microiontophoretic tests in that Purkinje cells responded reproducibly to norepinephrine and were routinely inhibited by it. When the staining methods indicated that Purkinje cells were the proper cell type to study, additional physiologic experiments indicated that norepinephrine slows Purkinje cells through interaction with a beta receptor, and that the slowing of

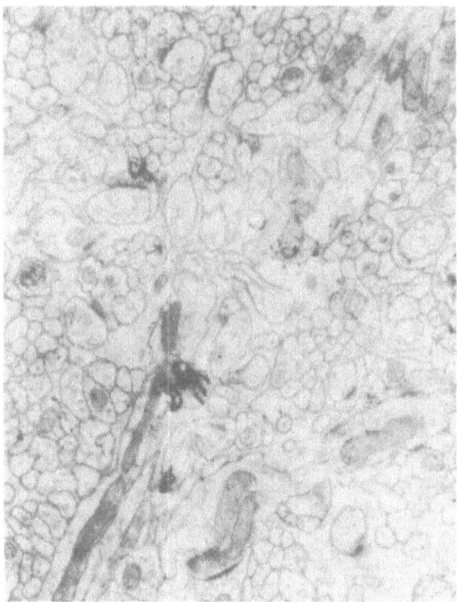

Figure 12.1 Electron-microscopic autoradiographic localization of H³-norepinephrine-containing axons and synaptic terminals in rat cerebellar cortex. Magnification = × 12,000. Modified from Bloom (1973).

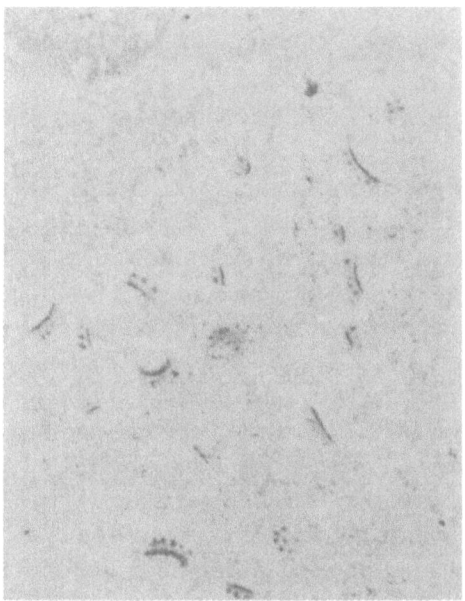

Figure 12.2   Cerebellar cortex stained with ethanolic phosphotungstic acid, showing electron density only in the synaptic contact points and chromatin, at upper left. Magnification = × 12,000.

the mean discharge rate is produced by prolonging pauses between bursts of single spikes without affecting climbing-fiber responses.

## Cyclic AMP and prostaglandin histochemistry

At this point we recognized that our choice of the cerebellar cortex as a point from which to study the central actions of norepinephrine was extremely fortunate, in that previous biochemical experiments (see Rall and Gilman, 1970; Rall, 1972) had demonstrated that this region of the brain was most rich in cyclic adenosine monophosphate (cyclic AMP). Cyclic AMP is the intracellular messenger that Sutherland and his colleagues had identified as the mediator of many norepinephrine actions in the sympathetic nervous system, especially at those receptors characterized pharmacologically as beta-adrenergic (see Rall, 1972). Furthermore, it was known that, of all the regions of the brain that had been challenged in vitro with neurotransmitters capable of stimulating cyclic AMP, the cerebellum exhibited the greatest incremental responses to norepinephrine. We were therefore quite captivated when microiontophoresis of cyclic AMP to Purkinje cells reproduced the actions of norepinephrine, both on discharge rates and patterns (Siggins, Hoffer, and Bloom, 1969), and on intracellular-membrane properties (Siggins, Hoffer, and Bloom, 1971a, b; Siggins et al., 1971). Furthermore, analogous to some of the norepinephrine-sensitive adenylate cyclase receptors of the sympathetic tissues, prostaglandins of the E series were able to block the responses of Purkinje cells to norepinephrine (Hoffer, Siggins, and Bloom, 1969). The actions of both norepinephrine and cyclic AMP on the Purkinje cells could be potentiated by microiontophoresis of drugs that blocked the catabolism of cyclic

AMP, i.e., the phosphodiesterase inhibitors such as methyl xanthines and papaverine (Siggins, Hoffer, and Bloom, 1969; Hoffer, Siggins, and Bloom, 1971; Hoffer et al., 1972).

At this point we again turned to cytochemical staining methods to test our hypothesis that the synaptic actions of norepinephrine in the cerebellar cortex could be mediated in Purkinje cells by the formation of cyclic AMP (Hoffer et al., 1972; Siggins, Hoffer, and Bloom, 1969, 1971a,b; Siggins et al., 1971). In all studies using the method of microiontophoresis, the question eventually arises as to whether the cells that are observed giving the response are actually those that directly respond to the drug ejected from the pipette, or whether the drug is really working at some electrophysiologically covert, but synaptically connected, presynaptic site. The electron-microscopic autoradiographs and the patterns of synaptic degeneration after 6-hydroxydopamine indicated that the norepinephrine synapses contacted the dendrites and cell body of Purkinje cells almost exclusively, and that these cells were therefore the physiologic target of the pathway (see Figures 12.3, 12.4). But, in proceeding to define the sites of action of the cyclic AMP, the phosphodiesterase inhibitors, and the prostaglandins, additional data were needed to eliminate the possibility that presynaptic actions underlie this pharmacology.

### Staining synapses with phosphotungstic acid

To gather such additional data we employed phosphotungstic acid, an electron stain first used by F. O. Schmitt (Hall, Jakus, and Schmitt, 1945) in his study of muscle fibrils, and later employed by E. G. Gray (1959) to add contrast to thin

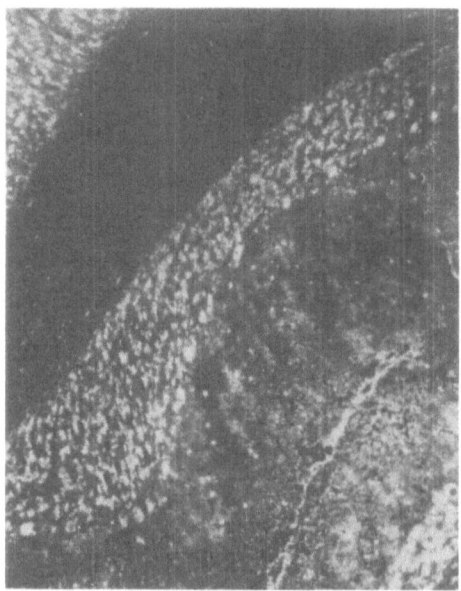

Figure 12.3 Immunocytochemical localization of cyclic AMP in the cerebellar cortex of rat-brain frozen biopsy material. Note the reactive staining in the granule and Purkinje cells, but no staining in the white matter. Magnification = × 600.

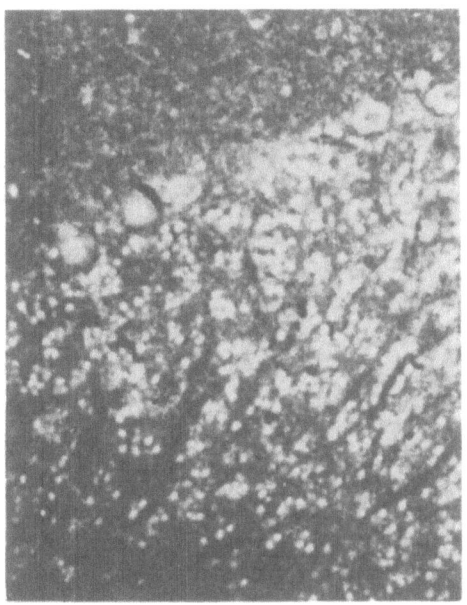

Figure 12.4   Higher magnification of a preparation similar to that shown in Figure 12.3. Staining for cyclic AMP can be seen within the cytoplasm, the nuclei of Purkinje cells, and the nuclei of granule cells. Magnification = × 5000.

sections of brain viewed by electron microscopy. Aghajanian and I had "rediscovered" this stain in our early attempts to enlarge the armamentarium of stains on synaptic chemicals (Bloom and Aghajanian, 1966, 1968a,b; Aghajanian and Bloom, 1968). We found that in the absence of other conventional treatments used to fix brain for electron microscopy, phosphotungstic acid (PTA) in absolute alcohol would selectively stain synapses. Furthermore, it stained them so selectively that the junctions could be readily identified in the electron microscope at low magnifications and the number of junctions quantified. This approach was used to study the time course of synaptogenesis in the rat cerebral cortex (Aghajanian and Bloom, 1967) and to complement our pharmacology of the cerebellar Purkinje cells.

When the PTA stain was applied to the region of the cerebellum in which our electrophysiological studies were performed, we found that no significant numbers of synaptic junctions were observed in our random thin sections until late on the third day after birth (Woodward et al., 1971), and this onset time correlated almost exactly with the onset of successful driving of Purkinje cells by surface activation of parallel fibers. Therefore, we reasoned that if the responses we were examining were direct, the pharmacological response patterns of the Purkinje cells to norepinephrine, cyclic AMP, prostaglandins, and phosphodiesterase inhibitors should be reproduced by the immature Purkinje cells, which had not yet received any synaptic connections. This is exactly what was observed, and these observations on the neonate, together with biochemical observations that rat cerebellar cortical adenylate cyclase was fully sensitive to norepinephrine (Schmidt

and Robison, 1971), strengthened the hypothesis that the synaptic actions of norepinephrine on the Purkinje cell could be mediated by formation of cyclic AMP.
In addition, a light-microscopic cytochemical method was used to demonstrate
that Purkinje cells exhibit the highest concentration in the brain of prostaglandin
dehydrogenase (Siggins, Hoffer, and Bloom, 1971a,b), again indicating that the
pharmacologic events we were studying on Purkinje cells were probably significant in the biology of these cells.

### Localizing the source of the noradrenergic synapses
In order to extend the significance of our work my colleagues and I realized that
it would be essential to determine which of the several nuclei of norepinephrine
neurons was the source of the axons whose synapses and postsynaptic receptors
we were analyzing in the cerebellum. We knew from the light- and electron-
microscopic studies that these axons belonged to neither of the two "classical"
afferent fiber systems to the cerebellum, i.e., they were neither climbing fibers
nor mossy fibers. Therefore, it was essential that we demonstrate the source of the
fibers so that the nucleus could be activated electrically and allow us to determine
the actions of the pathway that appeared to be mediated by norepinephrine.
Again the answer to the problem was provided by histochemistry. We applied
the fluorescence-histochemical methodology for catecholamines to young animals
subjected to hemicerebellectomy or to transections of the ascending projections of
the pontine norepinephrine nuclei. As initially described by Olson and Fuxe
(1971) and confirmed in our laboratory (Hoffer et al., 1973), the nuclei from which
the fibers could be traced to the cerebellum after these lesions were those of the
locus coeruleus. Furthermore, in unoperated adult animals, it has been possible
to employ autoradiography and the phenomenon of axoplasmic transport to
confirm the locus coeruleus–cerebellum Purkinje pathway. The locus coeruleus is
injected (stereotactically) with $H^3$-proline, and the labeled proteins formed from
this amino acid are traced into the nerve terminals (Cowan et al., 1972) in the
cerebellum (Segal, Pickel, and Bloom, 1973).

### Physiology of the noradrenergic projection to cerebellar Purkinje cells
With the anatomical source of the norepinephrine fibers to the cerebellum determined histochemically, the physiologic and pharmacologic effects of the pathway
could be compared to the previous results from microiontophoretic tests. These
comparisons disclosed that stimulation of the pathway inhibited Purkinje-cell
discharge, especially single-spike bursts, that the inhibitory effects of stimulating
the locus coeruleus required active synthesis of norepinephrine, and that no effects on cerebellar neuronal discharge were observed when the area of the locus
was stimulated in animals pretreated with 6-hydroxydopamine to eradicate the
adrenergic projection to the cerebellum. It was found, by intracellular recording
during the activation of the locus coeruleus, that Purkinje cells were hyperpolar-
ized, and this hyperpolarization was accompanied by a definite increase in the
resistance of the membrane. Pharmacologically, activation of the locus coeruleus
led to an inhibition of spontaneous discharge, which could be potentiated by local
iontophoresis of phosphodiesterase inhibitors onto the Purkinje cell, and could be

blocked by local iontophoretic administration of prostaglandins of the E series. All these results (Hoffer et al., 1973) supported the concept that this adrenergic projection could operate by producing transsynaptic elevation of cyclic AMP in Purkinje cells.

However, in order to probe fully the physiologic hypothesis involving cyclic AMP in the mediation of a synaptic response, we felt it necessary to demonstrate that norepinephrine and locus-coeruleus stimulation could, in fact, alter the cyclic AMP content of the Purkinje cell. To do so required development of yet another staining approach that would be specific and sensitive enough to detect intracellular cyclic AMP, since no biochemical method was likely to be successful at the cellular level. By applying an immunocytochemical method for cyclic AMP to cryostat sections (Wedner et al., 1972; Bloom et al., 1972), we observed that topical application of norepinephrine or electrical activation of the locus coeruleus elevated the number of Purkinje cells showing positive immunocytological staining for cyclic AMP, from resting frequencies of 5–15% to levels greater than 75% (see Siggins et al., 1973).

## Revelations and Tentative Conclusions

After this resumé of our last five years' work, it becomes painfully obvious that much of the gain in the analysis has been in the stain. It should be almost as obvious that determination of a discrete pathway of histochemically, physiologically, and pharmacologically defined chemical transmission can open relatively large vistas of research relevant to neuropsychopharmacology (Bloom, Hoffer, and Siggins, 1972) and to neurobiology.

We have, for example, examined the ontogeny of the locus coeruleus relative to the Purkinje cells and other postsynaptic macroneurons in the cortical regions to which the locus coeruleus projects (Bloom et al., 1974). The earlier "birth" of the locus coeruleus suggests that the norepinephrine system may participate in triggering differentiation in its postsynaptic target cells. We have examined the relationship between the firing patterns of norepinephrine-containing neurons of the cat locus coeruleus during unrestrained sleeping and waking behavior (Chu and Bloom, 1974). These data suggest that norepinephrine cells respond to, rather than trigger, shifts in cortical activity during sleep stages. We have repeated physiological and pharmacological studies similar to those on the norepinephrine projection to the cerebellum on the hippocampal pyramidal cells and have found that pyramidal cells respond almost exactly as described above for the cerebellar cortex (Segal and Bloom, 1973, 1974a,b). At the present time the staining and electrophysiological methods are being applied to the dramatic and unexpected plastic properties of partially transected central norepinephrine axons (Björklund et al., 1971; Katzman et al., 1971; Moore, Björklund, and Stenevi, 1971). These properties allow the axons to sprout in the adult and form increased numbers of axon terminals, as seen by fluorescence-histochemical (Pickel, Krebs, and Bloom, 1973) and autoradiographic indices (Segal, Pickel, and Bloom, 1973); they also show increased physiological effectiveness as measured by microelectrode studies (W. H. Krebs and F. E. Bloom, in preparation).

All of the methods applied above to the analysis of the norepinephrine synapses are capable of extension to the dopamine-containing neurons projecting to the caudate nucleus (B. J. Hoffer, G. R. Siggins, and U. Ungerstedt, in preparation), and to the serotonin projection systems as well (Bloom, Hoffer, and Siggins, 1972). The autoradiographic method of localization may also be able to reveal the distribution of amino-acid-containing terminals such as those that can accumulate gamma aminobutyrate or glycine (see Iversen and Schon, 1973).

The practical status of the problem, however, is perhaps not quite so rosy as the picture described above would suggest, when we look beyond the embellishment of the highly effective combination of histochemical stains and electrophysiological probes. While we can now begin to pursue the biological and behavioral implications of nerve circuits whose anatomy, pharmacology, and physiology are defined (in the case of catecholamines, serotonin, and certain amino-acid transmitters), the proportion of all synaptic systems in the brain that these chemicals mediate is unknown and is probably nowhere near a plurality of the chemical systems at large. Having studied almost all of the chemical systems in the mammalian central nervous system for which mammalian autonomic or invertebrate paradigms of physiology and pharmacology have been described (see Bloom, 1972b), we must begin to question the methods by which we might discover new transmitters when the chemical nature of the molecules to be sought is totally unknown. For example, if fluorescence-histochemical methods had not revealed the cellular location of the monoamine cells, it is unlikely that any of the cytochemical methods available today, including sophisticated energy-catabolism profiles (Sims et al., 1974), would have hinted at the nature of the transmitter synthesized by neurons of the locus coeruleus. It is to be hoped that an innovative screening method at the cytochemical level can be devised without additional delay, but no such success appears imminent.

### Significance of cyclic-nucleotide-mediated synaptic messages

While waiting for this momentous millenium of cytochemical tactics to arrive, I should like to offer a few terminal thoughts on the potential significance of the mediation of synaptic responses by cyclic AMP. In the cases of the cerebellar, hippocampal, and caudate neurons (mentioned above), the released catecholamine can initiate a cascade of effects by which the synaptic actions of the circuit are amplified. The neurotransmitter activates the adenylate cyclase to form the cyclic nucleotide, and the cyclic nucleotide can in turn activate the enzymes, known as protein kinases, which can phosphorylate a variety of endogenous brain proteins (Greengard, Kebabian, and McAfee, 1973). This capacity for amplification of the chemical mediators of the synaptic message may account for the marked potency of the activated norepinephrine pathway to the cerebellum (Hoffer et al., 1972, 1973) despite an apparent paucity of fibers synapsing on the Purkinje cells.

Two concepts flow directly from this cascade of actions. First, phosphorylation of a synaptic membrane protein (Johnson et al., 1972) might be expected to alter membrane permeability. It is possible that the increased membrane polarization and the decreased membrane permeability generated by the activation of adenylate cyclase are only epiphenomena of the process by which cyclic AMP concentration is increased. It seems worth considering, therefore, that the electrical effects of such synapses may be less important over their short-term interactions (i.e., with other synaptic potentials being generated simultaneously) than longer-term trophic effects on the properties of the postsynaptic cell (see Bloom, 1974). Second, the immunocytochemical staining pattern of Purkinje cells for cyclic AMP (see Bloom et al., 1972; Bloom, Hoffer, and Siggins, 1972; Siggins et al., 1973), indicates that the nucleotide concentration can be increased in the nucleus as well as in the cytoplasm, and this intracellular redistribution may correlate with the activation of protein kinases, which can phosphorylate nuclear proteins (Langan, 1970). Phosphorylation of nuclear histones could be expected to alter the extent to which certain genetic properties are being expressed, and this constitutes a mechanism by which a discrete synaptic input might eventually influence the entire biology of a neuron rather than merely the relative polarization levels of the plasma membrane. It is my hope that future developments along this line of attack will make the present gains in brain through stain seem plain.

## References

Aghajanian, G.K., and Bloom, F.E. (1967): The formation of synaptic junctions in developing rat brain: A quantitative electron microscopic study. *Brain Res.* 6:716–727.

Aghajanian, G.K., and Bloom, F.E. (1968): An osmiophilic substance in brain synaptic vesicles not associated with catecholamine content. *Experientia* 24:1225–1227.

Björklund, A., Katzman, R., Stenevi, U., and West, K.A. (1971): Development and growth of axonal sprouts from noradrenaline and 5-hydroxytryptamine neurones in the rat spinal cord. *Brain Res.* 31:21–33.

Bloom, F.E. (1971): Fine structural changes in rat brain after intracisternal injection of 6-hydroxydopamine. *In: 6-Hydroxydopamine and Catecholamine Neurons.* Malmfors, T., and Thoenen, H., eds. Amsterdam: North-Holland, pp. 135–150.

Bloom, F.E. (1972a): Localization of neurotransmitters by electron microscopy. *Res. Publ. Assoc. Res. Nerv. Ment. Dis.* 50:25–57.

Bloom, F.E. (1972b): Amino acids and polypeptides in neuronal function. *Neurosci. Res. Program Bull.* 10:122–251.

Bloom, F.E. (1973): Ultrastructural identification of catecholamine-containing central synaptic terminals. *J. Histochem. Cytochem.* 21:333–348.

Bloom, F.E. (1974): Dynamics of synaptic modulation: Perspectives for the future. *In: The Neurosciences: Third Study Program.* Schmitt, F.O., and Worden, F.G., eds. Cambridge, Mass.: The MIT Press, pp. 989–999.

Bloom, F.E., and Aghajanian, G.K. (1966): Cytochemistry of synapses: Selective staining for electron microscopy. *Science* 154:1575–1577.

Bloom, F.E., and Aghajanian, G.K. (1968a): An electron microscopic analysis of large granular synaptic vesicles of the brain in relation to monoamine content. *J. Pharmacol. Exp. Therap.* 159: 261–273.

Bloom, F.E., and Aghajanian, G.K. (1968b): Fine structural and cytochemical analysis of the staining of synaptic junctions with phosphotungstic acid. *J. Ultr. Res.* 22:361–375.

Bloom, F.E., Algeri, S., Groppetti, A., Revuelta, A., and Costa, E. (1969): Lesions of central norepinephrine terminals with 6-OH-dopamine. Biochemistry and fine structure. *Science* 166: 1284–1286.

Bloom, F.E., von Baumgarten, R., Oliver, A.P., Costa, E., and Salmoiraghi, G.C. (1964): Micro-electrophoretic studies on adrenergic mechanisms of rabbit olfactory neurons. *Life Sci.* 3:131–136.

Bloom, F.E., Costa, E., and Salmoiraghi, G.C. (1964): Analysis of individual rabbit olfactory bulb neuron responses to micro-electrophoresis of acetylcholine, norepinephrine and serotonin synergists and antagonists. *J. Pharmacol. Exp. Therap.* 146:16–23.

Bloom, F.E., Hoffer, B.J., Battenberg, E.F., Siggins, G.R., Steiner, A.L., Parker, C.W., and Wedner, H.J. (1972): Adenosine 3′-,5′-monophosphate is localized in cerebellar neurons: Immuno-fluorescence evidence. *Science* 177:436–438.

Bloom, F.E., Hoffer, B.J., and Siggins, G.R. (1972): Norepinephrine mediated cerebellar synapses: A model system for neuropsychopharmacology. *Biol. Psychiatry* 4:157–177.

Bloom, F.E., Krebs, H., Nicholson, J., and Pickel, V. (1974): The noradrenergic innervation of cerebellar Purkinje cells: Localization, function, synaptogenesis, and axonal sprouting of locus coeruleus. *In: Wenner-Gren Symposium in Dynamics of Degeneration and Growth in Neurons* (Symposium held in Stockholm, 16–18 May). Fuxe, K., ed. New York: Pergamon Press.

Chu, N.-S., and Bloom, F.E. (1974): The catecholamine-containing neurons in the cat dorso-lateral pontine tegmentum: Distribution of the cell bodies and some axonal projections. *Brain Res.* 66:1–21.

Cowan, W.M., Gottlieb, D.I., Hendrickson, A.E., Price, J.L., and Woolsey, T.A. (1972): The autoradiographic demonstration of axonal connections in the central nervous system. *Brain Res.* 37:21–51.

Curtis, D.R., and Eccles, R.M. (1958): The excitation of Renshaw cells by pharmacological agents applied electrophoretically. *J. Physiol.* 141:435–445.

Dahlström, A., and Fuxe, K. (1964): Evidence for the existence of monoamine-containing neurons in the central nervous system. I .Demonstration of monoamines in the cell bodies of brain stem neurons. *Acta Physiol. Scand.* 62 (Suppl. 232): 1–55.

De Robertis, E. (1966): Adrenergic endings and vesicles isolated from brain. *Pharmacol. Rev.* 18: 413–424.

Eccles, J.C., Ito, M., and Szentágothai, J. (1967): *The Cerebellum as a Neuronal Machine*. New York: Springer-Verlag.

Fuxe, K. (1965): Evidence for the existence of monoamine neurons in the central nervous system. III. The monoamine nerve terminal. *Z. Zellforsch. Mikrosk. Anat.* 65:573–596.

Glowinski, J., and Iversen, L.L. (1966): Regional studies of catecholamines in the rat brain. I. The disposition of [$^3$H] dopamine and [$^3$H] dopa in various regions of the brain. *J. Neurochem.* 13:655–669.

Gray, E.G. (1959): Axo-somatic and axo-dendritic synapses of the cerebral cortex: An electron microscope study. *J. Anat.* 93:420–433.

Greengard, P., Kebabian, J.W., and McAfee, D.A. (1973): Studies on the role of cyclic AMP in neural function. *In: Proceedings of the Fifth International Congress on Pharmacology.* Vol. 5. Acheson, G.H., and Bloom, F.E., eds. Basel: S. Karger, pp. 207–217.

Grillo, M.A., and Palay, S.L. (1962): Granule-containing vesicles in the autonomic nervous system. *In: Proceedings of the Fifth International Congress for Electron Microscopy.* Vol. 2. Breese, S.S., Jr., ed. New York: Academic Press.

Hall, C.E., Jakus, M.A., and Schmitt, F.O. (1945): The structure of certain muscle fibrils as revealed by the use of electron stains. *J. Appl. Phys.* 16:459–465.

Hoffer, B.J., Siggins, G.R., and Bloom, F.E. (1969): Prostaglandins $E_1$ and $E_2$ antagonize norepinephrine effects on cerebellar Purkinje cells: Microelectrophoretic study. *Science* 166:1418–1420.

Hoffer, B.J., Siggins, G.R., and Bloom, F.E. (1971): Studies on norepinephrine-containing afferents to Purkinje cells of rat cerebellum: II. Sensitivity of Purkinje cells to norepinephrine and related substances administered by microiontophoresis. *Brain Res.* 25:523–534.

Hoffer, B.J., Siggins, G.R., Oliver, A.P., and Bloom, F.E. (1972): Cyclic AMP-mediated adrenergic synapses to cerebellar Purkinje cells. *Adv. Cyclic Nucleotide Res.* 1:411–423.

Hoffer, B.J., Siggins, G.R., Oliver, A.P., and Bloom, F.E. (1973): Activation of the pathway from locus coeruleus to rat cerebellar Purkinje neurons: Pharmacological evidence of noradrenergic central inhibition. *J. Pharmacol. Exp. Ther.* 184:553–569.

Hökfelt, T. (1968): *In vitro* studies on central and peripheral monoamine neurons at the ultrastructural level. *Z. Zellforsch. Mikrosk. Anat.* 91:1–74.

Iversen, L.L., and Schon, F.E. (1973): The use of autoradiographic techniques for the identification and mapping of transmitter-specific neurones in CNS. *In: New Concepts in Neurotransmitter Regulation.* Mandell, A.J., ed. New York: Plenum Press, pp. 153–194.

Johnson, E.M., Ueda, T., Maeno, H., and Greengard, P. (1972): Adenosine 3', 5'-monophosphate-dependent phosphorylation of a specific protein in synaptic membrane fractions from rat cerebrum. *J. Biol. Chem.* 247:5650–5652.

Katzman, R., Björklund, A., Owman, Ch., Stenevi, U., and West, K.A. (1971): Evidence for regenerative axon sprouting of central catecholamine neurons in rat mesencephalon following electrolytic lesions. *Brain Res.* 25:579–596.

Langan, T. (1970): Phosphorylation of histones in vivo under the control of cyclic AMP and hormones. *In: Role of Cyclic AMP in Cell Function.* Greengard, P., and Costa, E., eds. New York: Raven Press, pp. 307–324.

Malmfors, T., and Thoenen, H., eds. (1971): *6-Hydroxydopamine and Catecholamine Neurons.* Amsterdam: North-Holland.

Moore, R.Y., Björklund, A., and Stenevi, U. (1971): Plastic changes in the adrenergic innervation of the rat septal area in response to denervation. *Brain Res.* 33:13–35.

Olson, L., and Fuxe, K. (1971): On the projections from the locus coeruleus noradrenaline neurons: The cerebellar innervation. *Brain Res.* 28:165–171.

Pickel, V.M., Krebs, H., and Bloom, F.E. (1973): Proliferation of norepinephrine-containing axons in rat cerebellar cortex after peduncle lesions. *Brain Res.* 59:169–179.

Poincaré, H. (1963): *Mathematics and Science: Last Essays.* Bolduc, J.W., trans. Reprint of 1913 edition. New York: Dover.

Rall, T.W. (1972): Role of adenosine 3′, 5′-monophosphate (cyclic AMP) in actions of catecholamines. *Pharmacol. Rev.* 24:399–409.

Rall, T.W., and Gilman, A.G. (1970): The role of cyclic AMP in the nervous system. *Neurosci. Res. Program Bull.* 8:221–323. Also in *Neurosciences Research Symposium Summaries,* Vol. 5. Schmitt, F.O., et al., eds. Cambridge, Mass.: The MIT Press, pp. 215–311.

Richardson, K.C. (1962): The fine structure of autonomic nerve endings of the rat vas deferens. *J. Anat.* 96:427–442.

Richardson, K.C. (1963): The fine structure of tissues following isolation in oxygenated saline for prolonged periods. *Anat. Rec.* 145:275.

Richardson, K.C. (1964): The fine structure of the albino rabbit iris with special reference to the identification of adrenergic and cholinergic nerves and nerve endings in its intrinsic muscles. *Am. J. Anat.* 114:173–205.

Richardson, K.C. (1966): Electron microscopic identification of autonomic nerve endings. *Nature* 210:756.

Salmoiraghi, G.C., and Bloom, F.E. (1964): Pharmacology of individual neurons. *Science* 144:493–499.

Salmoiraghi, G.C., Bloom, F.E., and Costa, E. (1964): Adrenergic mechanisms in rabbit olfactory bulb. *Am. J. Physiol.* 207:1417–1424.

Schmidt, M.J., and Robison, G.A. (1971): The effect of norepinephrine on cyclic AMP levels in discrete regions of developing rabbit brain. *Life Sci.* 10:459–464.

Segal, M., and Bloom, F.E. (1973): A projection of the nucleus locus coeruleus to the hippocampus of the rat. *In: Abstracts.* Third Annual Meeting, Society for Neuroscience, San Diego, Calif., p. 371.

Segal, M., and Bloom, F.E. (1974a): The action of norepinephrine in the rat hippocampus. I. Iontophoretic studies. *Brain Res.* 72:79–97.

Segal, M., and Bloom, F.E. (1974b): The action of norepinephrine in the rat hippocampus. II. Activation of the input pathway. *Brain Res.* 72:99–114.

Segal, M., Pickel, V.M., and Bloom, F.E. (1973): The projections of the nucleus locus coeruleus: An autoradiographic study. *Life Sci.* 13:817–821.

Siggins, G.R., Battenberg, E.F., Hoffer, B.J., Bloom, F.E., and Steiner, A.L. (1973): Noradrenergic stimulation of cyclic adenosine monophosphate in rat Purkinje neurons: An immunocytochemical study. *Science* 179:585–588.

Siggins, G.R., Hoffer, B.J., and Bloom, F.E. (1969): Cyclic adenosine monophosphate: Possible mediator for norepinephrine effects on cerebellar Purkinje cells. *Science* 165:1018–1020.

Siggins, G.R., Hoffer, B.J., and Bloom, F.E. (1971a): Studies on norepinephrine-containing afferents to Purkinje cells of rat cerebellum. III. Evidence for mediation of norepinephrine effects by cyclic 3′,5′-adenosine monophosphate. *Brain Res.* 25:535–553.

Siggins, G.R., Hoffer, B.J., and Bloom, F.E. (1971b): Prostaglandin-norepinephrine interactions in brain: Microelectrophoretic and histochemical correlates. *Ann. NY Acad. Sci.* 180:302–323.

Siggins, G.R., Oliver, A.P., Hoffer, B.J., and Bloom, F.E. (1971): Cyclic adenosine monophosphate and norepinephrine: Effects on transmembrane properties of cerebellar Purkinje cells. *Science* 171:192–194.

Sims, K.L., Kauffman, F.C., Johnson, E.C., and Pickel, V.M. (1974): Cytochemical localization of brain nicotinamide adenine denucleotide phosphate (oxidized)-dependent dehydrogenases: Qualitative and quantitative distributions. *J. Histochem. Cytochem.* 22:7–19.

Ungerstedt, U. (1971): Stereotaxic mapping of the monoamine pathways in the rat brain. *Acta Physiol. Scand.* (Suppl.) 367:1–48.

Wedner, H.J., Hoffer, B.J., Battenberg, E., Steiner, A.L., Parker, C.W., and Bloom, F.E. (1972): A method for detecting intracellular cyclic adenosine monophosphate by immunofluorescence. *J. Histochem. Cytochem.* 20:293–295.

Wolfe, D.E., Potter, L.T., Richardson, K.C., and Axelrod, J. (1962): Localizing tritiated norepinephrine in sympathetic axons by electron microscopic autoradiography. *Science* 138:440–442.

Wood, J.G., and Barrnett, R.J. (1964): Histochemical demonstration of norepinephrine at a fine structural level. *J. Histochem. Cytochem.* 12:197–209.

Woodward, D.J., Hoffer, B.J., Siggins, G.R., and Bloom, F.E. (1971): The ontogenetic development of synaptic junctions, synaptic activation and responsiveness to neurotransmitter substances in rat cerebellar Purkinje cells. *Brain Res.* 34:73–79.

# Adventures with Unexpected Observations

Berta Scharrer (b. 1906, Munich, Germany) is professor of anatomy at Albert Einstein College of Medicine. Her research, primarily with invertebrates, has dealt with neurosecretory systems, hormonal and nervous system interactions, and the ultrastructure of endocrine and neural tissues. She collaborated with her husband Ernst in the discovery of the neurosecretory cell.

# 13

# The Concept of Neurosecretion and Its Place in Neurobiology

## Berta Scharrer

The leitmotiv of this retrospective survey of a path of discovery is the phenomenon of neurosecretion. The gradual evolution of our current views on its place in neuroendocrinology and in the neurosciences will be discussed not in terms of one research career but of two, a husband-and-wife team that ended, after thirty years, with the death of Ernst Scharrer in 1965. As happens so often, the foundations for this investigative work were laid during student days, and its course was shaped by many events unforeseen at the time.

Ernst Scharrer, while working for his Ph.D. degree, made his discovery that certain hypothalamic neurons specialize in secretory activity to a degree comparable to that of endocrine gland cells. In 1928, when these results were published, I was a student in the same institute, the Department of Zoology of the University of Munich, headed by Karl von Frisch.

In recapturing the excitement of those early years, one must recall that German students, after their graduation from the very strict gymnasium, suddenly found themselves in an atmosphere of almost complete academic freedom. I remember my rising fascination with this new world of the intellect opening up before me, the spell cast by an outstanding group of academic teachers and investigators, among them Heinrich Wieland (chemistry) and Karl von Goebel (botany).

High standards of scholarship prevailed in the science faculty at that time, and the laboratory in which we undertook our first steps in biological research was no exception. Richard Hertwig was still at work at the old Zoological Institute, housed in a converted seventeenth-century monastery, our beloved "Alte Akademie," and, at our request, took part in our doctoral examinations. Sensory physiology was the keynote of the departmental research program, and the honey bee was the experimental animal that yielded the most spectacular results and brought von Frisch worldwide acclaim. We all came to appreciate the lucidity of his writing and the wide appeal of his lectures, but perhaps the most decisive imprint we received was that of the heuristic value of a broad comparative approach.

Those were happy, almost carefree days. Our love for and commitment to scientific research were firmly established by the time we received the "Doctor Hut."

But what was in store for us next? In the early thirties prospects for an academic career in Germany were bleak and, for a woman, virtually nonexistent. Ernst decided to acquire an additional degree, in medicine, and I to obtain a certificate for teaching in a German gymnasium. During those years we managed to keep laboratory fires burning, both of us having found working space at the Research Institute of Psychiatry in Munich, then under the inspired directorship of Walther Spielmeyer.

After our marriage in 1934, following Ernst's research appointment at the Edinger Institute of Neurology in Frankfurt am Main, the road seemed open for a joint effort to probe into the role of neurosecretion (Figure 13.1). We decided to divide the animal kingdom; Ernst would continue his studies on vertebrates, and I would set out to search for comparable phenomena among invertebrates. Several sojourns at the Stazione Zoologica, Naples, and a collecting trip around Africa yielded a wealth of material, as did the fauna of the Marine Biological Laboratory, Woods Hole, Massachusetts, at a later time.

Newcomers to the academic scene of the medical faculty of Frankfurt, we were warmly received and encouraged in our work by Albrecht Bethe, then codirector of the Neurological Institute, a man of boundless curiosity and youthful enthusiasm for new ideas. Other contacts that developed into lifelong friendships were with Tilly Edinger, daughter of Ludwig and a renowned palaeoneurologist, and with Wolfgang Bargmann who, in later years, was to contribute so much to the solution of the problem of neurosecretion.

The institute was well equipped and its library contained many treasures, among them a complete set of Ramón y Cajal's "Trabajos" and a monograph on the histology of the nervous system by Fridtjof Nansen. Work progressed beautifully in this deceptively sheltered milieu, but the political climate in Germany was becoming increasingly intolerable, and the outlook was grim.

When Ernst was granted a Rockefeller Fellowship in 1937, we set out for the University of Chicago in high spirits and, having had to leave behind all our scientific material, prepared for a new start. His sponsor was C. Judson Herrick, Ludwig Edinger's American counterpart in comparative neurology and also an early proponent of the concept of a structure-function relationship. We both benefited greatly from our contacts with him as with other members of the anatomy department (George Bartelmez, William Bloom, Robert Bensley, David Bodian), and of the zoology department next door (Paul Weiss, Carl Moore).

Then followed two years at the Rockefeller Institute (now Rockefeller University) in New York, under an arrangement made by Charles Stockard (Cornell University Medical School) who had taken a great interest in neurosecretion. Unfortunately, his untimely death cut short a collaborative study envisioned to

Figure 13.1   Ernst and Berta Scharrer preparing serial (celloidin) sections of hypothalamus at the Edinger Institute of Neurology, Frankfurt am Main, Germany, 1935.

become part of his overall program with purebred dogs. Subsequent moves took us to Ernst's first teaching position at Western Reserve University, Cleveland, Ohio, and then, in 1946, to the University of Colorado Medical School at Denver.

In 1955 we again shipped our growing collection of slides, books, and reprints, this time to the newly opening Albert Einstein College of Medicine, New York, where Ernst had accepted the chairmanship of the Department of Anatomy (Figures 13.2 and 13.3). It was here, no longer subject to the rule of nepotism, that I received my first regular academic appointment. This and other manifestations of the pioneering spirit of this fledgling institution provided us with a tremendous impetus.

Yet conditions that might appear to have been restrictive for me during the two preceding decades had their positive side. It was a privileged existence; free from official duties and other pressures, and much encouraged by my husband, I was allowed to develop and pursue my research program. The fact that much of it was carried out on a lowly laboratory animal, the cockroach, a choice originally dictated by the limited facilities available to a "guest investigator," likewise turned into an asset.

It is tempting to speculate on the variety of factors determining the gradual elucidation of the phenomenon of neurosecretion, which, in spite of various side excursions, remained our central interest throughout the years. The road was long and arduous, and for many years rather lonely. There was much uncertainty, but never any real doubt about the final outcome. In retrospect, there is every reason to be satisfied with the course of events that brought triumph during Ernst's lifetime.

It is quite remarkable that, from the very beginning, the sights set for this course of investigation had pointed in the right direction. The two initial propositions made by Ernst Scharrer (1928) in his first study on a teleost fish, *Phoxinus laevis*, turned out to be correct, i.e., the endocrine nature of special hypothalamic neurons, and their relationship with hypophysial function. This was a bold concept that did not fit into any existing mold, and it is not surprising that it was received with skepticism. Why should members of a class of cells as readily defined as neurons be capable of functioning as glands of internal secretion?

What is less understandable, however, is the almost universal rejection by the

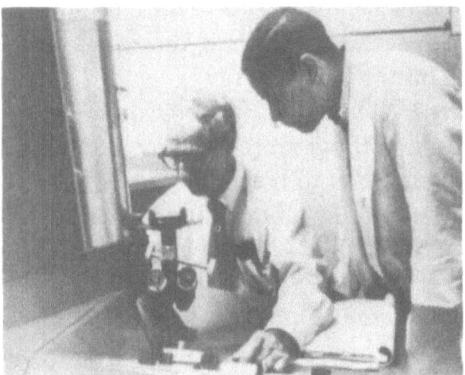

Figure 13.2    Ernst Scharrer and Stanley Brown at the electron microscope at the Albert Einstein College of Medicine, New York, 1960.

Figure 13.3 Ernst Scharrer, 1964.

scientific community of the validity of cytological evidence for the existence of a secretory process. I vividly remember Ernst's disappointment after an eagerly anticipated discussion with Professor Ranson in 1937 that was intended to correlate and interpret their respective results. As stated by another prestigious investigator, H.B. van Dyke (1936–1939): "The evidence that such cells secrete colloid and are to be considered a 'diencephalic gland' is morphological and does not deserve acceptance at this time."

Clearly, a more convincing approach, such as the search for the functional role of neurosecretory centers by classical endocrinological methods, was now called for. But, for a number of reasons, early attempts in this direction were unrewarding. Would this work have progressed more satisfactorily had the resources of Stockard's program not been ended through his untimely death?

One of the recurrent criticisms, according to which the cytological characteristics of neurosecretory cells were judged to be nothing more than manifestations of postmortem changes or pathological processes, had to be countered by demonstrating the very wide occurrence of such neurons in the animal kingdom. A search in the literature had yielded information on "glandlike" nerve cells in the spinal cord of skates, described and correctly interpreted by Speidel (1919), and on comparable cellular elements in various other animal phyla, reported by Hanström (1940, 1941). Ernst's studies soon encompassed representatives from all classes of vertebrates, and my own early contributions among invertebrates ranged from opisthobranch snails and annelids to arthropods. This broadly based search for cytophysiological correlates revealed a remarkable degree of analogy (Scharrer and Scharrer, 1944; E. Scharrer, 1956), and presented us not only with a wide choice of experimental animals, but with insights that could not have been obtained from mammalian material alone.

It is a matter of record that the first evidence for neurohormonal activities was the result of studies in invertebrates. As early as 1917, transplantation experiments in caterpillars had led the Polish biologist Kopeć to the conclusion that their brains furnish a "pupation hormone." Many years had to pass before the localization of this endocrine factor in implants of the dorsomedial cerebral area by Wigglesworth (1940) put the spotlight on a group of neurosecretory cells, first demonstrated by Weyer (1935) in the same part of the insect brain. Another case of circumstantial evidence resulted from two parallel studies in the horseshoe crab, *Limulus*, showing good correlation between the frequency of neurosecretory elements in different segments of the central nervous system (B. Scharrer, 1941), and the degree of chromatophorotropic activity in extracts prepared from corresponding segments (Brown and Cunningham, 1941).

But much still stood in the way of the primary goal, which was to elucidate mammalian hypothalamic function. Both the design and interpretation of comparable experiments among vertebrates were handicapped by the greater structural complexity of their neurosecretory centers and by the inadequacy of the histological methods then available for their visualization.

A lucky break occurred when Bargmann (1949) and his collaborators experimented with procedures originally designed by Gomori for the demonstration of the secretory product of pancreatic beta cells. By selectively staining the material elaborated in the perikarya of neurosecretory cells, these and comparable methods permitted its being traced throughout the entire neuron. In other words, a marker had been found that linked cells of origin with the special storage and release sites of their distinctive products. Examples of such structures are the neurohypophysis of vertebrates and the analogous corpus cardiacum of insects.

The functional implications of these spatial relationships became increasingly apparent and gave rise to the concept of "neurosecretory systems" as structural and functional units, whose prototype is the hypothalamic-hypophysial system of vertebrates.

The key to the correct interpretation of such systems was the realization that the terminals of the neurosecretory neurons forming the hypothalamo-neurohypophysial tract fail to establish synaptic contact with other neurons or nonneural effector cells. The redundancy of this nerve supply to the posterior lobe, long a source of puzzlement, and the close affiliation of the axon terminals with the vascular bed, suddenly made sense. Such an arrangement is designed for the discharge of special neurochemical messengers destined to become bloodborne in amounts appropriate for the control of multiple effector cells at some distance from the storage and release sites. Expressed in different terms, the stainable secretory material found in abundance within the neurohypophysis was now judged to be of hypothalamic origin and, more specifically, to be manufactured in the perikarya of the nuclei supraopticus and paraventricularis and their homologues.

Vasopressin and oxytocin, contained within this visible substance, thus became prototypes of a new class of neurochemical mediators, for which the designation of "neurohormones" is appropriate.

The posterior pituitary had lost the rank of an endocrine gland in its own right and was now demoted to that of a storage depot (Bargmann and Scharrer, 1951). The term "neurohemal organs," introduced by Knowles, gained wide acceptance, especially when analogous structures were identified in various arthropods. Interestingly, the best known among them, the corpus cardiacum of insects, harbors not only neurosecretory material of cerebral origin, but additional hormonal principles produced by intrinsic parenchymal cells.

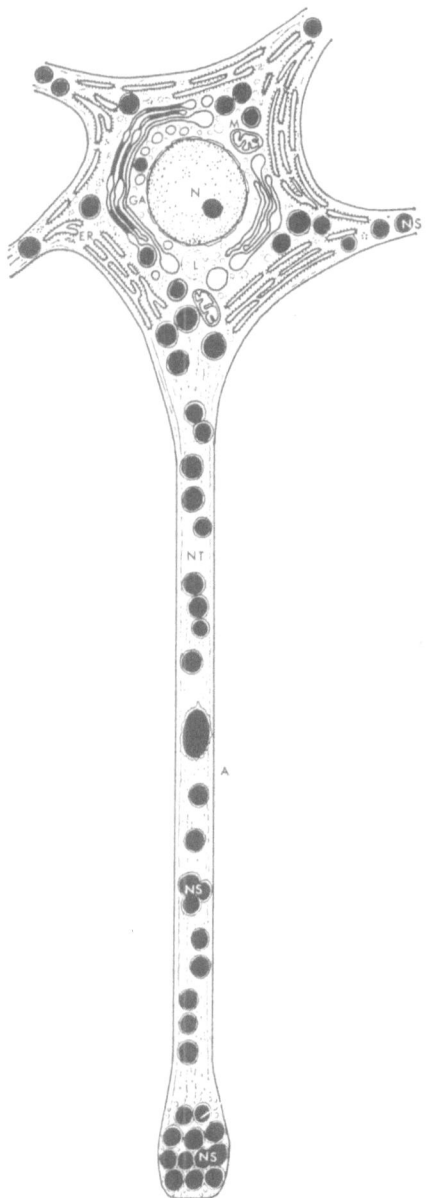

Figure 13.4   A diagram of the neurosecretory neuron showing organelles (ergastoplasm, ER, and Golgi apparatus, GA) involved in the production of neurosecretory material. Some of the neurosecretory granules (NS) in axon (A) have fused into larger units. From E. Scharrer (1966).

The correctness of this interpretation was confirmed by experiments such as the following: Evidence for the intraneuronal transport of extrinsic active principles, bound to carrier proteins of the neurophysin type, was obtained by severance of the neurosecretory pathway (Hild, 1951, mammals; B. Scharrer, 1952, insects), which resulted in an impressive accumulation of the material at the proximal stump. Furthermore, the pharmacologically determined degree of hormone activity of tissue extracts from such stumps, or from posterior lobes of dehydrated rats (vasopressin depletion), paralleled the amounts of visible neurosecretory material in these locations.

The time had come for the "neurosecretory neuron" to be accepted as a new and distinctive cell type with dual properties, i.e., neural and glandular (Figure 13.4). The major criterion for the separation of this special neuron from its conventional counterparts was that, instead of engaging in synaptic chemical transmission, it manufactures peptidergic neurohormones in substantial amounts, in a manner comparable to that observed in classical protein-secreting cells.

But this new insight, important as it was, did not provide an answer to the central question of why neurons should deviate so profoundly from the norm, if all they accomplish thereby is the dispatch of hormonal signals to terminal target cells, such as those of the kidney, the mammary gland, or the integument. As Ernst Scharrer postulated in 1952, the raison d'être for the neurosecretory neuron with its highly specialized dual properties lies in the need for effective communication between the neural and the endocrine regulatory centers, each of which operates in its own way. This important conceptual step marked the emergence of a new phase in neuroendocrine research, the main emphasis now being placed on the mechanism of control over the first way station of the endocrine apparatus, the adenohypophysis of vertebrates, and analogous structures in invertebrates. As

Figure 13.5   Some of the participants at the First International Symposium on Neurosecretion, Stazione Zoologica, Naples, 1953. From left: E. Scharrer, W. Bargmann, B. Scharrer, R. Dohrn, and J. Benoit.

will become apparent, the mechanisms comprising this "final common path" turned out to be more complex than originally anticipated.

By this time a respectable number of investigators here and abroad had developed an interest in neurosecretory phenomena, but, in part due to the Second World War, exchange of information and personal contacts were limited. A most auspicious occasion for an overview of the entire field, encompassing invertebrates and vertebrates, was the First International Symposium on Neurosecretion held at the Stazione Zoologica, Naples, in 1953 (Figure 13.5). Its twentieth anniversary was recently commemorated by the Sixth Symposium in London, the intervening conferences having taken place at approximately four-year intervals in Lund, Bristol, Strasbourg, and Kiel.

The proceedings of these symposia constitute a record of the history of progress in this field (Anon., 1954; Bargmann et al., 1958; Heller and Clark, 1962; Stutinsky, 1967; Bargmann and Scharrer, 1970). The topics featured in consecutive programs reflect an ever-growing spectrum of information, as well as shifts in focal areas that have been sparked by conceptual and methodological advances. It is quite evident that substantial progress has accrued from a combination of biochemical, neurophysiological, and pharmacological approaches. Some of the more recent contributions along these lines are concerned with the energetics of the intra-axonal transport and release of neurosecretory mediators; the differential rates of secretion characteristic of individual neurohypophysial hormones and their neurophysin carriers; the role of new hypothalamic polypeptides; and the special properties of putative aminergic neurosecretory elements.

But, throughout this modern period of investigation, morphological studies have continued to hold their own. This has been due primarily to the wealth of information derived from electron microscopy and, more recently, from fluorescence and immunoenzyme cytochemistry as well as the cobalt/axonal iontophoresis method.

But what have we learned from all this about the mode of operation of the hypothalamic-adenohypophysial axis?

The various cells of the anterior pituitary are under stimulatory and/or inhibitory control by neurochemical messengers originating in hypothalamic neurosecretory centers, such as the arcuate nucleus. In the face of mighty opposition by S. Zuckerman, G. W. Harris (1955) had convincingly demonstrated the role of the hypophysial portal system as a semiprivate, "directed" route for the dispatch of such messenger substances to their destinations. This concept was later substantiated by a detailed ultrastructural analysis of the respective release sites at the level of the median eminence.

These advances soon shared the spotlight with the fruits of concentrated efforts by two research teams to determine the chemical identities of several hypophysiotropic factors (see Blackwell and Guillemin, 1973; Schally, Arimura, and Kastin, 1973). Peptides of this nature, also commonly referred to as "releasing" or "regulating" factors, or hormones, or as "adenohypophysiotropins," are now recognized as a distinctive class of neurohormones. The adaptive features of their special mode of operation, which permit the dispatch of effective (i.e., simultaneous) neural signals to multiple endocrine effector cells without unnecessary detour, are readily apparent.

However, a different solution of the same problem appeared to be required in those invertebrates that lack capillary systems. Our attention was, therefore, directed to an endocrine organ that is analogous to the adenohypophysis of vertebrates, namely the corpus allatum of insects. This gland was found to possess a network of acellular stromal channels, which permits the exchange of various substances, including hormones, between the parenchymal cells and the general circulation (hemolymph).

But there is also an arrangement for bypassing the vascular route, whereby neural elements penetrate the organ through the same extracellular channels and deliver neurosecretory messengers of cerebral origin in close proximity to putative effector cells. More than one of these cells can be addressed simultaneously, where an extracellular layer of stroma intervenes between them and the axonal release site. Furthermore, "synaptoid" neurosecretory release sites were observed in direct contact with endocrine cells ("neurosecretomotor junctions"). In this case, as in conventional synaptic intervention, the signal not being shared with neighboring cells is thus equally "private."

Further examination of vertebrate neurosecretory systems revealed examples of comparable circumvention of the vascular route (see Vollrath, 1967). This was somewhat unexpected in view of the availability of the portal circulation, which seemed to be made to order. In fact, one might question the need for such a series of specializations characterized by a stepwise reduction of the extracellular pathway to the point where it is virtually nonexistent. Are some of these variants adaptive modulations, or are they signs of redundancy?

Be that as it may, a structural basis had now been demonstrated for the existence of modes of neurochemical communication that are neither neurohormonal nor neurohumoral, in the strict sense of the word, but somewhere in between. At this point the characterization of the "neurosecretory neuron" had to be modified, since one of its originally conceived criteria, namely the dispatch of neurohormones, no longer universally applied (see B. Scharrer, 1970, 1972).

The phenomenon of neurosecretion found its proper place within the larger framework of neuroendocrine communication which, in turn, became the central feature of the new discipline of neuroendocrinology (E. Scharrer and B. Scharrer, 1963; E. Scharrer, 1966). Neurology and endocrinology, having long followed their separate ways, had to take notice. The programs of neurobiological meetings of the past decade reflect a growing interest in this affiliation (see Kappers and Schadé, 1972). The same holds for endocrinology, in that 1953 marks the first appearance of "neurosecretion" in the annals of the Laurentian Hormone Conference (E. Scharrer and B. Scharrer, 1954).

A recent addition to our knowledge of neural commands to endocrine receptors, once more due to painstaking ultrastructural scanning, was the demonstration of conventional secretomotor junctions in both vertebrates and invertebrates (see Bargmann, Lindner, and Andres, 1967; B. Scharrer, 1970). However, in what manner and to what degree chemical synaptic transmission participates in the overall control mechanism has yet to be subjected to detailed experimental analysis.

In summary, what we end up with is a remarkably rich range of neurochemical signals for the endocrine apparatus to choose from, so as to permit every nuance in the orchestration of its performance. Is this impressive versatility on the part of the neuron a reflection of the special requirements of hormone-producing effector cells, or does it transcend the framework of neuroendocrine interaction?

A survey of nonendocrine receptors of nonconventional neural input reveals that there are indeed certain parallels with the patterns discussed thus far, except for differences in the frequency with which they occur. Again the mechanisms range from neurohumoral to neurohormonal.

Well-known examples of neurohormonal control are effects of the kind already discussed, among them that of vasopressin on the collecting ducts of the kidney. Aside from the conventional neuroeffector junctions, some striated muscles show signs of "innervation" by peptidergic fibers, which can be compared with those in salivary gland and corpus allatum cells on the one hand and mammalian pars intermedia cells on the other. Furthermore, certain elements of smooth and striated musculature reveal intermediate types of arrangement as defined previously. Here too, several cells can share the neural messenger substances released into an intervening stromal compartment by one axon *en passant*, the difference being that in this case the mediator is aminergic instead of peptidergic.

Perhaps the most unexpected outcome of this survey has been the detection of nonconventional neuron-to-neuron signals. Such departures from the standard pattern of chemical synaptic transmission, although they are very much in the minority, apply to situations where either one or both of the interacting neurons are of the nonconventional type. The major and, by definition, nonconventional response to afferent neural input is the release of neurohormone, and this input involves stimulatory and inhibitory conventional as well as neurosecretory neurons. Even more noteworthy is recent evidence suggesting that certain conventional neurons are capable of responding to neurohormonal stimuli. More specifically, it appears that dopaminergic cells in the mammalian central nervous system have become attuned to two hypothalamic neurohormones, TRF and MIF, with effects comparable to those of L-dopa (Plotnikoff et al., 1972a,b).

Recognition of the diverse modes of operation of neurosecretory neurons has elucidated the relationship of these cells to more conventional types. It has also directed attention to comparable modulations within the range of conventional neurons and has thus clarified the entire scope of neurochemical mediation, which may be viewed as follows:

Since all neurons share the capacity to synthesize and release distinctive chemical mediators, the existing dichotomy should be viewed as a matter of degree. It derives from the fact that, in the course of phylogeny, the classical "neurosecretory neuron" has developed its secretory activity to the point where it takes precedence over all of the cell's other functions. This specialization enables the nervous system to communicate by means of neurohormonal as well as neurohumoral signals, and the class of neurosecretory neurons takes its place on one side of a spectrum, the other side of which is occupied by conventional neurons. Furthermore, the different levels of specialization within the class seem to hold unequal rank in terms of functional significance as well as of frequency.

The position at the nonconventional end is taken up by the first-order systems in which neurohormonal commands reach terminal target cells directly via the general circulation. As mentioned earlier, in higher animals the necessity for this relatively primitive mechanism is not readily apparent, but its existence can be interpreted as a carry-over from an early state in the evolution of the endocrine system.

Next in line is the group of neurosecretory neurons that have risen to a key position commensurate with their dual capacities, i.e., that of serving as the "final common path" by which the nervous system accomplishes its liaison with the endocrine apparatus. The only question here is whether the several "semiprivate" hormonal and nonhormonal pathways used for this step represent necessary adaptations to special situations, or whether they are perhaps interchangeable. At any rate, their availability, together with that of some nonneurosecretory nervous input, appears to fulfill the complex requirements of the neuroendocrine axis.

An examination of the "conventional" side of the range reveals structural and functional digressions from the pattern of orthodox neural transmission which, although they are less prominent, parallel some of those first recognized within the group of neurosecretory neurons. Thus the two sides of the neurochemical spectrum are neither rigidly uniform nor separated by as clear-cut a line of demarcation as was originally supposed. Instead there is an intermediate zone where one neuron type gradually blends into another. The striking features distinguishing classical neurosecretory from conventional neurons have now become "part of a whole" and serve to underscore not merely the existence but the remarkable degree of flexibility inherent in neurochemical communication.

The classical neurosecretory cell retains its special position within this spectrum as a neuron that engages in secretory activity to a degree above and beyond that of other, more conventional neurons.

The concept of neurosecretion, once considered heretical, has reached its golden age. It has attained respectability and, in the process of entering the domain of modern biological thought, it is now approaching anonymity.

## References

Anon. (1954): Convegno sulla neurosecrezione. Riassunti. (Proceedings of the 1st International Symposium on Neurosecretion, Naples, Italy, 1953). *Pubbl. Staz. Zool. Napoli* 24 (Suppl.): 1–98.

Bargmann, W. (1949): Über die neurosekretorische Verknüpfung von Hypothalamus und Neurohypophyse. *Z. Zellforsch. Mikrosk. Anat.* 34:610–634.

Bargmann, W., Hanström, B., Scharrer, B., and Scharrer, E., eds. (1958): *Zweites Internationales Symposium über Neurosekretion* (Lund, 1–6 July 1957). Berlin: Springer-Verlag.

Bargmann, W., Lindner, E., and Andres, K.H. (1967): Über Synapsen an endokrinen Epithelzellen und die Definition sekretorischer Neurone. Untersuchungen am Zwischenlappen der Katzenhypophyse. *Z. Zellforsch. Mikrosk. Anat.* 77:282–298.

Bargmann, W., and Scharrer, B., eds. (1970): *Aspects of Neuroendocrinology* (Proceedings of the V International Symposium on Neurosecretion, 20–23 August 1969, Kiel). Berlin: Springer-Verlag.

Bargmann, W., and Scharrer, E. (1951): The site of origin of the hormones of the posterior pituitary. *Am. Sci.* 39: 255–259.

Blackwell, R.E., and Guillemin, R. (1973): Hypothalamic control of adenohypophysial secretions. *Annu. Rev. Physiol.* 35:357–390.

Brown, F.A., Jr., and Cunningham, O. (1941): Upon the presence and distribution of a chromatophorotropic principle in the central nervous system of *Limulus*. *Biol. Bull.* 81:80–95.

Hanström, B. (1940): Inkretorische Organe, Sinnesorgane und Nervensystem des Kopfes einiger niederer Insektenordnungen. *Kungl. Sven. Vetensk. Acad. Handl.* 18:1–265.

Hanström, B. (1941): Einige Parallelen im Bau und in der Herkunft der inkretorischen Organe der Arthropoden und der Vertebraten. *Lunds Univ. Årsskr. (N.F. Avd.)* 37(4): 1–19.

Harris, G.W. (1955): *Neural Control of the Pituitary Gland*. London: Edward Arnold.

Heller, H., and Clark, R. B., eds. (1962): *Neurosecretion* (Proceedings of the Third International Symposium on Neurosecretion, University of Bristol, September 1961). Memoirs of the Society for Endocrinology, no. 12. London: Academic Press.

Hild, W. (1951): Experimentell-morphologische Untersuchungen über das Verhalten der "Neurosekretorischen Bahn" nach Hypophysenstieldurchtrennungen, Eingriffen in den Wasserhaushalt und Belastung der Osmoregulation. *Virchows Arch. (Pathol. Anat.)* 319:526–546.

Kappers, J.A., and Schadé, J.P., eds. (1972): *Topics in Neuroendocrinology* (Progress in Brain Research, vol. 38). Amsterdam: Elsevier.

Kopeć, S. (1917): Experiments on metamorphosis of insects. *Bull. Int. Acad. Sci. Lett. Cracovie (Classe Sci. Math. Nat. B.)*:57–60.

Plotnikoff, N.P., Kastin, A.J., Anderson, M.S., and Schally, A.V. (1972a): Oxotremorine antagonism by a hypothalamic hormone, melanocyte-stimulating hormone release-inhibiting factor (MIF). *Proc. Soc. Exp. Biol. Med.* 140:811–814.

Plotnikoff, N.P., Prange, A.J., Jr., Breese, G.R., Anderson, M.S., and Wilson, I.C. (1972b): Thyrotropin-releasing hormone: enhancement of dopa activity by a hypothalamic hormone. *Science* 178:417–418.

Schally, A.V., Arimura, A., and Kastin, A.J. (1973): Hypothalamic regulatory hormones. *Science* 179:341–350.

Scharrer, B. (1941): Neurosecretion. IV. Localization of neurosecretory cells in the central nervous system of *Limulus*. *Biol. Bull.* 81:96–104.

Scharrer, B. (1952): Neurosecretion. XI. The effects of nerve section on the intercerebralis-cardiacum-allatum system of the insect *Leucophaea maderae*. *Biol. Bull.* 102:261–272.

Scharrer, B. (1970): General principles of neuroendocrine communication. *In: The Neurosciences: Second Study Program.* Schmitt, F.O., editor-in-chief. New York: Rockefeller University Press, pp. 519–529.

Scharrer, B. (1972): Neuroendocrine communication (neurohormonal, neurohumoral, and intermediate). *Prog. Brain Res.* 38:7–18.

Scharrer, B., and Scharrer, E. (1944): Neurosecretion. VI. A comparison between the intercerebralis-cardiacum-allatum system of the insects and the hypothalamo-hypophyseal system of the vertebrates. *Biol. Bull.* 87:242–251.

Scharrer, E. (1928): Die Lichtempfindlichkeit blinder Elritzen (Untersuchungen über das Zwischenhirn der Fische. I.). *Z. Vergl. Physiol.* 7:1–38.

Scharrer, E. (1952): The general significance of the neurosecretory cell. *Scientia* 46:177–183.

Scharrer, E. (1956): The concept of analogy. *Pubbl. Staz. Zool. Napoli* 28:204–213.

Scharrer, E. (1966): Principles of neuroendocrine integration. *Res. Publ. Assoc. Res. Nerv. Ment. Dis.* 43:1–35.

Scharrer, E., and Scharrer, B. (1954): Hormones produced by neurosecretory cells. *Recent Prog. Horm. Res.* 10:183–240.

Scharrer, E., and Scharrer, B. (1963): *Neuroendocrinology.* New York: Columbia University Press.

Speidel, C.C. (1919): Gland-cells of internal secretion in the spinal cord of the skates. *In: Papers from the Department of Marine Biology of the Carnegie Institution of Washington.* Vol. 13 (Publication No. 281). Washington, D.C.: Carnegie Institution, pp. 1–31.

Stutinsky, F., ed. (1967): *Neurosecretion* (IV International Symposium on Neurosecretion, Strasbourg, 25–27 July 1966). Berlin: Springer-Verlag.

Van Dyke, H.B. (1936–1939): *The Physiology and Pharmacology of the Pituitary Body.* Chicago: University of Chicago Press.

Vollrath, L. (1967): Über die neurosekretorische Innervation der Adenohypophyse von Teleostiern, insbesondere von *Hippocampus cuda* und *Tinca tinca*. *Z. Zellforsch. Mikrosk. Anat.* 78:234–260.

Weyer, F. (1935): Über drüsenartige Nervenzellen im Gehirn der Honigbiene, *Apis mellifica* L. *Zool. Anz.* 112:137–141.

Wigglesworth, V.B. (1940): The determination of characters at metamorphosis in *Rhodnius prolixus* (Hemiptera). *J. Exp. Biol.* 17:201–222.

Rita Levi-Montalcini (b. 1909, Turin, Italy) is professor of biology at Washington University, Saint Louis, and director of the Laboratory of Cell Biology in Rome. Her research in experimental neurology has dealt with problems in neurogenesis and with the isolation and study of the nerve growth factor.

# 14
# NGF : An Uncharted Route

## Rita Levi-Montalcini

"A scientist should never attempt to judge his own contributions, whether significant or not, but especially when not." (Lwoff, 1966)

## Introduction

A disclaimer of personal merit, such as phrased above by Lwoff, is not a disclaimer of the significance of a phenomenon that chance rather than calculated search has brought to one's attention, and for this reason I have accepted with pleasure the very flattering invitation to discuss the history of nerve growth factor (NGF). I am afraid, however, that the following account will not provide a unique glimpse into the paths of discovery that have shaped the course and content of neuroscience in recent decades. The NGF has in fact still not found its place in the broadening panorama of neuroscience, and, even worse, twenty years after its coming into existence this factor has disclosed only a few, perhaps the most trivial, of its traits. It keeps us wondering where it is heading, and whether its uncharted route has, indeed, any ending.

It is in the spirit of never-ending pursuit, which has characterized this search from its very beginning, that the present biography of the NGF is written by one who has watched with awe and wonder the birth of this "miracle" molecule from the sinister womb of malignant tissues.

Different moments of this experience were shared with three friends. Their names are Viktor, Stanley, and Piero. Their true identity, as well as their participation in this adventure, will emerge in the following pages.

### Biographical Sketch of the Biographer

#### Starting a career
The start of what was supposed to be a medical career took place in Turin, in northern Italy, in the early 1930s. Eugenia Lustig, my cousin, Salvador Luria, and Renato Dulbecco were my schoolmates and became my lifelong friends. We had in common a tremendous respect and fear of our teacher, Giuseppe Levi (Figure 14.1), a towering figure in the biology of that period, and shared a lukewarm interest in histology, the subject that he taught with great enthusiasm and unique knowledge. To be honest, I should confess that I hated histology and that up to this day I have never mastered even the most common staining techniques. I suspect that my three schoolmates nursed the same feelings, but so high was our esteem for "the master" that this dislike never came out in the open, nor prevented us from becoming *allievi interni* in the Istituto di Anatomia which he di-

Figure 14.1   Professor Giuseppe Levi.

rected with enormous energy and iron rules. It was my good luck that the master, realizing my lack of histological talent, decided that I should learn instead to culture different cell types in vitro, a field which was at that time just beginning. My first study, on the formation of reticular fibers in vitro from connective and epithelial tissues, pleased Levi and signaled the beginning of a master–pupil association that was to last until his death 33 years later.

In 1936 graduated I from medical school and specialized in neurology and psychiatry, equally attracted by the clinical profession and by a pure academic career in the footsteps of Levi. My perplexity was not to last too long: in June 1938 Mussolini issued the "Manifesto per la difesa della razza," signed by ten Italian "scientists," which barred academic as well as professional careers to non-Aryan citizens. Intermarriage between Aryan and Semitic citizens was prohibited to protect the pure Italian Aryan blood from contamination with that of inferior non-Aryan races. In 1939 I received and accepted an invitation from a neurological institute in Brussels, and I moved there a few months before the declaration of war between Germany, France, and England. However, when the invasion of Belgium appeared imminent, I returned to Italy to join my family in Turin. In the meantime the situation had greatly worsened: the only two alternatives to total stagnation were either to abandon the country and emigrate to the United States (there were no more safe places in Europe), or to pursue some activity that would need neither support nor connection with the outside Aryan world in which we lived. My family did not want to consider the first alternative, and I then decided to build a small research unit at home. Though I had never performed neuroembryological experiments, I was familiar with the literature (non-Aryan citizens were no longer allowed to consult the university libraries), and I was thrilled at the idea of this experience à la Robinson Crusoe.

I built a laboratory in my bedroom with few indispensable pieces of equipment, such as an incubator, a light, a stereomicroscope, and a microtome. The object of choice was the chick embryo, and the instruments consisted of sewing needles transformed with the help of a sharpening stone into microinstruments. My Bible and inspiration was an article by Viktor Hamburger dated 1934, which I had happened to read some years earlier. The title of this excellent classic study was "The effects of wing bud extirpation on the development of the central nervous system in chick embryos." Through my training with Levi I had become an expert in silver techniques, and I decided to reinvestigate the problem of the effect of the periphery on the developing nerve centers by making use of the specific Cajal–De Castro technique. The project had barely started when Levi returned from Belgium, where he had also moved soon after being discharged from the university. Levi asked to join me in this investigation, and he became, to my great pride, my first and only assistant.

**The trying years**
Looking back to the period 1940–1943, I wonder how I could have found so much interest in and have devoted myself with such a burning enthusiasm to the study of a small neuroembryological problem when all the values I cherished were being crushed and the triumphant advance of the Germans all over Europe seemed to herald the end of Western civilization. The answer may be found in the well-known refusal of human beings to accept reality at its face value, whether it be the fate of an individual, of a country, or of the whole of human society. Without this built-in defense mechanism life would be unbearable, not only to those doomed to impending death by fatal illness, but also to those who approach the physiological end of their lives with all the misery and suffering that are the companions of old age.

I believe that I inherited from my father an unusually efficient defense mechanism that was to be of great help during those years. This was strengthened by the association with Levi, then in his seventies. The old master followed with unfailing enthusiasm (comparable to and tuned to an even higher pitch than my own) the development of our experiments. I still hear his thundering voice, used to frighten assistants and legions of students, which now resounded even more powerfully in the narrow precinct of my small bedroom laboratory. During the period 1940–1942 the anti-Semitic campaign reached its peak, and the daily press found great pleasure in spreading all the more hideous slogans borrowed from the Nazis; but Levi and I ignored the threats and abuses, so absorbed were we in our work. My bedroom became the meeting center for old pupils and friends of Giuseppe Levi, who worshiped in him not only the great scientist but also the valiant and undaunted anti-Fascist. It should, in fact, be remembered that the majority of Italians rebelled vigorously against the racial laws and were not afraid of expressing dissent even at the risk of being themselves persecuted. In July 1942, however, the heavy bombing of Turin and other industrial cities by the Allies forced my family and me to move to a small country house, where I rebuilt my laboratory under conditions far worse than in my previous bedroom, while Levi moved to a small mountain village. Shortages of eggs (the eggs were used first for experimentation and then, five days later, upon removal of the operated embryo, as scrambled eggs for food) and of electrical power, which was cut off every few days, made the work extremely difficult, not to say impossible. Yet, to my great satisfaction, I was able to complete an experimental study on the acousticovestibular centers of the chick embryo that was to come out in print many years later in the United States (Levi-Montalcini, 1949).

In July 1943, on the verge of military disaster and the total collapse of the country, Mussolini was disavowed by his (up to then) most loyal followers and was jailed by the king. One and a half months later Italy was invaded by hordes of Nazis, and the small Jewish-Italian population became the object of ferocious hunting, mass killing, and deportation to the extermination camps. My family and I escaped capture by flying to Florence, where we mingled with hundreds of other refugees living, as we did, with false identity cards. Of the long months spent there in seclusion, listening secretly to the news broadcasts from London, in continuous danger of being discovered, I shall mention only one episode, which is perhaps more amusing in retrospect than when it actually occurred. One day the bell rang, and I heard the familiar voice of Professor Levi asking the landlady in his usual authoritative way to call me immediately. He did not know my new family name, but, with a wisdom that I would never have expected from him, he gave my first name: Rita. To the question, whom should she announce, he answered: "Professor Giuseppe Levi, ah no, I keep forgetting, Professor Giuseppe Lovisato." The landlady, who was not aware of our true identity, became from that moment very suspicious of us and of our absent-minded friend; but, being a gentle soul and not at all curious, she kept her suspicions to herself.

With the end of the war in May 1945, I returned to Turin and became the assistant of Levi, who had resumed his position as professor of anatomy at the university. One year later, in the spring of 1946, I received a letter from Viktor Hamburger, who had come across the 1942 article by Levi and myself on the correlations between periphery and developing nerve centers. This manuscript, refused by Italian journals in view of the non-Aryan names of its authors, had been accepted by the Belgian *Archives de Biologie* (Levi-Montalcini and Levi, 1942). Hamburger was intrigued by the different mechanisms of action that had been postulated by us and by him for the interaction between nonnervous and nervous structures. He invited me to work for a one- or two-year period in his laboratory in Saint Louis to collaborate with him in reinvestigating this problem. However, it was not until the fall of 1947 that I left Turin for Saint Louis, where I was to spend the next 26 years, the happiest and most productive years of my life.

## Viktor Hamburger, a Founding Father of Experimental Neuroembryology

If the beginning of my career was under the spell of Giuseppe Levi, the second period was under the influence of Viktor Hamburger, who had already played a key role in channeling my interests toward problems of growth and differentiation of nerve cells.

Viktor Hamburger was a former student and the "favorite son" of Hans Spemann, the great German embryologist who was awarded the Nobel Prize in 1934 for his discovery of the role played by induction in developmental embryonic processes. From his teacher Viktor had learned the trades and skills of the art, which requires both sharp thinking and sharp microinstruments. Spemann was a master in both fields; in fact he greatly enjoyed devising and forging his own microtools and, in the Renaissance tradition, passed on both skills to his pupil. In 1932, one year before Germany fell into the hands of Hitler, Viktor had come to the United States to spend one year in Chicago in the laboratory of Frank Lillie, a close friend of Spemann and himself a famous embryologist who had actually started experimental work on the chick embryo. The triumph of Hitler in 1933 persuaded Viktor to remain in the United States to pursue the analysis of the developing nervous system that he had started in Germany. In 1935 he moved from Chicago to Saint Louis where, in 1940, he succeeded F.O. Schmitt (who had moved to MIT) as chairman of the Department of Zoology of Washington University.

Viktor directed the department with rules that were quite different from those of Levi. Accustomed as I was to the thundering voice of Levi and to his explosive way of expressing dissent in political and scientific matters (a mild way of conveying his viewpoint to his interlocutor was: "I beg your pardon, but you are a perfect imbecile"), I was struck by the kindness and subtle dry humor of Viktor, who would never hurt other people's feelings nor show his disagreement with more than a few gentle remarks and a firm glance of disapproval. Working with him on the same problem that had absorbed so much of my time and thoughts in my secluded laboratory in Turin was a sheer pleasure. Instead of the sinister atmosphere of an Italian city during the fateful years 1940–1943, I was now surrounded by the cheerful environment of an American college. It was right after the war and the period of dissent was still far away; students strolled hand in hand on the university campus, which, to a European observer, seemed like the garden of Eden with no snake to tempt the naive inhabitants of such a Paradise. (But, alas, the snake was there.) But even more than these novel surroundings, it was my association with Viktor that I enjoyed. While I dearly loved the "old master," it had often been difficult to adjust to his temperamental fits and dogmatic way of thinking. With Viktor, however, there was no problem of this sort. What I liked most was the clarity of his thinking and his superb control of the English language. Writing a scientific paper was a new experience for me, and I concentrated on the effort of learning how to do it. Up to this day, everytime I write one, my first thought is: Will Viktor like it?

In the fall of 1948, one year after my arrival in Saint Louis (the work had progressed so well that my return to Turin was postponed indefinitely), Viktor showed me a short article by one of his former students (Bueker, 1948) which was to change entirely the direction of my research. In this article Bueker reported on the results of a bold and ingenious experiment that he had performed to test the ability of developing nerve fibers to innervate fast-growing tissues, such as neoplastic tissues, and to adjust their growth rate to that of a rapidly expanding tumor. He selected for these studies a mouse tumor, sarcoma 180, and to his satisfaction he found that the results agreed with his expectation. Fragments of this particular tumor (other mouse tumors did not produce the same effect), implanted into the body wall of three-day chick embryos, became established and innervated by nerve fibers growing out from adjacent sensory ganglia of the host. Histological studies performed three to five days later showed that the ganglia providing fibers to the tumor were larger than contralateral ganglia innervating the leg.

These results were so much in line with current concepts of the interrelation between nerve centers and end organs that Bueker overlooked other, more perplexing aspects of the growth response. A reinvestigation of this effect by Viktor and myself showed that the growth response of sensory ganglia innervating the tumor was, in fact, much more pronounced than that of the same ganglia innervating a supernumerary limb. Furthermore, while the motor somatic nerve cells in the spinal segment facing the tumor transplant were not increased (they actually decreased in number), the sympathetic ganglia were tremendously enlarged and contributed to the innervation of the neoplastic tissues to a much larger extent than did sensory ganglia. Sensory and sympathetic nerve fibers branching into the tumor established no contact with the neoplastic cells but wandered aimlessly among the cells. And yet, even conceding that the effect was much more marked than that produced by implantation of a supernumerary limb bud, and differed in many respects from it, we were still not prepared in 1951 (Levi-Montalcini and Hamburger, 1951) to see in this response any flagrant deviation from normality, so difficult is it to refute generally accepted concepts and to evaluate novel results with an unprejudiced mind.

It was a spring day in 1951 when the block was suddenly removed, and it dawned on me that the tumor effect was *different* from that of normal embryonic tissues in that the tumor acted by *releasing* a growth factor of unknown nature rather than by making available to the nerve fibers a larger-than-usual field of innervation. This hypothesis, which became a certitude with me long before I obtained supporting evidence in its favor, was suggested by the observation that the viscera of embryos bearing transplants of mouse sarcomas 180 or 37 (the latter proved to produce the same effect as sarcoma 180) were invaded by sympathetic nerve fibers at a stage when there was still no innervation apparent in controls. Furthermore, and this was an even more remarkable infraction of normal rules, sympathetic nerve fibers forced their way inside blood vessels of the host where they formed large neuromas coated with red blood cells. How many times during the past two years had I seen but not perceived this most intriguing routing of nerve fibers into the blood vessels?

While I was lost in the contemplation of this, for me, stupendous phenomenon, I heard the familiar thundering voice of the "old master," who had just arrived from Italy and was paying me his first and only visit in the United States. In the excitement of the moment of discovery I showed him under the microscope the nerve bundles filling the viscera and entering into the blood vessels. Levi shook his powerful leonine head, still covered with a thick red mane in spite of his eighty-one years: "How can you say such nonsense? Don't you see that these are collagenous and not nerve fibers? Did it take you such a short time to forget all that you learned in Turin?" Soon after Levi left the room, I showed the slide to Viktor, who immediately grasped the far-reaching significance of these findings and was enthusiastic. I obtained decisive evidence for my hypothesis by grafting fragments of sarcomas 180 or 37 onto the chorioallantoic membrane of four- to six-day chick embryos, in such a position that the tumor and the embryo shared the circulation but no direct contact was established between neoplastic and embryonic tissues (Levi-Montalcini, 1952; Levi-Montalcini and Hamburger, 1953). The results were the same as in intraembryonic transplants.

The excitement of this discovery was tempered by the realization that it would be tremendously difficult to identify the tumor factor by using the ordinary exceedingly laborious and time-consuming embryological experiments. What was needed was a much simpler and faster bioassay. The tissue-culture method that I had learned in Turin with Levi seemed to offer a possible approach to this problem. A dear friend of mine, Herta Meyer, a former associate of Fisher in Berlin and then research assistant with Levi in Turin, was now in charge of the tissue-culture unit in the Biophysics Institute of the Medical School of Rio de Janeiro, directed by Professor Carlos Chagas. After an exchange of letters with Professor Chagas and Herta and with their consent, I submitted a travel-grant application to the Rockefeller Foundation, which would permit me to perform these experiments in Rio de Janeiro. The proposal was approved, and in October 1952 I boarded a plane for Brazil.

### The NGF: Its Birth and Early Life History

The tumoral factor had given a first hint of its existence in Saint Louis; but it was in Rio de Janeiro that it revealed itself, and it did so in a theatrical and grand way, as if spurred by the bright atmosphere of that explosive and exuberant manifestation of life that is the Carnival in Rio.

I had come from Saint Louis with two mice bearing transplants of sarcomas 180 and 37 in my handbag. Immediately after arriving, I dissected small fragments of these tumors and cultured them in vitro in a hanging, semisolid drop of rooster plasma and embryonic extract; sensory or sympathetic ganglia from eight- to ten-day chick embryos were then transplanted in close proximity to neoplastic tissues. The first results were not only negative but worse: the ganglia cultured adjacent to the tumor produced less fibers than controls grown alone or in combination with embryonic chick tissues. It occurred to me that perhaps some other contaminants released from mouse tumors overshadowed or masked the tumor effect. I transplanted both tumors into chick embryos as intermediary hosts and then dissected out small fragments and cultured them in vitro, in proximity to sympathetic or sensory ganglia. Twenty years later I recall, as if it had just happened, the astonishment and wonder of that morning when, for the first time, I saw in the light microscope the outcome of these experiments. Figure 14.2 reproduces the drawing that I did with India ink (I did not have a camera and I could not wait long enough to search for one) and immediately sent in an express airmail letter to

Viktor, who returned it to me many years later. Nerve fibers had grown out in twelve hours from the entire periphery of sensory and sympathetic ganglia cultured in proximity to the neoplastic tissues and had spread out radially around the explants like the rays of the sun (Levi-Montalcini, Meyer, and Hamburger, 1954). The "NGF halo," as it soon became known, represents the most sensible and reliable index of the presence of this growth factor in any tissue or, as we shall see, in body fluids. Without this assay I very much doubt that we could ever have discovered and eventually identified this factor.

The news of the halo spread in no time throughout the Biophysics Institute and was celebrated at a party in the Chagases' most hospitable home. Three months later I returned to Saint Louis with my halo, a good dose of enthusiasm, and the naive belief that the identification and characterization of the tumoral factor would be only a matter of some months of solid work. How far we were from that goal we would learn in the subsequent twenty years; nor would we even today have solved the problem were it not for a most fortunate association with an outstanding biochemist and some fortuitous discoveries which, all of a sudden, made available far better sources of this factor than mouse sarcomas.

**Enter Stanley Cohen**

Stanley Cohen, or Stan as he became known to us from the very beginning, had never been interested in the nervous system. His first postdoctoral study was on the mechanism of urea excretion of the earthworm, and to this end he had spent long days collecting tons of them in the fertile ground around Denver, Colorado. Having solved this problem, he had then moved to Saint Louis, where he became associated with Martin Kamen.

I have wondered many times where we would stand now if Stan had not heard from Viktor about the halo effect and had not been tempted to work with us on this problem. Soon after joining the Department of Zoology as research associate, Stan fractionated the tumor homogenate and identified in the microsomal fraction the active tumoral principle. In 1954 this factor was christened "Nerve Growth Stimulating Factor" (Cohen, Levi-Montalcini, and Hamburger, 1954), a name shortened later to "Nerve Growth Factor" or, more simply, "NGF."

In a biological science, perhaps to a larger extent than in any other experimental

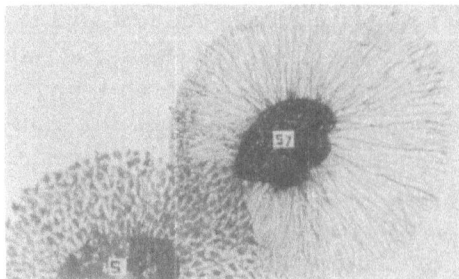

Figure 14.2 The "halo effect" at its first appearance in Rio de Janeiro in December 1952. This is an India ink drawing of a combined culture of sarcoma 180 (S) and a sympathetic ganglion (Sy) from an eight-day chick embryo after twelve hours of incubation.

science, chance and good luck play a notoriously great role. It is not only, as is so often stated, a matter of serendipity, or of the perception of a truth that is there all the time but goes unnoticed until the mind of the observer suddenly grasps it, but rather of a fortuitous stumbling into a cave of precious stones while hiking up a hill on a trail that is not expected to bring one anywhere but to the top of the hill.

When Stan, following the suggestion of Arthur Kornberg, made use of snake venom to purify further the tumoral factor, he did not, nor could he, anticipate the outcome of this routine experiment. The venom was used as a source of phosphodiesterase to degrade the nucleic acids present in the active fraction. If the nucleic acids were an essential component of the NGF, their enzymatic degradation would destroy the biological activity of this factor; if the activity remained, then the protein rather than the nucleic acids must be responsible for the growth effect.

It was again a spring morning (1956) when I inspected, as I used to do every morning, the cultures performed the day before and then, with no comment, asked Stan to do the same. Stan looked through the eyepiece of the microscope and mumbled: "Rita, I am afraid that with this we have used up all our good luck; we cannot count on it anymore." Fortunately he was wrong, but this will come out later.

The addition of the venom to the tumoral fraction endowed with NGF activity had so potentiated its effect as to transform the delicate fibrillar halo into one of tremendous density. Two alternatives were considered: either the venom had destroyed an inhibitor present in the tumoral fraction or it harbored the growth factor itself. Six hours later we knew the answer. The addition of minute amounts of snake venom to the culture medium evoked the same effect from sensory and sympathetic ganglia as when it was added in the presence of the tumoral fraction. Thus the venom was another most potent source of the NGF.

Between 1956 and 1958 Stan succeeded in purifying and characterizing the venom NGF (a feat that had not been possible with the tumoral NGF since in these tissues the factor is present in exceedingly small quantities) as a protein molecule of molecular weight 20,000. At the same time I studied the in vivo and in vitro effects of the venom NGF on sensory and sympathetic ganglia of the chick embryo. It soon became apparent that its biological activity was strikingly similar if not identical to that of the tumoral NGF, with only one important difference: the NGF factor was present in the venom at a concentration estimated as a thousand times higher than in the two mouse sarcomas (Cohen and Levi-Montalcini, 1956).

The finding of a protein molecule endowed with such a potent and selective biological activity in two unrelated sources, mouse tumors and snake venom, prompted a search for it in other tissues and body fluids. At the time of the discovery in Rio of the in vitro NGF effect of mouse sarcomas, other mouse tissues had been tested and I had found, to my dismay, that they also elicited a nerve growth effect similar to, but much milder than, that of the two sarcomas. This finding, reported in the 1954 article, seemed somewhat disturbing to the thesis that the production of this factor was the prerogative of some neoplastic tissues; but in the absence of clear-cut evidence against the hypothesis (other mouse tumors proved to elicit no growth effect), we did not pursue the study of normal mouse tissues any further.

The discovery of the venom NGF was a definite blow to the thesis and called for a reconsideration of the earlier findings. It was Stan who conceived the idea of testing the mouse salivary glands, a homologue of the venom gland, as another potential source of NGF. The results of in vitro experiments fully confirmed his guess: the mouse salivary glands proved to be a third and by far the most potent source of NGF (Figure 14.3). From 1958, when these experiments were first performed, to the present day, all our work and that of the other laboratories that have become interested in the NGF has centered on the purification and characterization of the salivary NGF. Stan identified the factor as a protein molecule of molecular weight 44,000, possibly a dimer of the venom NGF (Cohen, 1960). Extensive studies in newborn and adult mice and rats injected with the purified NGF gave evidence for the magnitude of the growth response elicited from sympathetic ganglia. While these ganglia remain receptive to the action of the NGF throughout their life, the sensory ganglia respond to it only during a restricted period of their embryonic development (Levi-Montalcini and Booker, 1960a).

In 1959 Stan made his last, but no less remarkable, contribution to the study of the NGF. He prepared an antiserum to the NGF and we tested its effects in vitro. When we found that it abolished the halo stimulated by the salivary factor around sensory and sympathetic ganglia, we injected the serum into newborn mice; treated and control animals of the same litter were sacrificed and inspected at the stereomicroscope twenty days later. On June 11, 1959, Stan and I saw for the first time in the stereomicroscope the outcome of this seemingly innocuous treatment. Control and injected mice were of the same size, and the vitality of the group injected daily with the antiserum to the NGF was in no way impaired; but the sympathetic chain ganglia, which in controls are easily detected at the two sides of the vertebral bodies, were no longer there. With considerable effort I succeeded in identifying them in two vanishingly thin translucent filaments. After dissection, fixation, and serial section, we saw in the light microscope what was left of each

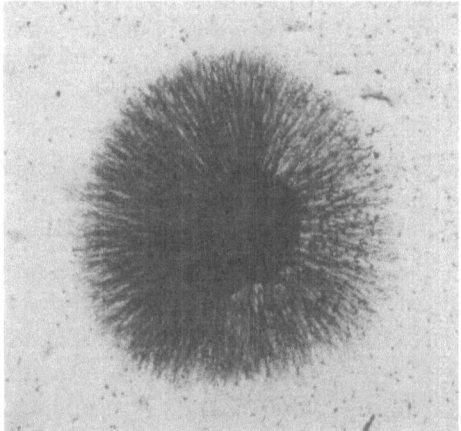

Figure 14.3   Fibrillar halo around a sensory ganglion from an eight-day chick embryo cultured in vitro in the presence of 0.01 μg/ml of the purified salivary factor. From Levi-Montalcini (1964).

ganglion: a population of glial cells with a few highly atrophic sympathetic neurons scattered among nonnervous cells. The entire para- and prevertebral population of sympathetic nerve cells was reduced to 3–5% of that of controls (Levi-Montalcini and Booker, 1960b).

These findings gave additional evidence for the key role played by the NGF in the life of the sympathetic adrenergic neuron, and at the same time provided a unique method of rearing mice and rats that are deprived of their sympathetic nervous system from birth. Subsequent studies showed that three to five injections of the antiserum to the NGF given in the first five days of life are sufficient to obtain the near-total destruction of sympathetic ganglia, and also showed that this effect is irreversible. This end result, which became known as "immunosympathectomy," has been utilized ever since 1959 in our laboratories and in several others to assess the role of the sympathetic nervous system in physiological and pathological conditions (Zaimis and Knight, 1972; Steiner and Schoenbaum, 1972).

### Exit Stan; Enter Piero Angeletti

Shortly after this last discovery, Stan, to my great regret, abandoned our joint pursuit and moved to Nashville, to Vanderbilt University, where he started the study of another factor, also isolated from mouse salivary glands, that is endowed with an equally potent and selective growth-promoting effect on epithelial cells. With the same ingenuity and talent that he had displayed in the NGF chase, he now succeeded in uncovering the properties of this new growth factor, which he named the "Epithelial Growth Factor" or, more simply, the "EGF" (Cohen, 1962). His departure from his small and not too clean office on the first floor of Rebstock Hall, where he had labored for six years to unveil the mysterious nature of the NGF, signaled the end of the most romantic and picturesque phase of this adventure. Stan used to spend the entire day and most of his evenings there, meditating with eyes half-closed, smoking his pipe, and playing the flute (his main talents were, however, not in this direction), while Smog, his gentle, dirty, and all-bastard dog, looked fondly at his master or slept peacefully at his feet.

Shortly after the departure of Stan, Piero Angeletti joined our group and for twelve years became my partner in this exploration. Piero moved into Stan's office and brought to the problem his youth and imagination combined with a strong scientific drive and remarkable talent. The NGF pursuit now took a somewhat different direction.

By the time Piero made his entrance we no longer entertained the naive belief that the NGF phenomenon could be explained with a few more months of work. The rules of the game were clear by now, and while the problem was all the more challenging, especially for a newcomer, they warned against hasty conclusions and prepared us to accept defeat gracefully.

The strategy of Piero was markedly different from that of Stan and reflected their different backgrounds, personal inclinations, and, most of all, the natural history of the NGF. Now that it had revealed itself, the search demanded a new course and suggested a different tactic. Rather than concentrating all his efforts on elucidating the nature of this molecule, Piero, at the head of a small team (in 1961 we built a new study center in Rome, which remained in close connection with that in Saint Louis through the continuous commuting of members of the two groups), guided the attack in several directions, aiming at uncovering not only the nature of the NGF but also its origin, significance, and mechanism of action.

The extraction procedure of the salivary NGF was considerably improved, and, as a result, a highly purified NGF became the object of studies that paved the way to its complete characterization (Bocchini and Angeletti, 1969). It was, however, only in 1971 that this goal was reached (R. H. Angeletti and Bradshaw, 1971). This achievement climaxed twenty years of work and settled one of the most debated and crucial aspects of the whole phenomenon, namely, the nature of the NGF molecule, which had been conceived by some investigators in the mystic light of a Trinity, or as a triune entity whose biological activity would result from the cooperative action of three different subunits (Varon, Nomura, and Shooter, 1967). The NGF, as unveiled by Ruth Angeletti and Ralph Bradshaw, was instead identified as a rather small dimeric molecule consisting of two identical

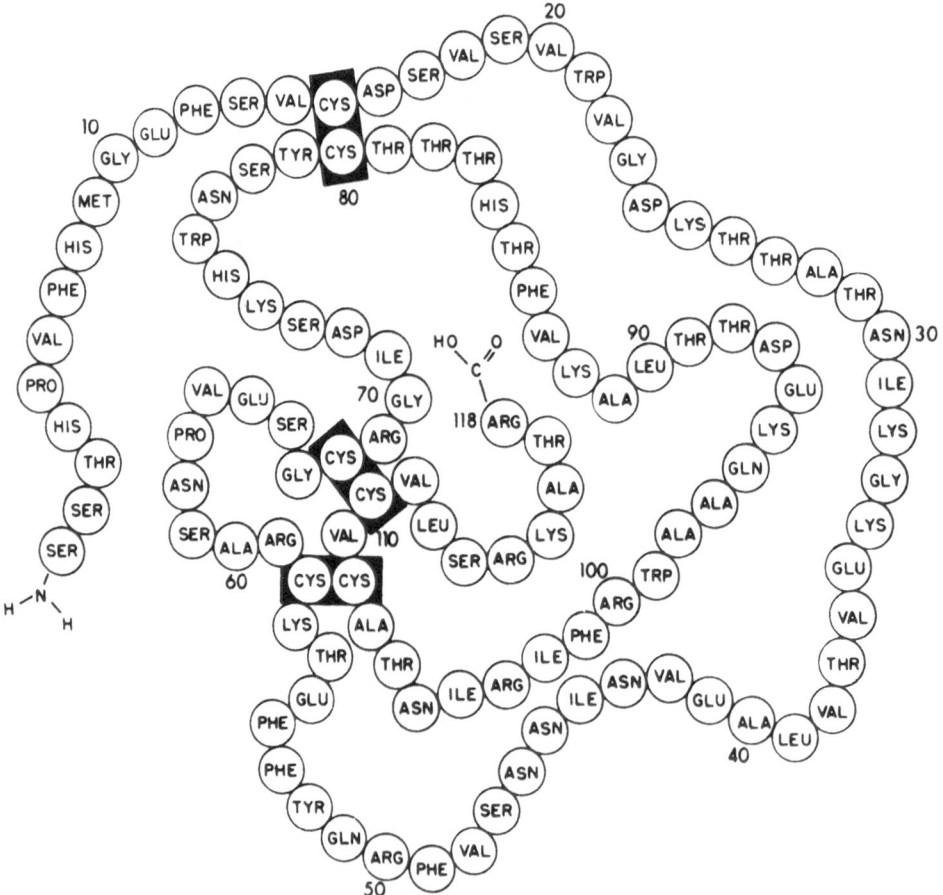

Figure 14.4   A portrait of the NGF: a schematic representation of the amino-acid sequence of the primary subunit of 2.5 S NGF from mouse submaxillary gland. From R. H. Angeletti and Bradshaw (1971).

subunits of molecular weight 13,259, each subunit consisting of 118 amino-acid chains held together by three disulfide bonds (Figure 14.4).

Among other advances that took place in the period 1960–1970 I shall mention here only the most significant ones. Biochemical and immunological studies gave evidence for the remarkable similarity of the venom and salivary NGF (Levi-Montalcini and Angeletti, 1968; Angeletti, Levi-Montalcini, and Zanini, 1971). The finding of such a potent growth factor in two exocrine glands such as the snake venom gland and mouse salivary gland raised the question of whether this factor is produced and released in a hormonal fashion from these glands, and should therefore be classified as a hormone, or whether growth factors such as the NGF and the EGF belong to a class of biologically active agents that is different from that of hormones. Extensive studies on the salivary NGF and its production and release mechanisms gave evidence for substantial differences between NGF and classical hormones (Levi-Montalcini, 1966; Levi-Montalcini and Angeletti, 1968); the question, however, is far from being settled, as indicated by the recent extensive reexamination of the problem by Hendry and Iversen (1973).

The mechanism of action of the NGF and the characterization of the growth response were studied at the metabolic and ultrastructural levels. Experiments in vitro showed that all anabolic processes are markedly stimulated in embryonic sensory and sympathetic cells; puromycin and cyclohexamide block the outgrowth of nerve fibers from the explanted ganglia, while actinomycin-D, at a concentration which inhibits RNA synthesis, does not entirely prevent the formation of the fibrillar halo, thus disproving the early hypothesis that this factor may act at the transcription level (Angeletti, Levi-Montalcini, and Calissano, 1968; Levi-Montalcini, 1964; Levi-Montalcini and Angeletti, 1971; Larrabee, 1972).

Electron-microscopic studies revealed that the most precocious and marked NGF effect is the production of neurofilaments and neurotubules, which fill the cell cytoplasm and from there are funneled into the axon (Levi-Montalcini et al., 1968). This effect, which is difficult to fit into the concept of a major role for neurotubules in intracellular transport processes (a role that would conceivably be called upon in the fully differentiated neuron rather than in nerve cells at their early inception), will be considered again in the last section.

While continuing these studies, our attention was focused on the sympathetic neuron, not only as the target cell of the NGF but also as a most convenient model of nerve-cell growth and differentiation. As is all-too-well known, we owe to this neuron the discovery of the chemical nature of nerve impulse transmission as well as the subsequent identification of the adrenergic neurotransmitter. These discoveries were soon followed by the elucidation of the metabolic pathways and of the enzymes involved in the synthesis and degradation of noradrenaline. The successful development of drugs that compete with noradrenaline for its storage sites or interfere with its release and uptake mechanisms opened a new field in neuropharmacology and was of immediate clinical significance. Among the drugs that were found to block the transmission of the nerve impulse from the sympathetic adrenergic neuron, a dopamine analogue, six-hydroxydopamine (6-OHDA), proved to be most effective. In 1968 Thoenen and Tranzer showed that, at variance with other adrenergic-neuron blocking agents, 6-OHDA suppresses the transmission of the nerve impulse by causing a selective destruction of the synaptic vesicles. The process is reversible, and four to eight weeks after discontinuation of the treatment both the integrity of these organelles and the sympathetic function are fully restored (Thoenen and Tranzer, 1968).

The extensive experience gained in our laboratory on the differential vulnerability of immature and mature sympathetic neurons (the former are destroyed by a specific antiserum to the NGF, while the latter are only temporarily impaired) suggested to Piero that we should assay 6-OHDA for its possible noxious effects in newborn animals. The dopamine analogue was injected daily in newborn mice and rats for a week, and the sympathetic chain ganglia were inspected by stereo and light microscopy weeks and months after discontinuation of the treatment. The results fully confirmed the hypothesis that immature sympathetic neurons are much more severely affected by this drug than are fully differentiated nerve cells. The 6-OHDA treatment resulted, in fact, in the destruction of 95–98% of neurons located in the sympathetic para- and prevertebral chain ganglia. The process is irreversible, as indicated by studies of treated animals one to two years later (Angeletti and Levi-Montalcini, 1970). Thus a new and most effective method to suppress the sympathetic system in newborn animals became available, known as "chemical sympathectomy."

Studies with the electron microscope showed that the lesions produced by 6-OHDA differ markedly from those caused by antibodies to the NGF. The latter are localized at first in the nuclear compartment and consist of disaggregation of the nucleolus and clumping of the chromatin (Levi-Montalcini, Caramia, and Angeletti, 1969), while the former consist of dilation and rupture of the cisternal lamellae and selective destruction of the synaptic vesicles in the nerve end terminals (Angeletti and Levi-Montalcini, 1972).

It seemed of interest to see whether NGF would prevent the lethal effects of 6-OHDA on the immature sympathetic neuron. Newborn mice and rats were injected daily for one- to four-week periods with NGF and 6-OHDA, and the ganglia were then dissected out, compared with those of controls or NGF-injected littermates, and examined at the optic and electron microscope. We expected that the ganglia would be equal in size to those of animals injected with only the NGF if the growth factor entirely obliterated the 6-OHDA effects, and of an intermediate size if the NGF and 6-OHDA effects added to and compensated for each other. Once again we were faced instead with a much more complex and, indeed, unpredictable result. The combined NGF and 6-OHDA treatment produced a dramatic enhancement of the NGF effects, as shown in Figure 14.5. Ganglia of rats injected with both agents undergo a further volume increase that amounts to three or four times that of NGF-injected littermates and about thirty times that of controls (Levi-Montalcini, 1974; Levi-Montalcini, Revoltella, and Calissano, 1974). Studies with light and electron microscopes showed that this effect is mainly due to an overproduction of axons by sympathetic neurons, which compare in size to those of NGF-treated rats. The mechanism and possible cause of this paradoxical effect are now under investigation.

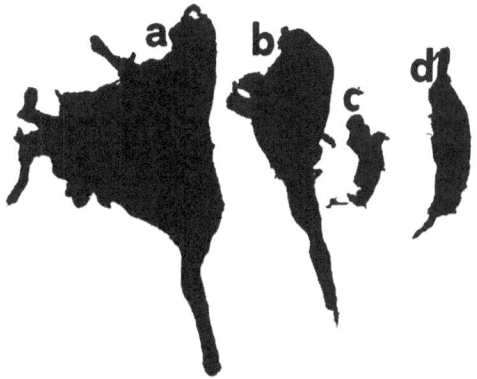

Figure 14.5   The paradoxical effect of NGF plus 6-OHDA. Whole mounts of the superior cervical ganglia of nineteen-day littermate rats injected since birth with: a, NGF plus 6-OHDA; b, NGF; c, 6-OHDA; d, saline. From Levi-Montalcini (1974).

## The NGF and the Forgotten Organelle

"Although the ubiquity in living cells of these fibrous structures (microtubules) is now established, little is known concerning their role; they may be regarded as organelles vital to cell function. The status of our knowledge about these organelles is comparable with that about mitochondria a generation ago" (Schmitt, 1968b). Thus F.O. Schmitt, in his usual forceful way, called the attention of the neurobiologist to the function of these forgotten organelles in this as well as in several other articles that appeared between 1967 and 1970 (Schmitt, 1968a,b, 1970).

Interest in neuronal fibrous proteins, first identified in neurofibrils, goes back to the nineteenth century, when these filamentous structures were first detected, in the cell cytoplasm and in the axons of nerve cells, by their strong and selective affinity for metal salts. Their ubiquitous presence in nerve but not other cells was correlated with the unique function of nerve cells: the production and transmission of the nerve impulse. The hypothesis was abandoned at the beginning of this century when the membrane theory replaced the neurofibril theory. Interest in neuronal fibrous proteins in nerve cells was revived in the fifties, mainly because of Schmitt and his classic studies on the physicochemical properties of these organelles (Schmitt, 1950, 1957; Schmitt and Davison, 1961; Schmitt and Geren, 1950). More recently, various authors have presented evidence that neurotubules play an all-important role in axoplasmic flow (Weiss, 1967, Dahlström, 1969), as well as in several other intracellular transport processes (Tilney, 1971). At the same time a much clearer distinction has been made between three different fibrous proteins: the neurofilament (present only in nerve cells); the microfilament or actinlike filament; and the microtubule (Fine and Bray, 1971; Shelanski, Gaskin, and Cantor, 1973; Yamada, Spooner, and Wessells, 1970); both microfilaments and microtubules are regular constituents of most eukaryotic cells.

What is the significance, if any, of the NGF in relation to these fibrous proteins? We shall consider only one of them, the microtubule.

It will be remembered that the most outstanding feature of the NGF response is the massive production by its target cells of neurotubules and neurofilaments. This effect, which was noticed at the beginning of the investigation in 1951, came into sharper relief in 1969 when we studied at the electron microscope the earliest changes evoked by the NGF in vitro (Levi-Montalcini et al., 1968). It was, however, not until the spring of 1973 that the overproduction of neurotubules by NGF-treated sensory and sympathetic nerve cells in vitro suddenly became the object of intensive efforts to elucidate the role of microtubules in the NGF growth response. The credit for this new turn (for, as we shall see, this novel approach has already showns its validity as well as its potential for further development) must go to a neurochemist, Piero Calissano, and an immunologist, Roberto Revoltella, both of whom joined our group and explored the NGF phenomenon with an unprejudiced eye and with new tools. In the following section I shall summarize both lines of investigation, even if the overall picture that emerges is still hazy and is likely to undergo changes as it comes into sharper focus.

Calissano, who had had extensive experience in studies on the specific proteins of nervous tissue, conceived the idea of exploring a possible interaction of NGF with some brain proteins. This inspired guess was fully justified by preliminary experiments which showed that a solution of brain proteins undergoes an almost instantaneous and marked turbidity upon addition of the NGF. Electrophoretic analysis of the pellet revealed that NGF coprecipitates with one single protein band, which exhibits a molecular weight and amino-acid composition strikingly similar to those of tubulin (Calissano and Cozzari, 1974).

While these experiments were in progress, Roberto Revoltella and his coworkers explored the binding of the NGF to its target cells with immunological and biochemical techniques as a possible model system to analyze the mechanism of interaction between a ligand and its structural membrane receptor. In preference to the sympathetic nerve cells they selected murine neuroblastoma (NB) cells (C 1300), which share many properties in common with sympathetic neurons but have the advantage of being much larger and of growing in suspension as well as in monolayer; they can therefore be harvested in large quantity and used for immunological and biochemical tests. Studies with $^{125}$I-NGF and with the rosette-forming technique showed that the NB cells bind NGF onto their surface with an avidity that is several orders of magnitude higher than that of a variety of other ligands. A binding capacity of the same order is exhibited by sympathetic but not by other normal or neoplastic cells (Levi-Montalcini, Revoltella, and Calissano, 1974).

Electron-microscopic studies of rosettes of NB- and NGF-coated sheep erythrocytes brought to light other features of the interaction between these two partners, strongly suggestive of a high specificity of the NGF binding for this "vicarious" cell target. At 37°C the NGF-coated erythrocytes are rapidly wrapped by folding of the NB-cell membrane and then phagocytized. At the contact point, immediately beneath the cell membrane, microtubules appear in large number. The tubulin polymerization, which is responsible for microtubule formation, is immediately followed by interiorization of NGF-coated erythrocytes into NB cells (Revoltella, Bertolini, and Pediconi, 1974; Revoltella et al., 1974). A protein extracted from

NB cells interacts and precipitates upon addition of the NGF. The amino-acid content, the electrophoretic pattern, and the molecular weight of this protein are remarkably similar to those of tubulin extracted from rat brain. These results suggest that, on the membrane of neuroblastoma and sympathetic cells (studies performed on the latter, although less extensive, are in agreement with those on NB cells), there is a protein that selectively interacts with NGF and exhibits physicochemical properties remarkably similar to those of tubulin.

These findings seem to indicate that microtubule proteins play a most prominent role in the chain of events triggered by NGF. Beside their generally acknowledged function in mediating intracellular transport processes, they would mediate the effects of agents such as the NGF at the membrane level. A similar role for neurotubule proteins has been suggested recently in many hormone-stimulated effects, in transduction processes, and in several immune reactions of lymphocytes or macrophages (Strom et al. 1973; Taylor et al., 1973; Taylor and Varela, 1971; Edelman, Yahara, and Wang, 1973; Inoué and Sato, 1967; Plaut, Lichtenstein, and Henney, 1973; Ukena and Berlin, 1972).

When compared with other similar or dissimilar systems, the one under investigation presents considerable advantages and recommends itself for this study. The target cell, the sympathetic neuron, has been the object of intense analysis, and its structural and functional properties are known to an extent unparalleled not only among nerve cells but even among other cells of simpler design than neurons. Hence, any deviation in growth and differentiation processes is easily noticed and tracked down. The microtubule proteins are unusually prominent in sympathetic nerve cells and lend themselves to exploration not only in this cell but also in its neoplastic counterpart, the neuroblastoma cell. Last, but not least, the NGF stands out today as one of the best known hormonal and growth factors, and its long-sought mechanism and site of action are slowly but steadily coming into focus.

**Twenty Years Later**

The long search for the identification of the NGF, which started under the sun of Rio de Janeiro in 1953, ended two years ago when the NGF with its small package of tightly coiled amino acids stood proudly in front of the photographer and of its admirers. But this achievement did not bring the exploration to an end.

In fact, the NGF, for all its feats, has still not found its place in the ever-changing game on the neuroscience chessboard, and this is, perhaps, the best sign of its vitality and of its potential rather than actual impact in a field that is itself in its most vigorous growth stage.

At the time of writing of this biography, the NGF, as has been its habit ever since it came to light, has driven its hunters into new surroundings. At variance with the past, however, and in line with prevailing trends, the new surroundings are not the open space of an unexplored world but, rather, the narrow precinct of a small field that has already been extensively explored but has still revealed only a few of its hidden treasures.

The chase is now taking place inside rather than outside the adrenergic neuron, a cell that many times in these last decades has generously rewarded its explorers. Thus, among other merits, the NGF deserves to be praised for once again bringing to the forefront this nerve cell that used to belong to the third world and is now asking for a long-overdue primary role on the widening stage of neurobiology.

Now that the NGF has come of age and the most picturesque and adventurous phase of its life is over, the biographer, who has had some part in the chase, entrusts it, with love, to younger and more skillful hands.

## References

Angeletti, P.U., and Levi-Montalcini, R. (1970): Sympathetic nerve cell destruction in newborn mammals by 6-hydroxydopamine. *Proc. Natl. Acad. Sci. USA* 65:114–121.

Angeletti, P.U., and Levi-Montalcini, R. (1972): Growth inhibition of sympathetic cells by some adrenergic blocking agents. *Proc. Natl. Acad. Sci. USA* 69:86–88.

Angeletti, P.U., Levi-Montalcini, R., and Calissano, P. (1968): The nerve growth factor (NGF): Chemical properties and metabolic effects. *Adv. Enzymol.* 31:51–75.

Angeletti, P.U., Levi-Montalcini, R., and Zanini, A. (1971): Immunochemical properties of the nerve growth factor. *In: Hormones in Development.* Hamburgh, M., and Barrington, E.J.W., eds. New York: Appleton-Century-Crofts, pp. 731–738.

Angeletti, R.H., and Bradshaw, R.A. (1971): Nerve growth factor from mouse submaxillary gland: Amino acid sequence. *Proc. Natl. Acad. Sci. USA* 68:2417–2420.

Bocchini, V., and Angeletti, P.U. (1969): The nerve growth factor: Purification as a 30,000-molecular-weight protein. *Proc. Natl. Acad. Sci. USA* 64:787–794.

Bueker, E.D. (1948): Implantation of tumors in the hind limb field of the embryonic chick and the developmental response of the lumbosacral nervous system. *Anat. Rec.* 102:369–389.

Calissano, P., and Cozzari, C. (1974): Interaction of NGF with the mouse brain neurotubule proteins. *Proc. Natl. Acad. Sci. USA* 71:2131–2135.

Cohen, S. (1960): Purification of a nerve-growth promoting protein from the mouse salivary gland and its neurocytotoxic antiserum. *Proc. Natl. Acad. Sci. USA* 46:302–311.

Cohen, S. (1962): Isolation of a mouse submaxillary gland protein accelerating incisor eruption and eyelid opening in the new-born animal. *J. Biol. Chem.* 237:1555–1562.

Cohen, S., and Levi-Montalcini, R. (1956): A nerve growth-stimulating factor isolated from snake venom. *Proc. Natl. Acad. Sci. USA* 42:571–574.

Cohen, S., Levi-Montalcini, R., and Hamburger, V. (1954): A nerve growth-stimulating factor isolated from sarcomas 37 and 180. *Proc. Natl. Acad. Sci. USA* 40:1014–1018.

Dahlström, A. (1969): Synthesis, transport, and life-span of amine storage granules in sympathetic adrenergic neurons. *In: Cellular Dynamics of the Neuron* (Symposia of the International Society for Cell Biology, vol. 8). Barondes, S.H., ed. New York: Academic Press, pp. 153–174.

Edelman, G.M., Yahara, I., and Wang, J.L. (1973): Receptor mobility and receptor-cytoplasmic interactions in lymphocytes. *Proc. Natl. Acad. Sci. USA* 70:1442–1446.

Fine, R.E., and Bray, D. (1971): Actin in growing nerve cells. *Nature [New Biol.]* 234:115–118.

Hamburger, V. (1934): The effects of wing bud extirpation on the development of the central nervous system in chick embryos. *J. Exp. Zool.* 68:449–494.

Hendry, I.A., and Iversen, L.L. (1973): Reduction in the concentration of Nerve Growth Factor in mice after sialectomy and castration. *Nature* 243:500–504.

Inoué, S., and Sato, H. (1967): Cell motility by labile association of molecules. The nature of mitotic spindle fibers and their role in chromosome movement. *J. Gen. Physiol.* 50 (Suppl.):259–288.

Larrabee, M.G. (1972): Metabolism during development in sympathetic ganglia of chickens: Effects of age, nerve growth factor and metabolic inhibitors. *In*: Zaimis and Knight (1972), pp. 71–88.

Levi-Montalcini, R. (1949): The development of the acoustico-vestibular centers in the chick embryo in the absence of the afferent root fibers and of descending fiber tracts. *J. Comp. Neurol.* 91:209–241.

Levi-Montalcini, R. (1952): Effects of mouse tumor transplantation on the nervous system. *Ann. NY Acad. Sci.* 55:330–343.

Levi-Montalcini, R. (1964): Growth control of nerve cells by a protein factor and its antiserum. *Science* 143:105–110.

Levi-Montalcini, R. (1966): The Nerve Growth Factor: its mode of action on sensory and sympathetic nerve cells. *Harvey Lect.* 60:217–259.

Levi-Montalcini, R. (1974): Control mechanisms of the adrenergic neuron. *In: Dynamics of Degeneration and Growth in Neurons* (Wenner-Gren Symposium Series, vol. 22). Fuxe, K., ed. New York: Pergamon Press, pp. 297–314.

Levi-Montalcini, R., Aloe, L., and Johnson, E.M., Jr. (1973): Interaction between the nerve growth factor (NGF), guanethidine and 6–hydroxydopamine in sympathetic neurons. *In: Frontiers in Catecholamine Research* (III International Catecholamine Symposium, Université de Strasbourg, 1973). Usdin, E., and Snyder, S.H., eds. New York: Pergamon Press, pp. 267–276.

Levi-Montalcini, R., and Angeletti, P.U. (1968): Nerve growth factor. *Physiol. Rev.* 48:534–569.

Levi-Montalcini, R., and Angeletti, P.U. (1971): Ultrastructure and metabolic studies on sensory and sympathetic nerve cells treated with the nerve growth factor and its antiserum. *In: Hormones in Development.* Hamburgh, M., and Barrington, E.J.W., eds. New York: Appleton-Century-Crofts, pp. 719–730.

Levi-Montalcini, R., and Booker, B. (1960a): Excessive growth of the sympathetic ganglia evoked by a protein isolated from mouse salivary glands. *Proc. Natl. Acad. Sci. USA* 46:373–384.

Levi-Montalcini, R., and Booker, B. (1960b): Destruction of the sympathetic ganglia in mammals by an antiserum to a nerve-growth protein. *Proc. Natl. Acad. Sci. USA* 46:384–391.

Levi-Montalcini, R., Caramia, F., and Angeletti, P.U. (1969): Alterations in the fine structure of nucleoli in sympathetic neurons following NGF-antiserum treatment. *Brain Res.* 12:54–73.

Levi-Montalcini, R., Caramia, F., Luse, S.A., and Angeletti, P.U. (1968): In vitro effects of the nerve growth factor on the fine structure of the sensory nerve cells. *Brain Res.* 8:347–362.

Levi-Montalcini, R., and Hamburger, V. (1951): Selective growth stimulating effects of mouse sarcoma on the sensory and sympathetic nervous system of the chick embryo. *J. Exp. Zool.* 116: 321–361.

Levi-Montalcini, R., and Hamburger, V. (1953): A diffusible agent of mouse sarcoma, producing hyperplasia of sympathetic ganglia and hyperneurotization of viscera in the chick embryo. *J. Exp. Zool.* 123:233–287.

Levi-Montalcini, R., and Levi, G. (1942): Les conséquences de la destruction d'un territoire d'innervation périphérique sur le développement des centres nerveux correspondants dans l'embryon de poulet. *Arch. Biol. (Liège)* 53:537–545.

Levi-Montalcini, R., Meyer, H., and Hamburger, V. (1954): In vitro experiments on the effects of mouse sarcomas 180 and 37 on the spinal and sympathetic ganglia of the chick embryo. *Cancer Res.* 14:49–57.

Levi-Montalcini, R., Revoltella, L., and Calissano, P. (1974): Microtubule proteins in the Nerve Growth Factor-mediated response (interaction between the Nerve Growth Factor and its target cells). *Recent Prog. Horm. Res.* 30:635–669.

Lwoff, A. (1966): The prophage and I. *In: Phage and the Origins of Molecular Biology.* Cairns, J., Stent, G.S., and Watson, J.D., eds. Long Island, N.Y.: Cold Spring Harbor Laboratory of Quantitative Biology, pp. 88–99.

Moran, D.T., and Varela, F.G. (1971): Microtubules and sensory transduction. *Proc. Natl. Acad. Sci. USA* 68:757–760.

Plaut, M., Lichtenstein, L.M., and Henney, C.S. (1973): Studies on the mechanism of lymphocyte-mediated cytolysis. III. The role of microfilaments and microtubules. *J. Immunol.* 110: 771–780.

Revoltella, L., Bertolini, L., and Pediconi, M. (1974): Unmasking of Nerve Growth Factor membrane-specific binding sites in synchronized murine C 1300 neuroblastoma cells. *Exp. Cell Res.* 85:89–94.

Revoltella, R., Bertolini, L., Pediconi, M., and Vigneti, E. (1974): Specific binding of Nerve Growth Factor (NGF) by murine C 1300 neuroblastoma cells. *J. Exp. Med.* 140:437–451

Schmitt, F.O. (1950): The structure of the axon filaments of the giant nerve fibers of *Loligo* and *Myxicola. J. Exp. Zool.* 113:499–515.

Schmitt, F.O. (1957): The fibrous protein of the nerve axon. *J. Cell. Comp. Physiol.* 49:165–174.

Schmitt, F.O. (1968a): The molecular biology of neuronal fibrous proteins. *Neurosci. Res. Program Bull.* 6:119–144. Also *In: Neurosciences Research Symposium Summaries.* Vol. 3. Cambridge, Mass.: The MIT Press (1969), pp. 307–332.

Schmitt, F.O. (1968b): Fibrous proteins, neuronal organelles. *Proc. Natl. Acad. Sci. USA* 60:1092–1101.

Schmitt, F.O. (1970): Molecular neurobiology: An interpretive survey. *In: The Neurosciences: Second Study Program*. Schmitt, F.O., editor-in-chief. New York: Rockefeller University Press, pp. 867–879.

Schmitt, F.O., and Davison. P.F. (1961): Biologie moléculaire des neurofilaments. *In: Actualités Neurophysiologiques*. 3ième série. Monnier, A.-M., ed. Paris: Masson et Cie., pp. 355–369.

Schmitt, F.O., and Geren, B.B. (1950): The fibrous structure of the nerve axon in relation to the localization of "neurotubules." *J. Exp. Med.* 91:499–504.

Shelanski M.L., Gaskin, F., and Cantor, C.R. (1973): Microtubule assembly in the absence of added nucleotides. *Proc. Natl. Acad. Sci. USA* 70:765–768.

Steiner, G., and Schoenbaum, E., eds. (1972): *Immunosympathectomy*. Amsterdam: Elsevier.

Strom, T.B., Garovoy, M.R., Carpenter, C.B., and Merrill, J.P. (1973): Microtubule function in immune and nonimmune lymphocyte-mediated cytotoxicity. *Science* 181:171–172.

Taylor, A., Mamelak, M., Reaven, E., and Maffly, R. (1973): Vasopressin: Possible role of microtubules and microfilaments in its action. *Science* 181:347–349.

Taylor, D., and Varela, F.G. (1971): Microtubules and sensory transduction. *Proc. Natl. Acad. Sci. USA* 68:757–760.

Thoenen, H., and Tranzer, J.P. (1968): Chemical sympathectomy by selective destruction of adrenergic nerve endings with 6-hydroxydopamine. *Naunyn-Schmiedbergs Arch. Pharmacol.* 261:271–288.

Tilney, L.G. (1971): Origin and continuity of microtubules. *In: Origin and Continuity of Cell Organelles* (Results and Problems in Cell Differentiation. A Series of Topical Volumes in Developmental Biology. Vol. 2). Reinert, J., and Ursprung, H., eds. Berlin: Springer-Verlag, pp. 222–260.

Ukena, T.E., and Berlin, R.D. (1972): Effect of colchicine and vinblastine on the topographical separation of membrane functions. *J. Exp. Med.* 136:1–7.

Varon, S., Nomura, J., and Shooter, E.M. (1967): Subunit structure of a high-molecular-weight form of the nerve growth factor from mouse submaxillary gland. *Proc. Natl. Acad. Sci. USA* 57:1782–1789.

Weiss, P. (1967): Neuronal dynamics and axonal flow, III. Cellulifugal transport of labeled neuroplasm in isolated nerve preparations. *Proc. Natl. Acad. Sci. USA* 57:1239–1245.

Yamada, K.M., Spooner, B.S., and Wessells, N.K. (1970): Axon growth: Roles of microfilaments and microtubules. *Proc. Natl. Acad. Sci. USA* 66:1206–1212.

Zaimis, E., and Knight, J., eds. (1972): *Nerve Growth Factor and Its Antiserum*. London: Athlone Press of the University of London.

Frédéric G. N. Bremer (b. 1892, Arlon, Belgium) has, since his "retirement" in 1962, continued active research at the Center for Brain Research of the University of Brussels. His areas of interest have included the physiology of spinal cord, cerebellum, reticular formation, hypothalamus, and acoustic and visual cortex, and neural mechanisms regulating sleep-wakefulness.

# 15
## The Isolated Brain and Its Aftermath

Frédéric G. N. Bremer

This essay tells how a problem, once encountered, can continue to impregnate, as a counterpoint, a life's work devoted to research in the neurosciences.

My activity in the twenties was centered on various topics. Their diversity reflected the wide-ranging curiosity of a young researcher encountering a neurophysiology that had just received its guiding principles from the neuron doctrine, established by Ramón y Cajal's monumental work, and from the illuminating concepts that Sherrington had patiently deduced from his analysis of spinal and brain functionings and had eloquently formulated in his Silliman Lectures, which would later be published as *The Integrative Action of the Nervous System*. The physiology of sleep, which had just begun to be investigated by Ranson and by Hess, formed part of my early interest because, during a stay as a foreign assistant in Pierre Marie's service at the Salpêtrière in Paris, I had been faced with the clinical and histological aspects of the lethargic encephalitis that raged during the years 1918 and 1919. In two publications I attacked the problems of sleep from its clinical-pathological (Trétiakoff and Bremer, 1920) and psychophysiological aspects.

Two years later, in Harvey Cushing's laboratory in Boston, my curiosity concerning the physiology of sleep was reawakened when I undertook, with my friend Percival Bailey, the study of the hypothalamic-pituitary basis of the adipose-genital syndrome and diabetes insipidus in the dog. We had hoped to observe that sleep problems followed our experimental lesions of the hypothalamus, but our expectations were not fulfilled: the hypothalamic lesions were never followed by a true lethargic state.

At the beginning of the thirties I was busy with physiological and pharmacological problems involving the basis of postural muscle tone and its cerebellar regulation. These experiments, done on decerebrate cat, required that the brainstem be fully functional. It occurred to me that it might be interesting to modify the ordinary method of the decerebrate preparation by leaving the forebrain in situ after the mesencephalic transection instead of destroying it. I thought that the danger of its deterioration by trauma and cooling would be reduced by this technical modification. A second, more important motivation for the change was my curiosity about how the forebrain functions after its disconnection from the caudal neuraxis. I suspected that the effect of this isolation might be not unlike, mutatis mutandis, the one that characterizes the change in the physiological condition of the spinal cord when it is separated from the frontal neuraxis by a transverse cut or by the annular cooling of a thoracic segment, as in Trendelenburg's experiment which I had often demonstrated to my students. Also, I thought that the mesencephalic transection might demonstrate the existence of a continuous facilitation of functional innervation of the forebrain resulting from the steady flow of ascending inputs from the spinal neuraxis and the brainstem.

With this idea in mind, then, I made the mesencephalic transection caudal to the third nerves, in order to preserve a behavioral ocular expression of the activity of the forebrain, the arterial circulation of which was protected by the use of a blunt instrument for the cut. When the initial ether narcosis had dissipated, I was struck by the intense contraction of the pupils and the downward and inward dropping of the eyes covered by the nictitating membrane, as in the normal sleep of the cat. I found that the fissured pupils could be made to dilate in response to an electrical stimulation of the internal capsule. This suggested that the miosis resulted from the lack of cerebrofugal impulses, which in the waking brain exert a continuous inhibitory action on the tonic activity of the Edinger-Westphal nuclei, and that the functional condition of the isolated forebrain was one of deep sleep. But, before I could venture this interpretation, I had to wait for the construction of an electronic amplifier and the adjustment of an optical device for the faithful display of the cortical potential waves.

When, for the first time, I observed simultaneously the ocular syndrome of the *cerveau isolé* and its electrocorticogram, which was characterized by regular bursts of its 7 to 10 Hz waxing and waning potential waves, I had the shock one experiences in the face of a novel natural phenomenon (Bremer, 1935 a,b).

I soon verified that this corticogram, with its monotonous ubiquitous "spindling," bore a striking resemblance to the EEG pattern of barbital anesthesia, and that the isolated forebrain, unlike the narcotized one, was capable of vigorous photic responses; these were, however, strictly limited spatially to the visual area. I found also that a high spinal transection at the $C_1$ level, disconnecting the whole encephalon from the spinal cord, was compatible with an alert waking condition of the brain, in spite of the low blood pressure of this *encéphale isolé* preparation.

If I had been more anatomically minded, I would have concluded from these experiments that between the cervical spinal section and the mesencephalic intercollicular one there lies a neural structure that is necessary for the maintenance of the waking condition of the forebrain. Instead I attributed the sleeplike ocular and EEG symptoms of the isolated forebrain to the extensive sensory deafferentation that abolishes the steady flow of cerebropetal impulses which, because they sustain the "cortical tone," are apparently indispensable for the waking state. The synchronizing and spindling effects exerted on the occipital cortex by a bilateral optic-nerve section (experiments performed in 1939 in my laboratory by E. Claes on the *encéphale isolé* cat) gave apparent support to this interpretation, as did the deep functional depression of a cortical area resulting from undercutting it. This type of EEG silence could, however, be interrupted by the local application of stimulating drugs to the undercut cortical area (Bremer, 1938).

Without losing sight of the sleep problem, I made use of these isolated-brain and isolated-encephalon preparations to elucidate the nature of the cortical potential waves, and to study the factors that control the regularity of their succession, their overall amplitude, and their homolateral and bilateral synchronism on the cortical mantle. These experiments, in conjunction with the study of spinal dorsal- and ventral-root potentials, led me to attribute the brain waves of the alpha-rhythm frequency to synchronized synaptic potentials evolving in the cortical neuronal network (Bremer, 1949). The autorhythmical character of their succession remained a problem. A plausible explanation had to await the analysis of Andersen and Eccles (1962), which clarified the interplay of inhibitory processes and post-inhibitory rebounds at the thalamic and cortical levels.

Meanwhile the two isolated-brain preparations had also proved useful for mapping and analyzing the responses of an unanesthetized cerebral cortex to volleys of afferent impulses and to direct electrical stimuli. Associated with these studies of the acoustic, vagal, cerebellar, and vestibular cortical projections are the names of Valentine Bonnet, Robert Dow, Percival Bailey, Earl Walker, Giuseppe Moruzzi, and M. Gerebtzoff, foreign guests of the laboratory. The EEG observations of Gerebtzoff (1940) on the cortical projections of the vestibular apparatus of the rabbit had, without our suspecting it, a direct bearing on sleep physiology and, more particularly, on the cause of cortical arousal, but this relationship became apparent only later, when the ascending reticular activating system was discovered. The strength of the effect of arousal revealed by these experiences remained, however, a very interesting fact in itself. This activity was interrupted for four years by the occupation of Belgium—and of my laboratory—by the German army.

I was still confident of the validity of our deafferentation hypothesis to explain the functional condition of the isolated brain when, in 1949, Moruzzi and Magoun published the results of their reticular-stimulation experiments. This was soon followed by the work of Magoun and coworkers on the neurosurgical analysis of the cerebral effects of brainstem transections, which led them to the conclusion that the lethargic state of the *cerveau isolé* was indeed the result of the blocking of the steady flow of ascending activating impulses (French and Magoun, 1952; Lindsley et al., 1950); however, these impulses originated from the mesencephalic reticular tegmentum, and the share of sensory impulses impinging directly on the cortical mantle was of negligible importance in this tonic energizing influence (Magoun, 1963).

Magoun and Moruzzi's discovery renewed and stirred my interest in the sleep problem and inspired experiments in my laboratory designed to study the physiological properties of the ascending reticular activating system. Among the findings of these studies were the following:

1. The demonstration of the reciprocity of the functional relations between the reticular formation and the cerebral cortex, and of the importance of this cortico-reticular feedback for the maintenance of the alert waking condition (Bremer and Terzuolo, 1953, 1954).

2. The powerful facilitation of cortical evoked potentials produced, in the *encéphale isolé*, by a conditioning reticular stimulation (the phenomenon was discovered independently by Bremer and Stoupel, 1958, and Dumont and Dell, 1958).

3. A facilitation whose explanation now includes the antagonistic action exerted by the ascending reticular impulses on intrathalamic and intracortical inhibitory

processes of endogenous origin (Bremer, 1970; Steriade, Apostol, and Oakson, 1971).

4. The cholinergic nature of ascending reticular terminals in the forebrain, a conclusion based on the strong cortical activation exerted by minute doses of acetylcholine in arterial injections (Bonnet and Bremer, 1937) and on the effect of anticholinesterases on the reticular facilitation of evoked potentials (Bremer and Stoupel, 1958).

5. The necessity, in observing the behavioral activating effects of anticholinesterases, of taking into account the simultaneous impact of the drug on the diencephalon and telencephalon (Desmedt and Franken, 1959).

The activating property of corticopetal reticular impulses had logically led to the idea that an essential factor in the initiation of sleep in mammals is the dampening of the activating reticular system, by whatever mechanism this dampening is produced. In 1954, in a report on the physiological problem of sleep at the Laurentian Conference, I discussed the role of the various factors that could contribute to the induction of sleep and pointed out that the notion of a hypnogenic structure or system still raised serious theoretical and technical difficulties. These objections were removed when Moruzzi and coworkers, in 1959, demonstrated the existence of a region of the caudal brainstem, in the immediate vicinity of the tractus solitarius, which could legitimately be called a hypnogenic structure. While its stimulation by low-frequency electrical pulses regularly resulted in a diffuse synchronized driving of cortical electrogenesis, its bilateral destruction or paralysis, or its separation from the mesencephalic tegmentum by a mediopontine transection, was regularly followed by a striking and lasting arousal of the forebrain. A few years later another hypnogenic structure with similar properties was discovered independently by Sterman and Clemente (1962) and by Hernández-Peón and Chavez-Ibarra (1963) in a basal preoptic area, a region which von Economo (1918) had already described as a "sleep center" on the basis of his anatomic and clinical study of encephalitis cases.

In the framework of the reticular interpretation of the sleep-waking cycle, a plausible explanation of the hypnogenic action attributed to these bulbar and diencephalic structures was that it resulted from the direct inhibition exerted by their efferent impulses on the reticular activating system. Starting from this assumption as a working hypothesis, I performed experiments on the *encéphale isolé* preparation which showed that the electrical stimulation of the preoptic area, by single or double shocks, within the spatial limits found by Sterman and Clemente for the cortical synchronizing and hypnogenic effects of its repetitive stimulation, regularly evokes in the whole ascending reticular system slow positive field potentials whose inhibitory properties can be demonstrated by various tests (Bremer, 1973a). These experiments are in agreement with the finding by De Armand and Fusco (1971) that in the behaving cat the sleep effect of the local warming of the same preoptic area is associated with a reduction of the spontaneous activity and responsiveness of the mesencephalic reticular formation.

Sterman and Clemente (1968) had already noted the possibility of a dissociation between peripheral manifestations of the hypnogenic action of preoptic stimulation and its cortical sleep-induction effect. Another striking instance of such dissociation can be observed. A *unilateral* high-frequency (100–200 Hz) stimulation of the basal preoptic area in the *encéphale isolé* cat results regularly in *bilateral* ocular sleep symptoms (miosis, downward dropping of the eyes and of the eyelids), which begin immediately after the onset of the stimulation, increase progressively up to a maintained maximum, and disappear slowly after stimulus cessation. This ocular response, which is unaccompanied by EEG symptoms of sleep, may be related to the feeling of heavy eyelids that warns us of the imminence of a nap that we are trying to resist. In their study of the hypnogenic effect of the stimulation of vagal afferents in the *encéphale isolé* curarized cat, Padel and Dell (1965) noted a similar but less complete dissociation: the miosis always preceded the appearance of spindles and cortical slow waves by at least ten seconds. In the case of the stimulation of the preoptic basal area, the fact that this stimulation immediately brings on an ocular sleep syndrome should be considered an additional proof for the existence of a sleep structure centered in this diencephalic region.

Finally, a recent episode of sleep research, in which I did not share, deserves to be told. It concerns the *direct* participation of sensory impulses in the maintenance of the tone of a cortical receiving area. After the discovery of the ascending reticular activating system, I had given up this idea. Yet experiments by Andersson, Sterman, and Rougeul and their associates in the last decade (for literature see Bremer, 1974) leave little doubt that sensory lemniscal impulses impinging without a reticular participation on the cat's somatosensory area contribute efficiently to the maintenance of its local tone, as indicated by a desynchronized electrogenesis. These observations do not endanger the reticular theory of arousal. The local cortex-activating effect of lemniscal afferent impulses could be considered as a corollary of the triggering and directing properties of tactile and proprioceptive impulses issued from the moving limbs; also significant is the absence of alerting significance of the same impulses. This would not apply to trigeminal impulses (Roger, Rossi, and Zirondoli, 1956) nor to visual impulses, which exert a powerful activation of the mesencephalic reticular formation resulting in part from a corticoreticular positive feedback. Bonnet and Briot (1972) have, in fact, observed in the *encéphale isolé* cat that the stimulation of the lateral geniculate body, even by a single electrical shock, regularly produces a strong activation of the mesencephalic reticular formation. This activation is linked to the reaction of the visual cortex because it is abolished immediately following the inactivation by freezing of the visual area ipsilateral to the geniculate nucleus that was stimulated. I was able to confirm this observation of Bonnet and Briot and to demonstrate (in studies not yet published) that this reticular activation of cortical origin is powerfully inhibited by a conditioning stimulation from the basal preoptic area.

The experimental data summarized in this essay, the autobiographical character of which was required by the theme of this symposium, can perhaps be considered as accounting satisfactorily for the regulation of arousal and for sleep induction. Yet, they leave unanswered the fundamental question of the mechanism by which the sleep state, once initiated, is maintained steadily for hours. Such terms as the

French *enclenchement*, used to describe the broad categorization of the phenomenon, illustrate by their naked descriptive verbalism the depth of our ignorance. It is here that the revelation of the physiological importance of the intracerebral concentration and turnover of monoamines, especially indoleamines and catecholamines (see Jouvet, 1972), may hasten the solution of the sleep problem and lead to a harmonious synthesis of the information collected by the traditional connectionist and electrophysiological methods and by the approaches that rely mainly on biochemical and pharmacological data (the "dry" and the "wet" physiologies, to use the humorous shorthand terms coined by Francis Schmitt).

## Postscript

Since the completion of this article I have been able to demonstrate the existence of a mutual tonic inhibitory interaction between the preoptic hypnogenic structure and the ascending reticular formation. A striking aspect of this interaction is the strong and lasting activation of the hypnogenic area, which follows the release from its tonic inhibition after the transection of the mesencephalic tegmentum. Another feature of this preoptic activation is its inhibition by light, under conditions suggesting the mediation of the photic effect by direct retino-preoptic fibers (*Brain Research*, in press).

## References

Andersen, P., and Eccles, J.C. (1962): Inhibitory phasing of neuronal discharge. *Nature* 196:645–647.

Bailey, F., and Bremer, F. (1922): Experimental diabetes insipidus. *Arch. Int. Med.* 28:773–803.

Batini, C., Moruzzi, G., Palestini, M., Rossi, G.F., and Zanchetti, A. (1959): Effects of complete pontine transections on the sleep-wakefulness rhythm: The midpontine pretrigeminal preparation. *Arch. Ital. Biol.* 97:1–12.

Bonnet, V., and Bremer, F. (1937): Action du potassium, du calcium, et de l'acétylcholine sur les activités électriques, spontanées et provoquées, de l'écorce cérébrale. *C.R. Soc. Biol. (Paris)* 126:1271–1275.

Bonnet, V., and Briot, R. (1972): Participation du cortex visuel à l'activation sensorielle de la formation réticulée. *C.R. Acad. Sci. (Paris)* 274:2341–2343.

Bremer, F. (1935a): Cerveau "isolé" et physiologie du sommeil. *C.R. Soc. Biol. (Paris)* 118:1235–1241.

Bremer, F. (1935b): Quelques propriétés de l'action électrique du cortex cérébral isolé. *C.R. Soc. Biol. (Paris)* 118:1241–1244.

Bremer, F. (1937): L'activité cérébrale au cours du sommeil et de la narcose. *Bull. Acad. R. Med. Belg.* (Series 6) 2:68–86.

Bremer, F. (1938): Effets de la déafferentation complète d'une région de l'écorce cérébrale sur son activité électrique spontanée. *C.R. Soc. Biol. (Paris)* 127:355–359.

Bremer, F. (1949): Considérations sur l'origine et la nature des "ondes" cérébrales. *Electroencephalogr. Clin. Neurophysiol.* 1:177–193.

Bremer, F. (1954): The neurophysiological problem of sleep. *In: Brain Mechanisms and Consciousness.* Adrian, E.D., Bremer, F., and Jasper, H.H., eds. Oxford: Blackwell, pp. 137–162.

Bremer, F. (1958): Cerebral and cerebellar potentials. *Physiol. Rev.* 38:357–388.

Bremer, F. (1970): Inhibitions intrathalamiques récurrentielles et physiologie du sommeil. *Electroencephalogr. Clin. Neurophysiol.* 28:1–16.

Bremer, F. (1972): La formation réticulaire activatrice et le problème physiologique du sommeil. *Acta Neurol. Belg.* 72:73–84.

Bremer, F. (1973a): Preoptic hypnogenic area and reticular activating system. *Arch. Ital. Biol.* 111:85–111.

Bremer, F. (1973b): Tonus cortical et afférences cérébrales. *Arch. Ital. Biol.* 111:462–467.

Bremer, F. (1974): Historical development of ideas of sleep. *In: Basic Sleep Mechanisms.* Petre-Quades, O. and Schlag, J.D., eds. New York: Academic Press, pp. 3–11.

Bremer, F., and Stoupel, N. (1958): De la modification des réponses sensorielles corticales dans l'éveil réticulaire. *Acta Neurol. Belg.* 58:401–403.

Bremer, F., and Terzuolo, C. (1953): Interaction de l'écore cérébrale et de la formation réticulée du tronc cérébral dans le mecanisme de l'éveil et du maintien de l'activité vigile. *J. Physiol. (Paris)* 45:56–57.

Bremer, F., and Terzuolo, C. (1954): Contribution à l'étude des mécanismes physiologiques du maintien de l'activité vigile du cerveau. Interaction de la formation réticulée et de l'écorce cérébrale dans le processus du réveil. *Arch. Int. Physiol.* 62:157–178.

Claes, E. (1939): Contributions à l'étude physiologique de la fonction visuelle. I. Analyse oscillographique de l'activité spontanée et sensorielle de l'aire visuelle corticale chez le chat non anesthésié. *Arch. Int. Physiol.* 48:181–237.

De Armand, S.J., and Fusco, M.M. (1971): The effect of preoptic warming on the arousal system of the mesencephalic reticular formation. *Exp. Neurol.* 33:653–670.

Desmedt, J.E., and Franken, L. (1959): Neurohumoral factors influencing the electrical activity of the cerebral cortex. *In: First International Congress of Neurological Sciences.* (Congress held in Brussels, 21–28 July 1957). Vol. III. *Electro-encephalography, Neurophysiology and Epilepsy.* Bogaert, L.V., and Radermecker, J., eds. London: Pergamon Press, pp. 356–360.

Dumont, S., and Dell, P. (1958): Facilitations spécifiques et nonspécifiques des réponses visuelles corticales. *J. Physiol. (Paris)* 50:261–264.

Economo, C. von (1918): *Die Encephalitis Lethargica.* Vienna: Deuticke.

Evarts, E.V. (1973): Motor cortex reflexes associated with learned movement. *Science* 179:501–503.

French, J.D., and Magoun, H.W. (1952): Effects of chronic lesions in central cephalic brain stem of monkeys. *Arch. Neurol. Psychiatr.* 68:591–604.

Gerebtzoff, M. (1940): Recherches sur la projection corticale du labyrinthe. I. Des effets de la stimulation labyrinthique sur l'activité électrique de l'écorce cérébrale. *Arch. Int. Physiol.* 50:59–99.

Hernández-Peón, R. and Chavez-Ibarra, G. (1963): Sleep induced by electrical or chemical stimulation of the forebrain. *Electroencephalogr. Clin. Neurophysiol.* (Suppl) 24:188–198.

Jouvet, M. (1972): The role of monoamines and acetylcholine-containing neurons in the regulation of the sleep-waking cycle. *Ergeb. Physiol.* 64:166–307.

Lindsley, D.B., Schreiner, L.H., Knowles, W.B., and Magoun, H.W. (1950): Behavioral and EEG changes following chronic brain stem lesions in the cat. *Electroencephalogr. Clin. Neurophysiol.* 2: 483–498.

Magoun, H.W. (1963): *The Waking Brain.* 2nd. ed. Springfield, Ill.: C. C Thomas.

Moruzzi, G. (1972): The sleep-waking cycle. *Ergeb. Physiol.* 64:1–165.

Moruzzi, G., and Magoun, H.W. (1949): Brain stem reticular formation and activation of the EEG. *Electroencephalogr. Clin. Neurophysiol.* 1:455–473.

Padel, Y., and Dell, P. (1965): Effets bulbaires et réticulaires des stimulations endormantes du tronc vago-aortique. *J. Physiol. (Paris)* 57:269–270.

Roger, A., Rossi, G.F., and Zirondoli, A. (1956): Le rôle des afférences des nerfs crâniens dans le maintien de l'état vigile de la préparation "encephale isolé." *Electroencephalogr. Clin. Neurophysiol.* 8:1–13.

Sherrington, C.S. (1906): *The Integrative Action of the Nervous System.* New Haven: Yale University Press.

Steriade, M., Apostol, V., and Oakson, G. (1971): Control of unitary activities in cerebellothalamic pathway during wakefulness and synchronized sleep. *J. Neurophysiol.* 34:389–413.

Sterman, M.B., and Clemente, C.D. (1962): Forebrain inhibitory mechanisms: Cortical synchronization induced by basal forebrain stimulation; sleep patterns induced by basal forebrain stimulation in the cat. *Exp. Neurol.* 6:91–102, 103–117.

Sterman, M.B., and Clemente, C.D. (1968): Basal forebrain structures and sleep. *Acta. Neurol. Lat. Am.* 14:228–244.

Trétiakoff, C., and Bremer, F. (1920): Encéphalite léthargique avec syndrome Parkinsonnien et catatonie. Rechute tardive. Vérification anatomique. *Rev. Neurol.* 7:772–775.

# Sensory and Motor Systems

William A. H. Rushton (b. 1901, London, England) has been, since his retirement from Cambridge University in 1968, Distinguished Research Professor at Florida State University. The two major phases of his research career have centered on nerve excitation and conduction (from 1927 to 1951) and then on visual pigments and the mechanism of vision (from 1951 to the present).

# 16
## From Nerves to Eyes

## William A. H. Rushton

My whole life has been fortunate much beyond my deserts. Those in a position to give me the opportunities for research seem to have had a higher expectation of its outcome than I had, and their confidence has been a great encouragement and a spur for me to do better than I thought I could. I am very sensitive to personal relations, and my colleagues have always given me the real friendliness I so much need, despite the ungovernable sharpness of my scientific criticism.

The school I went to was unusually orientated toward science (physics and chemistry). Though the practical work (of necessity) differed little from the instructions of a cookery class, and experiments were designed simply to verify what I had learnt from the book as Laws of Physics, yet somehow a different and very exciting idea got through. It was that an experiment could be devised that would answer a question, something not already known. The immortal spirit of Galileo moved among those routine operations and whispered and begat dreams of creative experimentation.

When I left my boarding school ($17\frac{1}{2}$ years old), my housemaster wrote to my father: "Your son, I understand, wishes to become a research scientist, but the science staff doubts that he has the ability. I suggest that he take up Law." (I used to argue with my housemaster, who was a poor arguer!) The view of the science staff was reasonable. I had then the characteristic—which I must confess I still have—of being much keener to invent my own experiments than to read and assimilate the experiments of others, though those might be greatly superior to mine. I did not in fact study properly my textbooks of physics and chemistry, and did not succeed in winning any kind of scholarship to the University.

As for my future, my father arranged an interview with a patient of his on the staff of University College, London. I can't remember his name, but his advise was excellent. "You are not likely to get the opportunity to do research unless you get a 'first' in your finals. Do you think you can manage that?" "I did not get a scholarship, so I suppose I shall not get a first." "Well then, you must plan for an alternative career." He suggested various possibilities, and I decided to study medicine with the hope (if I did well enough) of applying my interest in physics to the problems of physiology. That decision has determined my life.

Straightway I registered and studied elementary biology at the Middlesex Hospital in London with a view to going to Cambridge to continue with anatomy and physiology the next year. I hated biology—the dissection and the systematics —a hate which I carried over to anatomy. You have to piece together so much before that marvellous jig-saw puzzle begins to reveal its picture, and I have never known enough. But I managed to pass the First M.B. by the end of the year, and at Cambridge I passed the Second M.B. in a further four terms. I got a second class in the Tripos (Cambridge third year honours finals), taking Physics, Chemistry and Physiology, and then did advanced physiology in my fourth year; I managed to get a first class in it (eight candidates, two firsts).

My greatest inspiration at this time came from a man I have never seen, Keith Lucas, killed in 1916. His little monograph *The Conduction of the Nervous Impulse*, brought out after his death by E.D. Adrian, delighted me to intoxication. My ideal of physiological research is still the beautiful way in which Lucas dissected intellectually the components of excitation and conduction, and devised experiments and built simple but elegant equipment to distinguish between the possibilities raised. Occasionally people have said that they have detected something of this quality in my research and writing. If it is there I should be very pleased, for I have tried to copy this master.

On obtaining a first class, I was accepted for Ph.D. work by E. D. Adrian and received the Michael Foster Studentship. At this time the idea that narcotized nerve conducted with a decrement was being severely challenged by Kato and his Japanese colleagues. This was important because Lucas and Adrian had measured the size of the nerve impulse—the size of the "all" in all-or-none conduction—by finding the distance that it could travel through a narcotized region. I thought of an explanation to justify the Lucas–Adrian view and reconcile it with the Japanese results, and I proposed working for my Ph.D. to substantiate this explanation. Adrian said, "If the Japanese are right, it is no use trying to prove them wrong; if Lucas and I are right, we, not you, will reap most of the credit from your work. Think of an independent line of your own." This was generous advice, but it left me a bit in the air.

**Nerve Excitation: Cable and Chronaxie**

The most important contribution to the theory of nerve excitation at that time was that of Nernst, the famous physical chemist. He had worked out some simple relations rom the fact that the flow of current through nerve must change the concentration of ions close to membranes. Together with a few other assumptions, he had been able to obtain the time–intensity relations for nerve excitation. I asked the question "Whereabouts in nerve structure is Nernst's membrane?" and I hoped to get an answer by applying exciting currents in different places and in different directions when the nerve was immersed in a large bath of Ringer's fluid. I found that the efficacy of an oblique current was proportional to the component parallel to the nerve, and was disappointed to read that this nice result had already been noted by Galvani!

It was clear that the cable structure of nerve imposed a determinate distribution of current flow across the sheath and down the core at various points, depending upon the known potential distribution in the fluid outside. The steady-state solution is a very simple case of partial differential equations, but I had not met such an equation until I wrote it down in my attempt to specify this current distribution. The clumsy mathematics of my first paper (Rushton, 1927) was an expression of my total ignorance of this branch of mathematics, but the answer was correct.

And the experimental relation between threshold current and distance apart of the exciting electrodes corresponded to the calculation from cable theory and gave the value of the "characteristic length," which has since been recognized as an important nerve constant. The results showed that current excites by leaving the sheath, and excites most strongly where the current density leaving is greatest —the confirmation of an old conclusion.

While I was working on thresholds with various other arrangements of current distribution, showing that they too fitted the predictions from the cable formula (Rushton, 1928a,b), Lapicque published a lively account of his life-work in *L'Excitabilité en Fonction du Temps* (1926).

We may consider that a constant current builds up a state of excitation in nerve rather like the charging of a condenser, and the time of half-charge is found to vary greatly between one tissue and another (e.g., medullated nerve or smooth muscle). Practical ways (free from theory) of measuring this "half-charging" time had been proposed by Gildemeister (who called it the "*Kennzeit*"), by Lucas ("the excitation time"), and by Lapicque ("the chronaxie"). Of these three names for the same measure the chronaxie became best known because Lapicque made many very extended claims for the measure, such as that the chronaxie of any motor nerve was a fixed fraction of the contraction time of the innervated muscle!

One strong tenet of Lapicque was isochronism, that the chronaxie of a muscle was the same as that of its motor nerve. This had never been accepted in Cambridge. Lucas' beautiful experiments showed chronaxie disparity of ten-fold or more, and we in the advanced class under Adrian all repeated Lucas' experiments and confirmed his results.

Lapicque had to meet the challenge of Lucas' work, and in his book he suggested that this "engineer turned physiologist" could not carry out accurate experiments. I could not allow my idol to be slighted thus. I determined to vindicate him. "How," you may ask, "can competent experimenters get such very different results? They must be getting them by doing different things." That was the case. Lucas excited his muscle immersed in a large Ringer bath through which the current flowed; Lapicque excited his muscle in air by sticking a silver pin in it with some indifferent return. Lucas' results showed two excitabilities, one with the brief chronaxie that was the same as that of the nerve excited alone through separate electrodes, the other with a very much longer chronaxie, presumably due to the muscle directly excited. This was confirmed when, after adding the poison, curare, that blocked nerve–muscle conduction, the brief chronaxie (measured still by the muscle twitch) vanished, but the longer chronaxie remained unaltered.

The crucial question was "What tissue is directly excited (and hence responsible for the chronaxie measured) by Lapicque's silver pin which impales the muscle–nerve complex?" If nerve twigs were being excited at threshold, then of course the chronaxie would be that of nerve. And if the muscle chronaxie was quite different, this would be revealed when curare had abolished nerve–muscle conduction. Lapicque agreed that curare did reveal a much longer muscle chronaxie, but he put an entirely different interpretation upon the facts. He supposed that identity of chronaxie between nerve and muscle (isochronism) was a necessary condition for conduction from nerve to muscle, and that the specific action of curare was to increase the chronaxie of muscle (but not of nerve) so that it produced a paralysing *heterochronism*.

Though there was a good deal more to be said on both sides, I thought Lucas' case so clear that it should be easy to establish it beyond doubt, especially as my analysis of current distribution in tissues gave an entirely fresh way of identifying whether it was nerve or muscle that was being excited.

I performed these experiments mainly at the Johnson Foundation (University of Pennsylvania), to which D. W. Bronk invited me (and also Ragnar Granit) for two years when he started it in 1929. I wrote seven papers for the *Journal of Physiology*, nice, neat publications that were answered again by Lapicque. But I had no experience in controversy, and Lapicque was a formidable opponent, Napoleonic in the mobility with which he varied his fighting positions. If Lapicque's theories collapsed, it was not because I had disproved them, but because isochronism (the pseudo-resonance between muscle and nerve, of Monnier) could hardly survive the damping effect of Dale and Feldberg's secretion of acetylcholine at the resonance site. And speculation about the nature of excitation at the endplate was soon replaced by Kuffler and Katz's beautiful records of what actually happens there.

I was disappointed by the Lapicque polemic. What had started clear became muddier and muddier, and I found myself fighting not so much for the truth as for my own viewpoint. Though rather controversial by nature, I have never got myself into a long running fight since, and I hope I never shall.

I met Lapicque only once. It was in 1932 at the International Physiological Congress in Rome; I was introduced, I think, by Gasser. Lapicque was charming. He asked me to object to the paper he would be giving on chronaxie "and we shall have one or two *coups de box*, no knockouts, judged on points. And you," turning to Gasser and the rest, "you must be the judges."

In my first year with Bronk we went to the Federation Proceedings, which that year (1929) were in Chicago. There Ralph Gerard organized a dinner of "axonologists" (a word he had coined) to which I was invited. I still remember the thrill of suddenly meeting face to face so many of those who until then were simply the great names in the literature: Alex Forbes, Hal Davis, Gasser, Erlanger, George Bishop, "Iron Wire" Lillie, Newton Harvey, Ali Monnier, plus, of course, Det Bronk and Ragnar Granit from the Johnson Foundation, and I expect a few others whom at the moment I do not recall.

This experience was a great enlargement for me, and since then I have met again with pleasure not only Gerard but the majority of his axonologists, sometimes in their own homes, sometimes in mine, or in various parts of the world. The international comradeship of fellow experimentalists is one of the rich fruits of our trade.

## Clinical Studies: Death and Transfiguration

I returned to Cambridge in 1931 with a Research Fellowship at Emmanuel, my old College. Sir Joseph Barcroft, Professor of Physiology, said that he would give me a University Lectureship in his department if I would go to hospital and get qualified in medicine. (I had stopped my medical studies after qualifying in anatomy and physiology, when I managed to get grants to do research.)

This was rather a setback, but I took advice, gave up research, and went to University College Hospital, London. There, or across the road at the College, I met and got to know A. V. Hill, Wilfred Trotter, Sir Francis Walshe and Sir Thomas Lewis, men of great intellectual power and scientific or clinical insight. Each made a deep impression, and all were extraordinarily kind to me, in a slightly pitying way, for I was hopeless as a medical student and later failed four times in my finals and gave up. I failed, I think, not because I was stupider than the average man who passed, but because I was absolutely overwhelmed by the mystery of health and disease and by my incapacity to understand and cure. I could not bear to contemplate that someone should ever place his health or life in my hands for management, and I devoutly hoped that this situation would never arise. But, of course, it is only by continually seeing future responsibility in imagination that students can prepare themselves for the event and finally rise successful to the challenge. I was terrified of this challenge; my thought was paralysed and my little knowledge faded into oblivion.

At my utter failure, Barcroft took pity on me (bless him) and gave me the Lectureship all the same. He said I had got the medical slant on physiology whether I had satisfied the examiners or not. So I was established to teach students to get ultimately the degree I could not get myself. This I could not stand. Two years later I went back to Hospital during the Long Vacation, decided to make the fool of myself at the ward visits instead of at the examinations, and somehow got through the finals three months later.

Back in Cambridge I found that fear and inhibition had entered my conduct of experiments—fear that my experiments would not succeed in confirming the elegant ideas in my head. I felt like a painter bringing on to canvas a beautiful vision, while, because of some crack in the roof, a fall of mud may occur at any moment and ruin the picture. So did I regard the incidence of any experimental observation that did not accord with my pretty prediction.

I hope you are shocked at this attitude, which is infected with falsification and leads to paralysis. I do not think I published anything false, but my experimenting degenerated into a sort of sleep-walking routine.

When a man in his late thirties grinds to this kind of a halt, it is fairly safe to expect no more of him. If it be allowed that I have done some more—for instance, worked creatively in vision—my thanks must be to Dr. Susan Isaacs, the distinguished Freudian, who gave me about five years of psychoanalysis during the war years (when those of us who remained in Cambridge gave up research and taught full time in the compressed medical courses).

I do not understand what is meant to happen in psychoanalysis from all the talking by the patient. Dr. Isaacs said very little, yet that little made a great change in my attitude to work and to people. In particular I began to creep back to research, and to see that the important thing about an experiment is that it be well executed and lead to some definite conclusion. If it contradicts a preconception, it is the more likely to be non-trivial, but needs careful checking for errors. I began to enjoy experimenting as in the early days, and my joy was greatly enhanced by working with a newly graduated student, Cyril Rashbass. His subtle, acute and ingenious mind was a stimulus and a delight. We worked on the rather trivial question of the deformation of current flow by the connective tissue sheath around the nerve trunk. We enjoyed inventing some ingenious tricks to establish this and indulged our fancy in greatly over-proving the case that current is deformed by the connective tissue stocking worn by the nerve trunk (Rashbass and Rushton, 1949a).

Then, suddenly, Lorente de Nó brought out his two great volumes of research in which he claimed that no such deformation occurred and that his great array of results, which seemed to us obviously due to the connective tissue, were in fact the intrinsic properties of axons. Our over-proved case now had an antagonist, but I was not going to engage in a second polemic. We marshalled our evidence in one paper (Rashbass and Rushton, 1949b), and it appears to have been sufficient.

The year 1948–1949 was a great one for me. Hodgkin and I were elected Fellows of the Royal Society, and I took my first sabbatical year's leave. I went to work in Professor Granit's laboratory in Stockholm, because I had admired his work and his philosophy of science and life and had enjoyed his friendship when we were together in Philadelphia at Bronk's invitation (1929–1931). And he warmly welcomed my suggestion that I should come. I had intended experimenting on *mammalian* nerve, with which I was unfamiliar, but I soon changed my subject to vision.

When in 1935 I returned from Hospital to Cambridge, Alan Hodgkin was just starting research, and I had the privilege of giving tutorials to Andrew Huxley. I advised Hodgkin that the research he had carefully planned was too hard for a first attempt. He disregarded me and in the course of the year completed his first paper (Hodgkin, 1938), a masterpiece, proving that nerve propagation is caused by the action current spreading forward and stimulating the nerve region in front. This had often been suggested; but sound proof is something different, and it won for Hodgkin a Trinity Fellowship after only one year of research. His remarkable insight and experimental expertise makes me entirely endorse Gasser's dictum when Hodgkin went to work with him, "This is the reincarnation of Keith Lucas."

By the time I went to Stockholm, Hodgkin and Huxley already had in mind the analysis which won them the Nobel Prize. They asked me whether I would join them in this enterprise. I gave them the right reply with an absurd reason. I said, "I think I ought first to complete our work on the resistance of the connective tissue sheath." Hodgkin in a nice understatement said, "Perhaps the analysis we plan will turn out more important than the connective tissue sheath." It did. All the same, I could not have pulled any weight rowing in *that* boat. Rowing on my own, however, I have managed to move along.

## I Switch to Vision

In the Nobel Institute of Neurophysiology I got to know—besides the Granits—

the Donners, the Frankenhaeusers, Curt von Euler and others, and also C. G. Bernhard, Ulf von Euler, Rudolph Skoglund and Svaetichin from the Karolinska Institute opposite. And many a good dinner and delightful evening I enjoyed at the house of Yngve Zotterman, whom with his wife Brita I had met in 1925–1926 when he worked with Adrian in Cambridge. All these able physiologists I have enjoyed meeting not only in Stockholm at that time, but in other parts of the world on many occasions since.

I spent this year living in the guest room in Granit's laboratory, where the bed was long and narrow. Freud describes how many people sleep curled up in the ante-natal position; in that bed one slept in the post-mortem position!

I soon became fascinated with the work that Granit and Donner were doing, recording discharges from the retinas of cats. I wished to drop work on the theory of nerve excitation, which Hodgkin and Huxley would do so much better, and take up visual physiology. But I did not wish to follow Granit exactly, for I prefer a narrower and more analytical approach. I do not know how Granit received my request to switch at once from nerves to vision. He probably thought that it was a sign of restlessness and that I would soon be switching to something else. It must have been a disappointment too, as I could hardly be expected to publish much in an entirely new domain during what remained of my sabbatical year. And a research institute relies for its continued support upon a steady output of papers.

Whatever Granit may have thought, I received nothing but help and encouragement. I gave him almost nothing that year by way of publication, but for me it was a year of germination. Granit's sponsors will have seen no fruit from my visit, but Ragnar himself knows that the hundred papers I have written on vision since entering that wide sea at the age of fifty are the result of his successful launching of a somewhat ill-balanced craft.

On returning to Cambridge, I thought about vision while working with J. J. Lussier from l'Université de Montréal on excitation of de-sheathed nerves—some pretty little experiments, though backward-looking (Lussier and Rushton, 1951, 1952).

I found that two of my students had hatched out while I was away, so that our roles were now reversed and I began to learn from them, as I have done ever since. Giles Brindley was then working with E. N. Willmer, and in an experiment to measure by reflexion the density of the human macular pigment, they obtained evidence that rhodopsin could also be measured in that way in the living human eye, which was an inspiration both to me and to R.A. Weale. Horace Barlow had been recording from frogs' ganglion cells, and had established (in 1948–1949, although not published until 1953) the centre-surround inhibitory organization of the receptive field, which was independently discovered in the cat by Kuffler (1953).

**The Photochemical Theory**

The photochemical theory is a monument to Selig Hecht's determination to give simple chemical explanations to regular biological processes. Hecht spent the year 1925–1926 in Cambridge, my first year of research. He was working on hemoglobin with Barcroft and reading the old German papers of König, etc., to compare those fine measurements with his developing photochemical theory of vision. I was immensely impressed by Hecht from what I saw of him in the tea-room—not so much by his knowledge, which was often limited (his mathematics sometimes was shocking), but by his invincible and rugged determination to get what he wanted with the means at his disposal. Though I can see that he might have been an awkward colleague, he was always a kind friend to me (albeit a bit contemptuous of my ineffectiveness). I admired the directness of his approach and enjoyed his intolerance of vagueness in others.

Hecht, working first with clams and later with man, sought to find a chemical explanation of light and dark adaptation. He hit upon the (correct) idea that the recovery of sensitivity in the dark is due to the regeneration of rhodopsin that has previously been bleached by strong light adaptation.

Then Hecht succumbed to the common pitfall of the discoverer of a good theory: he tried to make it explain not only dark adaptation but every form of adaptation. The Weber-Fechner relation, describing the rise of threshold when a test flash falls upon a luminous background, was explained by the not unreasonable assumption that the background had bleached some of the rhodopsin away, so that there was less remaining to absorb light from the test flash. The expected relation from this assumption led to Fechner's Law. Other well-established relations that depended upon the level of illumination (e.g., acuity, flicker fusion) were also made to fit the photochemical theory by choosing suitable theoretical criteria for detection and assigning suitable values to the arbitrary constants of the bleaching-regeneration equilibrium equation.

In 1937 Hecht wrote up his theory in *Physiological Reviews*. By then he had realized that the theory that accounted for rod dark adaptation, with its half-hour's regeneration, could not also account for the Fechner relation, where adaptation of rods to different levels of background equilibrates almost instantaneously. Hecht had to drop rhodopsin regeneration as the explanation of one or the other of these adaptations. Unfortunately, in the *Reviews* paper he dropped dark adaptation (where the theory was correct) and retained field adaptation (where the explanation required a regeneration rate of rhodopsin a million times too fast).

In 1942 Hecht, Shlaer and Pirenne, in a paper that has become classic, proved that a human rod can be excited by the catch of a single quantum but that flash detection requires the cooperation of about five rods. This is incompatible with Hecht's photochemical kinetics. But he had no time to develop a new theory. His last few years were spent in assessing and describing aspects of the new and astounding fact that a change in a single rhodopsin molecule can excite a whole rod to activity.

R. D. Cohen and I proved that the bleaching of only 2% of rhodopsin raises the rod threshold not by 2% but ten-fold (Rushton and Cohen, 1954), and we demonstrated this to George Wald, who was surprised but at once confirmed it (Wald, 1954). The proponents of the photochemical theory had never made any attempt to measure the actual kinetics of the bleaching and regeneration of rhodopsin. The kinetics were simply invented to fit the theory, usually on the assumption that the rhodopsin level was inversely related to the visual threshold. This assumption is quite wrong, as Cohen and I showed, and it seemed to me important to develop a technique to measure rhodopsin physically in the living eye—ultimately the human eye—and thus to observe what the kinetics actually are.

I first measured rhodopsin by reflexion densitometry in the living cat, and after seeing my demonstration (Rushton, 1952), R. A. Weale took the matter up and pushed the investigation further (Weale, 1953). I then studied bleaching and regeneration of the rhodopsin in the albino rabbit, and was joined by W. A. Hagins who had come to work with me on earthworms (my wartime field of research: Rushton, 1945, 1946). I deflected him into vision, and with our densitometer he made on the rabbit some of the first observations on flash photolysis and the nature of the early photo-products of rhodopsin (Ph.D. Thesis, Cambridge University).

F. W. Campbell and I then addressed ourselves to the measurement of rhodopsin in man, which Brindley and Willmer had indicated was possible. We built a simple human reflexion densitometer with which we were able to prove by six criteria that what we measured was rhodopsin (Campbell and Rushton, 1955). The bleaching rate was proportional to the rate of quantal absorption by rhodopsin in a variety of conditions, and the pigment level recovered in the dark along an exponential curve that had the same time constant as the rod branch of the ordinary dark-adaptation curve.

In 1959, at NIH at the invitation of M. G. F. Fuortes, I was able to study a young woman who had practically no functional cones (Rushton, 1961). Measured physically by densitometry, her rhodopsin regenerated normally, and her rod dark-adaptation curve, uncontaminated by cones over a range of about six log units, showed throughout that the rise of log threshold was proportional to the fraction of rhodopsin still in the bleached (unregenerated) state.

Dowling (1960) independently found essentially the same relation in the rat. The log threshold for the $b$-wave of the electroretinogram was a linear function of the rhodopsin content of the partly regenerated retina measured directly by excision and extraction.

## Cone Pigments and Colour Defectives

Naturally I had the ambition to measure the visual pigments in cones and find whether there were in fact three, as would be expected from the trichromacy of colour vision, and whether these pigments regenerated as fast as was indicated by the cone branch of the dark-adaptation curve.

Previously cone pigment had been extracted only from the cone-rich retinas of fowls, and the general view was that in mammals the pigment density was immeasurably small. I was delighted therefore to find that the pigment density was about as great in cones as in rods, and that the kinetics of regeneration corresponded well with the cone branch of the dark-adaptation curve (Rushton, 1958).

It had long been known that one class of the common red-green colour defective called *protanope* lacks the red component of trichromatic vision. Is this because members of this class lack the red cone pigment *erythrolabe*, or does the absorption of red light by this pigment fail to generate the proper nerve response that should lead to a red sensation?

When these subjects were studied by retinal densitometry (Rushton, 1963a), they were found to be quite lacking in the red-photosensitive pigment erythrolabe that is present in normal eyes. Half-bleaching by either a red or a green light, which appear identical to these subjects, resulted in precisely the same difference spectrum. This could not occur if in the red-green spectral range they had more than one pigment in measurable amounts. It followed that protanopes lack erythrolabe and have only *chlorolabe* to catch quanta in the red-green spectral range. And since they have only one pigment, they have only one dimension of colour here, and all wave lengths can be matched by adjusting intensity alone. This has long been known to be true.

Deuteranopes, the other type of red-green dichromat, were similarly shown to have only erythrolabe, the red-sensitive cone pigment, and to lack chlorolabe, the pigment of the protanope (Rushton, 1965a).

The kinetics of bleaching and regeneration were investigated in some detail. For both protanopes and deuteranopes it was established that the rate of bleaching is proportional to the rate of quantal absorption by the pigment, i.e., to the light intensity multiplied by the fraction of pigment present. The rate of regeneration is proportional to the fraction of pigment bleached. These two rates contribute independently and add algebraically (Rushton, 1963b, 1965b).

Mitchell and I (1971a) confirmed my earlier work showing pigments so measured to be *visual pigments* (i.e., those responsible for vision), for the relative bleaching efficacy of lights of various wave lengths was determined by finding for each, the energy required at *equilibrium* to bleach the pigment down to 50% of its full dark-adapted value. The reciprocals of these energies give the action spectrum of the pigment. These bleaching lights were then matched in brightness with a fixed yellow light. (They all appeared to the dichromat the same in colour as the yellow light, and so a perfect match could be obtained.) Lights that appeared equally bright bleached the pigment equally fast, the error being within 0.1 log unit of intensity.

The action spectrum so determined is immune from the serious errors of stray light that contaminate the absorption spectrum, for light that has not passed through the cone pigment cannot bleach it, but it may add to transmitted light and falsify the absorption measurement. Also, it is the action spectrum, not the absorption spectrum, that seems applicable to the visual response.

Apart from the blue-sensitive cone pigment *cyanolabe*, which we have never been able to measure well by retinal densitometry (because so little blue light is reflected from the back of the eye), there remain two important questions on cone pigments in man. Was König right in supposing that normal green cones contain the pigment of the protanope, and red ones the pigment of the deuteranope? And what are the cone pigments of the commonest classes of colour defective? These are subjects who can distinguish between bright red and green and who need a mixture of three primary colours to match all other colours; but their mixture is abnormal, and so they are called "anomalous trichromats."

## The Last Problem*

How in the living human eye can we determine the spectral sensitivity of one class of cone pigment, since vision is the resultant of the outputs of different kinds of cone interacting in an unknown manner? A common practice is to "isolate" one type of cone by using a strong coloured background to which rival cones are more sensitive, in the hope that in this way the rival cone output will be reduced to zero, or to ineffectiveness. There is not much a priori ground to substantiate this hope, nor a posteriori ground to believe it true. Most who use this method find it sufficient to assert that they have "isolated" a single pigment, without any justification whatever of that assertion.

From my first entry into vision I had hoped that a solution might be achieved if, instead of exciting by suddenly presenting a light, a steady light was suddenly *exchanged* for another of different wave length and energy. For any given visual pigment P, the energies of the exchanged lights may be so adjusted that P catches quanta from the first at the same rate as from the second. We named this adjusted energy ratio the *isolept* ( = equally taken) of P for the two exchanging lights. It has been found that when one pigment is involved (e.g., rhodopsin in twilight vision, or chlorolabe in the red-green range with a protanope), lights adjusted to the isolept of that single pigment may be exchanged without visual effect. When its quantum catch is unaltered, the pigment does not "see" that any change has taken place. Consequently, when a normal eye (which contains chlorolabe) sees that the light has changed from red to green (the energies having been adjusted to the chlorolabe isolept), the signal cannot come from chlorolabe cones, and *must* be derived from erythrolabe cones. This is a method of "isolation" that can be trusted.

Stimulation by light exchange was what I tried in my first visual experiments on the cat in Stockholm, but my technique then was too poor to realize anything. In Cambridge, Donner and I (1959a,b) used exchanges more successfully on the frog in our analysis of receptor interaction. And in Tallahassee, Powell, White and I used it to measure in man the visual pigments of normal and anomalous eyes in the red-green range (Rushton, Powell and White, 1973a,b,c).

In our equipment a red light (640 nm) was exchanged for a green (540 nm), and the exchanging lights were projected upon a red or a green background. If the red and green light energies are set anywhere near the isolept for one cone pigment P, it must be the other pigment, Q, that detects the exchange. The presence of the steady green background naturally raised Q's threshold for exchange detection. Then the green background was exchanged for a red one whose intensity was adjusted to raise the Q threshold equally. In this condition the red

*"So far."

and green backgrounds must be equally absorbed by Q, and thus their ratio is the isolept for Q. Thus, assuming roughly the isolept for P, we may obtain rather exactly that for Q. Now, repeating the measurements the other way round and starting with the known Q isolept, we may find accurately the isolept for P.

The importance of measuring the red-green isolept for a pigment is this. If the red and green primaries in our "analytical anomaloscope" (Mitchell and Rushton, 1971b) are set to the isolept ratio for any cone pigment P, and a perfect Rayleigh match is made with the (red + green) mixture against $E_\lambda$ (a light of energy E and wave length $\lambda$), the spectral sensitivity of P at each $\lambda$ is inversely proportional to $E_\lambda$, which can be measured. Consequently we may obtain the spectral sensitivity of the two cone pigments in the red-green range in normal and anomalous eyes.

Figure 16.1 gives the results obtained. Curve 1 shows the log sensitivity curve of the normal "green" cone pigment. It coincides with that found in the protanope (Pitt, 1935), which corresponds accurately to the action spectrum of chlorolabe measured by densitometry (Mitchell and Rushton, 1971a). Curve 2 shows the log spectral sensitivity of the normal "red" cone pigment. This coincides with Pitt's curve for the deuteranope, which corresponds to the action spectrum of erythrolabe measured objectively.

Protanomalous subjects have pigments 1 and 3, which lie so close that they have escaped discrimination by partial bleaching. Deuteranomalous have 2 and 4.

The narrow difference between curves 1 and 3 and 2 and 4, as compared with that between 1 and 2, explains also why the anomalous trichromate is so weak in colour discrimination.

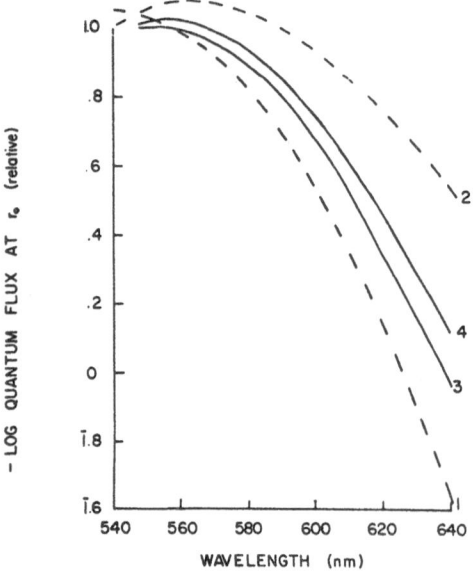

Figure 16.1   Log spectral sensitivity curves of four cone pigments. 1: chlorolabe. 2: erythrolabe. 3: protanolabe. 4: deutanolabe. The first two are the "green" and "red" pigments of normal eyes; the latter two are the pigments of anomalous eyes.

## Appointments and Travel

I came to Cambridge as a student in 1921 and have lived here ever since, but my appreciation of foreign colleagues and their techniques and my stimulus to research have been enlarged by generous opportunities for travel, some of which I should like to mention.

After my rebirth in Stockholm in 1949 I was invited in 1952 by Professor Carlos Chagas to spend the summer at his Biophysical Institute in Rio de Janeiro. I lectured on vision and experimented with electric eels.

In 1956 I had a grant from the Rockefeller Foundation (through the kind recommendation of Sir John Eccles) to travel and lecture in many universities in the United States. I had just measured cone pigments in normal and defective eyes by reflexion densitometry, and I was delighted with the generous and even enthusiastic reception that I received from George Wald and others, and with the opportunity to see many lovely parts of the United States and Canada.

I took 1959–1960 as sabbatical year and spent it at NIH in Dr. von Sallmann's department, working with Mike Fuortes and helped by Ralph Gunkel. I brought with me a "portable" reflexion densitometer, and we studied rhodopsin regeneration and dark adaptation, especially in the rod monochromat.

Seven years later (1966–1967) I was Visiting Professor at Johns Hopkins University in Ted MacNichol's laboratory. I spent most of the time in the library preparing lectures that I gave weekly to a very stimulating audience, which included MacNichol, Dowling and Wolbarsht among the professors, Steve Easter, Nigel Daw and Mark Dubin among the students.

In 1966, at the recommendation of Alan Hodgkin, I was made Professor of Visual Physiology at Cambridge. Two years later I resigned (one year before superannuation) and became Distinguished Research Professor at Florida State University, Tallahassee, at the recommendation of Professor Howard Baker. I was given space and equipment in the Institute of Molecular Biophysics by its director, Mike Kasha.

The opportunity to start afresh just when Cambridge was retiring me has been a tremendous stimulus. Mat Alpern and Shuko Torii came from Michigan for the first year to help me get started, and did so to such good purpose (Alpern, Rushton and Torii, 1970a,b,c,d) that the vigorous little research group then initiated has continued productive ever since, not least so during this last year (1972–1973) when I have had the honour to be Fogarty Scholar-in-Residence at NIH and my Florida group has worked largely on its own. And so it will be next year when I am invited to Australia.

This substantiates my opening remarks. Kind friends have always given me the opportunity for research, and by having a higher opinion of me than my own, they have encouraged me to do better than I thought I could.

And among my kind friends, the one who above all has helped my research throughout the past fifteen years is my able assistant, Clive Hood.

# References

Alpern, M., Rushton, W.A.H., and Torii, S. (1970a): The size of rod signals. *J. Physiol.* 206:193–208.

Alpern, M., Rushton, W.A.H., and Torii, S. (1970b): The attenuation of rod signals by backgrounds. *J. Physiol.* 206:209–227.

Alpern, M., Rushton, W.A.H., and Torii, S. (1970c): The attenuation of rod signals by bleachings. *J. Physiol.* 207:449–461.

Alpern, M., Rushton, W.A.H., and Torii, S. (1970d): Signals from cones. *J. Physiol.* 207:463–475.

Barlow, H.B. (1953a): Action potentials from the frog's retina. *J. Physiol.* 119:58–68.

Barlow, H.B. (1953b): Summation and inhibition in the frog's retina. *J. Physiol.* 119:69–88.

Brindley, G.S., and Willmer, E.N. (1952): The reflexion of light from the macular and peripheral fundus oculi in man. *J. Physiol.* 116:350–356.

Campbell, F.W., and Rushton, W.A.H. (1955): Measurement of the scotopic pigment in the living human eye. *J. Physiol.* 130:131–147.

Donner, K.O. (1959): The effect of a coloured adapting field on the spectral sensitivity of frog retinal elements. *J. Physiol.* 149:318–326.

Donner, K.O., and Rushton, W.A.H. (1959a): Retinal stimulation by light substitution. *J. Physiol.* 149:288–302.

Donner, K.O., and Rushton, W.A.H. (1959b): Rod-cone interaction in the frog's retina analysed by the Stiles-Crawford effect and by dark adaptation. *J. Physiol.* 149:303–317.

Dowling, J.E. (1960): Chemistry of visual adaptation in the rat. *Nature* 188:114–118.

Hecht, S. (1937): Rods, cones, and the chemical basis of vision. *Physiol. Rev.* 17:239–290.

Hecht, S., Shlaer, S., and Pirenne, M.H. (1942): Energy, quanta and vision. *J. Gen. Physiol.* 25:819–840.

Hodgkin, A.L. (1938): The subthreshold potentials in a crustacean nerve fibre. *Proc. R. Soc. B.* 126:87–121.

Kuffler, S.W. (1953): Discharge patterns and functional organization of mammalian retina. *J. Neurophysiol.* 16:37–68.

Lapicque, L. (1926): *L'Excitabilité en Fonction du Temps. La Chronaxie, sa Signification et sa Mesure.* Paris: Les Presses Universitaires de France.

Lorente de Nó, R. (1947): *A Study of Nerve Physiology.* Parts 1 and 2 (Studies from the Rockefeller Institute for Medical Research, Vols. 131 and 132). New York: The Rockefeller Institute for Medical Research.

Lussier, J.J., and Rushton, W.A.H. (1951): The relation between the space constant and conduction velocity in nerve fibres of the A group from the frog's sciatic. *J. Physiol.* 114:399–409.

Lussier, J.J., and Rushton, W.A.H. (1952): The excitability of a single fibre in a nerve trunk. *J. Physiol.* 117:87–108.

Mitchell, D.E., and Rushton, W.A.H. (1971a): Visual pigments in dichromats. *Vision Res.* 11: 1033–1043.

Mitchell, D.E., and Rushton, W.A.H. (1971b): The red/green pigments of normal vision. *Vision Res.* 11:1045–1056.

Pitt, F.H.G. (1935): *Reports of the Committee upon the Physiology of Vision. XIV. Characteristics of Dichromatic Vision with an Appendix on Anomalous Trichromatic Vision* (Medical Research Council Special Report Series, no. 200). London: His Majesty's Stationery Office.

Rashbass, C. (1949): The relationship between the slot excitability and the excitability due to a single pole. *J. Physiol.* 109:354–357.

Rashbass, C., and Rushton, W.A.H. (1949a): Space distribution of excitability in the frog's sciatic nerve stimulated by slot electrodes. *J. Physiol.* 109:327–342.

Rashbass, C., and Rushton, W.A.H. (1949b): Space distribution of excitability in the frog's sciatic nerve stimulated by polar electrodes. *J. Physiol.* 109:343–353.

Rashbass, C., and Rushton, W.A.H. (1949c): The relation of structure to the spread of excitation in the frog's sciatic trunk. *J. Physiol.* 110:110–135.

Rushton, W.A.H. (1927): The effect upon the threshold for nervous excitation of the length of nerve exposed, and the angle between current and nerve. *J. Physiol.* 63: 357–377.

Rushton, W.A.H. (1928a): Excitation of bent nerve. *J. Physiol.* 65:173–190.

Rushton, W.A.H. (1928b): Nerve excitation by multipolar electrodes. *J. Physiol.* 66:217–230.

Rushton, W.A.H. (1945): Action potentials from the isolated nerve cord of the earthworm. *Proc. R. Soc. B.* 132:423–437.

Rushton, W.A.H. (1946): Reflex conduction in the giant fibres of the earthworm. *Proc. R. Soc. B.* 133:109–120.

Rushton, W.A.H. (1949): The site of excitation in the nerve trunk of the frog. *J. Physiol.* 109:314–326.

Rushton, W.A.H. (1952): Apparatus for analysing the light reflected from the eye of the cat. *J. Physiol.* 117:47P–48P (abstract).

Rushton, W.A.H. (1958): Kinetics of cone pigments measured objectively on the living human fovea. *Ann. NY Acad. Sci.* 74:291–304.

Rushton, W.A.H. (1959): Excitation pools in the frog's retina. *J. Physiol.* 149:327–345.

Rushton, W.A.H. (1961): Rhodopsin measurement and dark-adaptation in a subject deficient in cone vision. *J. Physiol.* 156:193–205.

Rushton, W.A.H. (1963a): A cone pigment in the protanope. *J. Physiol.* 168:345–359.

Rushton, W.A.H. (1963b): Cone pigment kinetics in the protanope. *J. Physiol*. 168:374–388.

Rushton, W.A.H. (1965a): A foveal pigment in the deuteranope. *J. Physiol*. 176:24–37.

Rushton, W.A.H. (1965b): Cone pigment kinetics in the deuteranope. *J. Physiol*. 176:38–45.

Rushton, W.A.H., and Cohen, R.D. (1954): Visual purple level and the course of dark adaptation. *Nature* 173:301–302.

Rushton, W.A.H., Powell, D.S., and White, K.D. (1973a): Exchange thresholds in dichromats. *Vision Res*. 13: 1993–2002.

Rushton, W.A.H., Powell, D.S., and White, K.D. (1973b): The spectral sensitivity of "red" and "green" cones in the normal eye. *Vision Res*. 13:2003–2015.

Rushton, W.A.H., Powell, D.S., and White, K.D. (1973c): Pigments in anomalous trichromats. *Vision Res*. 13: 2017–2031.

Wald, G. (1954): On the mechanism of the visual threshold and visual adaptation. *Science* 119: 887–892.

Weale, R.A. (1953): Photochemical reactions in the living cat's retina. *J. Physiol*. 122:322–331.

Derek Denny-Brown (b. 1901, Christ-church, New Zealand), professor of neurology emeritus at Harvard Medical School, served after retirement as chief of the physiology section of the New England Primate Research Center from 1967 to 1972, and then became one of the Fogarty Scholars-in-Residence at the National Institutes of Health until 1973. His extensive laboratory and clinical work on the physiology of the nervous system has dealt with such subjects as the physiology of movement, neuropathology, and the histology of peripheral nerves.

# 17

# The Importance of Steady-State Equilibria in Small-Celled Reticular Systems

## Derek Denny-Brown

It is a special pleasure for me to join this testimonial to Frank Schmitt for several reasons, but chiefly because he and his associates at MIT have provided for those of us who work in Boston a direct view of the excitement and thrill of exploration of submicroscopic structure in relation to function. For those whose interest is primarily in the clinic, and who accordingly attempt to deal with neuronal mass effects, the submicroscopic has a peculiar unreality which the electron microscope and x-ray diffraction have not entirely dispelled. The behaviorist and clinician have more pressing problems in relation to functions that are of a larger order, but so diffuse that it is difficult to come to grips with them. Perhaps I might outline for you aspects of essential nervous integration that are not as yet soluble in terms of known spinal reflexes, yet are clearly necessary to all reflex function.

### Reflexes and Nerve Nets

My own introduction to the fundamentals of nervous function was the fortuitous offshoot of the claim of two Australians, Hunter and Royle, in 1924 to be able to relieve spasticity by the operation of sympathectomy. To the young anatomist and would-be surgeon that I was at that time, the proposition presented an irresistible challenge, which my professor of anatomy, W. P. Gowland, suggested could only be satisfied by going to work with Sherrington. With the aid of a Beit Fellowship this was eventually achieved, and I began work in the Oxford Laboratory of Physiology in the fall of 1925, inheriting a small unheated room that had just been vacated by John Fulton, containing a stretch-table and myograph and an old Einthoven string galvanometer. E. G. T. Liddell, who had earlier worked out the details of the stretch reflex with Sherrington, took me in hand, and we soon found that the sympathetic nervous system had no relation to postural reflexes. It was also clear that there was still much to be learned about the essential nature of postural reflexes, and indeed about nervous action in general. The next three years were the most exciting I have experienced. Sherrington, though past the usual age for retirement, was still most active, stimulating, and genial. Cooper and Creed had just joined the laboratory, and in the next two years Eccles and Granit also joined the research group.

We all, of course, read Sherrington's *Integrative Action* (1906a), though it was some years before the full impact of that remarkable book was to become clear to most of us. I would like to draw special attention to the long discussion in that work of what was then called "immediate and successive induction." Immediate induction was clear enough, for it is what is now understood by the term "facilitation." Successive induction was the swing-back to opposite effect, known also as "rebound," thought to imply that the stimulus was a mixed one containing both excitation and inhibition. The excitation had accumulated somewhere and rebounded, so to speak, after the inhibition had had its say. The effect was to provide an explosive recovery, like the bounce of a ball. But there were several difficulties. Why was the explosive aftereffect often more violent than that of the excitation when alone? Why was the rebound often very prolonged, lasting at times for four or five minutes, as in an experiment cited on page 208 of the *Integrative Action*? What sort of reflex mechanism could store its driving activity for so long a period after the stimulus? Sherrington noted that successive induction is a general phenomenon, comparable to the visual afterimage, but he could offer no further explanation.

Rebound of this type is often rhythmical, and in spinal reflexes the rhythm is clearly an alternating progression activity of the limbs (Sherrington, 1910). Such "walking" can be initiated in this way after thoracic-cord transection; it clearly involves a coordination of the muscles in both lower limbs and, hence, some kind of supranuclear "center." The nature of such a center has been most difficult to comprehend, for the coordinated activity of both lower limbs requires no particular nucleus. It is better coordinated if the upper lumbar segments are intact, and it becomes reciprocal with movements of the upper limbs in diagonal quadripedal patterns after cervical transection. The rhythm and pattern are no different in the decerebrate animal. In his original description of decerebrate rigidity, Sherrington (1898) noted the quadripedal progression pattern of movement elicited by a stimulus applied to any limb (Figure 17.1). This diagonal pattern of reflex spread is also prominent in labyrinthine ("alpha") rigidity where the gamma system is not involved (Denny-Brown, 1962). The local reflex response in any

A          B          C          D

Figure 17.1  Diagram of the pattern of decerebrate rigidity resulting from intercollicular section of the midbrain, with pattern of reflex response to stimulation applied to left forepaw, left hindpaw, and left ear, respectively. From Sherrington (1898), by combination of two of his figures.

limb is clearly a fragment of this quadripedal response. Its anatomical basis must likewise be a fragment of the whole, so that wherever the activity is set up it is automatically coordinated with whatever other parts of the total mechanism are available, as in a "nerve net," vaguely suggested by Sherrington in his Rede Lecture (1933).

This peculiar arrangement, coordination without any discrete coordinating "center," is also found in the scratch reflex, which Sherrington (1906b) showed to have a multitude of possible inputs over a wide reflexogenous area. Adequate stimuli applied concurrently at two different points can facilitate each other and adjust the location of the point scratched, but the remarkably constant rhythm and general nature of the response are unchanged. The motor apparatus is diffuse in the lumbar segments and has a corresponding postural component in curvature of the spine and abduction of the opposite limb. No particular segment is essential. As we ourselves found, the characteristic rhythm can be elicited in the ipsilateral hip flexor from cervical skin stimulation even after hemisection of the spinal cord at the D12 segment (Denny-Brown, 1966: p. 31).

Both the scratch reflex and automatic progression movement exhibit a natural rhythm, which is specific in each and surprisingly constant in the chronic spinal animal. In both instances deafferentation of one or both hind limbs does not alter the rhythm, showing that peripheral feedback for the effector part of the system is not an essential feature. The rhythm must, then, be "inherent" in the mechanism, a condition that seems to call for a central feedback built into the network.

Such coordination and integration requires the presence of some intercalated diffuse system, capable of storing the whole pattern of response and of integrating its individual parts. Sherrington's diagram of the scratch reflex (Figure 17.2) omits the intercalated neurons in the pathway. David Lloyd (1941, 1944) has demonstrated not only the extremely rapid conduction in some reticulospinal tracts, but also the slow polysynaptic transmission in propriospinal pathways.

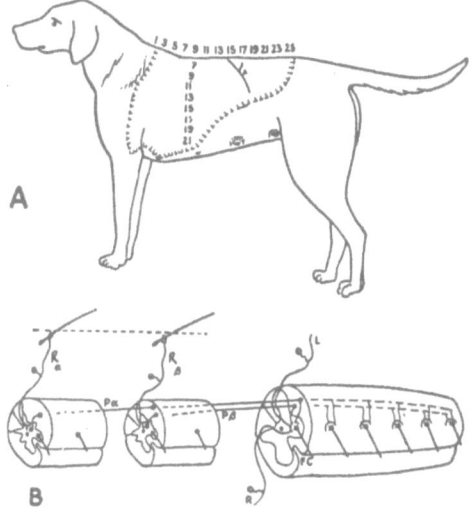

Figure 17.2   A diagram of the scratch reflex in the spinal dog. From Sherrington (1906b).

What type of intercalated neuron can serve such a function? The only clue seems to be the long forgotten figure of Ramón y Cajal (Figure 17.3) showing samples of small cells each with subdivided axons entering more than one funiculus. These clearly must belong to a primitive reticulum, a multichannel nerve net.

Current notions of the management of motor coordination usually require a suprasegmental center for the management of input to the gamma system. To say that reflex standing requires the activity of a pontine reticular center, however, does not take account of the eventual appearance of positive supporting reactions and intense stretch reflexes in the lower limbs of the chronic spinal dog (Denny-Brown and Liddell, 1927). In this situation other responses, such as a well-co-ordinated gallop response ("extensor thrust"), coitus reflex, and long sustained knee and ankle clonus, become prominent. In relation to the clonic response we found that a tendon tap elicited an inhibitory wave in synergic muscles, with marked rebound. The tendon reflex is in fact only the peak of a submerged iceberg in the form of a widespread subliminal reverberation throughout the isolated spinal segments, often initiating prolonged clonic spasms, for which the simplest explanation is the activity of an intercalated network. Thus the gamma system of the spinal cord is not completely dependent on the integrity of the pontine retic- ular nuclei.

**Patterned Supraspinal Responses**

In the supranuclear regulation of spinal reflexes the effect of the pyramidal system

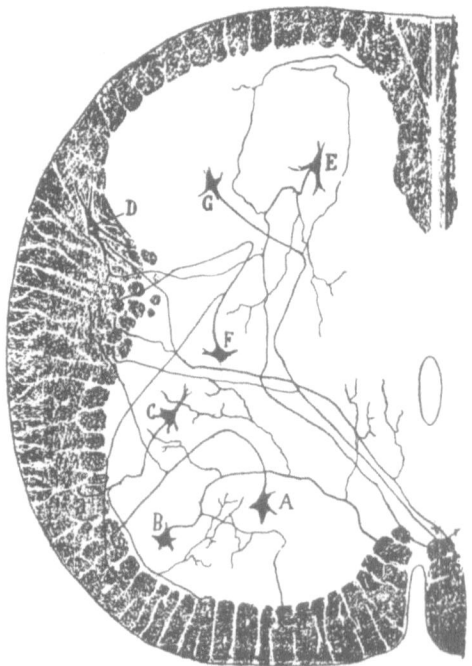

Figure 17.3   Figure of cells with axons entering two or more funiculi in the spinal cord of new-born mouse. Golgi method. From Ramón y Cajal (1909).

is well known. Knowledge of the effects of the descending reticulospinal pathways is most extensive in relation to the long reticulospinal paths. The effect of the descending path from the n. caudalis pontis and medial vestibular nucleus in facilitating gamma discharge is clear, though the input that drives their activity after intercollicular midbrain and eighth-nerve section is still obscure. The more obvious n. reticularis gigantocellularis (Brodal, 1957) in the lower medulla was found by Shimamura and Livingstone (1963) to be the source of a rapidly con-ducted volley of excitation to virtually all ipsilateral motor roots following a single volley to an afferent root at any level (the "spino-bulbo-spinal reflex"). Its latency was thus related to the distance of the test root from the medulla.

We found this reflex again in the form of myoclonus, a jerk of flexor muscles of the limbs, with extension of the neck, elicited by a single stimulus to any cutaneous nerve (Denny-Brown, 1968). This is the so-called myoclonic jerk, which becomes prominent after a variety of cortical, thalamic, basal ganglionic, or cerebellar lesions, indicating that all these areas normally suppress the spino-bulbo-spinal reflex. The response is the excitatory flexor counterpart of the inhibition of ex-tensor muscles, found to result from stimulation of the medullary reticular for-mation (Magoun and Rhines, 1948; Lindsley, Schreiner, and Magoun, 1949).

Of special interest to my present thesis is our finding that the maximal release of the myoclonic-jerk response, including its elicitation by visual and auditory stimuli, results from midline sagittal section of the pons, where all the descending suppressor pathways appear to decussate (Figure 17.4) In this situation the myo-clonic jerk is unchanged by intercollicular decerebration, or by deafferentation of the limb concerned, thus showing its independence of the regulation of the gamma system. It *is* abolished on the side of a paramedian section of medulla at the level of the obex since the descending tract from each n. gigantocellularis is ipsilateral near the midline. However, we found that seven to ten days after such a para-medial stab lesion the myoclonic jerk elicited by facial stimulus returns on the ipsilateral side. We have to conclude that it is then transmitted by a crossed path-way that recrosses at segmental level, as if the ipsilateral direct reticulospinal path had been only a *relative* canalization in a diffuse system.

Body contact righting and the tactile positive supporting reaction are the most elementary forms of postural response to the environment. Fragments of both are present in the chronic spinal animal, but these responses are better coordinated the greater the extent of brainstem below the level of transection. Only the midbrain preparation can successfully right, stand, and walk; yet no single structure can be said to be essential for this performance. Sagittal pontine section impairs contact reactions, leaving unstable myoclonic responses to any sudden change in tactile input and releasing automatic progression when the animal is suspended free of all body or limb contact (Denny-Brown, 1967). These reactions must reflect a progressively complex interaction at higher and higher levels of the reticular formation.

## The Modification of Quadripedal Progression by the Cerebral Cortex

There are multiple areas of cerebral cortex from which electrical stimulation

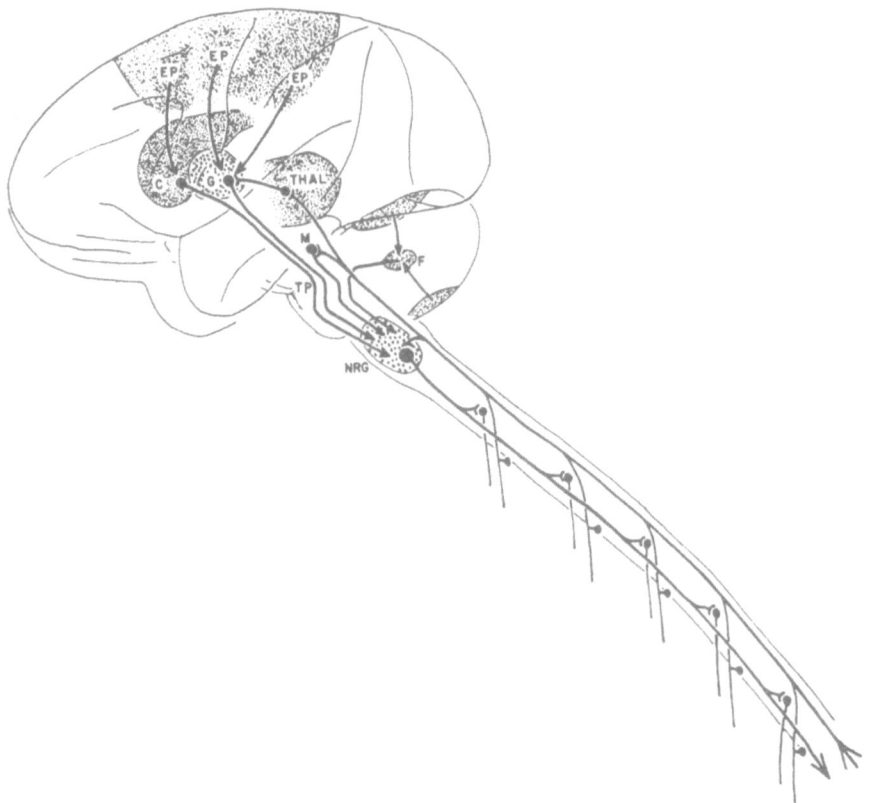

Figure 17.4  Diagram of the pathways involved in the spino-bulbo-spinal reflex and its control by extrapyramidal pathways from the cortex (EP), cerebellum, and midbrain tegmentum (M). Our own experiments indicate that the control pathways decussate in the pontine tegmentum (TP). The most important reflex structure is the n. reticularis gigantocellularis (NRG).

elicits movement of limbs after section of the pyramid in the medulla, as described by Tower, Hines, and others; these cortical areas are therefore called extrapyramidal motor areas. The responses obtained are bilateral and stereotyped, the common pattern elicited from the premotor area being extension of the contralateral upper limb, with flexion of the contralateral lower limb, and flexion of the ipsilateral upper limb (Denny-Brown, 1966). More medially, from that portion of the supplementary motor area that lies on the lip of the cingulate sulcus, the pattern is reversed. This reversed synergy, with flexion of opposite upper limb, is also obtained from the parietal extrapyramidal area (Denny-Brown, 1966). Following ablation of area 4 of Brodmann, or of the pre- and postcentral gyrus on both sides, the monkey learns to use the limbs to gain support on an approaching surface, or to paw repeatedly at a desired object in canine manner. The delicate, discrete movements of placing or reaching have then been replaced by a rhythmic quadripedal progression movement, of which the flexor or extensor synergy is the beginning of a cycle.

The most remarkable feature of this extrapyramidal performance is that, although it has a decussation in the pons, it is eventually regained in the limbs ipsilateral to a high cervical hemisection of spinal cord. In this circumstance stimulation of the premotor area of the cortex contralateral to the hemisection still elicits movement of normal pattern, in which the synergy begins with extension of the upper limb contralateral to the area stimulated (Denny-Brown, 1966). Stimulation of the precentral cortex ipsilateral to the hemisection produces the opposite pattern. This remarkable result points again to the presence of a bilateral conduction of specific patterns of motor response in the reticulospinal pathways. A second hemisection at any lower level on the opposite side reduces motor performance to the equivalent of a complete transection at the level of the lower lesion.

The importance of these cortical and subcortical extrapyramidal pathways is that their presence is necessary for the performance of discrete movements such as reaching, grasping, and placing, for which an intact pyramidal system is also essential. For example, a symmetrical lesion of the inner globus pallidus, if it is large enough, causes an akinetic state. The intact pyramidal system then seems powerless. A lesion of the whole putamen has the same result. A relative but less complete degree of akinesia of the same type is caused by bilateral ablation of the perirolandic cortex in the monkey, leaving the precentral gyrus intact (Denny-Brown, 1965, 1966). The extrapyramidal projections must supply the internuncial "set" upon which the pyramidal system applies its adjustments and facilitations.

### Diffuse Types of Dystonia

The physiologist working with human patients in the clinic is confronted by fascinating problems relating to motor control, but he can become greatly frustrated by his inability to simplify the reflex situation. The most interesting disorders of movement are related to anatomical disorders of extrapyramidal structures. The causal pathology is seldom clear-cut, being most often the result of some type of diffuse metabolic or viral disease. Yet enough is known to implicate some of the more obvious nuclei of the basal ganglia and brainstem. There are two very prominent features of such disorders: First, when the disorders are progressive, they begin by a disorder of movement, which gradually changes to a disorder of posture. Second, if the pathology is relatively sudden and well localized, there is an enormous capacity for compensation by remaining intact portions of the system. Conversely, if the causal damage is diffuse (as in metabolic or viral disease), a general instability of the whole system, in the form of myoclonus, results. Thus, in hypoglycemia, severe hepatic encephalopathy, or a subacute viral leucoencephalopathy, a period of spontaneous myoclonic jerking is eventually followed by a decorticate type of dystonic rigidity (Denny-Brown, 1968).

In such states not only is there an abnormal stereotyped response to stimulation, the myoclonic jerk, but in severe states spontaneous repetition of such jerking occurs. This commonly takes the form of a relatively localized fragment of the full myoclonic jerk, such as shocklike flexions of a portion of one limb or contractions of the abdominal wall, repeated regularly at the rate of approximately one a second ("epilepsia partialis continua") in acute disorders. In more slowly developing states, such as subacute leucoencephalopathy, the spontaneous twitch occurs only once every seven to twelve seconds. In most of these conditions there is also a widespread triphasic wave in the EEG of the two hemispheres, approximately but not exactly correlated with the myoclonic muscular twitch (Juul-Jensen and Denny-Brown, 1966; Denny-Brown, 1968). This results from an activation of the ascending portion of the reticular alerting system (French, Verzeano, and Magoun, 1953; French, 1960).

The rhythmical nature of these spontaneous myoclonic phenomena, both ascending and descending, indicates a reservoirlike feature of the sensory input for such disorders of the motor system, particularly evident in the long intervals encountered in patients with subacute encephalopathy. Rhythm is a general feature of extrapyramidal disorder, most commonly encountered in the form of parkinsonian tremor, or less frequently as the similar tremors at rest following damage to the mesencephalic tegmentum. We found it difficult to believe that there is a tremor-center, as has been proposed by some. Nor was it clear that any specific tremor circuits could be clearly defined. In one monkey we found that the tremor produced by paranigral lesion in the right upper limb, in all respects typical of such tremors as studied by others (Poirier, 1960, 1966), became reasserted in the same limb three weeks after ipsilateral hemisection of spinal cord at C3–4 level (Figure 17.5). Not only does this show that the pyramidal system has no essential part in such tremor, but it indicates once again that the indirect reticular connections of the spinal cord are polarized in fundamental patterns of movement.

## The Regulation of Sensory Input

In the last six years I have had the enormous privilege of becoming once again an experimental physiologist, at the New England Regional Primate Center. One of my interests there has been the investigation of the interrelationship of converging nerve roots that supply the same area of skin. With E. J. Kirk I examined the classical dermatomes of the monkey, using the Sherringtonian method of sectioning three roots above and three below the test root (Kirk and Denny-Brown, 1970). The results of intradural section were precisely those reported by Sherrington in his classic papers of 1893 and 1898. Like Sherrington and others since, we observed that the threshold of response was lower in the center of the remaining root area and highest at its caudal and cephalic edges, indicating some grading of the density of innervation of the skin. Nevertheless, the *area* from which responses were obtained remained surprisingly constant after the first few days and rarely varied more than two to three millimeters over periods of weeks or months.

Sherrington divided the roots intradurally between dorsal-root ganglia and the spinal cord, partly because they may be more clearly viewed here, and partly because no regeneration can ensue. We found, however, that if the roots were divided distal to their dorsal-root ganglia, the resulting dermatome area of the test root was greatly increased (Kirk and Denny-Brown, 1970). When the wound was

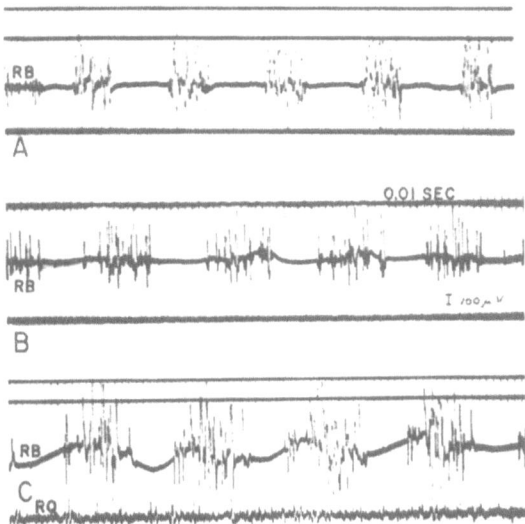

Figure 17.5　Three electromyograms of the right biceps muscle of a macaque monkey showing tremor at rest after a bilateral paranigral lesion. The tremor had persisted 63 days when recorded in A. After a further three weeks the right side of spinal cord was sectioned at C 3–4 level. Subsequent histology showed that the hemisection spared a small portion of right medial dorsal column but completely destroyed the remainder of the right side of the spinal cord and left ventral column. The tremor in the right side of the neck and right shoulder continued. After 8 days the tremor again involved the muscles of the right upper limb and at times the right lower limb. The electromyogram marked B was recorded 10 days after hemisection; C, 39 days later. The tracing RQ shows an occasional potential in right quadriceps at the height of a tremor beat. Myoclonic jerks in both lower limbs were obtained from stimulation of the right upper limb, and in the right upper limb also by the eighty-fifth day after hemisection.

reopened and the same roots were divided proximal to the ganglion, the result was surprising. The dermatome was at first still large, but then progressively shrank in the next three to five days to the area usually found after primary section proximal to the ganglia. These effects were the same for the general reaction of the animal as for the reflex response to the same stimulus.

Since the restriction of the field developed after a delay, it appeared that the change might be a functional one, related not to severance of a pathway, but rather to some change in the efficiency of conduction from the border areas of the dermatome where innervation was sparse. This was confirmed by giving a subconvulsive dose of strychnine (0.25 mg/kg) subcutaneously. This reversed the shrinkage within fifteen to thirty minutes for a period of an hour, following which the dermatome area returned to its previous extent. It would also appear that the mere presence of the ganglia of neighboring nerve roots contributed to the efficiency of conduction from the test root. More recently Kirk has found that the central processes of dorsal-root ganglion cells isolated from the periphery do in fact exhibit a slow spontaneous discharge after the first two or three days. This spontaneous activity would explain the facilitating effect of intact neighboring ganglia, even after section of their distal roots.

In a more recent paper we have pursued these effects further, showing that the facilitating effect of neighboring roots that innervate the same sensory field are conveyed by the medial division of the tract of Lissauer to reach the first synapse of the fibers of the test root (Denny-Brown, Kirk, and Yanagisawa, 1973). The effect is complicated by a powerful countereffect, conveyed by the lateral division of the tract of Lissauer, which can suppress conduction from the dermatome of the test root. Both effects must be relayed by the multisynaptic connections of the small cells in substantia gelatinosa, whose axons make up the bulk of the tract of Lissauer. When, following isolation of a test root on both sides by intradural section of three neighboring roots, the lateral division of the tract of Lissauer was sectioned on one side just above and just below the test root, the dermatome on that side was greatly expanded (Figure 17.6, upper charts), reaching a maximum in four to ten days in three different animals. When, instead, the medial division of the tract was sectioned above and below the test root on one side, the test dermatome became very narrow on the next day, shrank to two or three islands, and gradually disappeared in the following days (Figure 17.6, lower charts). Giving this latter animal a subconvulsive dose of strychnine completely restored the absent dermatome for an hour or more, together with the abdominal reflex response

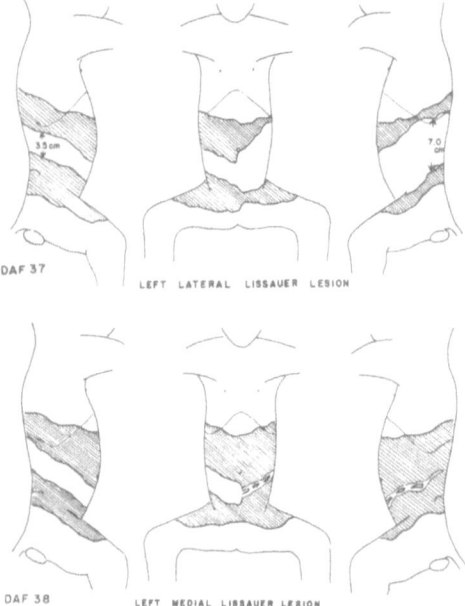

Figure 17.6   Two charts showing the first lumbar dermatome in two macaque monkeys isolated by bilateral section of D 10, 11, and 12 and L 2, 3, and 4 lumbar dorsal roots intradurally on both sides. The animal DAF 37 had, in addition, two small lesions of the lateral Lissauer tract on the left side, one above and one just below the isolated root. The corresponding dermatome enlarged greatly, reaching its maximum on the fifth day, then remaining constant. The lower charts are from animal DAF 38, in whom two small lesions were made in the medial portion of the left Lissauer tract, one just above and one below the isolated root. The left dermatome was only a very small, narrow band on the day after operation, and it was represented only by three small islands on the fifth day. Thereafter these islands were variably present or absent for many weeks. At any time in this period a subconvulsive dose of strychnine could restore for an hour or more the left dermatome to approximately the size of the control right side. From Denny-Brown, Kirk, and Yanagisawa (1973).

that is dependent upon its integrity. In some animals one or two small islands persisted for some days after the test dose of strychnine before the dermatome again became completely unresponsive.

Surprisingly, the same effect was produced by a moderate dose (50 mg/kg) of L-dopa given by mouth. Within half an hour the test dermatome began to expand, and it reached its full area in a little over an hour. The contraction and disappearance of the dermatome was then very gradual, some effect of a single dose of L-dopa persisting as long as five days. We thus found a most remarkable and persistent effect reversed for an unprecedented period of time by strychnine or L-dopa. In both cases there is no reason to suspect that a long-lasting transmitter is involved. Rather, it would appear that the enormous number of small, Golgi type-2 circuits in the substantia gelatinosa develop a peculiar momentum, still present for days after cessation of further input. The phenomenon is even more apparent in the results of section of the descending tract of the fifth nerve (Denny-Brown and Yanagisawa, 1973).

Now let us consider for a moment the implications of some of these findings. We have been able to effect a lasting change in the efficiency of conduction through the first sensory synapse, either in the direction of facilitation (hyperreactivity) or inhibition (absence of response), continuing for months in the monkey. Even when neighboring nerve roots have been sectioned, lesion of one or another part of Lissauer's tract still has an effect, demonstrating that both tonic E and tonic I have been active. We have therefore to suppose that, after section of the nerve roots and degeneration of their collaterals, not only do the cells of substantia gelatinosa take up to five days to reach a state of minimal tonic discharge, but some continued discharge, whether excitatory or inhibitory, continues indefinitely. It is this residual effect that is changed by subsequent section of the tract.

## Autonomous Activity in Small-Neuron Systems

Thus we come to recognize small residual fringe effects that continue without further input from the periphery. Indeed we begin to realize that without such fringe effects no function is possible. Thus the Schiff-Sherrington phenomenon (release of upper-limb reflexes following transection of thoracic cord) can occur, as Ruch (1936) has shown, even when all dorsal roots below the level of transection have been previously cut. The mechanism of intersegmental interrelation must, then, still be present after all input from the periphery has been abolished.

We therefore have to recognize mass effects from the conglomerations of polysynaptic short-axon chains of cells (or neuropil), such as substantia gelatinosa, that can continue prolonged tonic background activity. A number of those who have recorded from neurons in the dorsal horn have noted that those units that respond predominantly to C-fiber stimulation also show spontaneous activity. To this extent it may be supposed that the types of unit we have discussed here are among the less highly differentiated.

The substantia gelatinosa is only one conglomeration of neurons with a heavy synaptic network. The putamen and caudate nucleus are others where L-dopa may also be able to facilitate a general efficiency of reaction. In seeking the anatomical substrate for the prolonged effects and momentum of extrapyramidal management of movement and righting reflexes, we would suggest that similar mass effects are active.

The reticular formation of the brainstem forms a core that is continuous with the reticular substance of the spinal cord. Both are more obvious in the amphibian brain, where the primitive neuropil (reticulum) and its connections were extensively studied by C. Judson Herrick (1948). It is of special interest that the earliest motor development in the amphibian is the pattern of quadripedal progression movement for which the medullary and, at a later stage, the midbrain reticular formation are essential (Coghill, 1929; Detwiler, 1945). These primitive cell masses would appear to offer a better chance for exploration of fundamental neural organization than the more diffuse reticular systems of the mammalian brain.

## Conclusions

My necessarily sketchy survey of reflex activity has emphasized the presence of widespread subliminal mass effects, which provide a background for the classic phasic reflexes and exhibit their patterns as part of larger contexts. I would direct attention to the spontaneous activity of the primitive neuropil of the nervous system. The extensive networks of small neurons that comprise the substantia gelatinosa must be part of this fundamental mechanism. They provide a fringe of balanced excitation and inhibition, the equilibrium of which serves as a background for skin sensation. The time characteristics are extremely prolonged. It appears likely that a more diffuse spinal reticulum provides a motor neuronal counterpart, in which a similar long-lasting momentum of effect may occur.

In the motor performance of the nervous system there are reverberating effects that are seen most often as rebound following even the simplest transients, such as the tendon reflex, and lasting many minutes. It is suggested that these reflect the activity of similar systems of spontaneous, polarized, and patterned background activity in the central gray matter. At higher levels, as the result of facilitation of extrapyramidal mechanisms, there are released disequilibria that result in dystonia. Such states tend to stabilize after some days or weeks if there is sufficient remaining extrapyramidal mechanism, for we have found that the factors underlying compensation are nonspecific, being in terms of facilitation of general extensor or flexor activity. In relation to the function of the basal ganglia or cerebellum, for example, it is necessary to consider malfunction in terms of loss of efficiency in such reactions, rather than in terms of loss of specific movement patterns. It is in this area that disequilibrium of fundamental regulation of movement arises, leading to the tonic features of dystonia or, if acute and generalized, to the rhythmical instabilities of tremor or myoclonus.

## References

Brodal, A. (1957): *The Reticular Formation of the Brain Stem. Anatomical Aspects and Functional Correlations.* Edinburgh: Oliver and Boyd.

Coghill, G.E. (1929): The early development of behavior in amblystoma and in man. *Arch. Neurol. Psychiatr.* 21:989–1009.

Denny-Brown, D. (1962): The midbrain and motor integration. *Proc. R. Soc. Med.* 55:527–538.

Denny-Brown, D. (1965): The nature of dystonia. *Bull. NY Acad. Med.* 41:858–869.

Denny-Brown, D. (1966): *The Cerebral Control of Movement.* Springfield, Ill.: C. C Thomas.

Denny-Brown, D. (1967): The fundamental organization of motor behavior. *In: Neurophysiological Basis of Normal and Abnormal Motor Activities.* Yahr, M.D., and Purpura, D.P., eds. Hewlett, N.Y.: Raven Press, pp. 415–442.

Denny-Brown, D. (1968): Quelques aspects physiologiques des myoclonies. *Rev. Neurol. (Paris)* 119:121–129.

Denny-Brown, D., and Yanagisawa, N. (1973): The function of the descending root of the fifth nerve. *Brain* 96:783–814.

Denny-Brown, D., Kirk, E.J., and Yanagisawa, N. (1973): The tract of Lissauer in relation to sensory transmission in the dorsal horn of spinal cord in the macaque monkey. *J. Comp. Neurol.* 151:175–200.

Denny-Brown, D., and Liddell, E.G.T. (1927): The stretch reflex as a spinal process. *J. Physiol.* 63:144–150.

Detwiler, S. R. (1945): The results of unilateral and bilateral extirpation of the forebrain of amblystoma. *J. Exp. Zool.* 100:103–117.

French, J.D. (1960): The reticular formation. *In: Handbook of Physiology: Neurophysiology,* Vol. II, Chap. LII. Field, J., Magoun, H.W., and Hall, V.E., eds. Washington, D.C.: Am. Physiol. Soc., pp. 1281–1305.

French, J.D., Verzeano, M., and Magoun, H.W. (1953): An extralemniscal sensory system in the brain. *Arch. Neurol. Psychiatr.* 69:505–518.

Herrick, C.J. (1948): *The Brain of the Tiger Salamander.* Chicago: The University of Chicago Press.

Hunter, J.I. (1924): The postural influence of the sympathetic innervation of voluntary muscle. *Med. J. Aust.* 1:86–89.

Juul-Jensen, P., and Denny-Brown, D. (1966): Epilepsia partialis continua. *Arch. Neurol.* 15:563–578.

Kirk, E.J., and Denny-Brown, D. (1970): Functional variation in dermatomes in the macaque monkey following dorsal root lesions. *J. Comp. Neurol.* 139:307–320.

Lindsley, D.B., Schreiner, L.H., and Magoun, H.W. (1949): An electromyographic study of spasticity. *J. Neurophysiol.* 12:197–205.

Lloyd, D.P.C. (1941): Activity in neurons of the bulbospinal correlation system. *J. Neurophysiol.* 4:115–134.

Lloyd, D.P.C. (1944): Functional organization of the spinal cord. *Phvsiol. Rev.* 24:1–17

Magoun, H.W., and Rhines, R. (1948): *Spasticity: The Stretch Reflex and Extrapyramidal Systems*. Springfield, Ill.: C. C Thomas.

Poirier, L.J. (1960): Experimental and histological study of midbrain dyskinesias. *J. Neurophysiol.* 23:534–551.

Poirier, L.J. (1966): Neuroanatomical study of an experimental postural tremor in monkeys. *J. Neurosurg.* 24 (Suppl.): 191–193.

Ramón y Cajal, S. (1909): *Histologie du Système Nerveux de l'Homme et des Vertébrés*. Paris: A. Maloine.

Royle, N.D. (1924): A new operative procedure in the treatment of spastic paralysis and its experimental basis. *Med. J. Aust.* 1:77–86.

Ruch, T.C. (1936): Evidence of the non-segmental character of spinal reflexes from an analysis of the cephalad effects of spinal transection (Schiff-Sherrington Phenomenon). *Am. J. Physiol.* 114: 457–467.

Sherrington, C.S. (1893): Note on the knee-jerk and the correlation of action of antagonistic muscles. *Proc. R. Soc.* 52:556.

Sherrington, C.S. (1898): Decerebrate rigidity, and reflex co-ordination of movements. *J. Physiol.* 22:319–332.

Sherrington, C.S. (1906a): *The Integrative Action of the Nervous System*. New Haven: Yale University Press.

Sherrington, C.S. (1906b): Observations on the scratch reflex in the spinal dog. *J. Physiol* 34:1–50.

Sherrington, C.S. (1910): Flexion-reflex of the limb, crossed extension-reflex stepping and standing. *J. Physiol.* 40:28–121.

Sherrington, C.S. (1933): *The Brain and Its Mechanism*. Cambridge, Eng.: Cambridge University Press.

Shimamura, M., and Livingstone, R.B. (1963): Longitudinal conduction systems serving spinal and brain-stem coordination· *J. Neurophysiol.* 26:258–272.

Hallowell Davis (b. 1896, New York, New York) is emeritus professor of physiology at Washington University School of Medicine and emeritus director of research at the Central Institute for the Deaf in Saint Louis. His research has dealt with central nervous system and auditory physiology, psychoacoustics, and electroencephalography.

# 18
## Crossroads on the Pathways to Discovery

## Hallowell Davis

This will be a very personal account of the first half, the Harvard half, of my scientific career, with particular attention to certain conscious decisions that I made concerning its direction and several fortunate circumstances beyond my control that were equally important in determining my choices. To me, the crossroads of discovery are more interesting than the smooth, well-planned paths of systematic collecting of data or perfecting of a new technique.

During my school years I had never doubted that I would be an experimental scientist. I had a "laboratory" in either a cellar or an attic from the time I learned to saw wood and drive nails. In grammar school my pride was homemade electric batteries of carbon, zinc, and sulphuric acid. My first scientific hero was Michael Faraday, whose biography fell into my hands by chance. Later, in high school, a classmate and I home-built a full-sized Tesla coil that produced harmless ultra-high-frequency electric sparks ten inches long. These sparks would spread out on, but not penetrate, a sheet of clean window glass, but if the glass were smeared with oil, the spark would neatly puncture a pin-sized hole. This was my first scientific "discovery." I kept a fairly good notebook but never made a literature search or varied parameters systematically. In this era Hugo Gernsback, a colorful popular science writer, was a major source of both ideas and ways to do-it-yourself at home.

### Lawrence J. Henderson: Biochemistry

As a Harvard undergraduate I was an enthusiastic chemist. Why chemistry? Partly it was Faraday, partly the inspiring lectures of Elmer P. Kohler in elementary chemistry and later in organic chemistry. But my final hero was Lawrence J. Henderson, whose *Fitness of the Environment* (1913) made a lasting impression. I took his course in biological chemistry and was "hooked."

An early crossroad turned me to medical school rather than to a Ph.D. in chemistry. In 1917 I followed my mother's family tradition of the Society of Friends, and I was an outspoken pacifist before World War I. With the declaration of war I made a compromise with conscience and served in France for several months as a volunteer ambulance driver. When the U.S. troops arrived, the choice was to sign up as a member of the armed forces or return home. I returned and completed my senior year at college and enrolled at Harvard Medical School as a member of the Medical Enlisted Reserve.

At medical school Cecil K. Drinker turned me toward physiology. The study of medicine, however, became more and more attractive, and in my senior year I had the difficult choice of whether to become an internist, a neurologist, or an academic physiologist. Physiology won out, partly because I was allowed to spend the second half of my senior year almost entirely on individual experimental work in Cecil Drinker's laboratory under the immediate guidance of Wallace O. Fenn. Also, I had maintained personal contact with Lawrence J. Henderson by serving as "section man" for his college course in biological chemistry.

It was Professor Henderson who took the initiative in obtaining for me a Harvard Sheldon Traveling Fellowship, which I gladly accepted. My problem now was where to spend a year of foreign study (1922–1923), and my choice was Cambridge, England. Professor Henderson took it for granted that I would work there with Joseph Barcroft and later join Henderson's own group as a blood-chemistry physiologist. Some instinct told me, however, that this particular problem was already too far advanced to make it profitable to enter as a beginner, and I opted instead for electrophysiology under E.D. Adrian. My choice here was strongly influenced by Alexander Forbes and his promise of a place in his laboratory at Harvard when I should return.

Adrian suggested a timely problem, namely to try to resolve the contradictory results obtained by Louis Lapicque and Keith Lucas in regard to the chronaxie of muscle. The chronaxie is a time constant of the excitatory process. After a chemically oriented hypothesis relating to carbon dioxide had failed to resolve the issue, a physically oriented idea proved to be correct. It was the size of the electrode in contact with the tissue that gave a short or a long time constant.

The next year Walter B. Cannon welcomed me as an instructor in his Department of Physiology at Harvard Medical School. There, in Alexander Forbes's laboratory, I continued work on chronaxie but gradually lost enthusiasm as I realized that a proper treatment of the problem would require more physical chemistry and mathematics than I could command. I was ready for a crossroad.

## Alexander Forbes: The Nerve Impulse

The crossroad actually was a suggestion by Alex Forbes that the all-or-none law in nerve needed better proof than had been given by the small narcosis chambers of Adrian. Forbes proposed that we use a much longer stretch of mammalian nerve (peroneal nerve of cat) and measure the action potential at various positions along the nerve with his string galvanometer. David Brunswick and Ann Hopkins were my other collaborators. The experiment showed clearly that a nerve impulse could be considerably depressed in a region of partial narcosis (alcohol vapor), but that the depression was uniform throughout the narcotized stretch and not progressive as a function of the distance traveled; and that, finally, when the impulse emerged into the normal, unnarcotized region beyond the chamber, it returned to its original amplitude. Alex Forbes's prediction was fulfilled in complete detail, but we were two years late. Independently, the same experiment had been done in Japan by Genechi Kato (1924) and his numerous collaborators. Our experiments were more rigorous; theirs were more varied. The conclusions were identical. His monograph was in print some months before our report.

The next three years I spent on "mop-up" operations: further studies of the nerve impulse, using the same approach that we had developed for the all-or-none study. But I learned a great deal from my continued association with Alex Forbes—from both his thinking and his experimental style. He was perfecting and applying the vacuum-tube amplifier that he had designed after World War I on the basis of his experience with radio compasses in the navy. He had already developed a condenser-coupled amplifier, and a few years later his laboratory notebooks played a part in a patent dispute. An electrical engineer tried to patent the condenser-coupler amplifier circuit, but Forbes's notebooks established the proposition that the circuit was already prior art.

A few more words about Alex Forbes (Figure 18.1), because he was the single individual most important in shaping my career as a neurophysiologist. Alex was an innovator, oriented to applying electrical methods to the study of reflex activity, much as Keith Lucas had applied such methods to the study of peripheral nerve and muscle. Alex studied in England with Sir Charles Sherrington and, more briefly, with Keith Lucas. He brought back from Europe one of the first string galvanometers to be installed in New England. The ideas that he developed in his classic paper of 1922 on the interpretation of spinal reflexes set a style for one branch of American neurophysiology for a generation. I think he was also the first to use a vacuum-tube amplifier (homemade of course) in connection with a physiological preparation.

Alex was a delightful and cheery companion and raconteur in spite of his impaired hearing which, in those days before wearable hearing aids were available, put something of a strain on the vocal cords of his family and friends. He was a devotee of skating and skiing and above all yachting. He was a skillful navigator, and he also piloted his own airplane. Obviously he had independent financial resources, and these he used generously but anonymously to help support the

Figure 18.1  Lord Adrian, Sir Charles Sherrington, and Alexander Forbes. Photo taken in Ipswich, England, in June 1938 by Florence Forbes (Locke).

laboratories of physiology, both his own and Walter Cannon's. As the years went on he devoted more and more of his time to his outside interests, which culminated in expeditions to Labrador to map the northern coast by new techniques of aerial photography and mapmaking; but he still returned to do experiments and to stimulate the rest of us with his insightful discussions.

Alex Forbes may have been the first to use an amplifier in a neurophysiological experiment, but of course the great breakthrough came when Herbert Gasser and Joseph Erlanger used not only an amplifier but a cathode-ray oscilloscope to analyze the compound action potential of peripheral nerve. We at Harvard read avidly every word they published, and we ultimately saw their equipment in action when they brought it to Boston for the XI International Physiological Congress in 1929. The question before us was, of course, whether we should switch from string galvanometer to a cathode-ray oscilloscope, with its vastly superior time resolution, or whether there was another way. The great limitation of the CRO at that time was the poor visibility of the spot, which required a repetitive phenomenon with superimposed traces to obtain a readable photographic record. Even in 1929 it was necessary to get at least partially dark-adapted in order to see a single transit of the spot. But many of the phenomena that Alex and I wanted to explore in the central nervous system were not repetitive or closely time-locked to a stimulus. I skip our futile attempts to develop a high-speed string galvanometer and our abortive trials with a moving-mirror oscilloscope. At least by now we had ample laboratory space and were continuing to select experiments that could be handled by the good old Hindle string galvanometer.

### E. Glenn Wever: Auditory Neurophysiology

In 1930 I came to another crossroad when E. Glenn Wever and Charles W. Bray, members of the Department of Psychology at Princeton University, reported their famous experiment on the auditory nerve of cat (Wever and Bray, 1930). The action potentials picked up with a large electrode at the internal auditory meatus were amplified, and they activated a telephone receiver in a distant room. Words spoken to the cat's ear were clearly intelligible in the receiver. Control experiments, monitored by E. Newton Harvey, established the biological origin of the electrical signals. This was the Wever-Bray effect, and it posed a critical question: How can the auditory nerve deliver an impulse pattern at the high frequencies (above 2000 Hz) that are necessary for clear reproduction of speech?

Mammalian myelinated nerves such as the auditory nerve have absolute refractory periods of the order of a millisecond, or longer if high-frequency stimulation is sustained. Alexander Forbes had experimented on the medulla and brainstem of cat in 1926 using a string galvanometer and had found an upper limit to the frequency of auditory impulse volleys at about 200 Hz, as expected. Remember that Alex was hard of hearing, and therefore he did not *listen* to the output as Wever and Bray did. He simply concluded that the auditory nerve behaved like other myelinated sensory nerves. He did not expect to find anything different, but Wever and Bray did expect to find the high frequencies. Their theoretical model was not the place or resonance theory of Helmholtz but the "frequency theory" or "telephone theory" of hearing, first enunciated by Rutherford and further elaborated in the 1920s by E. G. Boring, professor of psychology at Harvard.

I recall a seminar in our physiology department about 1928 at which Boring and Forbes debated their rival models, or theories as we used to call them. Forbes pointed out that the place theory, with a postulated mechanical acoustic analyzer in the cochlea, was compatible with conventional properties of the auditory nerve. Boring developed an elaborate model for the representation of the auditory qualities of pitch, loudness, density, and volume, with pitch assigned to the frequency of impulses in individual nerve fibers, to be perceived as pitch by a postulated neural analyzer. He was forced to admit that he had to assume a rate of recovery in the auditory nerve at least ten times faster than had so far been measured in any nerve. He had faith in a special adaptation in the auditory system. At this point the physiologists and the psychologists parted company, amicably but mutually unconvinced.

Wever had studied under Boring, and he and Bray were putting Boring's theory to experimental test. The theory predicted that the high frequencies were in the nerve signal but that Forbes's string galvanometer had been too slow to detect them. With the telephone receiver the high frequencies were detected and, apparently, Boring was vindicated!

Immediately after the publication of Wever and Bray's report (1930), Alex and I arranged to visit the Princeton Laboratory. There Wever, Bray, and Harvey gave us a completely convincing demonstration. We were satisfied with the controls. We also heard Wever's "volley theory," which based the frequency theory of pitch on reasonable physiological properties of the auditory nerve. He did not assume, as Boring had done, that in response to a 10,000-Hz tone any one fiber carried up to 10,000 impulses per second, but rather that a group of twenty or more fibers cooperated by each responding as frequently as it could but always in phase with the 10,000-Hz tone. The fibers would be out of step with one another, but their synchronized volleys as a group would carry the frequency information.

Alex and I had to admit that the theory was tenable, particularly for frequencies below 1000 Hz. (Actually the volley theory is now an established part of the psychophysiology of hearing, as the basis of "periodicity pitch," but, as Alex and I suspected, the synchronization fades out progressively above 2000 Hz.) I decided almost immediately that I would concentrate my own research efforts on the auditory system. Tremendous advances had just been made in the electrical generation and measurement of sound. Sound as a stimulus was now under control, and good vacuum tubes and better amplifier circuits for recording action potentials had become available. A little later the phosphors of cathode-ray tubes were greatly improved, and single transients in the nerve-impulse time range could be photographed. It was my great good fortune that the Wever-Bray effect forced the auditory system on my attention just at this time.

Back at the laboratory Leon Saul and I started to follow the cat's auditory system by its impulse traffic in response to metronome clicks and pure tones. We used the string galvanometer at first but added an audio monitor. The auditory nuclei in the brainstem yielded fine responses to clicks; but words were no longer intelligible! Diction was clear only when the electrode was in, on, or very near the auditory nerve itself. Something generated in the cochlea seemed to spread up the nerve but to be rapidly attenuated with distance. The "spread" signal was particularly powerful in the auditory bulla at the round window. We had found the secret of the Wever-Bray effect: an unsuspected microphone-like transducer in the sense organ that we now call a "receptor potential." These potentials are local, they are continuously graded, and they have no refractory periods.

We were not really disappointed that Lord Adrian came to the same conclusion at the same time. Priority here is confused because Adrian gave the correct answer in a communication to the Physiological Society in 1931 but then seemed to retreat to "nerve impulses only" in a second communication that was nearly simultaneous with our first report. As in the Kato episode, here was another case of independent proofs of the same proposition: this time of a biological potential.

In order to exploit the new potential and to measure activity in central tracts and nuclei we needed better instrumentation. The audio monitor was not quantitative and the galvanometer was too slow. I succeeded in combining financial support from several sources (the DeLamar Mobile Research Fund, Harvard Medical School, the Josiah Macy, Jr., Foundation, the American Otological Society, and several anonymous donors) and in hiring a laboratory assistant with professional experience in electrical engineering, E. Lovett Garceau. He designed and constructed a complete audio stimulating system and sensitive amplifiers and installed my first cathode-ray oscilloscope. With it our progress in auditory neurophysiology, beginning with the cochlea, was rapid. Arthur W. (Bill) Derbyshire became my first Ph.D. graduate student.

### Hans Berger: The Electroencephalogram

I skip now to the next crossroad. Bill Derbyshire had a medical-student friend, Howard Simpson, who, for his second-year thesis, wanted to study the "convulsions" produced in cats by partial cerebral anemia. I allowed them to use my laboratory, and Simpson correctly identified the "convulsions" as decerebrate rigidity, not epilepsy. In the winter of 1933–1934, I recall clearly that Derbyshire and Simpson came to me with the story that a German psychiatrist named Hans Berger had reported in a German psychiatric journal that he could record from the head of his son and other subjects a spontaneous rhythmic electrical output with a frequency of ten per second (Berger, 1929). What did I think of this? I explained patiently that it must be a vibration in his equipment or other artifact because it was unthinkable that enough axons in the brain could be so synchronized in their activity as to yield such slow potentials. But, asked they, is there any reason why we shouldn't test his claim? Our equipment was sensitive enough and could pass a frequency of ten per second with only moderate attenuation, so I gave permission.

About three weeks later Bill and Howard came to my office again, looking a bit sheepish. "You are right, chief," said Bill. "We have stuck needles in each other's scalps. The base line is unsteady, but we can't see anything rhythmic on the scope. But come and see if we are doing it right before we say anything about it."

I went with them to the lab and Bill stuck needles into Howard's scalp. Howard sat in the shielded room and closed his eyes. The spot wobbled unsteadily across the scope. "That's what I thought," said I, "but three heads are better than two. Put the electrodes on my head." They did, and I sat in the room and closed my eyes. Immediately there were shouts outside: "There it is! There it is." It was indeed the Berger rhythm. It seems that I have very strong alpha waves. Bill's and Howard's are weaker, and they were excited, anxious, and perhaps more uncomfortable than I was. Other members of our staff volunteered, and they were divided about evenly into "Bergers" and "non-Bergers." We were convinced that Berger was right. (It was some time later that we learned that Adrian had already confirmed him. But at least my alpha rhythm was the first to be recognized as such in the Western Hemisphere.) I also soon realized that we were probably watching a new slow potential of neural origin.

Here was the crossroad. My decision again was immediate: to find out more about this slow, spontaneous cortical activity and its strange properties. This meant modifying our amplifiers to handle frequencies three octaves below the audio range and to develop a recording instrument that would give a permanent on-line record. Psychophysiological experiments could not be done properly with the delay required by photography, and the human memory was not good enough for more than qualitative observations on the oscilloscope.

It took Lovett Garceau about six months to develop our first inkwriting oscillograph: one pen and a $\frac{5}{8}$-inch-wide tape. He had obtained a Western Union "undulator," designed to record graphically the very slow signals from transatlantic submarine cables. Lovett put in stronger alnico magnets and stiffer springs and forced its natural period up above twenty per second. Damping was provided by the friction of the pen on the tape. Crude and nonlinear, but effective, it was the ancestor of the EEG inkwriters later developed by Albert Grass, Franklin Offner, and many others. Again we were early but not the first. By this time Tönnies in Berlin had developed a very fine multichannel inkwriter for animal experimentation by A. Kornmüller. My own greatest contribution to EEG technique was to introduce folding paper in 1938, thus creating a random-access memory.

We gave an early demonstration of an alpha rhythm, kindly provided by Donald Lindsley, in the Building C amphitheater at Harvard Medical School. I cannot identify the medical group for which we put on our show, but we have a picture dated 12/5/34 of our equipment and Don Lindsley and Bill Derbyshire, taken in the corridor before the relay racks were rolled back into the laboratory (Figure 18.2). I think it was this demonstration that led to the next step, the arrival of Fred and Erna Gibbs. Fred and Erna, dedicated to the study of epilepsy, were working at Boston City Hospital under William G. Lennox. Their chemical and circulatory hypotheses had not yielded a satisfactory explanation of the genesis of seizures, and Fred asked to join us and try our "brain-wave" instrument.

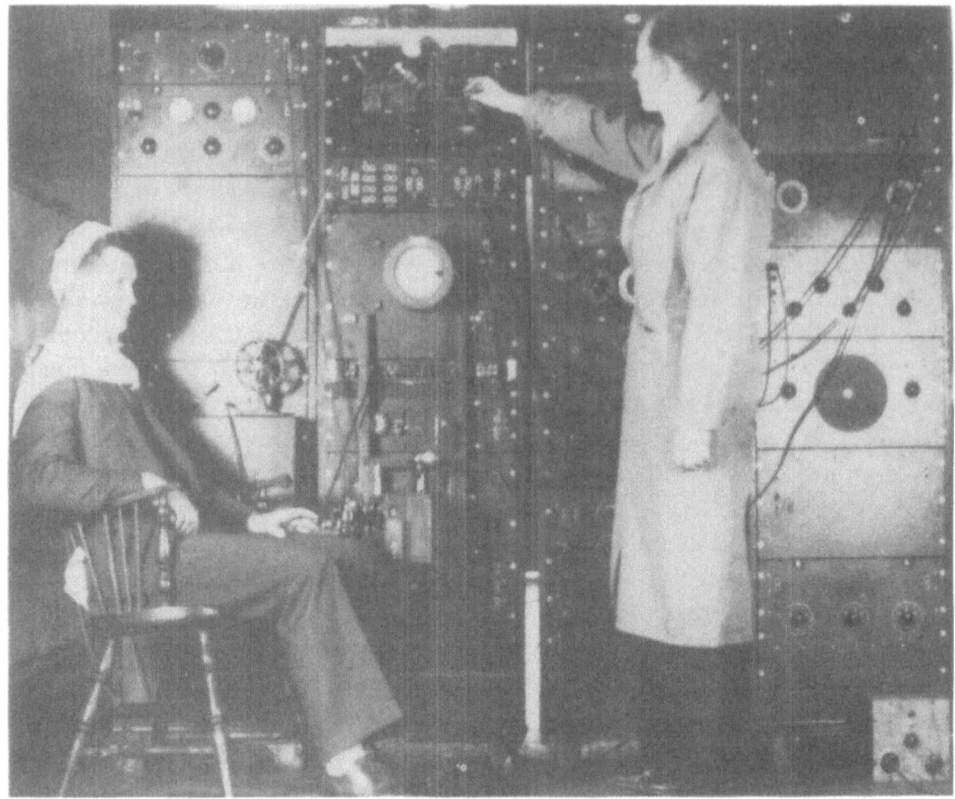

Figure 18.2   Donald B. Lindsley (left) and Arthur W. Derbyshire (right) demonstrating the alpha rhythm. Photo taken 12/5/34.

Fred had heard Berger mentioned in 1932, in very skeptical terms, by friends at the Johnson Foundation in Philadelphia, but he became a believer in our laboratory.

One evening in December 1934, Bill Lennox brought in one of his patients who suffered from very frequent petit mal attacks. Fred and Erna put on the scalp electrodes. Bill watched the patient in the outer room and reported the signs of the seizures while my wife Pauline operated our inkwriter in an inner room. I was the coordinator. Within twenty minutes we had the story and our first records of the spike-and-dome petit mal pattern. The most dramatic moment was when the patient said, "Please close that door. The noise from that room is giving me my seizures." She was hearing the scratching of the pen before she blanked out.

We published our report in December 1935, the second publication in the United States on brain waves. (Jasper and Carmichael were first.) Berger had seen the EEG pattern of petit mal, but he still thought it might be artifact. But does it matter who really was first? The important fact was that we had launched both clinical and experimental EEG work in the United States.

I had not given up the study of the cochlea and the auditory nerve, and 1934–1935 was my most exciting year in this area too. S. Smith (Smitty) Stevens worked with me in those days, relating place on the basilar membrane to audio frequency of the stimulus by means of small operative injuries. Moses H. Lurie provided the histological follow-up. This job, a high spot in our auditory program, and the identification of the petit mal pattern were simultaneous.

Never mind the later developments. For several years I concentrated chiefly on the EEG with my (first) wife Pauline. We studied sleep with Alfred Loomis in Tuxedo Park, but we failed in a long effort to relate the EEG patterns to characteristics of normal subjects or of neurotics or psychotics in mental hospitals. We got tantalizing partial correlations but no more.

## Robert Galambos: Single-Unit Activity

World War II brought me back abruptly to the auditory system. It became my task under the National Defense Research Committee to determine the tolerance of the ear for the loud sounds to which military personnel might be exposed. We gave the first proper description of "temporary threshold shift," but our work was heavily classified and we could not publish until 1950 (Davis et al., 1950). We found that men could tolerate, for short times at least, far more intense sound than anyone had believed, which proved to be critical for the operation of jet fighters from aircraft carriers. But through it all we managed to do a few purely scientific experiments, in evenings or on weekends. It was then that Bob Galambos, with micropipette electrodes, first succeeded in recording the activity of single units from the "auditory nerve" of cats (Galambos and Davis, 1943). He mapped out "response areas" and set the style for a generation of postwar studies. We could not do histological controls in our out-of-hours study, however. It was years later that we found to our dismay that Bob's electrodes, although within the internal auditory meatus, were actually in cell bodies in the cochlear nucleus, not in the axons of the nerve. This time we got wide acclaim for a "first" that we didn't achieve!

Let me revert finally to our confirmation of Berger's claims. It was years later when I asked Bill Derbyshire how he happened to find Berger's paper. He replied, "I didn't. You know very well I don't read the literature, particularly in German. *You* told me about it." I had to admit that he was right about his not reading German. Later I asked Howard Simpson the same question. He was astonished and, like Bill, said I had told them. He had duly included the Berger reference as the last one, apparently an afterthought, at the end of his thesis on the cat experiments, but he swears he didn't find it himself. Could it have been Fred and Erna Gibbs? Erna's native language was German, and Fred had heard of Berger, but he and I did not meet until a year after our initial trials. I didn't look up Berger's paper until after we had confirmed his claims; but we knew exactly what we were looking for in the laboratory that day when the subject closed his eyes.

I don't believe in ESP, but that is my story. Who told me? I believe that I shall never know. But whoever did it established a crossroad in my path of discovery.

## Acknowledgment

The preparation of this manuscript was supported by U.S. Public Health Service Research Grant NS 03856 from the National Institute of Neurological Diseases and Stroke.

## References

Adrian, E.D. (1931): The microphonic action of the cochlea: an interpretation of Wever and Bray's experiments. *J. Physiol.* 71:xxviii-xxix.

Adrian, E.D., Bronk, D.W., and Phillips, G. (1931): The nervous origin of the Wever and Bray effect. *J. Physiol.* 73:2P-3P.

Berger, H. (1929): Über das Elektrenkephalogramm des Menschen. I. *Arch. Psychiatr.* 87:527–570.

Boring, E.G. (1926): Auditory theory with special reference to intensity, volume and localization. *Am. J. Psychol.* 37:157–188.

Davis, H. (1923): The relationship of the "chronaxie" of muscle to the size of the stimulating electrode. *J. Physiol.* 57:lxxxi.

Davis, H. (1926): The conduction of the nerve impulse. *Physiol. Rev.* 6:547–595.

Davis, H., Forbes, A., Brunswick, D., and Hopkins, A. McH. (1926): Studies of the nerve impulse. II. The question of decrement. *Am. J. Physiol.* 76:448–471.

Davis, H., Morgan, C.T., Hawkins, J.E., Jr., Galambos, R., and Smith, F.W. (1950): Temporary deafness following exposure to loud tones and noise. *Acta Oto-Laryngol.* (Suppl.) 88:1–57.

Davis, H., and Saul, L.J. (1931): Action currents in the auditory tracts of the midbrain of the cat. *Science* 74:205–206.

Forbes, A. (1922): The interpretation of spinal reflexes in terms of present knowledge of nerve conduction. *Physiol. Rev.* 2:361–414.

Forbes, A., Miller, R.H., and O'Connor, J. (1927): Electric responses to acoustic stimuli in the decerebrate animal. *Am. J. Physiol.* 80:363–380.

Forbes, A., and Thatcher, C. (1920): Amplification of action currents with the electron tube in recording with the string galvanometer. *Am. J. Physiol.* 52:409–471.

Galambos, R., and Davis, H. (1943): The response of single auditory-nerve fibers to acoustic stimulation. *J. Neurophysiol.* 6:39–57.

Gasser, H.S., and Erlanger, J. (1922): A study of the action currents of nerve with the cathode ray oscillograph. *Am. J. Physiol.* 62:496–524.

Gibbs, F.A., Davis, H., and Lennox, W.G. (1935): The electro-encephalogram in epilepsy and in conditions of impaired consciousness. *Arch. Neurol. Psychiatr.* 34:1133–1148.

Gloor, P., ed. (1969): Hans Berger on the electroencephalogram of man. *Electroencephalogr. Clin. Neurophysiol.* (Suppl.) 28.

Henderson. L.J. (1913): *The Fitness of the Environment.* New York: Macmillan.

Jasper, H.H., and Carmichael, L. (1935): Electrical potentials from the intact human brain. *Science* 81:51–53.

Kato, J. (1924): *The Theory of Decrementless Conduction in Narcotised Region of Nerve.* Tokyo: Nankodo.

Stevens, S.S., Davis, H., and Lurie, M.H. (1935): The localization of pitch perception on the basilar membrane. *J. Gen. Psychol.* 13:297–315.

Wever, E.G., and Bray, C.W. (1930): Auditory nerve impulses. *Science* 71:215.

Ragnar A. Granit (b. 1900, Helsinki, Finland) shared the 1967 Nobel Prize with Haldan K. Hartline and George Wald for discoveries in the field of vision. Dr. Granit's research led him to an analysis of the electroretinogram, and to the discovery of inhibition in the retina and of color-specific modulators and broad-band dominators in single nerve fibers of lower vertebrates. In research on the motor system, he has described control mechanisms of the gamma motor neuron "loop." Dr. Granit is professor of neurophysiology at the Royal Caroline Institute, Stockholm.

# 19
# Half a Century in the Neurosciences: Personal Comments on Choices and Decisions

Ragnar A. Granit

If I were to meet the young man who emerged from the Swedish Normallyceum to matriculate at Helsingfors University in the spring of 1919, I am afraid I would find him a most disappointing companion. "Ripeness is all," and so profound are the changes brought about over the years that one would feel abashed at having to hail that youngster as an alter ego. Also, as a character he would seem to possess all one's faults without the insight and experience by which one cleverly tries to hide them.

The young man was interested in psychology—a widespread weakness in young people—and his gifted teacher, the philosopher Eino Kaila, introduced him to the experimental variety of that science. This was the time in Germany when Gestalt psychology rose to prominence, headed by a clique of talented enthusiasts. The best known were perhaps K. Wertheimer, W. Köhler, K. Koffka, and A. Gelb. The Gestalt school had started a spirited attack on what they held to be a dis-integrative and meaningless splitting-up of mental events by most of the experimental approaches of the time. Their leading thesis was that the experienced world consists of formed *(gestaltete)* entities, not further reducible by experimentation, though accessible to experiments properly designed to demonstrate Gestalt qualities or rules.

An example is needed to show how this attitude could be experimentally realized. I shall take an experiment that Gelb and I carried out in 1922 when I worked with him for two months in Frankfurt in a laboratory whose formal head was Professor F. Schumann (Gelb and Granit, 1922). We had a set of Maltese crosses photographically produced so as to make a given nuance and brightness of gray appear either as the cross or as its background. The threshold of a small spot of light was found to be higher when projected on the cross than when projected on the background between two of its arms. Crosses and backgrounds were alternated on separate charts. The experiment was designed on the Gestalt principle to check whether an integrated Gestalt might not create a greater resistance to an added disturbance than a background experience of the same physical qualities. It seemed that it did.

The Gestalt approach, whatever its theoretical value, has to its credit a number of important studies, such as those of Goldstein and Gelb on cases of head wounds from World War I and Köhler's behavioral experiments with chimpanzees. The latter became very well known and rendered him the nickname of *"Affenköhler."* The impending threat of Nazism broke up the school. Some of its most influential members emigrated to the United States, and the young man of whom I have spoken decided, on the advice of an uncle who was a practicing physician, that it would be necessary to study medicine seriously if he was to do something sensible in psychology. He therefore abandoned the field, and this is where I also quit him now. However, he may have left some trace in my upbringing because now, in my retirement, like a snake biting its tail I have begun to understand that psychological points of view could be used very profitably in the study of voluntary movement as a physiological process.

It is curious to reflect how young physiology is as a science. Those of us who were born at the turn of the century have had personal contact with pupils of the fathers of our science. While still interested in psychology, I listened—I believe in 1920—to lectures by Robert Tigerstedt in a course on the physiology of the special senses, and he was a pupil of Carl Ludwig. Tigerstedt was a true *Gelehrter*, a polyhistor such as our times do not produce. His textbook was read in many countries, but his reputation rests mainly on his large and authoritative monograph on circulation, containing everything published in that field up to the day of its appearance (Tigerstedt, 1921–1923). With Bergman, Tigerstedt discovered renin, a little too early for the physiology and biochemistry of his own time. As a young physiologist he moved from Helsingfors to the Caroline Institute in Stockholm, but in the end he left the chair at the institute and returned to Helsingfors University as professor of physiology. His son Carl succeeded him in this position.

When in 1926 I was made demonstrator *(Assistent)* at the Physiological Institute in Helsingfors, Robert Tigerstedt's creation, I still thought of physiology as something to be studied for the sake of a future career in experimental psychology, but the long-lasting medical course brought about a gradual change of attitude and interest. Clinical work was not without its temptations. However, I became more and more interested in the nervous system and the special senses, vision above all, for their own sake, and started to improve my knowledge in these disciplines.

At the time Helsingfors University, like other Scandinavian universities, was dominated by the German academic tradition, and most of the textbooks were in the German language. Alongside the course I read, first and foremost, Ewald Hering, then Helmholtz, Mach, König, von Kries, the special articles in the great German *Handbücher*, and some psychologists such as Ebbinghaus, Wundt, and G. E. Müller. I even read through Freud's *Vorlesungen über Psychoanalyse*, but apparently too late in the day relative to my own development, because I found them utterly unscientific, though highly entertaining as products of an original mind. Another quite original mind and great writer was William James, whose *Psychology* I studied at an early date with much admiration. Like Freud, James had had a medical education. But, though in published correspondence James made some very caustic remarks about the narrow-mindedness and bigotry of scientists, including experimental psychologists (naming Hugo Münsterberg), he himself stayed within the boundaries of scientific criticism when writing psychology.

I departed from the pattern of my own generation in feeling much attracted by the English language and civilization. For my first visit to a foreign country, in 1920 (thanks to the generosity of my father), I chose to go to a holiday course at London University in order to study English. This left me with a permanent feeling for Dr. Johnson's remark that the man who is tired of London is tired of life.

It is well known among physiologists that the decisive breakthrough of elee-tronics in physiology took place within the English-speaking world with papers by Adrian, Erlanger, Forbes, and Gasser, whose achievements are too familiar to need recapitulating. In spite of my academic background in German physiol-ogy, I was well prepared for drifting over into a slow Anglo-Saxon reeducation at a time when we, who had advanced to the Einthoven string galvanometer and the capillary electrometer, had to be reeducated anyway in preparation for the electronic era of neurophysiology. In point of fact, my personal reorientation toward physiology as a final career was the outcome of that process. It involved relearning physiology in a second language and, I felt, also with a change of emphasis.

Psychophysics of vision was my best subject and leading interest, but for a medi-cal man it seems natural to start with the retina and think of some sensible, fresh approach reaching beyond pure stimulus-response relationships of a more ele-mentary character. The study of Ramón y Cajal's work made me realize that it would be profitable to regard the retina as a nervous center. It impressed me a great deal to find him saying that the retina was his first love and that the study of its structure served as an introduction to his later work on other parts of the central nervous system. I still believe that this roundabout route to the brain is as good in physiology as its histological counterpart. For me the main question was, of course, how and where to begin. I made an effort with a study of interaction between the center and the periphery in the afterimage of movement (1927), but soon under-stood, on reading Sherrington's *Integrative Action of the Nervous System*, that I knew too little about the central nervous system to realize fully my general idea of translating psychophysical observations into what John Fulton was later to term "neurophysiology."

Sherrington at the time (1928) was surrounded by a lively group of well-trained young people engaged in teaching and experimenting, and it was my great fortune to be kindly received by him and the others—Sybil Cooper, R. S. Creed, Derek Denny-Brown, J. C. Eccles, and E. G. T. Liddell. Struggling alone for so long in order to reach a sensible attitude about the science of the eye, I felt happy when Sherrington immediately said that I had hit upon the right track and added that he had always wanted someone interested in the eye to work in his laboratory. Creed and I had many interests in common, including the eye; Eccles and I worked on the spinal cord, at times with Sherrington. What a pleasure not to be alone, to work for a change within a school where definite things had to be known and whose head was a wise and kindly old man of wide reading and culture, familiar with physiology virtually from its beginning. However, I have written about him elsewhere, in the first chapter of my appraisal of Sherrington's con-tributions to our field (Granit, 1966).

Sherrington wrote me a letter of introduction to Adrian, whom I visited for a day in Cambridge in 1928. At that time Adrian had carried out his pioneer work with Rachel Matthews on the optic nerve of the conger eel. Then, as on many later occasions, he gave me much of his time, though I never worked in his laboratory.

In the autumn of 1928, or it may have been in the spring of 1929, Alan Gregg of the Rockefeller Foundation visited Helsingfors and asked me to serve as his interpreter for visits to the university professors. His subsequent support and that of Sherrington sufficed to make D. W. Bronk appoint me a Fellow of the Eldridge R. Johnson Foundation of the University of Pennsylvania. Bronk had returned from a sojourn with Adrian at Cambridge and been offered the leadership of that brand-new institution. My wife and I came over in the autumn of 1929—our wedding trip—to a still virtually empty house; but this deficiency was soon remedied, and it became a place where everything one wanted could be obtained and where one could do what one wanted. This was my wife's and my introduction to the United States, another permanent association offering us new friends, new opportunities for work, a new way of living, and at the time also a country blazing in the brilliant coloring of October.

At the Johnson Foundation I met W. A. H. Rushton from Cambridge, initiating his experimental attack on the theory of isochronism, and a year later H. K. Hartline turned up from a long stay in Munich and started to build apparatus for his elegant studies of the retinae of *Limulus*, *Pecten*, and the frog. These two, and J. C. Eccles with whom I actually worked, are the friends of my own age group who have played the greatest role in my scientific development. A second period at Oxford, this time as a Rockefeller Fellow (1932–1933), completed my orientation toward the physiology of the nervous system.

I have made a long story short, believing that a person's start in science is nearly always of some interest but that readers must not be overdosed with individual fripperies. Looking back, I realize that these ripening years in neurobiology took a long time, where by "ripening" I mean arriving at a world of ideas broad and independent enough to sustain a durable interest in their development by experimentation. I have always believed that biological work presupposes a desire to advance understanding within a field, rather than a desire to express one's experimental cleverness. At the Johnson Foundation I made a systematic attempt at translating psychophysical observations into retinal neurophysiology, using the fusion of a flickering light as an index of excitability. My best coworker was the late C. H. Graham, finally professor of psychology at Columbia University. The most interesting results demonstrated varieties of spatial interaction (excitatory as well as inhibitory) by that index; and some of the findings were a close replication of results obtained by Adrian with the optic nerve of the conger eel (Granit, 1930; Granit and Harper, 1930; Graham and Granit, 1931).

From 1932 onwards I turned to electrophysiological methods. The results brought me invitations in 1940 to Harvard and to Stockholm, where in the end I settled down and founded an institute of my own as part of the Medical Nobel Institute (which belongs to the Royal Caroline Institute, Stockholm's medical university). The generous support of the Wallenberg Stiftelsen in Stockholm and of the Rockefeller Foundation made the move possible, and in 1946 the government created a chair in neurophysiology attached to the institute.

In the late forties I began to feel that I had been in the field of vision too long and needed a change. To this end I spent a term in my own laboratory repeating, for practice, most of the physiological experiments (on the cat) that others had done on the spinal cord after my time in Sherrington's laboratory fifteen years earlier. I concluded that a profitable opening would be the unduly neglected field of muscular reception in relation to spinal cord, muscular activity, and gamma control. Leksell had published his thesis on the gamma system in 1945, but I could not persuade him to follow it up; nor did the subject attract others to whom I offered it. From Leksell's point of view his attitude was sensible; he was a neurosurgeon, basically interested in his own discipline, and the spinal-cord aspect by itself required full-time laboratory work. At about that time (1948) Rushton turned up for a year of sabbatical leave in my laboratory, feeling, like myself, that he needed a change from his old interest in peripheral nerve. Rushton took up the subject of vision and both of us started a new scientific enterprise. Neither of us has regretted that decision.

An attractive aspect of motor work is that movement serves as an interpreter of what has been put in, whether an artificial electrical stimulus, an afferent message for a reflex, or a voluntary command. The completed integration is handed out to the observer or experimenter, and his role, apart from discovering general principles, is to investigate the mechanisms—their site, nature, and interaction—in order to reach a functional and structural interpretation of them. By "general principles" I mean statements of basic facts with some finality to themselves, such as that the pyramidal path increases in size from rat to man, that movement rather than conscious perception is disturbed by cerebellar lesions, and that muscle contractions are graded by firing and/or recruitment of fresh motoneurons.

This seems a suitable occasion for adding to the list of such principles two generalizations referring to the relation between the brain and its environment that most obviously are valid for higher organisms. One is that the greater the elaboration required for a percept or a motor act, the greater the number of cells engaged in it. "Elaboration" may not be the best possible term, but what I mean can be well illustrated by citing examples: delicate manipulation of objects, a high degree of sensory discrimination or generalization, and high levels of consciousness. I include in the last category acts and percepts that originally required a high degree of consciousness but which by practice have become automatized. Examples on the motor side are the large cortical areas devoted to thumb and finger movements (see the well-known figurines of Penfield and Rasmussen and of Woolsey); on the sensory side one could mention the great cortical expansion of the small fovea relative to the rest of the retina (Marshall and Talbot, Whitteridge).

The second generalization might be regarded as a corollary of the first, inasmuch as it is based on the common idea that the reason for a cortical expansion is the need for more contact points to facilitate the extensive combinations of data required for evaluation and anticipation. Hence the second generalization maintains that the greater the elaboration in the sense defined, the greater the number of sites in the brain engaged in the particular act or percept concerned. This is well illustrated by the large number of widely separated sites that are engaged in a motor act as defined by anatomical and clinical evidence and recently by Evarts's well-known physiological studies of wrist movement in the trained monkey. Another example is the steadily expanding number of visual areas, a minimum of nine at the moment (Cowan, Powell, Zeki).

These two generalizations are characteristically biological; they are as independent of physics or chemistry as the notion of evolution formulated by Darwin and Wallace. They deal with purposive responses of the organism to the environment and hence belong to the teleological aspects of physiology. I have been more explicit on teleological explanations elsewhere (Granit, 1972).

It must now be common knowledge that the organ of highest control, our brain, cannot be discussed as if it were a question of localization versus general representation. The whole problem is obsolete. It created a long-lasting dispute once upon a time, but this did not end in complete victory for either side. "We now realize that there are highly localized functions, even tricks—if I may say so— that only certain aggregates of cells can perform, as well as functions that require coactivation of several cortical areas widely apart. As we penetrate problems of localization, we tend to end up with problems of organization." (I quote myself from an unpublished essay on the nature of biological explanations.)

I have emphasized all these points in order to prepare a final confrontation with the question of whether this well-documented multicellular character of organized brain activity really can be understood by the single-cell techniques that in so many respects have been so rewarding. Having been one of the early birds in recording from single cells (retina), I have now been at the game long enough to have seen the whole field develop during my own active era of single-cell experimentation. The gain in understanding has been impressive, not only in terms of fundamental mechanisms at synapses and cell membranes but also from functional points of view in an integrative context. One might mention, for instance, the columnar organization of tactile specification in the cortex (Mountcastle), the processing of information within the retina (Barlow, Tomita, Dowling, and others), cells in the visual striate area responding to orientation (Hubel and Wiesel), the gamma system, the colony concept of pyramidal motor cells (Phillips), the different organizations for phasic and tonic motoneurons, much work on invertebrates, etc. These mechanisms form but a small selection from an overwhelming multitude of triggers, "detectors," "mandatory neurons" of different kinds, etc., discovered by the single-cell technique as used in an integrative context. And there is more to come, while, of course, this technique is and always will be indispensable for the analysis of specific synaptic events at cell membranes in chemical or physical terms. But this is another story.

Returning to the question of whether unicellular studies provide good enough insight into multicellular organizations, I do not think their importance in that regard should be exaggerated. In the first instance, one cannot but suspect that many aspects of coding intrinsically depend upon a combination of cells, which remains un-get-at-able by the single-cell approach, even if it is favored by an input extended in a plane (skin, retina). By the refined histological techniques of today many connections from primary projections to an ever-increasing number of sites in the brain have been discovered, as well as interconnections between them. At such levels of polysynaptic complication unicellular signals tend to degenerate into mere "spikes," detecting, to be sure, combinations of excitation and inhibition but not the relevance of the signal being transmitted. It is well known that in the primary afferent projections the discharge of a single cell often can be meaningfully interpreted as a "cue" or "detector." Similarly, on the motor side, its message can be evaluated as being significant for the regulation of movement in some way or other. In both these examples it is likewise possible to use the spike to analyze the organization of which it is an indicator. It is from these points of view that the unicellular response within a distant site loses its specific connotation, even though it is still capable of supporting lengthy papers on uninterpretable interactions. And when it comes to functions such as speech, understanding the significance of a sentence or a visual pattern, or deficiencies of motor control, few people would think of approaching them by the single-cell technique except in order to procure a useful hint or two.

This is where the "ablationists" come in with their behavioral tests and histological checks. As histology goes on improving, so does "ablationism," which can be regarded as its offshoot into the physiology and pathology of the nervous system. Ablation has always been regarded by physiologists as an equivocal technique because of compensatory processes, the risk for destruction of unwanted parts, and the fact that in many cases it merely tells us what the organism can do without the piece ablated, instead of telling us what the latter did in situ. But today my definite impression is that methods based on ablations have undergone great improvement, partly, as I said, because of the improved histology, but partly also because of improvement in the critical evaluation of behavioral tests. This is actually more of a difficulty than is histological verification. At its best a good ablation experiment becomes a salubrious antidote against overinterpretation of single-cell data. Of this, the field of vision provides excellent examples (see reviews by Doty, 1973, and Weiskrantz, 1972). It is truly surprising what an animal can do after ablation of the striate area.

However, I shall restrict my comments to the motor field, within which my reading and experimentation are of a more recent date. The now-available knowledge that many cells in many different sites take part in motor acts is apt to be discouraging. It suggests an inordinate number of local electrodes and a very optimistic faith in interval recording and correlation statistics. This experiment must be left to the believers in its future. But several electrodes could, indeed, be implanted without too much trouble and without an excess of machinery. The essential problem will then be to introduce constraints to limit the number of possibilities.

An analogy will make my meaning clear without involving technicalities. The central nervous system is capable of combining "anything" with virtually "anything else." As an example, take the word "horse." Sitting down thinking of my associations to "horse" for a while, I found them to be almost limitless. Yet I use the word "horse" only in a limited number of sensible contexts. Only when this is done does it become relevant either for my use of communication by means of the internal logic of language or for thinking about something. A "sensible context" in this example corresponds to a constraint applied to the electrophysiological analysis of a movement.

Introducing constraints into an experiment with a number of local electrodes takes us back into teleology, as it generally presupposes making use of some *purpose* of the motor act to be analyzed. Study of motor purposes begins with questions that sound conventionally physiological, but it ends high up in psychology with "demands" of a complex nature involving consciousness. Examples of conventionally physiological problems using teleological constraints are provided by a number of experiments trying to understand the role of recurrent inhibition in motor control or the different properties of tonic and phasic motoneurons in relation to fast and slow muscles and their employment in different tasks of motor regulation. A large number of familiar problems belong to this category. Further up in the hierarchy of movements, constraints will in general involve the volitional component that has the character of a "demand."

In the motor field I believe that far too little has been made of electrical recording of voluntary movement with demand as the leading variable. The experiments require men or monkeys as subjects, and a great asset of this approach is its close contact with the vast reserve of information from clinical neurology. In everyday life an exceedingly large number of motor acts are executed in response to a variation of demand. Surely it must be a relevant question to ask how different sites participate in such acts. Answers are bound to lead to some insight into basic principles of organization of "hardware" within the motor field.

This approach is not wholly untried. As indicated, clinical neurologists and neurosurgeons use demanded movements in their standard tests, and do so also in combination with local destructions, ablations, and case pathology. There are also electrophysiological studies making use of demanded movements. Buchwald and Eldred (1962) have shown that in a learning process the gamma motoneurons can be conditioned to respond to a cue more easily than the alpha motoneurons. Much experimentation by the technique of measuring H-reflexes in man shows that during the preparatory period of a voluntary act these reflexes undergo complex changes varying with demand and taking place also in motoneurons not participating in the final movement (e.g., Requin, 1969; Requin and Paillard, 1969). The preparatory period has also been studied in man and monkeys by electronic summation of potentials recorded electroencephalographically, though so far without sufficient attention devoted to demand as a variable. In the recording of the discharge of muscle-spindle afferents in man, a demand for a greater or lesser force of contraction is clearly reflected in the firing rate of the spindles (Hagbarth and Vallbo, 1969; Vallbo, 1970). Evarts's (1967) technique of recording from cells in the motor cortex of trained monkeys is loaded with possibilities for studying varying demands, and so is the electromyographic technique used by Marsden, Merton, and Morton (1972) for this very purpose. Some comments on these questions will be found in the *Report from the Conference on the Control of Movement and Posture* (Granit and Burke, 1973).

I am, of course, under no illusion that in the long run the electrophysiological approach will be omnipotent in solving all motor-control problems, but these comments are meant to emphasize that it still is capable of conquering new domains of understanding. My active time in neurophysiology has coincided with the electronic era, and so it seemed pertinent to discuss the role of electrophysiology in an integrative context. Many of the best known results of the microphysiological reductionist aspects of the period have consequently been neglected. These are the ones that have been most in the limelight, having been rewarded with ten Nobel Prizes.

## References

Buchwald, J.S., and Eldred, E. (1962): Activity in muscle-spindle circuits during learning. *In: Symposium on Muscle Receptors.* Barker, D., ed. Hong Kong: Hong Kong University Press, pp. 175–183.

Doty, R.W. (1973): Ablation of visual areas in the central nervous system. *In: Handbook of Sensory Physiology.* Vol. 7, Pt. 3. Jung, R., ed. New York: Springer-Verlag, pp. 483–582.

Evarts, E.V. (1967): Representation of movement and muscles by pyramidal tract neurons of the precentral motor cortex. *In: Neurophysiological Basis of Normal and Abnormal Motor Activities.* Yahr, M.D., and Purpura, D., eds. New York: Raven Press, pp. 215–251.

Gelb, A., and Granit, R. (1922): Die Bedeutung von "Figur" und "Grund" für die Farbenschwelle. *Z. Psychol.* 93: 83–118.

Graham, C.H., and Granit, R. (1931): Inhibition, summation and synchronization of impulses in the retina. *Am. J. Physiol.* 98: 664–673.

Granit, R. (1927): Ueber eine Hemmung der Stäbchenfunktion durch Zapfenerregung beim Bewegungsnachbild. *Z. Sinnesphysiol.* 58: 95–110.

Granit, R. (1930): On interaction between distant areas in the human eye. *Am. J. Physiol.* 94:41–50.

Granit, R. (1966): *Charles Scott Sherrington. An Appraisal.* London: Nelson.

Granit, R. (1972): In defense of teleology. *In: Brain and Human Behavior.* Karczmar, A.G., and Eccles, J.C., eds. New York: Springer-Verlag, pp. 400–408.

Granit, R., and Burke, R.E. (1973): The control of movement and posture. *Brain Res.* 53:1–28.

Granit, R., and Harper, P. (1930): Synaptic reactions in the eye. *Am. J. Physiol.* 95:211–228.

Hagbarth, K.-E., and Vallbo, Å.B. (1969): Single unit recordings from muscle nerves in human subjects. *Acta Physiol. Scand.* 76: 321–334.

Marsden, C.D., Merton, P.A., and Morton, H.B. (1972): Servo action in human voluntary movement. *Nature* 238: 140–143.

Requin, J. (1969): Some data on neurophysiological processes involved in the preparatory motor activity to reaction time performance. *Acta Psychiatr. Scand.* 30:358–367.

Requin, J., and Paillard, J. (1969): Depression of spinal monosynaptic reflexes as a specific aspect of preparatory motor set in visual reaction time. *Proc. Int. Symp. Bulg. Acad. Sci.,* pp. 391–396.

Tigerstedt, R. (1921–1923): *Die Physiologie des Kreislaufes.* Vols. 1–4. 2nd ed. Berlin: W. de Gruyter.

Vallbo, Å.B. (1970): Slowly adapting muscle receptors in man. *Acta Physiol. Scand.* 78: 315–333.

Weiskrantz, L. (1972): Behavioural analysis of the monkey's visual nervous system. *Proc. R. Soc. B.* 182:427–455.

# Behavior and Brain

Alexander R. Luria (b. 1902, Kazan, U.S.S.R.) is professor of psychology and head of the Department of Neuropsychology at Moscow University. His research has dealt with the neuropsychological analysis of brain injuries and the role of brain systems in complex processes including brain disorders and other human cognitive and behavioral phenomena.

# 20
# Neuropsychology: Its Sources, Principles, and Prospects

## Alexander R. Luria

Among that rapidly developing group of sciences of the nervous system which we now call the neurosciences, a special place is occupied by neuropsychology. Its roots go back through the centuries, yet at the same time it can be regarded as the youngest, the most recently founded, branch of neuroscience.

It differs from the other members of the group of neurological disciplines in that its concern lies not with the deep mechanisms of the molecular or biochemical bases of nervous activity, nor with the morphological structure or evolution of the nervous system, nor with the physiological mechanisms of nervous processes, but rather with the role of individual brain systems in the organization of human psychological activity.

In this respect neuropsychology stands astride the line separating the natural from the social sciences; it is one of the behavioral sciences, and it preserves the closest connection with such other branches of learning as neurology, psychology, and linguistics. In fact, progress in neuropsychology is largely dependent on progress in these other sciences; it can develop only on the basis of progress in scientific psychology, and, in turn, it must act as a powerful stimulus to further development in that field.

What are the sources of neuropsychology, the fundamental principles of its work, its theoretical and practical importance for psychology and medicine, and, finally, what are its prospects? The writer has been a direct participant in the creation of neuropsychology. In the pages that follow, theoretical propositions will therefore naturally be intermingled with personal reminiscences.

### The Sources and the Problem

The question of the role played by the different parts of the brain in psychological activity first aroused interest in the Middle Ages, and the first major attempt to provide an answer was made in the eighteenth and early nineteenth centuries by F. J. Gall, one of the founders of scientific brain anatomy. However, it was in-evitable that the manner of solution of the problems posed by what we now call neuropsychology would depend upon current philosophical and psychological concepts. In the seventeenth and eighteenth centuries the "theory of faculties" held a position of dominance in psychology (which at that time was still an integral part of philosophy). All psychological processes (perception and memory, voluntary action and abstract thought) were understood as "faculties" of the human mind, incapable of further subdivision. The analysis of psychological processes thus revolved around attempts to isolate these "faculties," and it was natural that an investigator such as Gall, who wished to discover their cerebral mechanisms, would strive to find brain structures that could be regarded as their organs or centers. That is how Gall came to produce his fantastic "brain map" on which all manner of different "faculties" (including "thriftiness" and "love of children") were ascribed to particular, narrowly localized areas of the brain.

The mythological basis of Gall's map was clearly realized long ago, and his phrenology has been rejected. However, the principle of the direct localization of complex psychological processes in circumscribed areas of the brain has persisted, virtually unchanged, on the basis of the clinical picture derived from the study of local brain lesions.

In 1861 Broca showed that a lesion of the posterior third of the inferior frontal gyrus causes a disturbance of motor speech but does not impair the ability to understand speech addressed to the patient. To Broca this was sufficient grounds to localize the "motor images of words" in this area of the brain. When Wernicke in 1873 described the opposite fact—a disturbance of the understanding of speech associated with a lesion of the posterior third of the superior temporal gyrus of the left hemisphere, in which the patient's motor speech remains intact—he felt fully justified in concluding that this area of the cortex is the center of localization of the "sensory images of words" *(Wortbegriff)*.

The view that complex psychological processes can be localized in circum- scribed areas of the cerebral cortex just like elementary processes (sensation, motor impulses) proved so attractive that, within a single decade (known in the history of neurology as the "splendid seventies"), scores of brain centers were discovered for such "functions" as reading, writing, calculation, and abstract concepts. The clinical pictures of their lesions, such as "agnosia," "apraxia," "alexia," "agraph- ia," "acalculia," and so on, also became firmly entrenched in clinical neurol- ogy. These views were followed up by investigations based on stimulation of individual points of the human cerebral cortex and observation of the resulting changes in behavior, which seemed to confirm the existence of cortical centers for complex psychological processes (Hoff and Pötzl, 1930; Foerster, 1936). It was later postulated that more complex disturbances of psychological processes could be interpreted as the result of the disconnection of these "centers." In the first decades of the present century theories about the direct localization of complex psychological functions (essentially indistinguishable from the idea of the direct localization of "faculties") reached their climax in the "cerebral pathology" of Kleist (1934) and in the doctrines of Nielsen (1946). Subsequently the concept that the most complex forms of psychological disturbances occurring in patients with local brain lesions result from the "disconnection" of the individual centers (the disconnection syndrome) acquired a firm foothold in modern neurology and was reflected by such writers as Geschwind (1965).

Very recently these views of the direct localization of complex psychological functions in circumscribed groups of cells received apparent confirmation from the brilliant researches initiated by Hubel and Wiesel (1962, 1963) and continued by many other investigators. These workers established to their satisfaction that isolated cortical neurons respond selectively to very specialized stimuli. They saw in this fact confirmation of the view that even very complex psychological processes can be "localized" in circumscribed groups of nerve cells and that there is no difference in principle between the localization of elementary functions and that of complex psychological processes. Some workers (for example, Konorski, 1967) saw here fresh grounds for returning to traditional views and for concluding that the secret of the cerebral organization of psychological processes in circumscribed areas of the brain had been discovered and that the brain mechanisms of psychological activity had at last been elucidated.

It thus seemed to some that neuropsychology had completed a cycle in its development, and that a new discipline, supported by a precise and finite system of facts, now existed in the family of the neurosciences.

## The Crisis

It would be wrong to suppose that the development of the ideas of the direct localization of complex psychological processes in circumscribed areas of the brain —or, as they are usually called, the ideas of "narrow localizationism"—marched forward like a triumphal procession without encountering any resistance and that they became fundamental and undisputed concepts for neuropsychology. History shows that whenever narrow localizationism has been propounded, opposing facts have been brought against it, and arguments have been presented that the brain works as a single entity.

At the beginning of the last century Gall's ideas were opposed by Flourens (1824), who showed that there are no firmly fixed centers in the bird's brain and that the specialization of the motor points of the brain can be changed by making the flexor muscles of the wing take over the function and innervation of the extensors. Some fifty years later Broca's views on "speech centers" were opposed by Hughlings Jackson (1884), who postulated that speech processes (like all complex types of psychological activity) are not localized in circumscribed areas of the brain and that their organization is based on the principle of hierarchical levels and dynamic systems. In the physiology of brain functions the same discussion was repeated between Fritsch and Hitzig (1870) and others who shared their views on the narrow specialization of cortical areas, and Goltz (1881), who postulated that the brain works as a whole.

In the first decades of this century a similar conflict was repeated in neurology and psychophysiology by leading investigators who were or are our contemporaries. Whereas Kleist (1934) logically defended the narrow localization of complex functions in the cortex, other eminent neurologists such as Monakow (1914) and Goldstein (1927, 1934), with equal conviction, defended the opposite view: only elementary functions can be localized in circumscribed areas of the brain; complex "semic" processes are always the result of the activity of large brain systems (Monakow, 1914); and complex psychological processes are dependent more on the mass of brain substance than on circumscribed areas of cerebral cortex (Goldstein, 1927, 1934).

This continuing conflict shows that the road of development of neuropsychological concepts is by no means easy, that the cerebral organization of higher forms of psychological activity is a far more complex phenomenon than has been realized, and that further searches must be made, perhaps with significant revision of our basic approaches, in order to break the vicious circle of controversy.

A century of history of what we might nowadays call the theoretical propositions of neuropsychology—the theory of cerebral organization and of human psychological activity—thus ended in deadlock. A crisis in the basic views of neuropsychology had arisen, and it would be pointless to pretend that it had not. It was far deeper than a conflict of views between individual scientists. Its bases were theoretical and factual and were concerned with fundamental aspects of our new science.

The theoretical basis of the crisis of concepts in classical neuropsychology centered on two groups of propositions which justifiably aroused the most serious misgivings. First, it could hardly be imagined that complex psychological processes, such as logical perception or voluntary action, not to mention speech, writing, reading, and calculation, could be understood as elementary and even as indivisible "faculties." Such views, which might have been acceptable in the Middle Ages or even at the dawn of the modern era, had lost all their justification by the middle of the last century, when these activities began to be understood as complex systems of more elementary psychological processes. And now, when higher forms of psychological activity are regarded as a product of development taking place under the combined influences of maturation and of exposure to the formative action of the social environment (Vygotskii, 1956, 1960), they are quite simply untenable. Such complex forms of psychological activity, formed under different conditions and to different degrees, carried out by completely different methods, and having different structures in different persons, cannot in any sense of the term be regarded as elementary inborn "faculties," and under no circumstances can they be understood as functions of a strictly localized group of nerve cells.

On the other hand, the crisis of basic views in neurology inevitably implicated fundamental views of the brain—the highest apparatus of psychological activity. With the modern development of the neurosciences it is impossible to imagine the brain as a homogeneous mass, for whose activity the mere volume of brain tissue is of decisive importance, or as a group of isolated microorgans, each of which possesses the highest degree of specialization and can function independently of neighboring cell formations.

Moreover, the specificity of the work of single neurons (admittedly, often responding selectively to very specialized stimuli) has been shown to depend not so much on their isolated qualities as on the work of the "functional ensembles" discussed at various times by such eminent workers as Monakow in Switzerland, Ukhtomskii and Anokhin in the Soviet Union, and Hebb in Canada. They described them variously as "functional constellations," "cell ensembles," and "functional systems." But whatever term they used, these investigators always implied that the specific character of the activity of a group of nerve cells is determined chiefly by the architectural entity into which they fit.

These considerations have raised fundamental doubts about whether the cerebral mechanisms of complex psychological activity can be approached either in terms of narrow localizationism, or in terms of viewing the work of the whole brain as an undifferentiated entity. What was acceptable as a hypothesis a century ago is no longer acceptable today, and the present, profound theoretical crisis of concepts in classical neurology is the result.

This crisis has another, factual aspect. The view that complex psychological functions are "localized" in circumscribed areas of the brain originated from the clinical observation that a lesion in a localized area of the cortex can lead to the loss of isolated psychological processes such as motor speech, the understanding of speech, spatial representation, reading, writing, calculation, and so on. However, this original viewpoint, on which all the ideas of classical localizationism have been based, has proved to be misleading, if not completely incorrect.

In any deductive activity, and particularly in science, the law we can call the "law of disregard of negative information" holds good: facts that fit into a preconceived hypothesis attract attention, are singled out, and are remembered; facts that are contrary to it are disregarded, treated as "exceptions," and forgotten.

This explains all those "facts" that gained a secure foothold in the textbooks of neurology and were repeated in them for almost 100 years. Lesions of the lower part of the premotor area (Broca's area) are known to cause disturbances of motor speech; but the fact that lesions of the lower part of the postcentral region can cause similar speech disturbances (although with different mechanisms) has only rarely attracted attention and has usually been dismissed or interpreted as an unexplained "exception." Lesions of the anterior zones of the occipital region are known to cause a disturbance of complex visual perception and "optical agnosia." The fact that a disturbance of perceptual activity can be found in patients with other lesions (for example, lesions of the frontal cortex) was not in accord with the customary views and so almost failed to attract its due attention. Finally, disturbances of writing, which, in the classical view, should be found in lesions of "Exner's center" (the middle zones of the premotor cortex of the left hemisphere), can in fact arise as a result of lesions in many different zones of the cortex of the dominant hemisphere (although, admittedly, their psychological structure is totally different). These facts, which I have repeatedly pointed out (Luria, 1962a, 1963a, 1970d, 1971a, 1973a,c), have usually been ignored or disregarded.

A situation of profound *factual* crisis was thus created. The facts on which the classical theory of the direct localization of complex psychological processes in circumscribed areas of the brain had been constructed proved to be insufficiently precise, and the theory itself appeared to be invalid. A radical revision of the classical views became necessary. This task of revision has been undertaken by neuropsychology in the last forty years, and in the process this field of knowledge has become one that meets the strictest demands presented to a properly based scientific discipline.

Let us now consider this revision in its most general features.

## The Way Out of the Crisis: A Revision of the Concept of "Localization of Functions"

A solution to the problem of the cerebral mechanisms of psychological activity was possible only after a radical revision of all the basic principles on which the classical neurologists had based their assumptions and which had led to the equally false notions of "narrow localizationism" and "antilocalizationism."

This work of revision, without which the further development of scientific neuropsychology was impossible, required a radical reexamination of the two underlying concepts of the theory of the "localization of brain functions": the concept of "function" and the concept of "localization."

The term "function" can be understood in two completely different ways. It can refer to the "function" of a particular tissue. In this sense the function of the pancreas is to secrete insulin, the function of muscle tissue is to contract, and the function of the retina is to decompose visual purple. Such functions are specific, and even indivisible, and are strictly "localized" in the particular tissues.

However, the term has another meaning, as has been made clear in particular by the work of the Soviet physiologist P.K. Anokhin (1935, 1949, 1955, 1968): a "function" can be understood as a complex adaptive activity aimed at the performance of some vitally important task, or, in other words, as a complex functional system that pursues its constant (invariant) aim by complex and variative methods.

Perhaps the best and the simplest examples are the "functions" of respiration and locomotion. The function of respiration is to supply air to the alveoli of the lungs. However, the fulfillment of this task requires expansion of the chest, and this act in turn requires the participation of a system of muscles, particularly the muscles of the diaphragm. However, if the muscles of the diaphragm are paralyzed, the person does not die from asphyxia: the intercostal muscles are brought into action, and to a large extent they take over the function of the muscles of the diaphragm. If for some reason the intercostal muscles are also out of action, the person can swallow air and utilize the apparatus of the pharynx and larynx, not previously concerned with the act of respiration.

The situation is the same with respect to the "function" of locomotion. As several workers have shown, it is impossible to regard locomotion as a fixed chain of motor innervations. It was Lashley (1937) who showed originally that if the cerebellum is extirpated or both halves of the spinal tract are divided at different levels in a rat trained to walk through a maze, the animal cannot perform any of its previous movements; yet it will reach its goal by falling head over heels or by moving in totally different and unusual ways.

Respiration and locomotion must thus be regarded, not as functions of a fixed tissue, but as complex "functional systems" in which the constant (invariant) task leads to a constant (invariant) effect through the use of interchangeable (variative) methods, as Figure 20.1 illustrates.

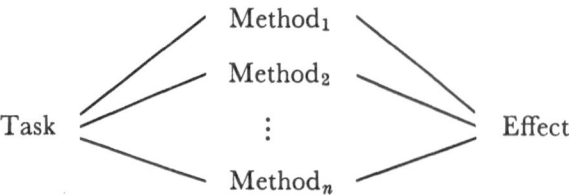

Figure 20.1.

What I have said above applies with even greater force to higher psychological processes, which are not "functions" in the first meaning of the word, but, rather, complex and variable functional systems. The famous Soviet psychologist L. S. Vygotskii (1934, 1956, 1960) made the decisive contribution to the development of scientific psychology when he stated that psychological processes are not elementary and inborn "faculties," but are, rather, formed during life in the process of reflection of the world of reality, that they have a complex structure, utilizing different methods for achieving their goal, which change from one stage of development to the next. He considered that the most important feature characterizing higher psychological functions is their mediated character, the fact that they rest on the use of external aids (tools for movements and actions, language for perception, memory, and thought), and that the structure of these complex psychological processes changes radically in ontogeny. He pointed out, for example, that, whereas a small child thinks by recalling, an adult may recall by thinking, and that even the structure of these processes varies during development. I have shown by special investigations that during the development of memory not only its structure changes, but also its relationship to its genotypic prerequisites, and thus that even such a comparatively simple process as memory cannot be regarded as elementary and unchanging (Luria, 1936, 1962b).

Further investigations by Leontiev (1959) and Gal'perin (1969) have shown the decisive role of the active manipulation of objects in the development of complex human psychological processes, and the complexity of both the pathways of formation and the structures of these processes. We thus reach the indisputable conclusion that complex forms of psychological processes cannot in any way be regarded as functions incapable of further subdivision (resembling the "faculties" of the older writers), but that they are functional systems, using different methods and having a complex structure.

This revision of the concept of "function" leads inevitably to a radical revision of the concept of "localization." When speaking about the "center of respiration," Pavlov stated that the old, classical concepts of the narrowly localized "respiratory center" had recently been radically modified. Whereas "from the very beginning we thought that it was something the size of a pinhead in the medulla, now, however, it has proved extremely elusive, climbing up into the brain and sinking down into the spinal cord, so that now nobody knows its precise boundaries" (Pavlov, 1949: Vol. 3, p. 127).

Exactly the same comment can be made about the "center for locomotion" and, with even greater justification, about the "centers" of complex psychological processes. If the higher psychological processes are complex functional systems, any attempt to "localize" them in circumscribed groups of cells is meaningless. Instead of trying to localize psychological functions in narrowly circumscribed areas of the cortex or subcortex, we must change our task and instead consider the arrangement of complex functional systems in the cortex and lower formations, the role of each part of the brain in the activity of complex functional systems, and the contribution each makes to their performance. This has become the basic problem in neuropsychology, the branch of science whose purpose it is to analyze the cerebral organization of complex forms of human psychological activity.

**The Way Out of the Crisis: Revision of the Concept of "Symptom"**

Revision of the concept of "localization of functions" has made it necessary for us to revise a third concept that plays a key role in the history of neuropsychology—the concept of "symptom." As I said above, much of our neuropsychological evidence was obtained by observations made on patients with local brain lesions. If a local brain lesion led to a clearly defined loss of function or, in other words, if it caused a definite symptom, observers concluded that that particular area of the brain was the "center" in which the corresponding function was "localized." The idea that the concepts of function and symptom coincide, even though they are opposite in sign, gave basic support to the idea of the direct localization of functions.

However, despite the apparent simplicity and attractiveness of this idea, it began long ago to arouse fundamental misgivings, and the idea that the concept of "symptom" is by no means equivalent to the concept of "function" gradually began to penetrate deeper and deeper into classical neurology.

The neurologist who did more than anyone else to shed light on the complexity of the true state of affairs was Kurt Goldstein, one of the most brilliant intellects of our time. It was in 1925 that he published his well-known paper "The symptom, its origin and significance," which can with complete justification be regarded as a seminal work of modern scientific neuropsychology.

The basic idea expressed by Goldstein is simple. If a local brain lesion causes a certain disturbance of a complex form of psychological activity such as perception, thought, speech, writing, or reading—or, in other words, if it leads to the appearance of the "symptom" of agnosia, aphasia, alexia, or agraphia—this by no means implies that this complex function is localized in that particular part of the brain.

A careful neurological (we should now call it neuropsychological) analysis shows that the symptom is complex in character. It may be based upon a primary, basic disturbance (*Grundstörung*) which by itself does not account for the whole range of the symptom, but which merely removes the fundamental condition required for the particular "function" to be performed. The actual symptom arising as the result of the local brain lesion, in Goldstein's opinion, is only the secondary result of the basic disturbance, and it can in no way be regarded as the direct consequence of a focal disturbance or as the result of the fact that the particular "function" is directly "localized" in the corresponding area of the brain.

In order to understand the true mechanisms that lead to the disturbance of a psychological process, a careful analysis must be made of the structure of the resulting symptom, the primary disturbance at its base must be distinguished, and the ways in which the secondary disturbances that constitute the directly observed pattern of the disorder arise must be shown. Inability to draw an object in a conventional picture, for instance, may be the result of the fact that the patient cannot discontinue a concrete activity and change to the act of conventional representation; inability to repeat a sound or word is by no means necessarily the result of an audioverbal disturbance, but may arise because the patient has lost the particular "abstract set" that he needs in order to change from the cognitive activity of spoken communication to the artificial act of repeating groups of sounds (Goldstein, 1948).

An externally manifested "symptom" is completely different from the primary defects that lie behind it. Careful analysis of the "psychological qualification of the symptom" is thus required to progress from the externally observed picture of disturbances of behavior to the mechanism lying at their basis. This was shown by Goldstein and his colleague Gelb in a series of meticulous investigations described under the general heading of *Psychological Analyses of Cases of Brain Pathology* (Gelb and Goldstein, 1920). This careful analysis of the internal logic of the changes in psychological activity arising as the result of local brain lesions can be regarded as the main contribution of these eminent authors and as the true foundation of scientific neuropsychology.

No matter that the basic principles from which Goldstein set out were incorrect, and that there are few nowadays who would agree with his view that every brain lesion (almost regardless of its localization) causes, as its "basic disturbance," the disintegration of the "abstract set" or of "categorical behavior." The facts gathered by subsequent generations of investigators have confirmed neither this view nor the hypothesis that the extent of brain damage plays a more important role in the genesis of a disturbance of higher psychological functions than the localization of the focus. The facts established by neuropsychology have proved to be incomparably richer and more varied than these examples of the "finest abstractions."

Neuropsychology must be indebted to Gelb and Goldstein for having introduced the idea of the existence of "basic" or "primary" disturbances and associated secondary disorders, and for having shown that, instead of a superficial description of symptoms, it is essential to give a careful "psychological qualification of the symptom" and to distinguish the "basic disturbance". This way (and not the superficial statistical assessment of symptoms) is the only possible highway of neuropsychological investigation.

Qualification of the symptom and identification of the basic disturbance form the first stage in every neuropsychological investigation. The second, no less important stage is the comparison of all the symptoms presented by the patient and their synthesis into a single picture or, in other words, establishment of the syndrome arising because of the local brain lesion. This is the final stage, and a special examination of the logic behind it is called for.

American workers, including Hans-Lukas Teuber (1959), have formulated a "principle of double dissociation," which has been accepted in neuropsychology as a basic principle of its techniques. Its essence can be stated as follows: If a local brain lesion leads primarily to the loss of a particular factor directly connected with the work of that part of the brain, all types of psychological activity that incorporate this basic factor will be disturbed, whereas all types of psychological processes that do not incorporate this factor will remain intact. An example is the comparative analysis of the pattern of disturbance of psychological processes in lesions of the temporal and parieto-occipital region of the left hemisphere. The first of these lesions directly disturbs complex phonemic hearing; processes that include this factor (for example, the understanding of speech, the recalling of words, writing) must therefore inevitably disintegrate, whereas processes that do not incorporate this factor (for example, orientation on a map, written calculation) remain intact. The opposite situation is found in lesions of the parieto-occipital region of the left hemisphere. The primary result of such lesions is a disturbance of visuospatial synthesis; types of psychological activity that do not incorporate this factor (for example, the understanding of speech, writing) therefore remain intact, while those for which the preservation of spatial synthesis is an essential condition (orientation on a map, calculation) disintegrate.

The analysis of whole syndromes, the description of their structure, and their comparison thus constitute the final stage of neuropsychological investigation; they are its most important component for the study of the role played by individual brain systems in the construction of psychological activity.

## Modern Approaches

Fifty years have passed since the publication of Goldstein's first paper, which was one of the original sources of neuropsychology. These have been years of intensive development of the whole range of neurosciences: neuromorphology, neurophysiology, neurochemistry, neuropathology. We have acquired much new and valuable information, leading to a new system of ideas in the light of which many of the factual statements by neurologists such as Goldstein can no longer stand up to criticism, although, as I have stated, their original theoretical propositions still have great importance.

Important new facts have been obtained in the last few decades in neuro-anatomy and neurophysiology. The work of many eminent investigators, among them such authorities as Moruzzi and Magoun (1949), Jasper (1954), and Lindsley (1960), has shown that neither the waking state of the cerebral cortex nor its precise, selective activity can exist without the participation of the nonspecific activating system of the brainstem and basal ganglia, and that any interruption of the structural continuity of the ascending and descending activating formations must lower the working tone of the cortex or interfere with the control of its waking state.

A no less important field of research was initiated by the classical studies of Olds (1956) on the one hand, and those of Penfield and Milner (1958) and Scoville and Milner (1957) on the other. These studies showed that structures of the thalamo-hypothalamic region, hippocampus, and caudate nucleus, as well as structures of the limbic system, participate in the activity of the more complex zones of the neocortex. They also revealed that these structures, which are nonspecific in character, play the decisive role in the comparison of actions, the retention of traces, and the production of responses to novelty, all of which are essential conditions for the performance of any type of complex psychological activity.

These investigations later underwent substantial development as a result of progress in new techniques, culminating in morphological studies of single neurons (Hubel and Wiesel, 1962, 1963). These later studies showed that, besides non-

specific neurons, which form a large part of the neuron population of deep structures and of the limbic region, there are other neurons with a surprisingly high
level of specialization in some parts of the cortex, neurons that respond only to
features exhibiting the finest selectivity and that are evidently the elementary
systems for the reception of information from the outside world.

Finally, work begun on the initiative of Grey Walter (1953, 1966, 1973; Walter
et al., 1964), allowing the formation and spread of slow electrical phenomena to be
investigated through the use of implanted electrodes, demonstrated the role of the
frontal cortex in the regulation of complex forms of activity. Another important
contribution was made by Livanov (1972; Livanov, Gavrilova, and Aslanov, 1964,
1966, 1973), who studied the spatial spread of excitation over the cortex and
discovered new criteria for the objective evaluation of its organized work. These
advances, even at this early stage, could not fail to have made great changes in the
information available to modern neurology.

There have also been developments in our knowledge of the morphophysiology
of cortical structures. Two generations ago we had only theoretical views on the
hierarchic organization of cortical structures, the foundations of which were laid
by Campbell (1905), Brodmann (1909), and the Vogts (1919, 1927). It is now
possible to add the wealth of information obtained by morphologists, such as
Filimonov (1940) and Polyakov (1956, 1962, 1966) in the Soviet Union and Nauta
(1958, 1964, 1971) in the West, who have described not only the details of the
neuronal structure of the cortex at different genetic levels, but also the connections
that actually exist between the various parts of the cerebral hemispheres and that
are fundamental to the activity of the living brain.

A field of particularly intensive and important research has been that connected
with hemisection of the brain, which developed rapidly following the work of
Sperry (1966, 1967, 1968) and Gazzaniga (1970). This research has provided
completely new information on the work of the isolated hemispheres and has
compelled us to revise many views that seemed hitherto to be soundly based.

A rapid flow of new information also has begun to arrive in recent decades from
investigations conducted at the molecular level, and it now forms the material for
an important subdivision of the neurosciences.

**Medical Advances in the Development of Neuropsychology**

In the preceding section I have mentioned the theoretical issues of modern
neuropsychology. However, the practical sources of its development are perhaps no
less important.

Recent years have seen great developments in neurology and neurosurgery.
With the introduction of new methods of contrast investigation, such as angiography, with the improvement of techniques for studying the cerebral blood flow,
and with the use of methods involving radioactive isotopes, the scope of investigation of the living brain has become significantly widened. We can now localize a
pathological focus very precisely; whereas only two generations ago it was a matter
of pure guesswork, we can now rely on objective verification. The considerable

extension of the frontiers of operative surgery, methods of stereotaxic neurosurgery, and methods of influencing the dynamics of the cerebrospinal fluid have provided even greater opportunities for the objective verification and study of local brain pathology. It is therefore not surprising that progress in this field of medicine has provided extraordinarily favorable conditions for the development of neuropsychology at its new level. That is why in the last few decades rapidly developing centers of neuropsychology have been created in many countries. It is hard to exaggerate the contribution made to this field of knowledge by workers such as Pribram, Teuber, Benton, and others in the United States, Milner in Canada, Zangwill, Critchley, and Weiskrantz in England, Hécaen in France, De Renzi in Italy, and many others.

It is difficult and, indeed, it would be superfluous for me to summarize the facts my colleagues and I have obtained on these fundamental problems in neuropsychology, which I began to investigate originally forty years ago in conjunction with L. S. Vygotskii. These facts have been published collectively in such books as *Traumatic Aphasia* (1947, English edition, 1970), *Higher Cortical Functions in Man* (1962, English edition, 1966), *The Human Brain and Psychological Processes* (Volume 1, 1963, English translation, 1966; Volume 2, 1970), and *Fundamentals of Neuropsychology* (English edition: *The Working Brain*, 1973), and also in the survey "Neuropsychological Research in the USSR" published in the *Proceedings of the National Academy of Sciences of the U.S.A.* in 1973. In the remainder of this article I will concentrate only on the fundamental principles underlying these investigations.

All the facts I have obtained have led me to conclude that the higher forms of human psychological activity and all human behavioral acts take place with the participation of all parts and levels of the brain, each of which makes its own specific contribution to the work of the functional system as a whole.

With a little approximation we can say that the apparatus of the brain comprises three principal functional units, each of which participates in the organization of psychological activity in its own particular way.

The first unit, which can be thought of as the unit providing the tone or "energy" of the brain, is connected with the structures of the brainstem, the thalamo-hypothalamic region, and part of the old limbic system. It is this unit which, through the intermediary activity of the ascending reticular formation, exerts activating influences on the cortex and maintains its tone, and in turn is controlled by descending influences from the cerebral cortex. Careful observations on patients with lesions (usually tumors) of the deep structure of the brain, which I shall discuss again below, have shown how the cortical tone is sharply reduced when corresponding impulses cease to reach the cortex, how the state of vigilance fluctuates in them, how their level of consciousness varies, and how quickly they become exhausted by any activity. Although they remain nonspecific in character, these influences participate in the course of all psychological activity and maintain the necessary level and background for primary psychological activity.

The second unit can be described as the unit for the reception, analysis, and storage of information reaching the body from the outside world. It comprises the posterolateral zones of the cerebral hemispheres and is subdivided into parts possessing high modal specificity: occipital (visual), temporal (auditory), and postcentral (general sensory). The cortex of each of these zones has a hierarchic structure, with primary (projection), secondary (projection-association), and tertiary (integrative) areas (the last are sometimes called "zones of overlapping"), and their structural plan is one of "diminishing specificity" or "increasing integrative action." A lesion of the secondary zone of any of these cortical areas leads to the corresponding modality-specific disorder which, in turn, leads to manifestation of one of the forms of disturbance of analyticosynthetic activity known in classical neurology as optical, acoustic, or tactile agnosia. A lesion of the tertiary zones of the cortex of this second unit (the parieto-temporo-occipital region or zones of overlapping) inevitably leads to disintegration of the combined activity of these areas and gives rise to primary disorders of simultaneous and spatial organization of incoming information.[1]

The third functional unit of the brain, which so far has been studied less than the other two, can be described as the unit for programming, regulating, and verifying complex forms of activity. It comprises the anterior zones of the hemispheres and, in particular, the frontal lobes of the brain. Many investigations (Luria, 1966a,b, 1969, 1973a; Luria and Homskaya, 1966; Pribram and Luria, 1973) have shown that lesions of the zones of the brain composing this unit neither give rise to modality-specific disturbances nor cause disintegration of any special operations. Such lesions do lead invariably to disturbance of the regulation of the state of activity (Homskaya, 1973). They also cause disintegration of the complex organizations of goal-directed activity and loss of that process of comparison of the effect of an action with the original intention which Anokhin (1955, 1968) has described by the term "acceptor of the results of action" and which is a fundamental mechanism for the control of an individual's behavior and the awareness of his defects.

Careful analysis has shown that all three functional units of the brain participate in all types of activity, but each unit, as I have already said, makes its own particular contribution to the organization of psychological processes. It is therefore natural to find that lesions of these units lead to disorders of higher cortical functions that are totally different in character from one another. It is this feature that has made it possible to describe the complex cerebral organization of human behavior and that has provided an approach to the study of the psychophysiological composition of the more complex forms of human conscious activity.

It was along these lines that such complex psychological processes as perception and manipulative action, speech, writing, reading, calculation, and problem solving have been analyzed (Luria, 1962a, 1963a, 1966a,b, 1970b, 1973a; Luria and Homskaya, 1966; Luria and Tsvetkova, 1966, 1967). These analyses have shown how complex is the cerebral organization of these forms of human conscious activity and how many different components are involved in it. They demonstrate how far our ideas have progressed both from the "faculties" and

1. Besides the sources mentioned above I have also made a detailed analysis of this syndrome in a patient described in *The Man with the Shattered World* (1971a).

narrow localizationism of classical neurology and from the notion that such activity is based on the "abstract set" and on "categorical behavior," functions of the "whole brain" whose mechanisms we are not yet able to unravel.

These investigations led to the discovery of the functional processes that are the working units of the secondary zones of the cerebral cortex and yielded descriptions of the disturbances of phonemic hearing that arise in patients with lesions of the left temporal region; the disintegration of simultaneous visuospatial syntheses that results from lesions of the parieto-occipital zones of the cortex; and the disturbance of the synthesis of movements into "kinetic melodies" that is observed following lesions of the premotor area.

They also yielded descriptions of the complex disorders arising as the result of lesions of the tertiary zones of the cortex. Special attention has been paid to analyzing disturbances of the ability to consecutively convert incoming information into simultaneously observable syntheses, such as are found in patients with lesions of the parieto-temporo-occipital zones of the left hemisphere and are manifested equally during orientation in space and in such complex symbolic operations as calculation and the understanding of complex, "reversible" grammatical constructions. Finally, I have been occupied for many years with the detailed analysis of the function of the frontal lobes, lesions of which lead to the disintegration of programmed action and of the sustained verification of its results. All these investigations are described in the long series of publications cited above, and I shall not discuss them in more detail in these pages.

Since Broca and Wernicke first laid the foundations of neuropsychology a century has passed; since Goldstein first attempted a scientific interpretation of its fundamental facts only fifty years have elapsed. However, it is hard to appreciate two fundamental changes that have taken place during this period and that are characteristic of modern neuropsychology, primus inter pares of the neurosciences: the radical change in the basic approach to the cerebral organization of complex human psychological activity, and the radical replacement of the original propositions. At the same time it would be difficult to exaggerate the contribution made by the data of neuropsychology to modern scientific psychology, to the theory of knowledge, and to linguistics. (I deal specially with this last problem in a book entitled *Basic Problems in Neurolinguistics*, 1975.) I shall not deal further with the discovery of the relevant facts in this field, but will devote the remaining pages to a brief account of facts not previously published and to problems that have remained unsolved and require further investigation.

## Current Research and Its Prospects

For many years I have been engaged in an examination of the role of individual brain systems in cognitive processes and in an analysis of perception, speech, writing, reading, calculation, and problem solving. A no less important aspect of my research has been the analysis of mechanisms that are at the basis of voluntary action and conscious activity. However, during the many years spent on these investigations I had to put aside some problems which seemed to me difficult to submit to neuropsychological analysis, and to which I hoped to devote my full attention later.

## The neuropsychology of memory

When studying patients with local lesions of the convexital part of the cerebral hemisphere, I never observed a single case of a patient disoriented in space and time who exhibited an oneiroid state of consciousness and massive, modality-nonspecific disorders of memory resembling the picture of Korsakov's syndrome. In order to make a neuropsychological analysis of those disturbances it was necessary to change the test object, to study patients with deep lesions of the brain.

Investigations of memory in the last decade have become a favorite topic in physiological and psychological laboratories. However, although the study of memory at the molecular and neuronal level has introduced an enormous flow of new information, in most cases it has suffered from a serious defect: memory has been understood by most investigators as the simple imprinting, storage, and reproduction of traces, and the whole wealth and variety of the different levels of organization of memory and its different components have usually been disregarded.

The great wealth and complexity of the structure of memory is found not only in psychological investigations, which have numbered in the hundreds in the last decade, but also in observations on disturbances of modality-nonspecific forms of memory arising in patients with deep brain lesions. These investigations, which began with the classical observations of Penfield, Scoville, and Milner (see Milner, 1958, 1962, 1966, 1970), showed that lesions of both hippocampi can give rise to lasting and modality-nonspecific disorders of memory that have the character more of difficulties of reproduction than of difficulties of impression. Later investigations, conducted at the neuronal level, showed that this region in fact possesses a particularly powerful system of nonspecific neurons, which respond to every change in incoming stimuli and which can be regarded with every justification as the nervous mechanism of comparison or as "neurons of memory" (Vinogradova, 1970).

All these facts compelled me to make a careful study of the changes in memory that result from local lesions of the deep brain structures that form the "circle of Papez"[2] and that affect hippocampal structures and lead to marked general disturbances of memory and sometimes also to disturbances of consciousness. I considered three fundamental questions: What is the character of these disturbances of memory? At what functional level of organization of mnestic activity do they occur? What is the mechanism of the pathologically increased forgetfulness which these patients show?

The results of these investigations are described in the two volumes of my *Neuropsychology of Memory* (1974), the contents of which I shall summarize here only very briefly.

The answer to the first question was quite evident: modality-specific disturb-

---

2. Circle of Papez: hippocampal nuclei, fornix, and mammillary bodies.

ances of memory can be produced only by lesions of the convexital zones of the hemisphere, and in their character they are more prolongations of gnostic disorders than true disturbances of memory. Conversely, lesions of the deep zones of the brain lead to modality-nonspecific disturbances of traces which, if the lesion extends to the anterior zones of the brain, assume the character of disturbances of mnestic activity and may be accompanied by distinctive disorders of consciousness.

The answer to the second question was equally evident: in massive brain lesions in different situations, modality-nonspecific disturbances of memory can differ in structure and can occur at different functional levels. For instance, tumors of the hypophysis, extrasellar in their location and influencing the region of the hippocampus, often evoke only disturbances of memory for a chain of isolated elements (sounds, traces, numbers, pictures); but the change to a higher level of organization (the remembering of phrases or stories) compensates for these difficulties (Kiyashchenko, 1973). Conversely, massive tumors of deep parts of the brain, spreading to the thalamic and hypothalamic nuclei and also affecting the hippocampal structures, disturb equally the recalling of discrete series of elements and of well-organized material (Popova, 1972; Luria, 1971b, 1974). Finally, lesions of the deep parts of the brain extending to the frontal regions lead to disturbances of mnestic activity in that the active goal-directed recalling of any organized material is generally impossible and can be replaced either by the uncontrollable production of random associations or by the inert reproduction of the same rigid stereotype (Luria, 1971b, 1974).

Perhaps the most interesting answer was that to the last question: What are the physiological mechanisms of the pathologically increased forgetfulness which so often occurs in local brain lesions?

Some psychologists have explained all increased forgetfulness by trace decay, whereas others have regarded it as the result of the inhibitory influence of interfering factors. Numerous observations on patients with local brain lesions have shown conclusively that the pathologically increased forgetfulness observed in such cases is not so much the result of direct trace decay as of pathologically increased inhibitability of traces by random interfering factors, and that this mechanism lies at the basis both of the modality-specific disturbances of memory arising in local lesions of the convexital zones of the cortex and of the modality-nonspecific disturbances of memory observed in patients with deep brain lesions (Luria, 1971b, 1974). This fact is proved by a simple test in which the patient easily retains a given series of traces (words, phrases, the meaning of a story) after an "empty" pause of 1–1.5 minutes, but immediately "forgets" them if the pause is filled by any type of irrelevant, interfering activity (see Peterson and Peterson, 1959).

Facts such as these clearly demonstrate the complex character of memorizing and recalling, and they show the great variety of phenomena concealed behind such an apparently simple concept as "memory."

## The neurodynamics of the disturbance of psychological processes in local brain lesions

The facts I have just mentioned lead us to the last region of neuropsychology, one which has so far received very little attention: the study of the neurophysiological mechanisms of disturbance of psychological processes arising through local brain lesions.

The classical neurologists were fond of saying that the result of a local brain lesion is the loss of a certain function. However, Monakow (1914) first emphasized that a pathological focus can cause only temporary blocking (diaschisis), spreading over the cerebral cortex, of a particular function, and Pavlov established a series of important neurodynamic laws, which are reflected in the function of the pathologically changed cortex.

Although more than half a century has elapsed since the description of these laws of the "pathophysiology of higher nervous activity," we are fully justified in turning to them once again to see whether they can explain the real forms of change in psychological processes which neuropsychologists observe in patients with local brain lesions. The phenomenon of forgetfulness, which I understand as a pathologically increased inhibitability of traces by interfering factors, is one example of this neurodynamic explanation of symptoms; the replacement of selective trace recall by reproduction of a pathologically inert stereotype is another example.

The question naturally arises: Can we explain many pathological phenomena in the working of the brain by pathophysiological mechanisms such as these? I shall examine only one decisively important series of phenomena. The course of psychological processes in the normal cerebral cortex is known to be characterized by very high selectivity and mobility: the necessary systems of connections are easily formed, and irrelevant random systems of connections are just as easily inhibited. The opposite situation applies in oneiroid states and in pathological states of the cortex: in such cases both the required and the irrelevant connections arise with equal probability, and the process of choosing from different alternatives (or, as it is now expressed in psychology, the process of decision making) is severely disturbed.

The explanation of this fact evidently lies in the altered cortical neurodynamics produced by the pathological focus. As Pavlov showed many years ago, factors such as fatigue or a pathological focus can reduce the cortex to a pathological state, which is reflected in a disturbance of the normal "law of strength" on the one hand, and of the normal mobility of nervous processes on the other. Whereas in the normal state of the cortex strong (or important) stimuli evoke strong responses and weak (or unimportant) stimuli evoke weak responses, in a pathological state of the cortex the situation is different: in the first, "equalizing" phase strong and weak stimuli (or their traces) begin to evoke equal responses, while in the second, more severe, "paradoxical" phase responses to weak stimuli are actually stronger than those to strong stimuli. The main condition of selectivity of psychological processes is thus naturally disturbed, all connections begin to arise with equal probability, and the organized course of psychological processes becomes impossible.

In deep brain lesions this loss of selectivity of psychological processes may be global in character (Luria et al., 1967); in local lesions of the convexital zones of the cortex this pathological state of the cortex can be localized or regional in character and can lead to the "equalization of excitability of traces" either in the visual or in the audioverbal sphere (Luria et al., 1967). The difficulties that a patient with aphasia experiences when attempting to find an essential word, and the phenomena of literal and verbal paraphasias that arise with the same probability as the desired word, can very likely be explained by this equalizing phase of the pathologically changed speech area of the cortex (Luria, 1972, 1975).

A factor that is just as important for the neurodynamic explanation of pathological forms of brain activity is a disturbance of the normal mobility of nervous processes, observed particularly clearly in lesions of the anterior zones of the brain. It is manifested as perseverations, which may assume different forms and may arise at different levels (Luria, 1965) and which disturb the normal course of psychological activity. The role of pathological inertia in the clinical picture of local brain lesions is very great, and this phenomenon has been studied in detail in many neuropsychological investigations (Luria, 1966a,b, 1969, 1970c, 1971a, 1973a; Luria and Homskaya, 1966).

All I have said goes to show that the introduction into neuropsychology of criteria elaborated in the pathology of higher nervous activity is one of our most important tasks, but one that has so far received inadequate attention.

## The prospects

I have discussed very briefly the sources, principles, and basic facts of neuropsychology. It now remains for me to say a few words about its prospects. As in any new branch of science the facts at our disposal are only the beginning, and to use Pavlov's expression, "the mountain which remains unknown is immeasurably larger than the small heap taken from it which is known." I shall attempt to summarize this mountain of the unknown in a few short words.

The first and the longest period of time was occupied in formulating the basic principles of neuropsychology as a science, in describing the symptoms found in lesions of the basic zones of the human brain, and in characterizing the principal neuropsychological syndromes arising in local lesions of this part of the brain.

The last few years have brought with them the urgent necessity of studying the pathophysiological mechanisms lying at the basis of these syndromes. This work has only just begun and in all probability it will occupy a whole generation, linking neuropsychological observations with neurophysiological analysis. Workers in my laboratory, including E. D. Homskaya and N. A. Filippycheva, have already set out along this road, and much can be expected from their investigations.

The next step, without which any further development in neuropsychological research would be impossible, is a careful clinical analysis of the nature of the facts observed. Whereas in the first stages it was sufficient to indicate the location of the focus and to analyze the symptoms produced by it, it is now necessary, in the present state of the science, to correlate the discoveries obtained by modern methods of investigation of the living brain with the data of angiography, with a careful investigation of the compensatory capillary circulation, and with an analysis of the relations between the local and general cerebral factors that are involved in every local brain lesion. Without this work no further development of neuropsychology could take place, and its results would rest on an imperfect clinical basis. However, the work has now begun and much is expected from it.

The third and equally important prospect of neuropsychological research is connected with the widening of the field of its activity. Until recently nearly all the information we obtained was concerned with the function of the left, dominant hemisphere; the function of the nondominant, right hemisphere (like the function of the deep brain structures) has remained largely a closed book. Only very recently, with the work of Sperry and his collaborators on the "disconnected" brain and with their observations during stereotaxic operations, has the attempt begun to make this unexplored field accessible for investigation.

In recent years we have begun to gather material from which we can hope that the functions of the right hemisphere will eventually be revealed to us. We know already that the line dividing the functions of the two hemispheres does not necessarily follow the line of separation of speech and nonspeech processes, that the right hemisphere is intimately connected with the direct realization of its activity and with the automation of its course, that in lesions of the right hemisphere anosognosia and deautomation of the processes can arise, that residual gnostic and speech processes may become uncontrollable in such lesions, and that the right hemisphere participates in a unique way in the consolidation of traces.

However, although these observations are still purely preliminary in character, perhaps it will not take a complete generation to collect the more informative material we require. We can look hopefully at the abundant literature now appearing and at the systematic work begun only a few years ago in my laboratory by E. G. Simernitskaya and her collaborators.

It now only remains for me to mention the last prospect for the development of neuropsychology as a science, and this time it is more technical in character. During the years I have spent on this work I have recorded many observations that have helped to provide the foundation for a theory of neuropsychology. However, these facts still require verification of detail, careful evaluation, and, where necessary, quantification in order to establish their complete reliability. I am far from thinking that quantitative measurement is the only true criterion of science. I am not too greatly impressed by the usual methods of medical statistics, which usually amount to nothing more than stating a number of cases showing a given symptom. In a special communication published not very long ago (Luria and Artemieva, 1970), I expressed the view that in neuropsychology, as in other fields of science where the material is comparatively limited in amount, attention should be focused on the intercorrelation of symptoms, and that a mathematical apparatus should be specially developed for use in such cases.

The time has come for mathematical analysis of the vast amount of material gathered as a result of my observations over the last forty years, observations that lie at the basis of the neuropsychological syndromes of local brain lesions and that give these syndromes their essential reliability. I have no doubt that this work will take more than a generation, but an inspired look into the future is sometimes just as important as a sober evaluation of the present.

## Acknowledgment

This paper was translated from the Russian by Basil Haigh.

## References

Ajuriaguerra, J., and Hécaen, H. (1960): *Le Cortex Cérébral*. Paris: Masson et Cie.

Anokhin, P.K., ed. (1935): *Reports on the Problem of Center and Periphery in the Physiology of Nervous Activity. Experiments with Anastomosis of Nerves*. (In Russian.) Gor'kii: Gosizdat.

Anokhin, P.K. (1940): Problems of localization from the viewpoint of systematic ideas on nervous functions. *Nevropatol. Psikhiat.* 9.

Anokhin, P.K. (1949): *Problems in Higher Nervous Activity*. (In Russian.) Moscow: Izd. Akademii Meditsinskikh Nauk.

Anokhin, P.K. (1955): New data on the properties of the afferent apparatus of the conditioned reflex. (In Russian.) *Vopr. Psikhol.* 6:16–38.

Anokhin, P.K. (1968): *Biology and Neurophysiology of the Conditioned Reflex and Its Role in Adaptive Behavior*. (In Russian.) Moscow: Meditsina. (See also Anokhin, 1974.)

Anokhin, P.K. (1972): *Fundamental Problems in the General Theory of Functional Systems*. (In Russian.) Moscow: Izd. Akademii Nauk. SSSR.

Anokhin, P.K. (1974): *Biology and Neurophysiology of the Conditioned Reflex and Its Role in Adaptive Behavior*. Oxford: Pergamon Press.

Bernstein, N.A. (1935): Problem der Wechselbeziehunger der Koordination und der Lokalisation. (In Russian with German summary.) *Arkh. Biol. Nauk.* 38:1–34.

Bernstein, N.A. (1947): *The Structure of Movements*. (In Russian.) Moscow: Medgiz.

Bernstein, N.A. (1957): Some forthcoming problems concerning the regulation of motor acts. (In Russian.) *Vopr. Psikhol.* 6:70–90.

Bernstein, N.A. (1966): *Outlines of the Physiology of Movements and the Physiology of Activity*. (In Russian.) Moscow: Meditsina. (See also Bernstein, 1967.)

Bernstein, N.A. (1967): *The Co-ordination and Regulation of Movements*. Oxford: Pergamon Press.

Broca, P. (1861): Remarques sur le siège de la faculté du langage articulé. *Bull. Soc. Anthropol.* 6.

Brodmann, K. (1909): *Vergleichende Lokalisationslehre der Grosshirnrinde in ihren Prinzipien dargestellt auf Grund des Zellenbaues*. Munich: J.A. Barth.

Campbell, A.W. (1905): *Histological Studies on the Localization of Cerebral Functions*. Cambridge, Eng.: Cambridge University Press.

Filimonov, I.N. (1940): The localization of function in the cerebral cortex. (In Russian.) *Nevropathol. Psikhiat.* 9.

Flourens, P. (1824): *Recherches Expérimentales sur les Propriétés et les Fonctions du Système Nerveux, dans les Animaux Vertébrés.* Paris: Crevot.

Foerster, O. (1936): Symptomatologie der Erkrankungen des Gehirns. *In: Handbuch der Neurologie.* Vol. 6. Bumke, O., and Foerster, O., eds. Berlin: Springer-Verlag.

Fritsch, G., and Hitzig, E. (1870): Ueber die elektrische Erregbarkeit des Grosshirns. *Arch. Anat. Physiol. Wiss. Med.* 37:300–332.

Gall, F.J. (1825): *Sur les Fonctions du Cerveau et sur Celles de Chacune de ses Parties.* Vols. 1–6. Paris: Baillière.

Gal'perin, P.Y. (1969): Stages in the development of mental acts. *In: A Handbook of Contemporary Soviet Psychology.* Cole, M., and Maltzman, I., eds. New York: Basic Books, pp. 249–273.

Gazzaniga, M.S. (1970): *The Bisected Brain.* New York: Appleton-Century-Crofts.

Gelb, A., and Goldstein, K., eds. (1920): *Psychologische Analysen Hirnpathologischer Fälle.* Leipzig: Springer-Verlag.

Geschwind, N. (1965): Disconnexion syndromes in animals and man. *Brain* 88:237–294, 585–644.

Goldstein, K. (1925): Das Symptom, seine Entstehung und Bedeutung für unsere Auffassung vom Bau und vom der Funktion des Nervensystems. *Arch. Psychiatr. Nervenkr.* 76:84–108.

Goldstein, K. (1927): Die Lokalisation in der Grosshirnrinde nach den Erfahrungen am kranken Menschen. *In: Handbuch der normalen und pathologischen Physiologie.* Vol. 10. *Spezielle Physiologie des Zentralnervensystems der Wirbeltiere.* Bethe, A., von Bergmann, G., Embden, G., and Ellinger, A., eds. Berlin: Springer-Verlag.

Goldstein, K. (1934): *Der Aufbau des Organismus; Einführung in die Biologie unter besonderer Berücksichtigung der Erfahrungen am kranken Menschen.* The Hague: M. Nijhoff.

Goldstein, K. (1948): *Language and Language Disturbances. Aphasic Symptom Complexes and Their Significance for Medicine and Theory of Language.* New York: Grune and Stratton.

Goltz, F. (1881): *Über die Verrichtungen des Grosshirns.* Bonn: Strauss.

Hebb, D.O. (1949): *The Organization of Behavior. A Neuropsychological Theory.* New York: John Wiley.

Hécaen, H. (1972): *L'Introduction à la Neuropsychologie.* Paris: Flammarion.

Hoff, H. (1930): *Die zentrale Abstimmung der Sehsphäre.* Berlin: S. Karger.

Hoff, H., and Pötzl, O. (1930): Über die Grosshirnprojektion der Mitte und Aussengräuzen des Gesichtsfeldes. *Jahrb. Psychiatr. Neurol.* 52.

Homskaya, E.D. (1973): *The Brain and Activation.* (In Russian.) Moscow: Izd. Moskovskogo Universiteta.

Hubel, D.H., and Wiesel, T.N. (1962): Receptive fields, binocular interaction and functional architecture in the cat's visual cortex. *J. Physiol.* 160:106–154.

Hubel, D.H., and Wiesel, T.N. (1963): Receptive fields of cells in striate cortex of very young, visually inexperienced kittens. *J. Neurophysiol.* 26:994–1002.

Jackson, J.H. (1932): Evolution and dissolution of the nervous system (Croonian Lecture). *In: Selected Writings of John Hughlings Jackson.* Vol. 2. London: Hodder and Stoughton.

Jasper, H.H. (1954): Functional properties of the thalamic reticular system. *In: Brain Mechanisms and Consciousness* (Symposium organized by the Council for International Organizations of Medical Sciences, Ste. Marguerite, Quebec, 1953). Delafresnaye, J.F., ed. Oxford: Blackwell, pp. 374–401.

Kiyashchenko, N.K. (1973): *Disturbances of Memory in Local Brain Lesions.* (In Russian.) Moscow: Izd. Moskovskogo Universiteta.

Kleist, K. (1934): *Kriegsverletzungen des Gehirns in ihren Bedeutung für die Hirnlokalisation und Hirn-pathologie.* Leipzig: J.A. Barth. (Also published with title: *Gehirnpathologie, vornehmlich auf Grund der Kriegeserfahrungen.*)

Konorski, J. (1967): *Integrative Activity of the Brain. An Interdisciplinary Approach.* Chicago: University of Chicago Press.

Lashley, K.S. (1929): *Brain Mechanisms and Intelligence. A Quantitative Study of Injuries to the Brain.* Chicago: University of Chicago Press.

Lashley, K.S. (1937): Functional determinants of cerebral localization. *Arch. Neurol. Psychiatr.* 38:371–387.

Leontiev, A.N. (1959): *Problems of Mental Development.* (In Russian.) Moscow: Izd. Akademii Pedagogicheskhikh Nauk.

Lindsley, D.B. (1960): Attention, consciousness, sleep and wakefulness. *In: Handbook of Physiology.* Sect. 1. *Neurophysiology.* Vol. III. Field, J., editor-in-chief, and Magoun, H.W., ed. Washington, D.C.: American Physiological Society, pp. 1553–1593.

Livanov, M.N. (1972): *The Spatial Spread of Excitation over the Cerebral Cortex.* (In Russian.) Moscow: Nauka.

Livanov, M.N., Gavrilova, N.A., and Aslanov, A.S. (1964): Cross-correlation between various cortical points in man during mental work. (In Russian with English summary.) *Zh. Vyssh. Nerv. Deiat.* 14:185–194.

Livanov, M.N., Gavrilova, N.A., and Aslanov, A.S. (1966): Correlation between biopotentials in the frontal zones of the human cerebral cortex. *In: The Frontal Lobes and Regulation of Psychological Processes.* (In Russian.) Luria, A.R., and Homskaya, E.D., eds. Moscow: Izd. Moskovskogo Universiteta, pp. 176–189.

Livanov, M.N., Gavrilova, N.A., and Aslanov, A.S. (1973): Correlation of biopotentials in the frontal parts of the human brain. *In: Psychophysiology of the Frontal Lobes.* Pribram, K.H., and Luria, A.R., eds. New York: Academic Press, pp. 91–107.

Luria, A.R. (1936): The development of mental functions in twins. *Character Person.* 5:35–47.

Luria, A.R. (1947): *Traumatic Aphasia.* (In Russian.) Moscow: Izd. Akademii Meditsinskikh Nauk. (See also Luria, 1970c).

Luria, A.R. (1948): *Restoration of Function after Brain Injury.* (In Russian.) Moscow: Medgiz. (See also Luria, 1963b.)

Luria, A.R. (1962a): *Higher Cortical Functions in Man.* (In Russian.) Moscow: Izd. Moskovskogo Universiteta. (See also Luria, 1966a.)

Luria, A.R. (1962b): On variability of mental functions in the process of child's development. (In Russian with English summary.) *Vopr. Psikhol.* 3:15–22.

Luria, A.R. (1963a): *The Human Brain and Psychological Processes.* Vol. I. (In Russian.) Moscow: Izd. Akademii Pedagogicheskhikh Nauk. (See also Luria, 1966b.)

Luria, A.R. (1963b): *Restoration of Function after Brain Injury.* Oxford: Pergamon Press.

Luria, A.R. (1965): Two kinds of motor perseverations in massive injury of the frontal lobes. *Brain* 88:1–10.

Luria, A. R. (1966a): *Higher Cortical Functions in Man.* New York: Basic Books.

Luria, A. R. (1966b): *The Human Brain and Psychological Processes.* New York: Harper & Row.

Luria, A. R. (1969): Frontal lobe syndromes. *In: Handbook of Clinical Neurology.* Vol. 2. *Localization in Clinical Neurology.* Vinken, P. J., and Bruyn, G. W., eds. Amsterdam: North-Holland, pp. 725–757.

Luria, A. R. (1970a): The origin and cerebral organization of man's conscious action. *In: 19th International Congress of Psychology* (London, 27 July–2 August 1969). Moscow: Izd. Akademii Pedagogicheskhikh Nauk.

Luria, A. R. (1970b): *The Human Brain and Psychological Processes.* Vol. II. (In Russian.) Moscow: Izd. Prosveshchenie.

Luria, A. R. (1970c): *Traumatic Aphasia. Its Syndromes, Psychology, and Treatment.* The Hague: Mouton.

Luria, A. R. (1970d): The functional organization of the brain. *Sci. Am.* 222:66–78.

Luria, A. R. (1971a): *The Man with a Shattered World.* (In Russian.) Moscow: Izd. Moskovskogo Universiteta. (See also Luria, 1973b.)

Luria, A. R. (1971b): Memory disturbances in local brain lesions. *Neuropsychologia* 9:367–375.

Luria, A. R. (1972): Aphasia reconsidered. *Cortex* 8:34–40.

Luria, A. R. (1973a): *The Working Brain. An Introduction to Neurophysiology.* New York: Basic Books.

Luria, A. R. (1973b): *The Man with a Shattered World.* New York: Basic Books.

Luria, A.R. (1973c): Neuropsychological studies in the USSR. *Proc. Natl. Acad. Sci. USA* 70:959–964, 1278–1283.

Luria, A. R. (1974): *The Neuropsychology of Memory.* 2 volumes. (In Russian.) Moscow: Izd. Pedagogika.

Luria, A.R. (1975): *Basic Problems of Neurolinguistics.* (In Russian.) Moscow: Izd. Moskovskogo Universiteta.

Luria, A. R., and Artemieva, E. Y. (1970): Two ways towards reliability of psychological investigations. (In Russian.) *Questions of Psychology* 3:105–112.

Luria, A. R., and Homskaya, E. D., eds. (1966): *The Frontal Lobes and Regulation of Psychological Processes.* (In Russian.) Moscow: Izd. Moskovskogo Universiteta.

Luria, A. R., Homskaya, E. D., Blinkov, S. M., and Critchley, M. (1967): Impaired selectivity of mental processes in association with a lesion of the frontal lobe. *Neuropsychologia* 5:105–117.

Luria, A. R., Konovalov, A. N., and Podgornaya, A. Y. (1970): *Disorders of Memory in the Clinical Picture of Aneurisms of the Anterior Communicating Artery.* (In Russian.) Moscow: Izd. Moskovskogo Universiteta.

Luria, A. R., Sokolov, E. N., and Klimkowski, M. (1967): Towards a neurodynamic analysis of memory disturbances with lesions of the left temporal lobe. *Neuropsychologia* 5:1–11.

Luria, A. R., and Tsvetkova, L. S. (1966): *Neuropsychological Analysis of Problem Solving.* (In Russian.) Moscow: Izd. Prosveshchenie.

Luria, A. R., and Tsvetkova, L. S. (1967): Les troubles de résolution des problèmes. *In: Analyse Neuropsychologique.* Paris: Gauthier-Villars.

Magoun, H. W. (1952): The ascending reticular activating system. *Res. Publ. Assoc. Res. Nerv. Ment. Dis.* 30:480–492.

Magoun, H. W. (1958): *The Waking Brain.* Springfield, Ill.: C. C Thomas.

Milner, B. (1958): Psychological defects produced by temporal lobe excision. *Res. Publ. Assoc. Res. Nerv. Ment. Dis.* 36:244–257.

Milner, B. (1962): Les troubles de la mémoire accompagnant des lésions hippocampiques bilatérales. *In: Physiologie de l'Hippocampe* (Colloques Internationaux du Centre National de la Recherche Scientifique, Montpellier, 24–26 August 1961). Paris: Centre National de la Recherche Scientifique, pp. 257–272.

Milner, B. (1966): Amnesia following operation on the temporal lobes. *In: Amnesia.* Whitty, C. W. M., and Zangwill, O. L., eds. London: Butterworth.

Milner, B. (1970): Memory and the medial temporal regions of the brain. *In: Biology of Memory.* Pribram, K. H., and Broadbent, D. E., eds. New York: Academic Press, pp. 29–50.

Monakow, C. (1914): *Die Lokalisation im Grosshirn und der Abbau der Funktion durch kortikale Herde.* Wiesbaden: Bergmann.

Moruzzi, G., and Magoun, H. W. (1949): Brain stem reticular formation and activation of the EEG. *Electroencephalogr. Clin. Neurophysiol.* 1:455–473.

Nauta, W. J. H. (1958): Hippocampal projections and related neural pathways to the mid-brain in the cat. *Brain* 81:319–340.

Nauta, W. J. H. (1964): Some efferent connections of the prefrontal cortex in the monkey. *In: The Frontal Granular Cortex and Behavior.* Warren, J. M., and Akert, K., eds. New York: McGraw-Hill, pp. 397–409.

Nauta, W. J. H. (1971): The problem of the frontal lobe. A reinterpretation. *J. Psychiatr. Res.* 8:167–187.

Nielsen, J. M. (1946): *Agnosia, Apraxia, Aphasia. Their Value in Cerebral Localization.* New York: Paul B. Hoeber.

Olds, J. (1956): Physiological mechanisms of reward. *In: The Nebraska Symposium on Motivation.* Jones, M. R., ed. Lincoln, Neb.: University of Nebraska Press.

Olds, J., and Olds, M. E. (1958): Positive reinforcement produced by stimulating hypothalamus with iproniazid and other compounds. *Science* 127:1175–1176.

Pavlov, I. P. (1947–1949): *Complete Works.* Vols. 3 and 4. (In Russian.) Moscow and Leningrad: Izd. Akademii Nauk. SSSR.

Penfield, W., and Milner, B. (1958): Memory deficit produced by bilateral lesions in the hippocampal zone. *Arch. Neurol. Psychiatr.* 79:475–497.

Peterson, L.R., and Peterson, M.I. (1959): Short-term retention of individual verbal items. *J. Exp. Psychol.* 58:193–198.

Polyakov, G. I. (1956): Relations between the principal types of neurons in the human cerebral cortex. (In Russian.) *Zh. Vyssh. Nerv. Deiat.* 6:469–478.

Polyakov, G. I. (1962): Modern data on the structural organization of the cerebral cortex. *In: Higher Cortical Functions in Man.* (In Russian.) Luria, A. R. Moscow: Izd. Moskovskogo Universiteta. (See also Polyakov, 1966.)

Polyakov, G. I. (1966): Modern data on the structural organization of the cerebral cortex. *In: Higher Cortical Functions in Man.* Luria, A. R. New York: Basic Books, pp. 39–69.

Popova, L. T. (1972): *Disturbances of Memory in Local Brain Lesions.* (In Russian.) Moscow: Izd. Meditsina.

Pribram, K. H. (1954): Towards a science of neuropsychology. *In: Current Trends in Psychology and the Biological Sciences.* Patton, R. A., ed. Pittsburgh, Penn.: University of Pittsburgh Press.

Pribram, K. H. (1959): *On the Neurology of Thinking* (Behavioral Science Series, no. 4). New York: Harper & Row.

Pribram, K. H. (1960): The intrinsic systems of the forebrain. *In: Handbook of Physiology.* Sect. 1. *Neurophysiology.* Vol. II. Field, J., editor-in-chief, and Magoun, H. W., ed. Washington, D. C.: American Physiological Society, pp. 1323–1344.

Pribram, K. H. (1963): The new neurology. Memory, novelty, thought, and choice. *In: EEG and Behavior.* Glaser, G. H., ed. New York: Basic Books, pp. 149–173.

Pribram, K. H. (1971): *Languages of the Brain. Experimental Paradoxes and Principles in Neuropsychology.* Englewood Cliffs, N. J.: Prentice-Hall.

Pribram, K. H., and Luria, A. R., eds. (1973): *Psychophysiology of the Frontal Lobes.* New York: Academic Press.

Scoville, W. B., and Milner, B. (1957): Loss of recent memory after bilateral hippocampal lesions. *J. Neurol. Neurosurg. Psychiatry* 20:11–21.

Sperry, R. W. (1966): Brain bisection and mechanisms of consciousness. *In: Brain and Conscious Experience* (Study Week of Pontificia Academia Scientiarum, 28 September–4 October 1964). Eccles, J. C., ed. New York: Springer-Verlag, pp. 298–313.

Sperry, R. W. (1967): *Mental Unity Following Surgical Disconnections of the Hemispheres*. New York: Academic Press.

Sperry, R. W. (1968): Hemisphere deconnection and unity in conscious awareness. *Am. Psychol.* 23:723–733.

Sperry, R. W., and Gazzaniga, M. S. (1967): Language following surgical disconnection of the hemispheres. *In: Brain Mechanisms Underlying Speech and Language* (Proceedings of a conference held at Princeton, N. J., 9–12 November 1965). Darley, F. L., ed. New York: Grune & Stratton, pp. 108–121.

Sperry, R. W., Gazzaniga, M. S., and Bogen, J. E. (1969): Interhemispheric relationships: The neocortical commissures; syndromes of hemisphere disconnections. *In: Handbook of Clinical Neurology*. Vol. 4. *Disorders of Speech, Perception, and Symbolic Behavior*. Vinken, P. J., and Bruyn, G. W., eds. Amsterdam: North-Holland, pp. 273–290.

Teuber, H.-L. (1959): Some alterations in behavior after cerebral lesions in man. *In: Evolution of Nervous Control from Primitive Organisms to Man* (Symposium organized by Section on Medical Sciences of American Association for Advancement of Science, New York, 29–30 December 1956). Bass, A. D., ed. Washington, D. C.: American Association for the Advancement of Science, pp. 157–194.

Teuber, H.-L. (1966): Alterations of perception after brain injury. *In: Brain and Conscious Experience* (Study Week of Pontificia Academia Scientiarum, 28 September–4 October 1964). Eccles, J. C., ed. New York: Springer-Verlag, pp. 182–216.

Tsvetkova, L. S. (1972): *Rehabilitative Training in Local Brain Lesions*. (In Russian.) Moscow: Izd. Pedagogika.

Ukhtomskii, A. A. (1945): *Outlines of the Physiology of the Nervous System. (Collected Works*. Vol. IV. In Russian.) Leningrad.

Vinogradova, O. S. (1970): Registration of information and the limbic system. *In: Short-Term Changes in Neural Activity and Behaviour* (A Conference Sponsored by King's College Research Centre, Cambridge). Horn, G., and Hinde, R. A., eds. Cambridge, Eng.: Cambridge University Press, pp. 95–140.

Vogt, C., and Vogt, O. (1919): Allgemeinere Ergebnisse unserer Hirnforschung. *J. Psychol. Neurol.* 25:279–462.

Vogt, O. (1927): Architektonik der menschlichen Hirnrinde. *Allg. Z. Psychiatr.* 86:247–274.

Vygotskii, L. S. (1934): *Thought and Language*. (In Russian.) Moscow: Sotsekgiz. (See also Vygotskii, 1962.)

Vygotskii, L. S. (1956): *Selected Psychological Investigations*. (In Russian.) Moscow: Izd. Akademii Pedagogicheskhikh Nauk.

Vygotskii, L. S. (1960): *The Development of Higher Psychological Functions*. (In Russian.) Moscow: Izd. Akademii Pedagogicheskhikh Nauk.

Vygotskii, L. S. (1962): *Thought and Language*. Cambridge, Mass.: The MIT Press.

Walter, W. G. (1953): *The Living Brain*. New York: W. W. Norton.

Walter, W. G. (1966): Human frontal lobe function in the regulation of active states. *In: The Frontal Lobes and Regulation of Psychological Processes*. (In Russian.) Luria, A. R., and Homskaya, E. D., eds. Moscow: Izd. Moskovskogo Universiteta, pp. 61–81.

Walter, W. G. (1973): Human frontal lobe function in sensori-motor association. *In: Psychophysiology of the Frontal Lobes*. Pribram, K. H., and Luria, A. R., eds. New York: Academic Press, pp. 109–122.

Walter W. G., Cooper, R., Aldridge, V. J., McCallum, W. C., and Winter, A. L. (1964): Contingent negative variation: An electric sign of sensorimotor association and expectancy in the human brain. *Nature* 203:380–384.

Wernicke, C. (1874): *Der aphasische Symptomencomplex. Eine Psychologische Studie auf anatomischer Basis*. Breslau: Max Cohn & Weigert.

Zangwill, O. L. (1964): Neurological studies and human behaviour. *Br. Med. Bull.* 20:43–48.

Eliot Stellar (b. 1919, Boston, Massachusetts) was appointed provost of the University of Pennsylvania in 1973 after having served since 1965 as director of the Institute of Neurological Sciences. His principal research areas have been the neural mechanisms of motivation and drive, especially the regulation of hunger in animals and humans.

# 21
# Physiological Psychology: A Crossroad in Neurobiology

## Eliot Stellar

### Background: Influences and Colleagues

In a rapidly developing interdisciplinary field, it is the good fortune of a few scientists to stand at the crossroad. In that position one has the opportunity to become a point of intersection for the older generation of scientists, representing the classical disciplines, and the younger generation of one's own students, representing the new interdisciplinary field.

I was fortunate enough to be at such a crossroad in neurobiology at the right time, just when the traffic was beginning to roar. The time was 1940, the place Harvard. I was an undergraduate psychology student, and I stood there with my teacher, Clifford T. Morgan, a man only a few years my senior. We were in the laboratory of Karl S. Lashley, the leading physiological psychologist, a man who had combined the strengths of behaviorist psychology and Gestalt psychology, and who had already promoted a coalescence of neurology and psychology.

These were exciting days for a young man. Bob Galambos and Don Griffin were down the hall, watching bats dodge piano wires by echolocation. Don was just beginning to track homing seagulls in his airplane, and he asked me to try and follow the gulls from the rear passenger seat. (It was only when I got to the airport carrying an envelope from Don's desk that I realized it contained that daring young man's license to fly passengers.) We followed the gulls as they were released one at a time from the ground in western Massachusetts, and we tracked six of them to a nearby river only to lose them in the glare off the water. An overcast day proved better for further experiments.

Just before I left Harvard to go to graduate school at Brown, Roger Sperry showed up on a fellowship with Lashley. He had funny rats with nerves to the extensor and flexor muscles of the forelimb crossed so that they curled their legs up as they walked. I was terribly impressed that the rats' walking never improved with practice.

During this time Morgan was writing his book *Physiological Psychology*, first published by him in 1943 and revised by the two of us in 1950. The book was dedicated to Lashley and Curt P. Richter, a student of behaviorist J. B. Watson and a young colleague of Adolph Meyer, the psychiatrist who called himself a psychobiologist. (Richter later became a colleague of great significance to me.) It contained the work of neuroanatomists, neurophysiologists, and physiological psychologists, many of whom later had personal impact on me and my students: Philip Bard, H. Keffer Hartline, Clinton Woolsey, E. F. Adolph, Frank Beach,

Neal Miller, Carl Pfaffmann, and Donald Lindsley. Interestingly enough, although the book was only about 500 pages long, it covered just about everything in the field. Compare that to current multivolume handbooks of neurobiology and you have a measure of our explosive growth in three decades.

Through this expanding, multidisciplinary environment passed my students, first in the psychology department at Johns Hopkins and later in the anatomy department and the Institute of Neurological Sciences at the University of Pennsylvania. Morgan had invited me to join his department at Hopkins in 1947, and while he was the only other physiological psychologist there, I soon found a growing number of colleagues who were interested in neurobiology and the biology of behavior. On the Homewood campus there was V. G. Dethier, the invertebrate neurophysiologist, who later became my coauthor in the publication of *Animal Behavior*, a little book that has found its way into ten languages and three English editions so far (Dethier and Stellar, 1961, 1964, 1970). There was also W. D. McElroy, the biochemist and student of bioluminescence, who later became my collaborator in studying the possible role of glutamic acid in learning (Stellar and McElroy, 1948).

At the Medical School, thirty minutes across town, were Curt Richter, Philip Bard, David Bodian, Clinton Woolsey and his student Vernon Mountcastle, Louis Flexner, James Sprague, Jerzy Rose, Steve Kuffler, and their younger colleagues and students. In 1950 Detlev Bronk became president of Hopkins and brought a whole biophysics department to the Homewood campus from the Johnson Foundation of the University of Pennsylvania, including Keffer Hartline, Frank Brink, Martin Larrabee, and their younger colleagues and students. It was a golden era, but Hopkins made a fatal error and refused our plans to form a neurological institute to keep it all together. For this reason (and others to be sure) the people scattered: Woolsey and Rose to Wisconsin; Bronk, Hartline, Brink, and company to Rockefeller; Flexner, Sprague, Dethier, and myself to Pennsylvania; Kuffler to Harvard. These were the seeds of neurological institutes and neurobiology departments that scattered around the country and took root and grew.

At Pennsylvania the Institute of Neurological Sciences was Louis Flexner's brainchild. He started it to give two of his young colleagues in anatomy, James Sprague and W. W. Chambers, more time for research, better access to graduate and postdoctoral students, and an opportunity for collaboration in the study of behavior. I joined in 1954 as the behaviorist and found that the Institute was exactly the right mechanism for providing day-to-day interdisciplinary and interdepartmental interactions among both faculty and students. It was good for physiological psychology, and it was the best thing that ever happened to me.

### Two Themes: Cognitive and Affective Functions

From the beginning there were two general themes in my scientific life: the minor one was the study of *cognitive functions* such as learning, memory, and perception, and the major one was the study of *affective functions* such as instinct, emotion, and motivation. This dichotomy was very much in the air at Harvard when I was there, deriving from the writings of William James and from the teachings of Wilhelm Wundt to Edward Bradford Titchener to E. G. Boring, the director of Harvard's psychological laboratories in my day. Lashley, of course, pursued both the "cognitive" and "affective" themes, and this was the most significant influence on me, even though at that time most of it filtered to me through Morgan.

Lashley's 1938 paper, "An experimental analysis of instinctive behavior," was already a classic in the field, even though it represented his minor scientific interest. In a very direct style he posed the major questions that should be asked of the nervous system by students of instinct, emotion, or motivation:

Where in the brain is there an excitatory process, like Sherrington's "central excitatory state," that can be the basis of such biological drives as sex, maternal behavior, emotion, hunger, thirst?

Is there an inhibitory or satiety mechanism?

How do such neural mechanisms add up the effects of sensory input and of such influences of the internal environment as sex hormones, variations in body temperature, or osmotic pressure?

How can instinctive behavior, once launched, be patterned in intricate sequences of responses, such as those seen in nest-building and web-spinning, often with only minimal external guidance?

I tried to answer the first three questions in 1954 in the light of recent experimental evidence about the hypothalamus (Stellar, 1954). The last question was much more elusive, but clearly implied a "central engram" that could be triggered to run its course. The ethologists later called this engram a "fixed-action pattern" (Tinbergen, 1951), and still later the students of invertebrate neurophysiology called it the "central score" (Wilson, 1970).

**First Experiment: Cognitive Theme**

By my time, Lashley (1929) had already made legendary contributions to the study of neural mechanisms in sensation, perception, learning, and memory. Because of this I began my work in his laboratory on cognitive functions, in a classic Lashley-type study of the cortical localization of kinesthesis in the rat. Originally I trained rats to do a weight discrimination in a Lashley jumping stand, but I later shifted to a more manageable distance discrimination in a linear maze. The rat simply had to run four feet and turn left in a situation where all visual, auditory, tactile, and olfactory stimuli were mixed up from trial to trial so that only kinesthesis was consistent in giving the correct cue. Suffice it to say that I was able to find a small area in the dorsal part of the rat's frontal cortex whose integrity, judged by thalamic degeneration and direct cortical damage, was critical for the distance discrimination (Stellar, Morgan, and Yarosh, 1942). (This was recently confirmed in work by Konorski's colleague Lukaszewska (1973) at the Nencki Institute in Warsaw.) Later one of my students, John Zubek (1951), did a similar study of somesthesis by training rats to do a roughness discrimination in a two-lever Skinner box; weight discrimination could also be done in the same box.

**Affective Theme: Motivated Behavior**

I learned much about behavioral testing, experimental surgery, and neuroanatomy from Lashley. But I changed themes when I went to graduate school at
Brown, even though the initial attraction at Brown was W. S. Hunter, the world's
leading authority on animal learning. What brought about this change at Brown
was the offer of a chance to work with a younger man, J. McV. Hunt, who was interested in experimental testing of the Freudian hypothesis that infantile frustration causes abnormal adult behavior. I was also pleased at the opportunity to work
with a fellow graduate student, Dick Solomon, who was to be another research
assistant for Hunt. Part of the excitement of our study was that the "abnormal"
behavior was to be food hoarding, which I thought would provide an excellent
"model" of motivated behavior for investigating physiological mechanisms of the
sort Lashley had discussed in 1938. I had a fancy theory that there was a metabolic
deficit at the root of hoarding, and I did both my masters and Ph.D. theses
investigating carbohydrate and energy metabolism, without success (Stellar,
1943, 1951). The infantile-frustration idea didn't work out too well either, and I
was quite depressed. Fortunately Professor Clarence H. Graham saved me from
total decline when he told me that his Ph.D. thesis had been even worse than
mine. Another professor, Harold Schlosberg, helped even more, for he convinced
me that all of these behaviors like hoarding, hunger, and sex were under multifactor control, and that my experiments had failed because I was thinking in
terms of single-factor control.

When I went to my first job at Hopkins in 1947, I made the important decision
that I could not survive on negative results. This meant that I would have to be
more empirical and work on behavior that might be easier to understand than
hoarding. Hence, in the quest of success, I shifted to the study of hunger and
thirst. I also abandoned fancy hypotheticodeductive theory and became quite
empirical, using inductive theorizing when I had something to go on. I was lucky.
Everything was ripe. The limbic system and hypothalamus had been identified as
the neurological target for investigation in the classic work of Papez, Ranson,
Brobeck, and Bard. The stereotaxic method was available. Since I didn't have
money to buy an instrument, but did have access to a machinist, Nelson Krause,
we built one for the rat with a few improvements (Stellar and Krause, 1954) and
made lesions to study the hypothalamus. We also stimulated the brain, for Hess
had already shown that chronic implantation of electrodes (and therefore pipettes)
was possible.

Most important of all, however, was the conceptual contribution to my thinking
that Richter (1942–1943) made when he spoke of hunger, thirst, specific hungers,
nest-building, etc., as self-regulatory behaviors, and thus tied our work conceptually
to the great work of Cannon on homeostasis and of E. F. Adolph on physiological
regulation. It was in this scientific setting that my students worked and began
their own specialized approaches to problems of physiological psychology and
neurobiology. R. A. McCleary (1953) started out working on sugar preferences,

but later shifted to identifying the role of excitatory and inhibitory mechanisms in active and passive avoidance conditioning (McCleary, 1966). Philip Teitelbaum (1955) worked on hypothalamic hyperphagia and the lateral hypothalamic starvation syndrome (Teitelbaum and Epstein, 1962). He and I showed that there was recovery of function after lateral hypothalamic lesions (Teitelbaum and Stellar, 1954), and he went on to show how recovery recapitulates ontogenetic development of feeding (Teitelbaum, Cheng, and Rozin, 1969). Alan Epstein started out as an undergraduate with me, doing a study of salt hunger in the adrenalectomized rat (Epstein and Stellar, 1955). Later he collaborated with Teitelbaum to study the eating and drinking deficits produced by lateral hypothalamic lesions and pioneered with intrahypothalamic cannulae (Epstein, 1971). More recently he has focused his interest on thirst (Epstein, 1973); here he has made it clear that there is a double depletion in thirst, osmotic and volemic, and he has shown that they can be separated in the preoptic area and that the renin-angiotensin system is important in volemic thirst.

Early in this period Harry Hill and I developed the "electronic drinkometer" by simply making the rat's tongue the switch in an electronic counter during drinking (Hill and Stellar, 1951). This technique revealed that the tongue lapped at a constant 6 to 7 times a second and netted between 0.004 and 0.005 ml of water per lap. The drinkometer opened the door to a wide range of studies of ingestive behavior, taste, reward, etc. Somewhat later my student Douglas Mook (1963) perfected the esophageal-fistula preparation in the rat that I had first tried to develop at Harvard; he later went on to study the role of hormones in hunger (Mook et al., 1972). John Corbit (1965) did his thesis on intravenous self-injection of water and salt solutions in a study of thirst in the rat; later he investigated the neural mechanisms involved in thermoregulatory behavior (Corbit, 1969).

I tried to pull my neurological ideas together in a paper entitled "The physiology of motivation" (Stellar, 1954). These included separate excitatory and inhibitory mechanisms in the hypothalamus, multifactor control of sensory input and internal-environment changes, and the hypothesis that hunger, thirst, emotion, sex, temperature regulation, maternal behavior, aggression, etc., all had basically the same underlying neurological and physiological mechanism. By 1960, in an update for the *Handbook of Physiology*, I was able to borrow from Neal Miller, Philip Teitelbaum, and James Olds and argue that all of these behaviors could be placed in the general category of motivated behavior since animals would learn arbitrary instrumental acts to achieve their natural goals (e.g., food, water, mating, warmth) (Stellar, 1960a). This was getting pretty close to having a measure of "voluntary" behavior in animals and to thinking in terms of rewards, reinforcements, and hedonic processes. Certainly James Olds's important discovery of self-stimulation in the rat pushed my thinking in this direction.

**A Return to the Cognitive Theme**

I'll come back to hedonic processes at the end, but I need to make a brief digression to show how I returned to the study of cognitive functions several times during this period. The first shift occurred while I was still at Hopkins, when I thought I could really make headway if I switched from the stimulus-bound and stereo-typed rat to the monkey. Since I didn't have the facilities to deal with rhesus monkeys, I tried my hand at marmosets. Suffice it to say that I learned how to keep them alive in the laboratory, devised a diet for them, and showed that they could solve such complex problems as matching-from-sample (Stellar, 1960b). But I never got to study the effects of brain lesions, for I moved to Pennsylvania during that period and got involved in research on cats with my new colleagues Jim Sprague and Bill Chambers. We had prepared a cat with bilateral reticular-formation lesions to demonstrate somnolence to the medical students. To our surprise the cat recovered wakefulness within a couple of months, before we could demonstrate it to the whole class. Keen on serendipity, we decided to investigate the matter further. To this end we prepared cats with control lesions in the lateral lemniscal regions, sparing the midbrain reticular formation. The controls, of course, proved the most interesting, for they showed no sense of danger, very flat affect, and maladaptive behavior reminiscent of autistic children; they also de-monstrated a marked sensory neglect reminiscent of patients with parietal-lobe lesions (Sprague, Chambers, and Stellar, 1961; Sprague et al., 1963).

To study the visual neglect more thoroughly, we prepared additional cats with split optic chiasms, and before we knew it my students and I were hooked on in-vestigating transfer of training in "split-brain" cats. Tom Meikle showed that interocular transfer of brightness discrimination occurs in cats that do not transfer pattern discrimination (Meikle and Sechzer, 1960); Jeri Sechzer (1964) found that cats do show interocular transfer of pattern discrimination when they are trained under shock-avoidance motivation, but not when food-reward training is used (as had been the case in all previous studies). We thus demonstrated that what pathways are involved in the interhemispheric transfer of training crucially depends on both the type of task and the type of motivation used to measure be-havior.

Not long after this Louis Flexner suggested that we pick up the thread of in-vestigation of the biochemical basis of learning and memory that Bill McElroy and I had dropped after our glutamic-acid failure. Flexner and I had decided to wait until we had an agent that could make a profound and specific difference in brain chemistry; we found it in puromycin, which had been shown by our col-league Gabriel de la Haba to inhibit protein synthesis (Yarmolinsky and de la Haba, 1959). Flexner had long thought that the very high rate of protein synthesis in the brain might have something to do with such complex functions as learning and memory, and here was a good way to test the idea. Sure enough, puromycin blocked memory and did so in proportion to the degree to which it inhibited protein synthesis (Flexner, Flexner, and Stellar, 1963). Flexner and his wife Josefa then went on to investigate many facets of the memory problem and showed among other things that the memory block is reversible and that the basis of the block is not inhibition of protein synthesis after all, but more probably some altera-tion of synaptic chemistry.

## Motivation and Hedonism in Man

Meantime, I was pulled back to the investigation of hunger and the regulation of food intake because I now saw an opportunity to become involved directly in the study of man. It had always perplexed me that almost all the papers and books in the literature on hunger in man were based on subjective reports of hunger and food intake. Very few contained objective measures of food intake, which was the main method used in the study of animals and which had proven so instructive (Stellar, 1967). When Henry Jordan, a young psychiatry resident, came to see me about doing studies of animal food intake, I proposed that we look at man instead. He jumped at the idea, and we set out to measure food intake in single experimental meals just as we had done in animals (Jordan et al., 1966). Later we discovered that Hashim and van Itallie (1964) had started, at about the same time, to look at 24-hour liquid-food intake in patients at St. Luke's Hospital in New York. But we were taking a different track with our investigation of single meals of Metrecal, pumped by the press of a button into the mouth through straws from an unseen reservoir in the next room or pumped via stomach tube in voluntary intragastric ingestion, which bypassed taste, smell, etc. (Jordan, Stellar, and Duggan, 1968).

A. J. Stunkard, chief of psychiatry, was our collaborator, and we looked at obese patients, anorexic patients, and patients on intravenous hyperalimentation. In addition we studied hedonic processes in normal subjects. For the first time we were able to establish ratings of hunger, satiety, and pleasure as we studied eating behavior and food intake. With our student Ray Hawkins, we also looked at thermal comfort and discomfort in man and thermal pleasure and pain experience as a function of body temperature, which was manipulated by immersion in a hot or cold bath. Whether we looked at ratings of the pleasure experienced in dipping one hand into water of different temperatures or the shower temperature a subject selected as the most pleasurable, the results turned out to be the same: when core temperature was lowered, hand and shower temperatures that had previously been intolerably hot were found to be most pleasurable, and vice versa (Hawkins, 1975). Essentially the same results were found behavioristically in the rat by John Corbit (1969) when he cooled or warmed the preoptic area of the brain.

So I have made a full circle, going several times through both of my scientific themes and always coming back to my major concern with affective processes. It was refreshing and exciting to take flyers at the study of cognitive functions, but I am afraid I can't claim much on this side. So far the problems of learning and memory haven't really yielded. Perhaps we are asking the wrong questions of the nervous system by trying to make the brain conform to the demands of such behavioral concepts as association, reinforcement, storage, retrieval, and interference. Perhaps, though, experiments of the sort the Flexners are doing will teach us to ask the right questions, and even tell us what to look for in the brain and where.

I believe we are in much better shape on the affective side (Stellar and Jordan,

1970; Stellar and Corbit, 1973). First, I think that we can have a general theory dealing with a common brain mechanism for hunger, thirst, sex, maternal behavior, temperature regulation, fear, aggression, territorial defense, etc. Second, I believe the same basic brain mechanisms serve at three levels of complexity of function: physiological regulation, motivated behavior, and hedonic experience. Third, and most encouraging, the field is in a rapidly accelerating phase of development; we have trained many new neurobiologists in the study of motivated behavior, and they have powerful new tools and concepts. In fact, we have a whole new breed of investigators who go under the names of physiological psychologists, neuropsychologists, behavioral biologists, psychobiologists, psychophysiologists, behavioral neurobiologists, and even neuroscientists. They are scientists skilled in behavioral analysis, but also in brain surgery, electrophysiology, neuroanatomy, and neurochemistry. Perhaps it is developing this new breed that has been our major accomplishment in neuroscience.

From the standpoint of the physiological psychologist, the field is now on the threshold of significant new insights and new syntheses. In particular, I think we are on the verge of understanding how the brain works in the generation of human subjective experience (Stellar, 1974). Up to now we have made some headway in learning about the neural codes involved in sensory experience but only a little in the memory code. I believe, however, that we are much closer, in neuroscience, to understanding the basis of hedonic experience in both man and animal and that this insight will provide the model of brain function for the eventual understanding of other subjective experiences.

## References

Corbit, J. D. (1965): Effect of intravenous sodium chloride on drinking in the rat. *J. Comp. Physiol. Psychol.* 60:397–406.

Corbit, J. D. (1969): Behavioral regulation of hypothalamic temperature. *Science* 166:256–258.

Dethier, V. G., and Stellar, E. (1961, 1964, 1970): *Animal Behavior. Its Evolutionary and Neurological Basis.* Englewood Cliffs, N.J.: Prentice-Hall.

Epstein, A.N. (1971): The lateral hypothalamic syndrome: Its implications for the physiological psychology of hunger and thirst. *In: Progress in Physiological Psychology.* Vol. 4. Stellar, E., and Sprague, J. M., eds. New York: Academic Press, pp. 263–317.

Epstein, A. N. (1973): Epilogue: Retrospect and prognosis. *In: The Neuropsychology of Thirst: New Findings and Advances in Concepts.* Epstein, A. N., Kissileff, H. R., and Stellar, E., eds. Washington, D.C.: V.H. Winston and Sons. pp. 315–332.

Epstein, A. N., and Stellar, E. (1955): The control of salt preference in the adrenalectomized rat. *J. Comp. Physiol. Psychol.* 48:167–172.

Flexner, J. B., Flexner, L. B., and Stellar, E. (1963): Memory in mice as affected by intracerebral puromycin. *Science* 141:57–59.

Hashim, S.A., and van Itallie, T.B. (1964): An automatically monitored food dispensing apparatus for the study of food intake in man. *Fed. Proc.* 23:82–84.

Hawkins, R. (1975): Ph.D. Thesis, University of Pennsylvania, Philadelphia, Pa.

Hill, J. H., and Stellar, E. (1951): An electronic drinkometer. *Science* 114:43–44.

Jordan, H. A., Stellar, E., and Duggan, S.Z. (1968): Voluntary intragastric feeding in man. *Commun. Behav. Biol. A.* 1:65–68.

Jordan, H. A., Wieland, W. F., Zebley, S. P., Stellar, E., and Stunkard, A. J. (1966): Direct measurement of food intake in man: A method for the objective study of eating behavior. *Psychosom. Med.* 28:836–842.

Lashley, K. S. (1929): *Brain Mechanisms and Intelligence. A Quantitative Study of Injuries to the Brain.* Chicago: University of Chicago Press.

Lashley, K. S. (1938): An experimental analysis of instinctive behavior. *Psychol. Rev.* 45:445–471.

Lukaszewska, I. (1973): Distance discrimination in frontopolar rats. *Acta Neurobiol. Exp.* 33:523–526.

McCleary, R. A. (1953): Taste and post-ingestion factors in specific-hunger behavior. *J. Comp. Physiol. Psychol.* 46:411–421.

McCleary, R. A. (1966): Response-modulating functions of the limbic system: Initiation and suppression. In: *Progress in Physiological Psychology.* Vol. 1. Stellar, E., and Sprague, J. M., eds. New York: Academic Press, pp. 209–272.

Meikle, T. H., and Sechzer, J. A. (1960): Interocular transfer of brightness discrimination in "split-brain" cats. *Science* 132:734–735.

Mook, D. G. (1963): Oral and postingestional determinants of the intake of various solutions in rats with esophageal fistulas. *J. Comp. Physiol. Psychol.* 56:645–659.

Mook, D. G., Kenney, N. J., Roberts, S., Nussbaum, A. I., and Rodier, W. I., III (1972): Ovarian-adrenal interactions in the regulation of body weight by female rats. *J. Comp. Physiol. Psychol.* 81:198–211.

Morgan, C. T. (1943): *Physiological Psychology.* New York: McGraw-Hill.

Morgan, C. T., and Stellar, E. (1950): *Physiological Psychology.* New York: McGraw-Hill.

Richter, C. P. (1942–1943): Total self regulatory functions in animals and human beings. *Harvey Lect.* 38:63–103.

Sechzer, J. A. (1964): Successful interocular transfer of pattern discrimination in "split-brain" cats with shock-avoidance motivation. *J. Comp. Physiol. Psychol.* 58:76–83.

Sprague, J. M., Chambers, W. W., and Stellar, E. (1961): Attentive, affective, and adaptive behavior in the cat. *Science* 133:165–173.

Sprague, J. M., Levitt, M., Robson, K., Liu, C. N., Stellar, E., and Chambers, W. W. (1963): A neuroanatomical and behavioral analysis of the syndromes resulting from midbrain lemniscal and reticular lesions in the cat. *Arch. Ital. Biol.* 101:225–295

Stellar, E. (1943): The effect of epinephrine, insulin, and glucose upon hoarding in rats. *J. Comp. Psychol.* 36:21–31.

Stellar, E. (1951): The effects of experimental alterations of metabolism on the hoarding behavior of the rat. *J. Comp. Physiol. Psychol.* 44:290–299.

Stellar, E. (1954): The physiology of motivation. *Psychol. Rev.* 61:5–22.

Stellar, E. (1960a): Drive and motivation. *In: Handbook of Physiology.* Sect. 1. *Neurophysiology.* Vol. III. Magoun, H. W., ed. Washington, D.C.: American Physiological Society, pp. 1501–1527.

Stellar, E. (1960b). The marmoset as a laboratory animal: Maintenance, general observations of behavior, and simple learning. *J. Comp. Physiol. Psychol.* 53:1–10.

Stellar, E. (1967): Hunger in man: Comparative and physiological studies. *Am. Psychol.* 22:105–117.

Stellar, E. (1974): *Brain Mechanisms in Hunger.* Philadelphia: American Philosophical Society (in press).

Stellar, E., and Corbit, J.D. (1973): Neural control of motivated behavior. *Neurosci. Res. Program Bull.* 11:295–410.

Stellar, E., and Jordan, H. A. (1970): Perception of satiety. *In: Perception and Its Disorders* (Proceedings of the Association for Research in Nervous and Mental Disease, New York, 6–7 December 1968). Hamburg, D. A., Pribram, K. H., and Stunkard, A. J., eds. Baltimore: Williams and Wilkins, pp. 298–317.

Stellar, E., and Krause, N. P. (1954): New stereotaxic instrument for use with the rat. *Science* 120:664–666.

Stellar, E., and McElroy, W. D. (1948): Does glutamic acid have any effect on learning? *Science* 108:281–283.

Stellar, E., Morgan, C. T., and Yarosh, M. (1942): Cortical localization of symbolic processes in the rat. *J. Comp. Physiol. Psychol.* 34:107–126.

Teitelbaum, P. (1955): Sensory control of hypothalamic hyperphagia. *J. Comp. Physiol. Psychol.* 48:156–163.

Teitelbaum, P., Cheng, M.-F., and Rozin, P. (1969): Development of feeding parallels its recovery after hypothalamic damage. *J. Comp. Physiol. Psychol.* 67:430–441.

Teitelbaum, P., and Epstein, A.N. (1962): The lateral hypothalamic syndrome: recovery of feeding and drinking after lateral hypothalamic lesions. *Psychol. Rev.* 69:74–90.

Teitelbaum, P., and Stellar, E. (1954): Recovery from the failure to eat produced by hypothalamic lesions. *Science* 120:894–895.

Tinbergen, N. (1951): *The Study of Instinct.* Oxford: Clarendon Press.

Wilson, D.M. (1970): Neural operations in arthropod ganglia. *In: The Neurosciences: Second Study Program.* Schmitt, F.O., editor-in-chief. New York: Rockefeller University Press, pp. 397–409.

Yarmolinsky, M. B., and de la Haba, G. L. (1959): Inhibition by puromycin of amino acid incorporation into protein. *Proc. Natl. Acad. Sci. USA* 45:1721–1729.

Zubek, J. P. (1951): Studies in somesthesis. I. Role of the somesthetic cortex in roughness discrimination in the rat. *J. Comp. Physiol. Psychol.* 44:339–353.

James Olds (b. 1922, Chicago, Illinois) is Bing Professor of Behavioral Biology at the California Institute of Technology. His research has dealt with reward, punishment, drive, and learning, especially as investigated by the electrical self-stimulation experimental method he helped pioneer.

# 22
## Mapping the Mind onto the Brain

## James Olds

Models of the mind, and the hope that these could be tied to neuroanatomical structures by brain maps, have motivated much of my research.

Introspective modeling was my spontaneous response to the psychological wonderland I entered when, as a first-year graduate student, I changed from philosophy to join a mixed department whose aim was to understand human action. Models of the brain based on introspection have been a source of ambivalence. On the one hand, ideas in this area are for me a game preferred above all others. On the other hand, my evaluation of such ideas oscillates: at the high point I believe them to be of dubious value, and at the low point I am convinced of their utter worthlessness. Yet, for all of this, I have never been able to do anything of an experimental nature that has not been tied in my secret consciousness to these dubious or valueless ideas. I hope therefore that the high-point evaluation is correct.

If such models are of value, it is because there is a second and almost magic source of information about the brain. It is a pathway that leads through the looking glass of the mind.

The high point of my graduate career came when Edward Chase Tolman, then a grand old man among psychologists, was a visiting professor. He and Clark L. Hull represented two poles of an argument that was at the crux of things (see Tolman, 1949, and Hull, 1943). The question was: How do motives work? Why does an animal "come back for more"? Hull held that reward "stamps in" a connection between stimulus and response at the time of its original occurrence. Later the animal comes back for more because he cannot help it; the stamped-in connection forces him. Tolman put more emphasis on information processing that goes on later, at the time the animal is deciding whether to go back. The first time around the animal learns where things are and how they "taste." These events are recorded, as if on a tape recorder. Later the animal plays back his tapes, at first vicariously. If he finds something he likes on one of the tapes, he then plays it back a second time for real. Thus, for Tolman, remembered reward did the main job. Tolman had created ingenious experiments to buttress his position and he really had the best of the argument (or so it seemed to me). Anyway, Tolman's presence caused me to modify my imagined brain machines so they would do the information-processing jobs he talked of.

My Ph.D. dissertation was directed by a younger psychologist, Richard L. Solomon, who was noteworthy for his insistence that ideas be cast in a form that could be tested by experiments. He guided me through a series of experiments that were obliquely related to my rather grandiose ideas. But in his view the obliqueness was improper. If I wanted to talk about the brain, I should work on it.

In retrospect, it seems that all of my ideas about the brain were derived at either first or second hand from Donald O. Hebb. I had a set of ideas about circling or reverberating neuronal activity that were in the air at that time. I found later that Hebb had put them in the air. I then read Hebb's book *The Organization of Behavior* (1949) and rearranged my theories so they explicitly matched his.

The wisdom of Solomon eventually sent me to be an apprentice in Hebb's laboratory. There Hebb again had direct and indirect influences. His direct communications were neither numerous nor verbose, but his ideas were pervasive and I readily accepted many of them as my own.

I arrived in the summer of 1953. I spent the summer mainly talking about my ideas and his with Seth Sharpless, a graduate student and one mainstay of Hebb's laboratory at that time. Sharpless introduced me to the discovery of the reticular activating system in the core of the brain. The great names concerned with this event were Moruzzi and Magoun (1949), Lindsley (1951), and Jasper (1949).

After a summer of discussions, I spent the early fall learning methods to implant probes, to record, and to stimulate electrically in the rat brain. I learned these from Peter M. Milner, another graduate student and a second mainstay of Hebb's laboratory at that time.

## The Reward Experiments

Late that fall I stumbled accidentally onto the discovery that rewarding effects could be produced by electrical stimulation of the rat brain. The discovery was influenced indirectly by Hebb and more directly by Neal Miller, who had recently reported negative reinforcing effects caused by electrical stimulation of the cat's brain (Miller, Roberts, and Delgado, 1953). I was not thinking about a possible discovery of rewarding effects when I did the experiment: my only purpose was to test whether the stimulus would be aversive, because aversive stimulation would interfere with some other things I had planned.

I made the tests in a relatively open situation which permitted the animal to show both approach and avoidance. A rat was free to move around in a fairly large tabletop enclosure (3 × 3 feet or more). A pair of stimulating wires was planted in the brain; it was aimed to affect the arousal or activating system, but it missed and landed in an area near the hypothalamus. I applied a brief electrical stimulus whenever the animal entered one corner of the enclosure. Surprisingly, the animal did not stay away from that corner, but came back quickly after a brief sortie that followed the first stimulation, and even more quickly after a briefer sortie that followed the second stimulation. By the time the third stimulation had been applied the animal seemed indubitably to be "coming back for more."

Later this same rat furthered my view that the electrical stimulation was rewarding by moving toward any part of the enclosure chosen by an independent observer, provided only that I turned on the switch whenever the animal took a step in the right direction. Still later the animal learned to run rapidly to that arm of an elevated T-maze in which the electrical stimulation was regularly applied.

Attempts to repeat the experiment by implanting probes in similar places in other animals did not readily meet with success. Some animals evidenced ambivalence or aversive reactions, and it soon became clear that a careful and systematic mapping would be required in order to find and exploit the exact brain areas that had yielded the original phenomenon. It was this problem that began my first brain-mapping exercise.

Milner and I developed the self-stimulation experiment to provide a measure of reward that might be correlated with the different brain points where electrical stimulation could be applied. In this experiment the animal was provided with a very large lever in a very small "Skinner box." (B. F. Skinner (1938) had invented the box as a measure of one kind of learning in the rat. Had he had more tolerance for ambiguous and subjective ideas, he would have seen that he had invented a measure of pleasure.) Our whole mechanism consisted of the box, the pedal, an opening in the top of the box for the stimulating leads to pass through, and a very simple electrical stimulator that was activated by the animal's depression of the pedal. Because a very small box was used, and because the large pedal was placed so that it would be depressed whenever the animal peered through the only aperture, initial response rates ranged as high as 60 responses in the first ten-minute period, even in cases when no reinforcement at all was used. Because of this very high initial response rate, a very low pedal rate could be taken as evidence of aversive effects of the electrical brain stimulus, and very high rates (ranging from 200 to 1000 responses in a ten-minute period) could be taken as evidence of rewarding effects. This self-stimulation experiment provided the method needed for tracking the brain reward to its neuroanatomical lair (Olds and Milner, 1954), making it possible to reproduce the original phenomenon, as is now done regularly in many brain and behavior laboratories.

Before offering my own interpretation of brain-reward research, I will review the further findings from our laboratories and from other laboratories that seem to me to be of most lasting value.

**Anatomy**

Eight different olfactory forebrain areas project bundles into a tubular structure at the base of the brain, called the medial forebrain bundle (see Figure 22.1). Stimulating in six of these bundles is rewarding (M.E. Olds and J. Olds, 1963; Olds, Travis, and Schwing, 1960; Olds, 1956b; Routtenberg, 1971). Several different areas in the midbrain, pons, and medulla project bundles upward into the same tubular structure (Fuxe, 1965). Stimulation in most of these is rewarding (Routtenberg and Malsbury, 1969; Crow, Spear, and Arbuthnott, 1972; Crow, 1972). Many or most of these ascending fibers carry a particular kind of chemical with a generally inhibitory influence in the central nervous system; the chemicals involved are the monoamines: noradrenaline, serotonin, and dopamine (Fuxe, 1965). The monoamine fibers project not only into this tube but through it to the eight forebrain structures (and to other structures besides). Stimulation of the tubular structure itself is very strongly rewarding, more so than stimulation anywhere else (M. E. Olds and J. Olds, 1963).

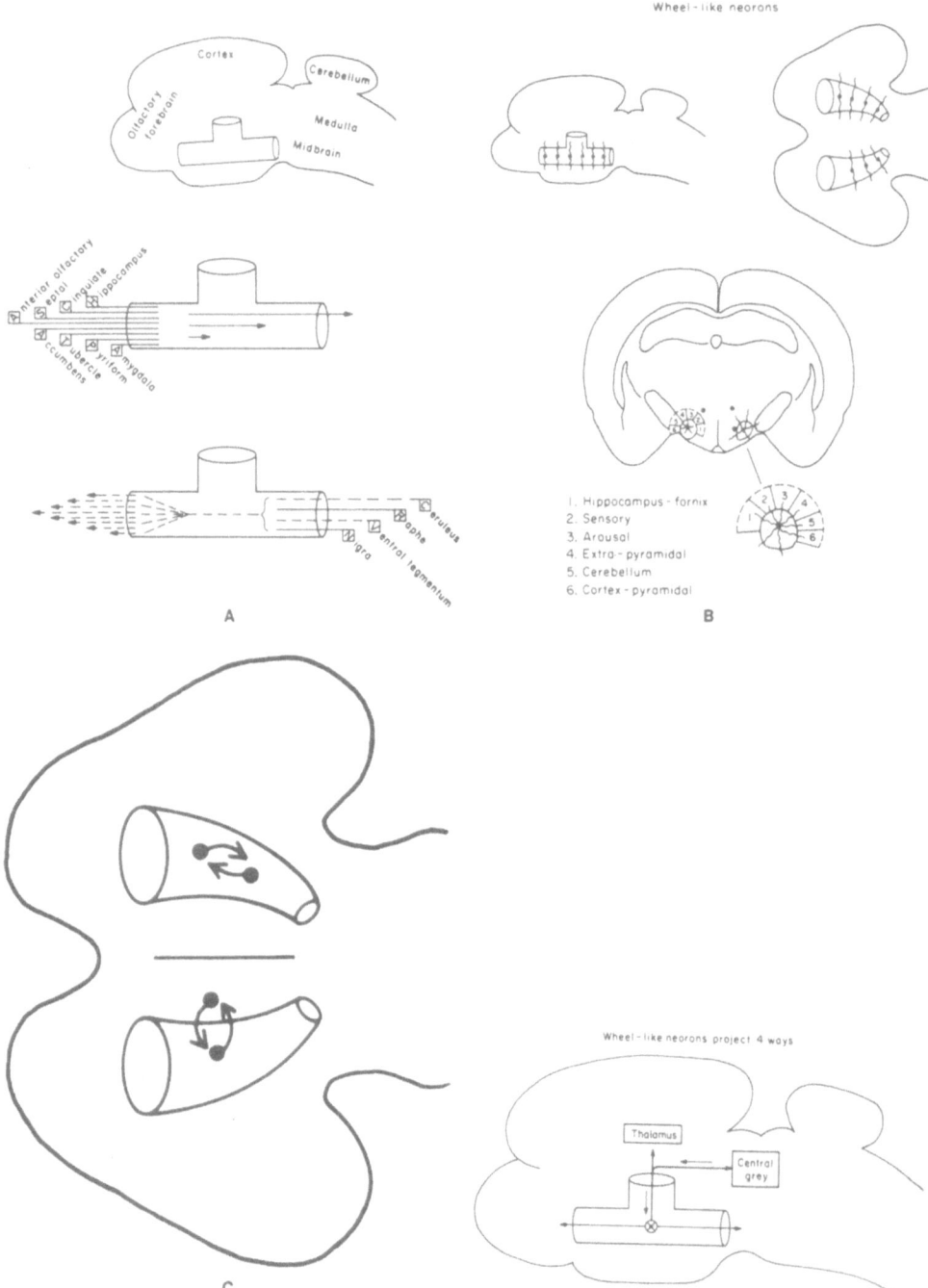

Figure 22.1 (A) Medial forebrain bundle shown as a tube at the base of the brain, penetrated by eight fiber systems (identified by their sources) from the front, and four from the rear. (B) The wheellike neurons in the path of the medial forebrain bundle with dendrites probing into six fiber systems (identified by their sources). (C) Interaction of the wheellike neurons with their neighbors. (D) Projections of the wheellike neurons. From Millhouse (1969).

In the tube, besides the numerous bundles, there is a family of neurons with dendrites like the spokes of a wheel, spread as if to monitor all the bundles in the tube, plus another set of bundles that forms a partial shell around it (Millhouse, 1969). In the shell there is a great concourse of sensory and motor systems: the three main motor pathways (from the cortex, extrapyramidal system, and cerebellum) pass here; there are special sensory fibers from olfactory and gustatory receptors (Scott and Pfaffmann, 1967; Norgren and Leonard, 1973); there are collaterals from the other sensory systems (Dafny, Bental, and Feldman, 1965); there is a bundle of fibers from the "arousal" system; and there is a special bundle from the computerlike hippocampus.

The spokes of the wheellike neurons penetrate into these paths and must therefore receive afferents from many if not all of them. They also receive messages from one another and from a closely related neighbor family in the medial hypothalamus, which seems to monitor hormone levels and blood constituents. The output from the large wheellike neurons goes in four or five directions: (1) back to the many forebrain structures, (2) back to the several midbrain and lower structures, (3) back to the blood monitors; but the main outputs seem to be (4) up toward the thalamus, and (5) back toward the central gray (Millhouse, 1969). From both of these last two the messages are relayed through "nonspecific" pathways to the cortex and probably also to outgoing motor centers.

All the structures that yield rewarding effects to electrical stimulation seem to have at least two things in common. First, they project to and receive from the large wheellike neurons in the tubular medial forebrain bundle. Second, they are penetrated by the monoamine fibers that project upward through this structure to all of its connected systems.

One interesting hypothesis is that the wheellike neurons are mainly responsible for motives, both drives and rewards. An equally interesting alternative is that the monoamine neurons, as inhibitors, do the rewarding and that the wheellike neurons act as drives, inhibited by the rewards. It seems, therefore, that careful studies of the tubular structure and its wheellike neurons might help unravel the conundrum of drive and reward.

**Behavior**

Three behavioral features divided the brain-reward behavior into two syndromes. One was the amount of conflict evoked by the stimulation; this was between reward and escape behaviors, both of which were in the conflicted cases evoked by the same electrical stimulus. The second was the dependence on or independence from antecedent brain stimulations: in some cases the behavior was fragile and compulsive, depending for its continuance on readily available and rapidly repeated brain stimulations; in other cases it was durable, being sustained even during long interreward intervals. The third was the amount of behavioral activation or pacification. There was a tendency for conflict, stimulus-dependence, and activation to go together with stimulation in the tube, but the correlation was far from perfect.

*Conflictedness.* In most cases the animal not only turned on the electric stimulation but worked to turn it off (Roberts, 1958). With probes in some places the on-behavior was stronger, in other places the off-behavior was stronger. When the off-behavior was very strong, the animal would escape from the test chamber if a way out was provided, but would self-stimulate at a substantial rate if locked in. Supporting the hypothesis of a mixed negative factor was the fact that stimulation of the axons of the wheellike neurons, after they had departed from the tube toward the thalamus and central gray, caused escape behavior without any sign of reward (M. E. Olds and J. Olds, 1963). Continuous trains of stimulation applied to these axons halted self-stimulation behavior via probes in the tube (M. E. Olds and J. Olds, 1962). The character of this negative effect was unclear. The animal would not make anticipatory avoidance responses, but would wait until the stimulus began and then shut it off. It was attractive to suppose that stimulation of these axons was neither rewarding nor aversive but "driving"; that is, after the stimulation had continued for a moment, the behavior to turn it off became compelling.

When rewarding stimulation was applied to some reward areas outside the tube, the off-behavior was less apparent and in some cases absent altogether (e.g., cingulate cortex).

*Priming.* The stimulus-dependent character of some brain-reward behavior was best evidenced by the need for priming (Gallistel, 1973; Deutsch, 1963). The animal sometimes needed a prior brain stimulation or some special signal to get the behavior started. In these cases rapidly repetitive behaviors were self-sustaining, but if the animal had to wait between stimuli the behavior died out. Therefore the behavior appeared compulsive, the more so because the animal could usually not be distracted while self-stimulating. As a consequence, many features of reward behavior were absent: the animals would not respond many times for a single reward, and they would not run mazes or cross obstructions.

It was attractive to suppose that in these cases the only component of reward was the behavior-repetition one, with the striving feature absent.

With probes in other places no priming was required (Olds, 1956b). Animals would strive to move toward the locations and the tools of brain reward. They crossed obstructions and solved mazes (Olds, 1962), and error-free maze performance occurred many hours, sometimes even days, after the last previous brain stimulation.

Thus, while there were some places where only the behavior-repetition component of reward behavior was present, there were other places where object-striving was also present and a much more normal pattern appeared.

*Activation.* When stimulation was applied in those parts of the tube with a strong negative factor, there was behavioral activation during stimulation. But often the animal's disposition became quiet and gentle when the stimulus was withdrawn or turned off. When stimulation was applied in some rewarding places outside the tube, there was arrest of behavior during the stimulation, often followed by a rebound of heightened activity containing components of fear and aggression when the stimulus was turned off (Olds, 1962).

Thus electrical stimulation in the tube caused a mixture of effects that was less pronounced or absent altogether when stimulation was applied elsewhere. Stimulation in the tube produced activity (like a drive), caused behavior repetition (but not striving), and was mixed with a negative counterforce (not quite like pain). Outside the tube the stimulation was quieting. It caused both behavior-repetition and striving, and there was little or no negative counterforce.

## Drives

The relation of the basic drives to brain reward supported the view that stimulation in the tube was having two effects that were partly synergistic and partly opposed.

The parts of the brain designated "feeding" and "satiety" centers on the basis of other mapping experiments made up a small subset of the areas in the lateral tube and the neighboring medial hypothalamus where electric stimulation caused reward (Hoebel and Teitelbaum, 1962; Margules and Olds, 1962). A thermostat-like mechanism had been discovered in the medial and lateral hypothalamus (Anand and Brobeck, 1951). The medial (satiety) area appeared to detect an upper limit of food stores and cut off feeding, while the lateral (feeding) center appeared to detect a lower limit and turn on feeding. The main data were that medial lesions caused animals to become obese and that some lateral lesions caused animals to starve. It was generally supposed that the satiety region acted by inhibiting the feeding center. But there was evidence that both centers were independently connected to a more remote "food-reinforcement" center where medial messages caused food to be aversive and lateral messages caused it to be attractive (Ellison, Sorenson, and Jacobs, 1970).

Stimulation in both centers caused escape and reward behavior. In the medial satiety center the current caused somewhat more escape; the medial reward was moderate, but it had a striving component (like other regions outside the tube). In the lateral feeding center stimulation caused more reward behavior, but it was the active and compulsive kind characteristic of stimulation in the tube.

Fitting in with the view that the lateral area was a feeding center, stimulation here often caused eating behavior (Miller, 1957). This combined with the rewarding effect to present a paradoxical picture: the same brain stimulation had both the driving effect of hunger and the rewarding effect of food. In some cases this caused two apparently different drives to alternate rapidly. One, directed to food, occurred when a continuous train was applied. The other, directed to places and pedals of brain reward, occurred whenever the train was stopped. The drive appeared equally high in both cases, but there was rapid change of object when the electrical train was stopped (Gallistel, 1973). The pursuit of places and pedals of reward was learned, of course, but it was originally thought that the stimulated feeding behavior was unlearned.

Valenstein, however, presented strong evidence that learning was also importantly involved in the stimulus-bound feeding behavior and in all the other drive behaviors that could be evoked by stimulating parts of the tube (Valenstein, 1973). He showed that if the brain stimulus was applied in interrupted trains in the presence of a drive object, the onset of the train would gradually come to evoke the appropriate drive. At the least it would evoke consummatory and instrumental behaviors directed to the drive object. If the object was food, the stimulus appeared to evoke hunger; if the object was water, the stimulus appeared to evoke thirst.

Because this learning occurred when the train was repeatedly turned on and then off, it was not clear whether the onset, the offset, or the continuation of the train was most important for this special kind of learning. It seemed possible, however, that when the cessation of the brain-stimulated drive followed regularly after feeding, this caused it to be pointed toward feeding. It would be reasonable for such cessation to cause food to become an object of the drive evoked by the brain stimulus.

If so, then two features of reward, striving vs. behavior-repetition, were dissociated by electric stimulation in the tube. During stimulation there was striving aimed at objects associated with stimulus-cessation. During periods without stimulation there was repetition of behaviors that led to objects and responses associated with the onset of the stimulus. Thus the part of the reward that caused behavior-repetition was applied at the onset of the brain signal, and the part of the reward that caused object-striving was applied at the offset of the signal. When stimulation was applied outside the tube, there was no evidence of this dissociation.

The drive experiments therefore furthered the view that brain stimulation in the tube might well have two influences: one driving, the other rewarding. But they added the notion that the rewarding effect of stimulus onset might cause mainly repetition of actions, and the rewarding or drive-reducing effect of stimulus offset might cause object-striving.

### Pharmacology

A proposal of Larry Stein (1964a, 1968) that noradrenaline might be especially involved in brain reward has continued to gain ground despite setbacks and good counterarguments. Antipsychotic drugs, such as chlorpromazine and reserpine, do counteract brain rewards and have known actions against noradrenaline (Olds, Killam, and Bach-y-Rita, 1956). However, this fact does not point very strongly to a relation between noradrenaline and brain reward because the spectrum of effects is broad: the drugs are equally potent against almost all other purposive behavior, even avoidance, and they have other actions than those against noradrenaline.

Similar arguments apply to experiments of Stein and others with amphetamine and drugs that amplify noradrenaline messages (Stein, 1964b). When brain-reward behavior is weak or slow (at threshold), amphetamine has a large potentiating effect, but this does not establish a strong relation between noradrenaline and reward because other slow purposive behaviors are also greatly accelerated by amphetamine (Dews, 1958). The argument could be made that all purposive behaviors involve the reward system, and the question thus arose whether non-purposive behaviors would be equally augmented. The answer was no. Pedal behavior was augmented if it was "rewarded" with subthreshold brain stimulation, but not if there was no reward at all (M. E. Olds, 1972).

Marianne Olds (1972) also showed that amphetamine could reverse a brain-

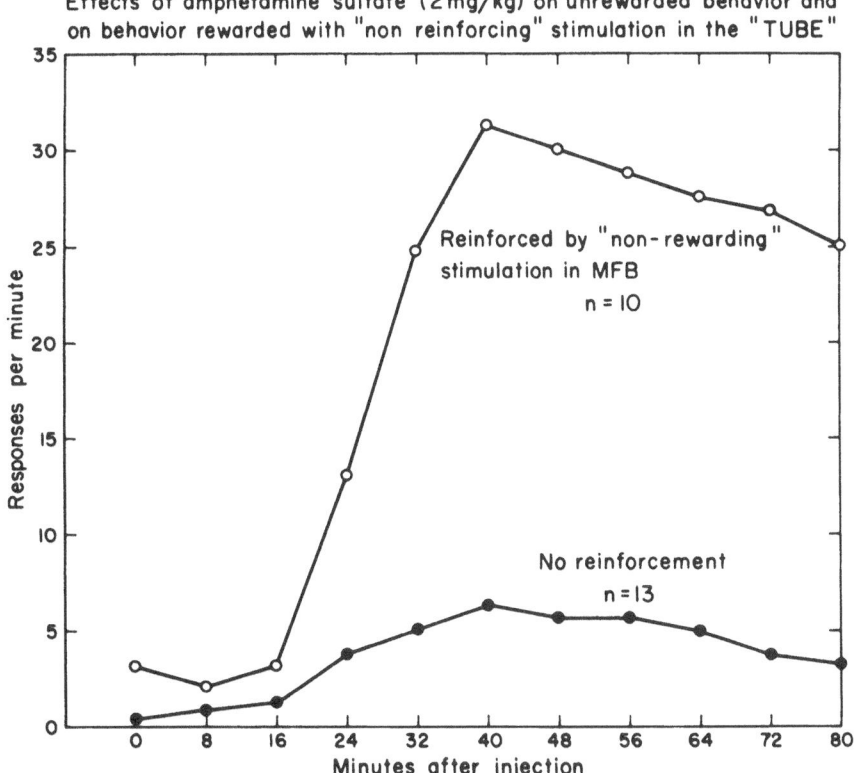

Figure 22.2   Effects of amphetamine (d-amphetamine sulfate, 2 mg/kg, i.p.) on pedal rates at various times after injection. Open circles represent data from ten animals reinforced by electrical stimulation in the medial forebrain bundle (MFB) (0.25-sec trains, 60-Hz sine-wave stimulation, 50 microamperes rms); these animals were selected because, even though the probes were "correctly" placed, there was no sign of reward behavior prior to the drug. Closed circles represent data from thirteen control animals that were not reinforced. Redrawn from M.E. Olds (1972).

reward deficit (see Figure 22.2). This deficit was evidenced when brain-reward tests failed even though the probes were correctly placed. It was as if some constitutional deficiency had caused the failure. In all cases the failure was reversed by the drug. Therefore, it is possible to suppose that a chemical constituent of the noradrenaline system might sometimes be in marginal supply, causing the brain-reward test to fail. Amphetamine would counteract this condition.

Support for the noradrenaline theory also came from the anatomical evidence of Fuxe (1965) showing noradrenaline-containing fibers to be present in the brain-reward pathways. This evidence was not specific to noradrenaline because other monoamine fibers were also present in most of the locations. The case for noradrenaline was supported by the rewarding effects of stimulation in the medulla at or near the site of origin of noradrenaline fibers (Crow, Spear, and Arbuthnott, 1972). Evidence for the role of another monoamine, dopamine, was also offered, showing that stimulation at the origin of dopamine neurons was also positive (Routtenberg and Malsbury, 1969; Crow, 1972).

My wife has bolstered the noradrenaline position still further by showing that animals with lesions in the noradrenaline sectors of the medulla appear to suffer the brain-reward deficit described above. They failed brain-reward tests even though the probes were correctly placed, but the failure was reversed by amphetamine. Thus the same kind of deficit that sometimes occurred naturally could be induced by deleting a large number of noradrenaline-containing fibers from the bundle.

Other evidence pointing even more directly to a role for noradrenaline in the brain-reward phenomenon came from direct application of this substance in the ventricle. Potentiating effects, such as those of amphetamine, were produced by noradrenaline itself. Because this substance is stopped by the blood-brain barrier, it did not act when applied by normal routes of administration. But it did work when it was pipetted in very small quantities into the lateral ventricle of the brain. This effect, reported first by Stein's group (Wise, Berger, and Stein, 1973), is difficult to demonstrate because the quantities of noradrenaline are critical: if they are raised too high, the effect is reversed. Although the experiments have been repeated successfully in a skeptical laboratory—my wife's—I still have a reservation. It might be that the noradrenaline, applied as it was in the lateral ventricle, inhibited the adjacent septal area and thereby released the reward behavior (lesions in the septal area also have this effect, as demonstrated by Keesey and Powley, 1968). Nevertheless, it seems likely to me that Stein has pointed correctly to some special role for noradrenaline.

Some indirect corroboration came from a different source. Proacetylcholine drugs such as physostigmine caused brain-reward behavior to be interrupted (Stark and Boyd, 1963). It could then be restored by antiacetylcholine drugs such as atropine. The negative influence of excessive acetylcholine furthered the view that different neurohumors might have general roles in emotional processes. Previous evidence of antagonism between acetylcholine and noradrenaline in the peripheral nervous system made these findings compatible with Stein's hypothesis of a positive role for noradrenaline.

I originally disagreed with the theory because direct application of noradrenaline in the tube suppressed behavior (Olds et al., 1964). But I have changed my mind. Noradrenaline often suppresses neuronal activity, and this may be its main function in the central nervous system; if it has a role in reward, this may be its mechanism of action. Directly applied in the tube it might suppress too many things, but applied in the ventricle it might have a more nearly normal influence.

My view now is that noradrenaline, acting as an inhibitory neurohumor, may well play a critical part in the control of behavior by reward. If it is true that there are both drive and reward elements in the tube, the pharmacological data make it attractive to suppose that the inhibitory noradrenaline fibers carry the reward message, and that one of their most important actions is to bring inhibition to play upon the wheellike "drive" neurons.

## Electrophysiology

The wheellike neurons were not turned on by rewards (Hamburg, 1971). They were turned off during consummatory behavior, except in special cases when there was some other behavior at the same time. If the animal was both striving and eating at the same time, then there was activity of these units. But if the striving stopped, and only eating continued, these elements became silent.

Muneyuki Ito (1972) later discovered that a surprisingly frequent suppression of these neurons was caused by "rewarding" brain stimulation when this was applied near the site of recording (see Figure 22.3).

Ito and my wife then showed that when one of the antinoradrenaline drugs, tetrabenazine, halted self-stimulation, it also interfered selectively with the suppression of these neurons. It left other recorded effects of the brain stimulation

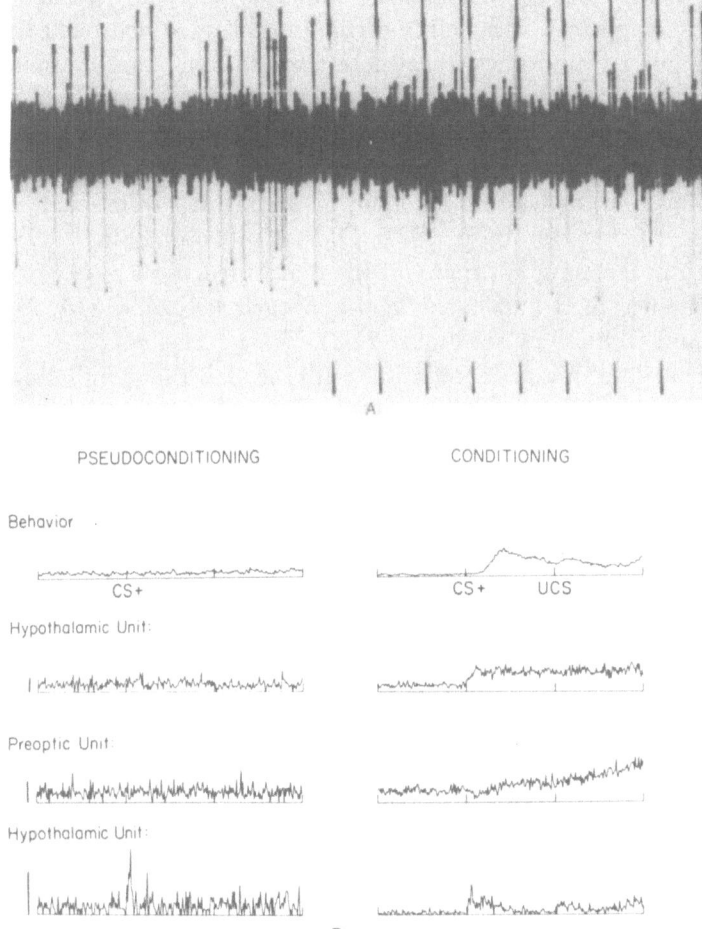

Figure 22.3  (A) Action potentials from path neurons (spikes on the left of the picture) stopped by a train of rewarding stimuli (larger spikes on the right). The stimuli were biphasic square waves, 300 microsec negative, 300 microsec interphase, 300 microsec positive (4 volts, approximately 500 microamperes), repeated at 40 Hz. The electrical stimulation was self-administered by a pedal response. The large spikelike stimulus artifacts obscured some parts of the recording, and one or more spikes often appeared between the stimuli of a train. However, the generally suppressive influence of the rewarding brain shocks on the action potentials was clearly observed. From Ito (1972). (B) Unconditioned (left) and conditioned (right) responses for gross behavior (top line) and three path neurons (lower three lines). The path neurons often had no response prior to conditioning and substantial response when the auditory signal signified food (see line 2). The base lines are 3 sec in length (marked in 1-sec intervals). The amplitude calibration at the left of the unit tracings stands for a spike rate of 5/sec. The behavior is gross movement measured automatically in arbitrary units. From Linseman and Olds (1973).

more or less intact (M. E. Olds and Ito, 1973). This made it appear that suppression of the local neurons was important. However, another antinoradrenaline drug, chlorpromazine, which also halted self-stimulation, left the suppression of nearby neurons unaffected.

Until further notice I assume that the work of Hamburg and of Ito and Marianne Olds taken together establishes a presumption that there are some neurons in the tube (some of the wheellike neurons) that are turned off by rewarding brain stimulation and also turned off during normal consummatory behavior (particularly if instrumental driving ceases). Maybe the turning off of these neurons can by itself be reinforcing, but I would guess that there are also other mechanisms involved.

The question arises, what turns on these wheellike neurons? We do not yet know what "unconditioned stimulus" is involved. Often they are active in a hungry animal that is looking for food, or waiting for food, or performing instrumental behaviors directed toward food (Hamburg, 1971; Olds, Mink, and Best, 1969). We do know one important fact: a great many of them are turned on by conditioned stimuli that announce food and cause food-approach behaviors (M. E. Olds, 1973; Linseman and Olds, 1973; see Figure 22.3).

The electrophysiological data thus are compatible with the view that conflicting systems coinhabit the tube. The large wheellike neurons were stopped by stimulating the fiber systems that engulf them. They were also silenced when striving stopped and consummatory behavior began. They were turned on by sensory signals, but these worked well only after associative training made them into conditioned stimuli that evoked anticipation and striving toward goals.

## A summary of my view of the brain-reward research

All of the structures where electrical stimulation has proved rewarding have two things in common: they send messages toward wheellike neurons in a tubular structure at the base of the brain, and they contain a set of noradrenaline fibers that have mainly inhibitory effects in the central nervous system. It seems possible, therefore, that inhibition of the wheellike neurons by noradrenaline axons might be at the root of reward.

In the tube the stimulation is often rewarding *and* aversive, the behavior provoked is often compulsive rather than purposive, and the animal is activated rather than quieted. Outside the tube the rewarding stimulation is less aversive, there is purposive striving in maze experiments, and the animal is quieted.

Stimulation in the tube causes paradoxical drive and reward behaviors and a dissociation of behavior-repetition and object-striving. Behaviors are repeated if they accompanied the onset of the stimulus, and drive objects are pursued if they accompanied its offset.

Both real consummatory behaviors and brain rewards are correlated with a cessation of firing in the wheellike neurons of the tube. The activity of these neurons is accelerated, however, by conditioned stimuli that trigger striving behaviors.

These data lead me to suppose that the wheellike neurons are a special kind of drive neuron; you might call them "learned-drive" neurons. These neurons, according to my theory, would become coupled on their dendrite side to motor axons so that the correlated actions would gain a source of power. On their axonal side they would become attached to sensory-system dendrites so that the correlated sensory stimuli would become objects of striving (see Figure 22.4). As part of a "self-wiring" system, these neurons would be actively coupled on input and output sides to ongoing brain processes. The rule would be that any sudden deceleration in the drive-neuron activity would cause them to couple in both directions. If the axons of such neurons were mixed in the tube with those of their monoamine inhibitors, then the electrical brain stimulus would have a paradoxical influence. It might silence the somatodendritic surface of the wheellike elements by activating the monoamine fibers, while sending false positive signals down their axons by activating these directly. Thus, at the onset of the train, spiking at the dendrite end

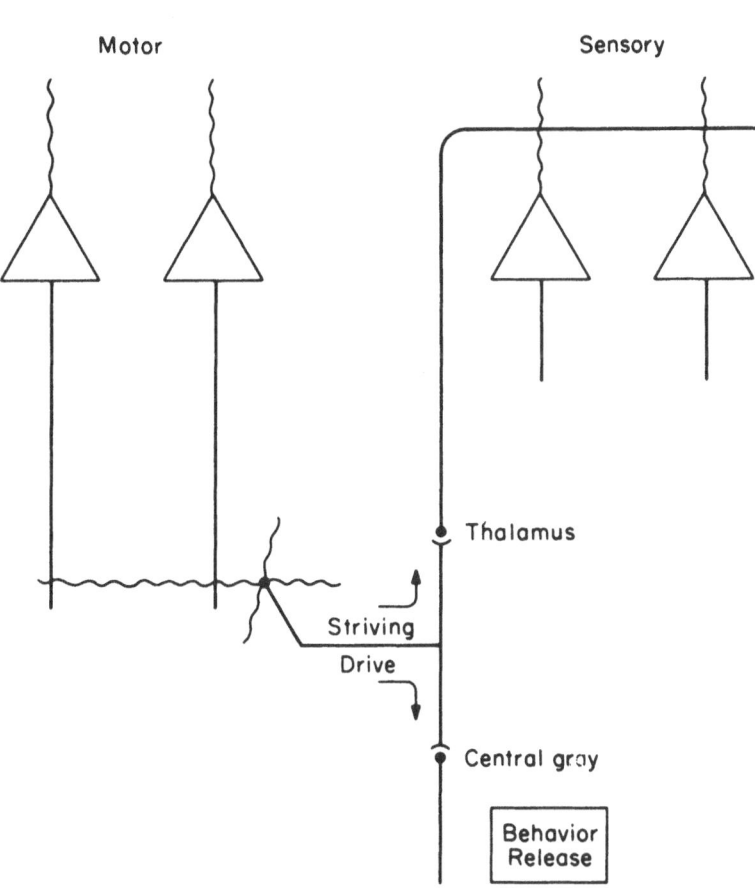

Figure 22.4   A theory of drive-neuron function. Connection of motor axons to drive-neuron dendrites, and of thalamus afferents to sensory dendrites, are considered to be plastic. When motor or sensory activity coincides with a deceleration in drive-neuron activity, these connections are reinforced. Later motor neurons with these connections can muster "drive support" for their correlated behaviors; the correlates of sensory neurons with these connections become objects of striving when the drive is active.

of the drive neurons would be decelerated, and at the offset of the train, spikes at the axonal end would subside. This *could* cause a compulsive behavior-repetition of actions accompanying the onset, and an object-striving toward objects accompanying the offset.

## Learning Experiments

My second mapping enterprise has been to some degree independent of the brain-reward research. It is part of a continuing effort in neuropsychology to make learning and cognitive processing of information amenable to study by means of implanted brain probes.

The method has been to apply an auditory signal with a sharp onset, and to track the arrival of its message, in successive time frames, through the various stations of the brain (Olds et al., 1972). By this method it is possible to establish a message map portraying the course of the signal through the brain in time and space. In my experiments I make one message map for a signal when it is habituated and meaningless during a control experiment. Then I make successive series of message maps for the different phases of a Pavlovian conditioning experiment, during a period when the meaning of the auditory signal is changed by applying it one second prior to food on a regular basis (see Figure 22.5). Because the animal is hungry, the auditory signal begins to elicit instrumental behaviors directed to the food.

By overlaying successive message maps it is possible to trace out a family of changes that succeed one another during the course of training. For each of these changes it is possible to gain some indication of where the message branched from its old pathway into a new one as a consequence of the training procedure.

These experiments have been my main concern for four years, and their underpinnings reach back at least another four. Nevertheless, they are still "just getting under way." The slow start is no cause for pessimism. To develop a careful method aimed at characterizing a large-scale network is a goal worth a great deal of time, and it is a goal that now seems within our grasp. I will give some tentative findings that need more replication to be completely verified. What they already show is that the method can get at the critical facts about this kind of network.

### Phases of learning

For the method to work, one requirement is that it separate the different learning events in a neuronal chain by showing them to occur at different phases of learning. Not only would this be an interesting finding in itself, but methodologically it would separate out the different links in the chain to show where the plastic coupling points lie.

Learning was found to occur in a sequence of phases. At first the animal continued to ignore the habituated signal. Later the animal paused when the stimulus was applied and looked toward it. After this the signal began to cause overt behavior directed in a slow and awkward way toward the food tray. Then this movement became fast and well directed. Finally, after obvious behavior improvement was completed, the behavior became so ingrained it was difficult to untrain in later procedures.

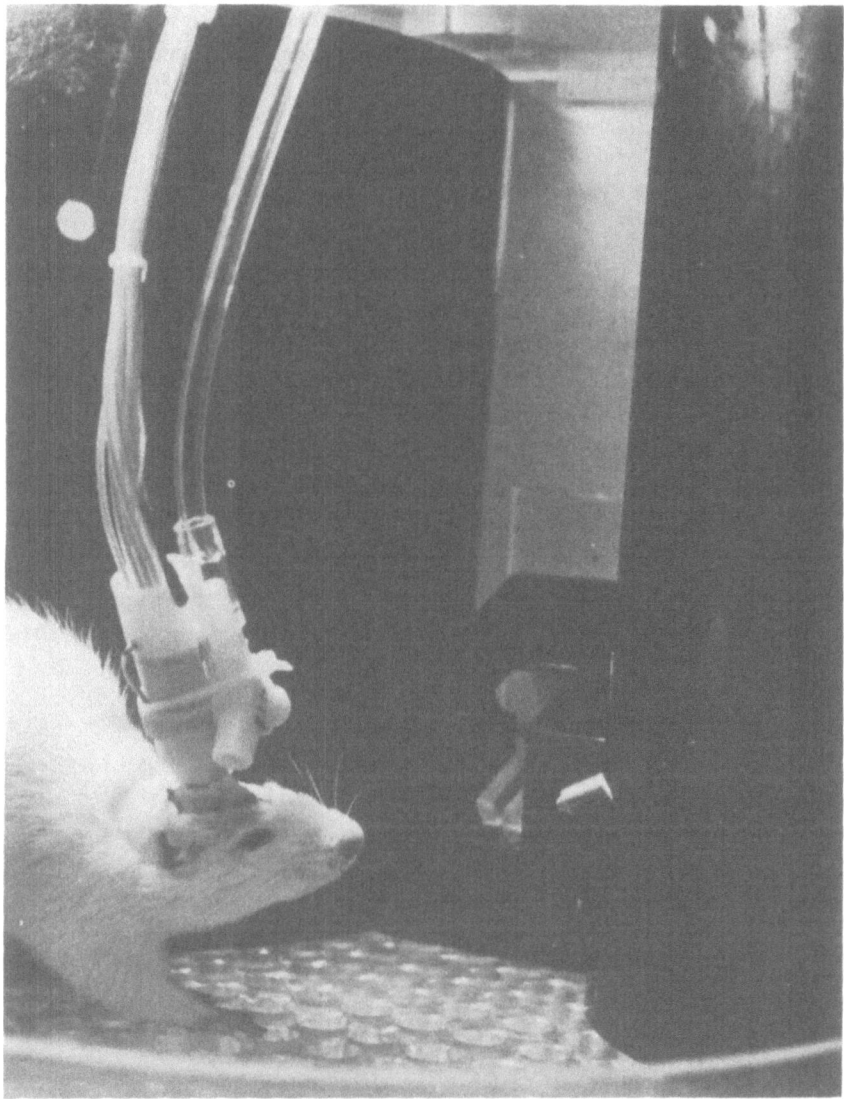

Figure 22.5   A rat prepared for the learning experiments. Cables carry unit responses to recording devices; there is also a noisy cable that detects gross movements. For a "stabilized auditory image," the clear plastic tube brings auditory signals down to the Y-shaped horn; thus the animal's movements cannot change its relation to the auditory signal source. On the right is the food chute, where reward is automatically presented, and the water bottle spout which is continuously available.

Our data suggested that different parts of the brain were involved in these phases (Olds, 1973a). First, while renewed interest was attaching itself to the signal, the only new brain responses to appear were observed in the hypothalamus and the related "emotion" centers of the brain. The changes in these centers came before there were any signs of changed behavior.

Second, when the animal began to look toward the source of the signal, there were new responses in the brain's "activating system"—the arousal systems of Moruzzi and Magoun (1949), Jasper (1949), and Lindsley (1951). At this time there were also new responses in parts of the thalamus related to movement.

Third, when the behavior became purposive, there were new accelerations of neuron activity in the central motor systems of the brain (i.e., the extrapyramidal systems) and also new responses in certain "association" centers of the thalamus.

Fourth, when behavior became faster and better organized, new responses appeared for the first time in the auditory cortex. Finally, some time after behavior improvement was completed, new responses appeared for the first time in the frontal lobes.

While these data validated some older ideas about the different functions of different areas, their main importance was to point out one feature of the method: it required that different brain areas "learn" at different rates. They did so, and the differences were large.

**Auditory tuning**

Training caused the signal to branch off into new pathways at almost every auditory station. It also caused preexisting responses in the auditory stations to be modified, usually to be amplified or enhanced (J. F. Disterhoft and D. K. Stuart, unpublished observations; J. Olds, unpublished observation; see Figure 22.6). There was also a substantial change in the background firing rate, the so-called

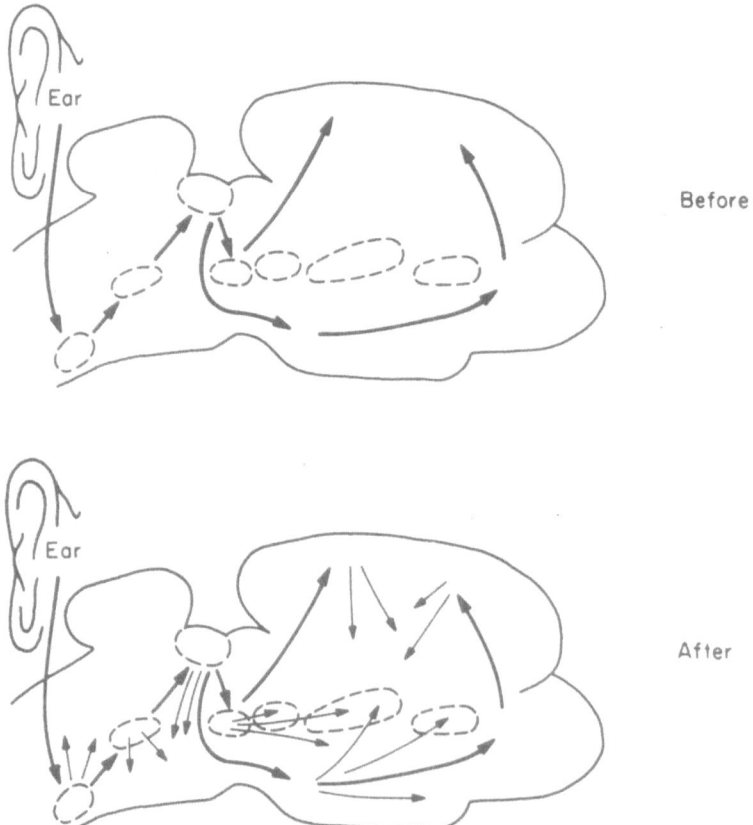

Figure 22.6  Paths of the auditory message in the rat brain before and after conditioning. The smaller arrows in the lower picture show that the message branches out into adjacent "nonspecific" centers from all the primary auditory stations.

spontaneous discharge rate of neurons, in some of the auditory stations (Disterhoft and Olds, 1972).

These enhancements extended at least as far down as the cochlear nucleus (see Figure 22.7) and may have even characterized recordings from the eighth nerve (J. Olds, unpublished observation). They showed up in the very first component of the cochlear-nucleus response, which appeared one to three milliseconds after the auditory signal reached the ear. They did not result merely from changing the relation of the ear to the auditory sound source because we used something like a

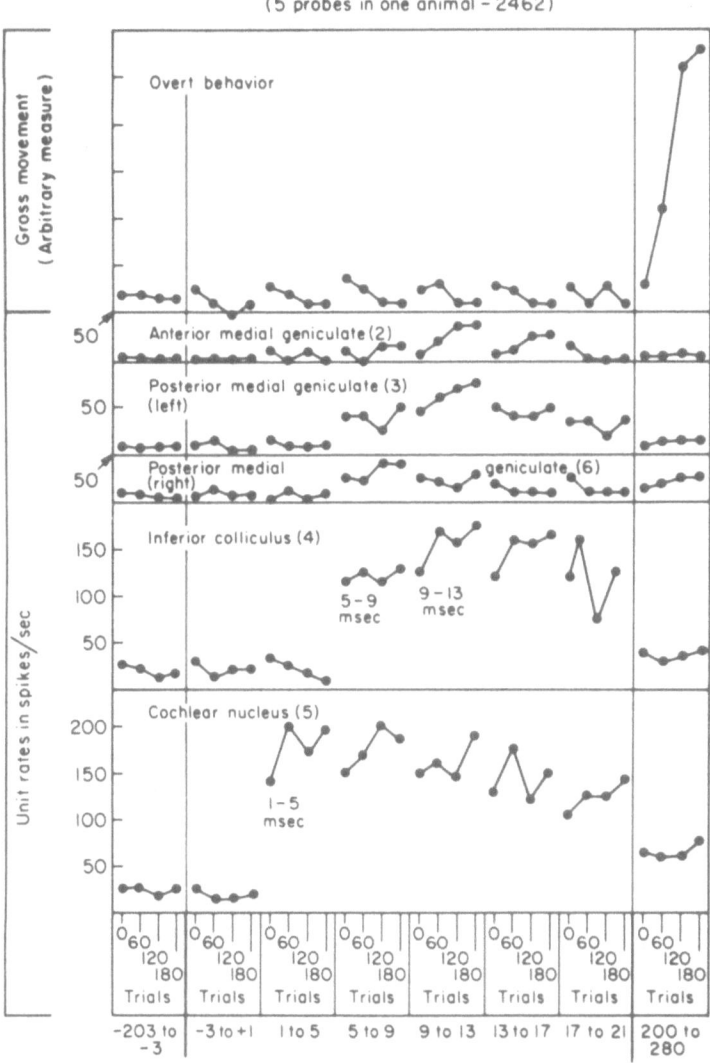

Figure 22.7   Receptor tuning. Spike rates in successive 4-msec time bins after the onset of the auditory CS+. The response occurred before the fifth msec in the cochlear nucleus, before the ninth msec in the inferior colliculus and medial geniculate. Training caused a 30% increase in the very first part of the cochlear-nucleus response, about a 50% increase in the second part of the colliculus response, and a de novo response in the anterior part of the medial geniculate. In other cases not shown here very large response changes appeared in the earliest part of the inferior-colliculus response. From unpublished observations of J.F. Disterhoft and D.K. Stuart.

stabilized auditory image (see Figure 22.5). It therefore appeared that the auditory system was ready for, that is, tuned for, these signals before they arrived at the ear. It was appealing to suppose that this readiness was maintained by a dynamic memory process in the cortex. Similar changes (with different latencies, of course) occurred at the several stations of the auditory pathway. While it was possible that the higher-auditory-center changes were merely sharpened versions of the lower ones, several features of the changes made this difficult to accept. First, the changes in higher centers were much larger in proportion to the total response; and second, the changes in different centers often occurred at different stages of conditioning. In some experiments changes in the inferior colliculus and cochlear nucleus were completed during the first sixty trials while changes in the medial geniculate continued into the second and even the third set of sixty trials. I guess, therefore, that the changes in the higher auditory centers were not directly caused by the smaller changes in the lower centers, or even by the same dynamic change in the cortex that modified the lower centers. Nevertheless, because there were changes in cortex background firing rate at this time, it was fair to suppose that "dynamic memory processes" in the cortex caused many of the response changes in the lower centers.

### Dynamic and structural changes in cortex

The changes in cortex caused by training were complex (see Figure 22.8). In most of the lower centers training caused brief and insignificant changes in background firing rates and left them well within control ranges. The main effect of training in

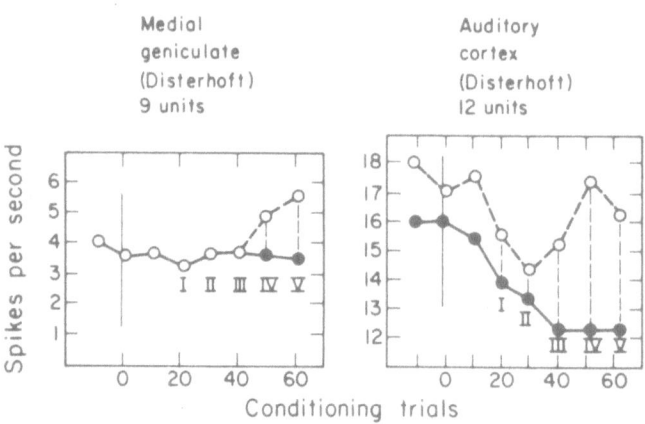

Figure 22.8   One kind of learning in the medial geniculate and two kinds in the auditory cortex. Filled circles show background firing rates (when these deviated from rates after stimulation). Open circles show rates after stimulation (during the 1-sec interval between conditioned and unconditioned stimulus). In the geniculate cases shown here, the background rate was equal to the rate after stimulation until trials 40–50, when a conditioned response began to appear. In the auditory-cortex cases there was an unconditioned response to the auditory signal (amounting to about 12% of the background rate). Training caused a concurrent change in background and stimulated response rates: there was a steady decline starting in the first 10 trials and continuing through trial 40. This formed the first kind of learning. When this was almost complete, the second kind began: a widening of the percentage gap between the background rate and the stimulated rate. From Disterhoft and Olds (1972).

these cases was to cause a widening gap between the firing rate produced directly by stimulation and the background firing rate. Only in the cortex was there a stable and sustained change in the background firing rate itself. It is interesting that during the first two stages of training, this was the only change observed in the cortex. There was about a 25% decline in the average spiking rate of these auditory cortex neurons. This decline appeared before and after stimulus onset. Then, after training had progressed, the responses to the auditory signal began to rise, even though the background rates remained depressed. It was appealing to suppose, therefore, that these neurons in the auditory cortex first reduced their background rate, or perhaps their connection to other irrelevant noises, and then increased their connection to the special new auditory signals. The decreased spontaneous discharge rate might be part of an early and temporary memory process involved in causing changed neuronal responses in other parts of the brain. At the same time as it caused this background change, the training procedure might plant the seeds of a new structural connectivity, so that later this newly significant auditory signal would influence a larger set of cortical neurons.

To get at the hypothesis of structural growth during time, it was necessary to have an experiment in which behavior not only improved during training, but continued to improve during an extended time-out period. We found that if we first trained animals to respond positively to one auditory signal, called the CS +, and to ignore a second auditory signal, called CS −, and then reversed the significance of these two, the behavior followed an appropriate course (J. R. Disterhoft, M. Segal, and J. Olds, unpublished observation). After the switch, behavior improved in response to the new CS + and deteriorated in response to the old one. But at the end of an eight-hour training series, the responding to the two signals was at a middle level, about equal. If an eight-hour time-out period was then interpolated, there was great improvement. The response to the new CS + was greatly augmented, and the responses to the old one disappeared almost altogether. The changes that occurred in cortex during the time-out interval substantiated the view that there was some kind of growth process going on there. The response of neurons to the new CS + that had been "planted" by training was substantially augmented by the time-out period, just as was the behavioral response. Meanwhile the response to the old CS + that had been to some degree extinguished during training sprang partly back to life, like a weed. Thus, while the cortex changes could account for the behavioral improvement in response to the CS + during the time-out, it could not account for the equally adaptive behavior loss in response to the CS −. A reduction in response to the CS −, without any changed response to CS +, occurred during the time-out period in neurons of the hippocampus and nonspecific thalamus in the same experiment. This experiment suggested, therefore, that in this particular case some structural consolidation of a newly acquired positive response occurred in the cortex, and some similar consolidation of extinction might have occurred in the hippocampus or the thalamus or both.

Many repetitions are still required before the "facts" about dynamic and structural changes become "true." But these experiments showed that there is a way to define the two kinds by unit-recording experiments.

### Reinforcement learning in the hippocampus

Reward and punishment enter the problem of learning in many ways. For higher learning the role of reward is at least twofold: reward or some alerting event is required to "turn on the learning machine," and reward enters again to determine what behaviors will be repeated and what ideas will be rehearsed.

The hippocampus ties processed information of the association cortex to attentional and motivational centers of the lower brain. Therefore it may be especially involved in critical interactions between motivation and learning. Our experiments tracking a conditioned stimulus through the hippocampus fit this view. One particular family of neurons in the hippocampal system seems to be involved mainly in turning on the hippocampus as a learning machine during training. The order of firing in the fully trained animal would then be compatible with the view that the same family is later involved in causing remembered behaviors to be performed.

The main family of neurons in the hippocampus, the CA-3 elements, are arranged in a fashion that matches a computer memory grid. Because they also have marked conditioned responses, I call them the memory neurons. This grid of elements is fed by four different warps of fibers, possibly bringing information to be remembered. Three of these come from, respectively, the cortex, the drive system, and the arousal system; the fourth comes from a neighboring family, the dentate granules. Because their main input is from the drive system, I will call them the "motive" set of neurons. The drive-system messages have direct access to the memory grid, but the main drive information is relayed through the motive neurons. The latter send their message only to the memory grid and nowhere else.

Our experiments suggest that the activity of the motive set is necessary to turn on the memory grid, to make it record, and that later, during playback, memories may need to trigger this motive system in order to evoke behavior.

There were three findings that pointed in this direction. The first was that the motive set learns first (Segal and Olds, 1972). Early in training the signal came to cause a briefly delayed acceleration in the motive neurons. Only after the conditioned stimulus began to turn on these motive neurons did it begin to influence the neurons of the memory grid. This intimated that these neurons might play a role in turning on the hippocampal learning machine. This hypothesis was soon supported by the further finding that when the auditory signal is associated with punishment instead of reward, the motive set of neurons becomes inhibited instead of accelerated (Segal, Disterhoft, and Olds, 1972). In this case the memory elements failed to acquire any new response at all. This finding not only corroborated the view that dentate neurons might be required to turn on the hippocampal learning machine, but added to this concept the hypothesis that because the dentate neurons are turned on only by reward signals, these neurons represent, or reflect into the hippocampus, the promise of reward.

The third finding was that after training is done, the dentate elements (the motive neurons) fire with a longer latency than the CA-3 units (the memory set) (Segal, 1973). The anatomical arrangement made this a surprising finding. The dentate neurons project only to the CA-3 neurons. The CA-3 neurons fired first, apparently acknowledging message #1 from the conditioned stimulus. Then they seemed to receive a second or reconfirming message relayed through the dentate gyrus. Why was the second message needed? What did it add?

One guess was that dentate activity represented in both stages of training some "hope for reward." Early in conditioning this is possibly provoked by diffuse pathways from the conditioned stimulus, through cingulate cortex and septal area. At this time the "hope" makes the animal attentive and turns on the hippocampal learning device. After learning has progressed, the message from the conditioned stimulus might go first to the memory device to select behavior plans; some of these (the ones previously followed by reward) might then promote themselves by their afferent connections to the dentate.

Now the message from the conditioned stimulus would feed through the "remembered behaviors" of CA-3 to the "hope of reward" in the dentate gyrus. The memory plus the hope or promise would cause the remembered behaviors to occur.

**Summary of the learning experiments**
This study of learning has not yet pushed through to the solid outcomes I expect of it, but it has guided guesses about the way the outcomes will fit together. When the number of cases is small, this kind of research gives "iffy" outcomes; it requires a very large number of repetitions before the observations can be accepted as real. The work so far has, however, validated this method of attack on the higher functions of the mammalian brain. It has done this by fitting with and adding to several ideas we had before we started. The first idea was that different brain areas perform different learning operations for different functions: hypothalamus for interest, reticular formation for arousal, extrapyramidal system for purpose. The second idea was that by this method changed responses in lower centers could be traced to active memories in the cortex (if indeed this turns out to be the mechanism involved). The third was that if training does plant the seeds of connection in some areas and then requires a growth process to make a lasting structural memory, this growth process could be tracked down by unit recordings. The fourth idea was that the order of learning and the order of firing in a "learning machine" such as the hippocampus could help to trace functions such as reward, first to turn on the attentional mechanism and later to cause behavior when training is done.

**Models of the Mind**

If maps can promote understanding, it is because the detailed wiring of a structure can help us understand a process that has been "mapped into it." This really has not happened yet, but it is worthwhile to keep in mind how it might work. The two poles that were at the crux of things when I was a graduate student can be taken to illustrate this. The question was: How do motives work? Hull's answer (that rewards work at the time of learning to stamp in connections between stimuli and responses, and that the animal later repeats a rewarded behavior because the stamped-in connections force the issue) implied a three-wire grid. A family of sensory axons was needed to carry a picture of the world. These had to intersect a family of motor dendrites so that every sensory representation would have access to every motor act. A coincidence of sensory axonal activity with motor dendrite activity would establish a very temporary connection between them; this connection would be fixed or prolonged if reward was applied to it during its moment of life. To get the reward message to the right place at the right time, a third wire was required. Interestingly, the cerebellum does indeed provide the kind of three-wire grid that this speculation requires; its detailed structure has been used by Marr (1969) in this kind of theory. The parallel fibers, which carry sensory messages, pervade the Purkinje dendrites, which influence motor output. Rising to all the intersection points from a small focal area in the medulla (or a small set of focal areas, possibly including the noradrenaline centers) are the climbing fibers. They entwine the dendrite trees like vines, bringing them into close proximity with all the "sensorimotor" synapses. They could therefore bring some general influence to bear on synapses that have just been "livened up" by a coincident activity of pre- and postsynaptic elements. It could be a positive influence connecting the two, as was suggested by Marr, or a negative influence disconnecting them. The learning of skills seems to involve a gradual but more or less automatic stamping in of sensorimotor integrations on the basis of satisfactory outcomes. Thus there is the possibility that Hull was right for skills and for the cerebellum. It is, of course, a very tenuous possibility, but it shows the way in which the mapping of a function into a structure would help.

Tolman's idea (that sensorimotor processes are recorded as if on a continuously recording tape and that if later playbacks reveal a reward in the succession of events, a second playback occurs with the recorded behaviors actually happening) can be modeled by a computer core memory that is a special kind of two-wire grid (Olds, 1972). It is as if the successive sensorimotor events were written into the successive memory addresses. If the computer memory were "associative" or "content-addressable," addresses could be reactivated by some part of their content. A reactivated address could reactivate its successors, and the computer could inspect them for reward recordings. Finding one, it could easily cause the correlated output sequence to occur. For a biological system based on this general principle to work, a pair of grids rather than a single core memory would be required. This is because neurons are unidirectional in a way that a core memory is not. In the biological machine sensory axons would need to cross through memory dendrites, and memory axons would need to cross through output dendrites. This arrangement would permit sensory input to reactivate memories and memories to reactivate motor output. Interestingly, the hippocampus provides this kind of double grid. I have used its structure as a basis in setting forth this kind of theory (Olds, 1972). As I said earlier, input to hippocampus arrives in four warps of axons from associative, arousal, and drive systems of the brain, and from a neighbor set of motive neurons. The memory set of neurons in hippocampus sends its dendrites into these four warps in the appropriate fashion. Axons leaving the memory set compose a similar gridlike arrangement making, *en passant*, synapses with a very large number of output elements. It would therefore be possible for the hippocampus to perform a function of this type (i.e., to remember a limited amount of material as a sequential temporal recording). The memory of the daily agenda for yesterday, today, and tomorrow seems to work in a manner like this. Thus the possibility can be entertained that Tolman was right for agenda items, and for the hippocampus.

One should mark ideas of this kind clearly to separate them from the experimental facts and working hypotheses that make up the main body of the science of the brain, and it is appropriate for one's attitude toward them to oscillate. At the low point it is clear that most of the ideas will not turn out to have any important organizing value because evidence will not converge to give them value and they will not become focal concepts among disparate research trends. At the high point each of them should be briefly treated as doubtful but possible. Even doubtful ideas are candidates for brief consideration because the path our field will travel on the road to organization is not at all clear in advance. It may be a highly informative series of experiments that marks the turning point; or it may be a set of organizing ideas with doubtful beginnings. Mendeleev pushed through to the periodic table by carefully determining atomic weights, which then put themselves in order. Watson and Crick decided to force an issue by thinking their way through to order. In any event, mere data collection is not enough to get us central ideas about the brain. The lack of a generally accepted set of focal concepts has kept our field in a backward state. But I do not agree with those who hold our field to be so basically complicated that it might not be put into much better shape by good ideas that are relatively easy to express and to understand.

## References

Anand, B.K., and Brobeck, J.R. (1951): Hypothalamic control of food intake in rats and cats. *Yale J. Biol. Med.* 24:123–140.

Crow, T. J. (1972): A map of the rat mesencephalon for electrical self-stimulation. *Brain Res.* 36:265–273.

Crow, T. J., Spear, P. J., and Arbuthnott, G. W. (1972): Intracranial self-stimulation with electrodes in the region of the locus coeruleus. *Brain Res.* 36:275–287.

Dafny, N., Bental, E., and Feldman, S. (1965): Effect of sensory stimuli on single unit activity in the posterior hypothalamus. *Electroencephalogr. Clin. Neurophysiol.* 19:256–263.

Deutsch, J. A. (1963): Learning and electrical self-stimulation of the brain. *J. Theor. Biol.* 4:193–214.

Dews, P. B. (1958): Studies on behavior. IV. Stimulant actions of methamphetamine. *J. Pharmacol. Exp. Ther.* 122:137–147.

Disterhoft, J. F., and Olds, J. (1972): Differential development of conditioned unit changes in thalamus and cortex of rat. *J. Neurophysiol.* 35:665–679.

Ellison, G.D., Sorenson, C.A., and Jacobs, B. L. (1970): Two feeding syndromes following surgical isolation of the hypothalamus in rats. *J. Comp. Physiol. Psychol.* 70:173–188.

Fuxe, K. (1965): Evidence for the existence of monoamine neurons in the central nervous system. IV. Distribution of monoamine nerve terminals in the central nervous system. *Acta Physiol. Scand.* (Suppl.) 247:37–84

Gallistel, C. R. (1973): Self-stimulation: The neurophysiology of reward and motivation. *In: The Physiological Basis of Memory.* Deutsch, J. A., ed. New York: Academic Press, pp. 175–267.

Hamburg, M. D. (1971): Hypothalamic unit activity and eating behavior. *Am. J. Physiol.* 220: 980–985.

Hebb, D. O. (1949): *The Organization of Behavior. A Neuropsychological Theory.* New York: John Wiley.

Hoebel, B. G., and Teitelbaum, P. (1962): Hypothalamic control of feeding and self-stimulation. *Science* 135:375–377.

Hull, C. L. (1943): *Principles of Behavior.* New York: Appleton-Century-Crofts.

Ito, M. (1972): Excitability of medial forebrain bundle neurons during self-stimulating behavior. *J. Neurophysiol.* 35:652–664.

Ito, M., and Olds, J. (1971): Unit activity during self-stimulation behavior. *J. Neurophysiol.* 34: 263–273.

Jasper, H. (1949): Diffuse projection systems: The integrative action of the thalamic reticular system. *Electroencephalogr. Clin. Neurophysiol.* 1:405–420.

Keesey, R. E., and Powley, T. L. (1968): Enhanced lateral hypothalamic reward sensitivity following septal lesions in the rat. *Physiol. Behav.* 3:557–562.

Lindsley, D. B. (1951): Emotion. *In: Handbook of Experimental Psychology.* Stevens, S. S., ed. New York: John Wiley, pp. 473–516.

Linseman, N. A., and Olds, J. (1973): Activity changes in rat hypothalamus, preoptic area, and striatum associated with Pavlovian conditioning. *J. Neurophysiol.* 36:1038–1050.

Margules, D. L., and Olds, J. (1962): Identical "feeding" and "rewarding" systems in the lateral hypothalamus of rats. *Science* 135:374–375.

Marr, D. (1969): A theory of cerebellar cortex. *J. Physiol.* 202:437–470.

Miller, N. E. (1957): Experiments on motivation. *Science* 126:1271–1278.

Miller, N. E., Roberts, W. W., and Delgado, J. M. R. (1953): Learning Motivated by Electrical Stimulation of the Brain (Motion picture shown by N. E. Miller at the Experimental Division of the American Psychological Assn., September 1953).

Millhouse, O. E. (1969): A Golgi study of the descending medial forebrain bundle. *Brain Res.* 15:341–363.

Moruzzi, G., and Magoun, H. W. (1949): Brain stem reticular formation and activation of the EEG. *Electroencephalogr. Clin. Neurophysiol.* 1:455–473.

Norgren, R., and Leonard, C. M. (1973): Ascending central gustatory pathways. *J. Comp. Neurol.* 150:217–237.

Olds, J. (1956a): A preliminary mapping of electrical reinforcing effects in the rat brain. *J. Comp. Physiol. Psychol.* 49:281–285.

Olds, J. (1956b): Runway and maze behavior controlled by basomedial forebrain stimulation in the rat. *J. Comp. Physiol. Psychol.* 49:507–512.

Olds, J. (1958): Self-stimulation of the brain. *Science* 127:315–324.

Olds, J. (1962): Hypothalamic substrates of reward. *Physiol. Rev.* 42:554–604.

Olds, J. (1972): Learning and the hippocampus. *Rev. Can. Biol.* 31 (Suppl.): 215–238.

Olds, J. (1973a): Brain mechanisms of reinforcement learning. *In: Pleasure, Reward, Preference. Their Nature, Determinants, and Role in Behavior.* Berlyne, D. E., and Madsen, K. B., eds. New York: Academic Press, pp. 35–63.

Olds, J. (1973b): The discovery of reward systems in the brain. *In: Brain Stimulation and Motivation.* Valenstein, E. S., ed. Glenview, Ill.: Scott, Foresman, pp. 81–99.

Olds, J., Disterhoft, J. F., Segal, M., Kornblith, C. L., and Hirsh, R. (1972): Learning centers of rat brain mapped by measuring latencies of conditioned unit responses. *J. Neurophysiol.* 35:202–219.

Olds, J., Killam, K. F., and Bach-y-Rita, P. (1956): Self-stimulation of the brain used as a screening method for tranquilizing drugs. *Science* 124:265–266.

Olds, J., and Milner, P. (1954): Positive reinforcement produced by electrical stimulation of septal area and other regions of rat brain. *J. Comp. Physiol. Psychol.* 47:419–427.

Olds, J., Mink, W. D., and Best, P. J. (1969): Single unit patterns during anticipatory behavior. *Electroencephalogr. Clin. Neurophysiol.* 26:144–158.

Olds, J., Travis, R. P., and Schwing, R. C. (1960): Topographic organization of hypothalamic self-stimulation functions. *J. Comp. Physiol. Psychol.* 53:23–32.

Olds, J., Yuwiler, A., Olds, M. E., and Yun, C. (1964): Neurohumors in hypothalamic substrates of reward. *Am. J. Physiol.* 207:242–254.

Olds, M. E. (1972): Comparative effects of amphetamine, scopolamine and chlordiazepoxide on self-stimulation behavior. *Rev. Can. Biol.* 31 (Suppl.): 25–47.

Olds, M. E. (1973): Short-term changes in the firing pattern of hypothalamic neurons during Pavlovian conditioning. *Brain Res.* 58:95–116.

Olds, M. E., and Ito, M. (1973): Noradrenergic and cholinergic action on neuronal activity during self-stimulation behavior in the rat. *Neuropharmacology* 12:525–539.

Olds, M. E., and Olds, J. (1962): Approach-escape interactions in rat brain. *Am. J. Physiol.* 203: 803–810.

Olds, M. E., and Olds, J. (1963): Approach-avoidance analysis of rat diencephalon. *J. Comp. Neurol.* 120:259–295.

Roberts, W. W. (1958): Both rewarding and punishing effects from stimulation of posterior hypothalamus of cat with same electrode at same intensity. *J. Comp. Physiol. Psychol.* 51:400–407.

Routtenberg, A. (1971): Forebrain pathways of reward in *Rattus norvegicus. J. Comp. Physiol. Psychol.* 75:269–276.

Routtenberg, A., and Malsbury, C. (1969): Brainstem pathways of reward. *J. Comp. Physiol. Psychol.* 68:22–30.

Scott, J. W., and Pfaffmann, C. (1967): Olfactory input to the hypothalamus: Electrophysiological evidence. *Science* 158:1592–1594.

Segal, M. (1973): Flow of conditioned responses in limbic telencephalic system of the rat. *J. Neurophysiol.* 36:840–854.

Segal, M., Disterhoft, J. F., and Olds, J. (1972): Hippocampal unit activity during classical aversive and appetitive conditioning. *Science* 175:792–794.

Segal, M., and Olds, J. (1972): Behavior of units in hippocampal circuit of the rat during learning. *J. Neurophysiol.* 35:680–690.

Skinner, B. F. (1938): *The Behavior of Organisms.* New York: Appleton-Century-Crofts.

Solomon, R. L., and Wynne, L. C. (1954): Traumatic avoidance learning: The principles of anxiety conservation and partial irreversibility. *Psychol. Rev.* 61:353–385.

Stark, P., and Boyd, E. S. (1963): Effects of cholinergic drugs on hypothalamic self-stimulation response rates of dogs. *Am. J. Physiol.* 205:745–748.

Stein, L. (1964a): Amphetamine and neural reward mechanisms. *In: Animal Behaviour and Drug Action* (Ciba Foundation Symposium). Steinberg, H., ed. Boston: Little, Brown, pp. 91–118.

Stein, L. (1964b): Effects and interactions of imipramine, chlorpromazine, reserpine, and amphetamine on self-stimulation: Possible neurophysiological basis of depression. *In: Recent Advances in Biological Psychiatry.* Wortis, J., ed. New York: Plenum Press, pp. 288–308.

Stein, L. (1968): Chemistry of reward and punishment. *In: Psychopharmacology. A Review of Progress, 1957–1967.* Efron, D. H., editor-in-chief. Washington, D. C.: U. S. Government Printing Office, pp. 105–123.

Tolman, E. C. (1949): *Purposive Behavior in Animals and Men.* Berkeley: University of California Press.

Valenstein, E. S. (1973): Commentary. *In: Brain Stimulation and Motivation.* Valenstein, E. S., ed. Glenview, Ill.: Scott, Foresman, pp. 162–172.

Wise, C. D., Berger, B. D., and Stein, L. (1973): Evidence of a-noradrenergic reward receptors and serotonergic punishment receptors in the rat brain. *Biol. Psychiatry* 6:3–21.

# The Brain: Neural Object and Conscious Subject

Herbert H. Jasper (b. 1906, La Grande, Oregon) is professor of neurophysiology at the University of Montreal, where he directs a coordinated, multidisciplinary program of research on all aspects of the function of the central nervous system. His research career has been associated closely with the study of the electrical activity of the brain. He founded and was first editor of the *International Journal of Electroencephalography and Clinical Neurophysiology* and served as first executive secretary of the International Brain Research Organization of UNESCO.

# 23
## Philosophy or Physics—Mind or Molecules

## Herbert H. Jasper

It is with particular pleasure that I take the opportunity to make a contribution to this symposium in honor of Frank Schmitt, for he is one of a small but distinguished group of neuroscientists who have had a considerable influence upon my own career over the past forty years. Frank (and his brother Otto) were among the few physical scientists in the early 1930s who were introducing more sophisticated techniques to analyze the structural properties of nerve membranes at the molecular level and to understand their electrical properties in terms of changes in impedance and specific ion flux. Little did I know at the time that Frank and I shared the same ultimate objectives, my path starting from philosophy and his from physics. These objectives have become abundantly manifest in the work of the Neurosciences Research Program, expressed by Frank in his introduction to *The Neurosciences: Second Study Program* as follows:

> If the physical basis of brain function were better understood, substantial progress could be made in the alleviation of mental ills and in the search for an understanding of the nature of man as a cognitive individual. Concomitantly, new dimensions of mental capability would be available to solve the pressing survival problems facing man today and to open up unexpected opportunities of human accomplishment. . . .
> One of the NRP's purposes is to scan the horizon for physical and biological theories that could lead to key discoveries that may resolve some of the mysteries and complexities of neuroscience, including those of the relationship of molecules to mind. (Schmitt, 1970)

The title of this account of my own rather devious path of exploration in the neurosciences is intended to represent our common concern, approached originally from different points of view: Philosophy or Physics, Mind or Molecules.

Among others who had a great deal to do with the shaping of my career in the neurosciences in the early 1930s were George Bishop, Herbert Gasser, Joseph Erlanger, Lorente de Nó, Lord Adrian, Sir Charles Sherrington, Sir John Eccles, Det Bronk, A. V. Hill, Ralph Gerard, K. C. Cole, Alex Forbes, Hal Davis, Ali Monnier, Alfred Fessard, and Louis Lapicque. Adding spice to this exciting era of neuroscience was the young renegade Rushton, enthusiastically attacking Lapicque's theory of "chronaxie," the deathblow being delivered by Sir Henry Dale and Feldberg with the establishment of chemical transmission at the neuromuscular junction and at synapses in sympathetic ganglion. Heated debates between Jack Eccles and Henry Dale enlivened all meetings of the Physiological Society at the time of my initiation into neurophysiology. These were the teachers and colleagues who had most to do with the shaping of my early career, but they had little to do with its beginning. We have been asked to try to trace the origins and influences that formed and directed our course in the neurosciences over the years. For this purpose I shall have to go back further. My career in the neurosciences was determined long before I had the good fortune to be associated with these stars and renegades of the early 1930s. In keeping with the macromolecular bias of Frank Schmitt, I shall even try to suggest some of the genetic determinants that propelled me inevitably into the neurosciences as a way of life.

## The Formative Years: Genetics and Imprinting

The genetic pool from which I derived strong behavioral determinants came from restless pioneer stock on both sides of the family. On my father's side were ancestors who pioneered the early settlements in America, dissatisfied with restrictions on life and thought in Europe, willing to venture across the Atlantic in a sailing ship to start a new life in a strange and savage country. Once life in the Eastern United States became somewhat more settled and stable, the pioneer genes expressed themselves once again. My great-grandfather joined the long trek by covered wagon to the Oregon Country, where I was born two generations later.

On my mother's side I have been able to trace my ancestry to the French Huguenots, that hardy band of much persecuted French Protestant reformists of the sixteenth and seventeenth centuries who, in spite of being massacred by the thousands, persisted in their fight for freedom of thought and religion and against the tyranny and oppression of the established government of the time. Some survived the Saint Bartholomew's Day massacre to fight on against overwhelming odds, and finally, after revocation of the Edict of Nantes in 1695, many fled for their lives to other countries. My own ancestors escaped to Switzerland, where they established an isolated community in the high Alps, safe from attack.

The Huguenots gained freedom and security in Switzerland. Two hundred years later one of their offspring, still living in a small alpine village with his wife and nine children, felt impelled to seek a better life and future for his family in the "promised land" of America. Eventually they also arrived in the great Northwest territories. Here the genetic strain of Pilgrims and covered-wagon pioneers mixed with the Huguenot strain, and I was one of the offspring. I feel sure that the genetic determinants resulting from the combination of Oregon pioneer stock with the militant reformists of France, fighting the "establishment" of the time, were of considerable importance in directing me toward the neurosciences, especially when considered together with early environmental influences.

My father sought satisfaction for his restless spirit in religion and intellectual pursuits, excelling in mathematics and philosophy as well as in practical engineering. He was a thoroughly dedicated minister, a living example of his teaching. I became convinced that money was at the root of all evil and that there was a purity in poverty to be sought after. "Seek ye first the kingdom of Heaven and all things else will be added unto you." These have proven to be valuable guidelines to a career in the neurosciences, particularly the purity of poverty!

With this genetic background, then, and with the strong influence of my father's example and teaching, I started my intellectual career, beginning with philosophical studies coupled with scientific curiosity and an interest in engineering. I don't remember ever making a deliberate decision about the direction of this career. To the despair of my parents, I had no plans to be a professional of any kind, certainly not a stuffy university professor. I was considered a drifter. I wanted only to learn how the brain works and to understand more of the meaning of life itself, wherever the search would lead. This was considered a futile fancy fifty years ago, as it is by many today.

When I entered Willamette University in 1923, I had already decided to concentrate on philosophy and psychology, though, like my father, I was fascinated also by physics, chemistry, and the biological sciences. (Unfortunately, I lacked his genius in mathematics.) I soon became disillusioned with the purely philosophical approach to the nature of the mind, reality, and the meaning of life. Psychology seemed to have more promise, especially since experimental methods were being devised that seemed to make possible the application of a few of the methods of physical science, mathematics, and engineering to studies of sensation, perception, learning, motivation, and even the higher intellectual processes—problem solving and "intelligence."

This was before the days of vacuum-tube amplifiers and oscilloscopes. Little was known precisely about nerve and muscle action currents, and no one suspected the chemical nature of synaptic and neuromuscular transmission. Still, it seemed to me that philosophers had reached a dead end in their attempts to understand the mind through intellectual reasoning alone. But I was not yet convinced that the methods of the physical sciences had much to offer in terms of a direct approach to the investigation of brain function. Psychophysics had the advantage of providing a quantification of subjective experience, but the central nervous system and brain were still largely being treated as a "black box." And so it was with many theories and systems of psychology, including psychoanalysis, at the time. Knowledge of the brain mechanisms underlying these mental and behavioral functions was inadequate and considered unnecessary. How to close the gap between stimulus and response was the question, and what did the brain itself contribute independently of specific stimuli in the immediate environment. Lurking in the background was the eternal mind-body (or mind-brain) problem, which had been treated from various points of view in my philosophical studies. Was it still possible that they were two forms of reality, closely related but not identical? Would it ever be possible to understand the mind on the basis of knowledge of the intimate and intricate complexities of brain function? With some reservation, I concluded that one would never know without trying. Growing up in the rugged Far West and working on a ranch in eastern Oregon had prepared me for a pragmatic approach to such problems, following the philosophy of John Dewey and Charles Peirce more than the idealism of Plato, Kant, Hegel, and Berkeley.

Searching through my memory stores of fifty years ago, I am impressed with certain experiences that stand out above others and that seem to have had a continuing effect on my path in the neurosciences. I have tentatively labeled these "imprinting" experiences, by analogy with the imprinting mechanisms that have been described for early experiences in animals. One of these occurred during my first two years at Willamette University, when I was working part-time in the State Mental Hospital to help pay for my college expenses. Through acquaintance with the attractive daughter of the superintendent at the hospital, I was allowed special privileges to attend staff conferences and to interview and live with many of the patients themselves. This was my first experience with the effect of brain disease upon the mind, and I found it quite disturbing. It also impressed me vividly with the rather tenuous nature of mental stability. My philosophical notions of the relationship between brain and mind would have to be reconsidered, since the intellectual processes and mental world of these patients appeared so different and yet so similar to our own. I was also deeply affected by the suffering of these patients and our inability to provide them with effective treatment. I was very much impressed by the value of deviations in mental activity for the understanding of normal brain function, and this impression or imprinting of experiences has remained with me to bias my entire career.

During this time these impressions were greatly fortified by a personal experience involving one of my closest college friends and classmates, one of the most brilliant students at Willamette University at the time. This promising young man became so disturbed and confused in his attempt to come to terms with university life and thinking, and in particular the conflicts between religious beliefs and philosophical teaching, that he was unable to face the intellectual problems before him and committed suicide. These experiences motivated my first thesis, at Reed College in Portland, Oregon, which attempted an objective study of the causes of such tragedies in student life. Entitled "Optimism and pessimism in college environments," it was published two years after my graduation in 1927, in the *Journal of Sociology*, as my first quasi-scientific publication (Jasper, 1929). These experiences imprinted on my developing interest in the neurosciences a permanent bias toward the social and individual importance (and potential tragedy) of mental disorder and disease, and toward the value of studying mental disorders for understanding the mechanisms of the brain and the mind.

I experienced another enduring imprint while a student at Reed College. I had a classmate and colleague in the Department of Psychology, Lewis Goodman, who was immersed in the study of psychoanalysis while I was immersed in the behavioral and reflex psychology in vogue at the time. Lewis Goodman and I were nearly expelled from Reed College for some extracurricular psychological experiments we conducted on the effects of certain hallucinogenic drugs, using ourselves as subjects. Before their abrupt termination, these experiments produced unforgettable experiences in both of us. The imprinting must have been equally effective in Lewis Goodman, who went on to become one of America's outstanding pharmacologists. For my part, I was most impressed with the effects that very small quantities of chemical substances could have upon the whole perceptual and intellectual organization of the mind. This imprint has been another bias in my career, although it did not come to fruition until forty years later, when I ventured into the fields of nenrochemistry and neuropharmacology as part of my attempt to understand the synaptic mechanisms of the brain and in search for chemical correlates of states of consciousness.

**From Psychology to Electrophysiology**

Beginning graduate studies at the University of Oregon in 1927, I joined an active group of graduate students in experimental psychology, among whom was Ted Ruch. The head of the department at the time was Edmund Conklin, a specialist in abnormal psychology who was to have considerable influence in fixing the impressions and imprinting of my earlier days in this direction. I began with studies of the visual perception of movement, afterimages, eidetic imagery, the phenomenon of reversibility in form and perception, and perseveration in behavior (Jasper, 1931), and even engaged in an attempt to quantify the illusion of movement of the phi phenomenon. I also became initiated into experimental methods for the study of learning and recall in human subjects as well as in experimental animals.

Ted Ruch and I had many long discussions far into the night with our colleagues at the University of Oregon Graduate School concerning the future of experimental psychology. We both came to the conclusion that more direct methods had to be developed for studying the mechanisms of the brain. Neither of us could be satisfied with the phenomenology of psychology at the time, in spite of improved methods of quantification. Upon leaving Oregon, Ted Ruch went to Yale to join John Fulton in neurophysiology, and I went to the University of Iowa from 1929 to 1931 to study under Lee Edward Travis, who was just establishing an electrophysiology laboratory in the Department of Psychology.

At Iowa, vacuum-tube amplifiers and rapidly moving mirror oscillographs were available, so that finally we were able to record with reasonable accuracy the electrical activity of nerves and muscles. Unfortunately, our amplifiers built for recording nerve and muscle activity failed to reveal the alpha rhythm when we made attempts to record the electrical activity of the brain. Little did we know at the time that Hans Berger in Germany had just published his first article, "Über das Elektrenkephalogramm des Menschen" (Berger, 1929). It was two years later that I first learned of Berger's work, brought to my attention by Adrian.

From my vantage point in the Midwest, I was soon in touch with George Bishop and Howard Bartley at Saint Louis, who were making their first observations on the electrical activity of the brain. There also were Gasser and Erlanger, who were introducing the cathode-ray oscilloscope, which gave the first accurate pictures of the nerve impulse and the compound action potential of peripheral nerve. Studying in their laboratories at that time was a young Frenchman by the name of Ali Monnier, who was going from physics to physiology, a course opposite to my own. We met for the first time at an American Physiological Society meeting in Chicago. I liked the cut of his jib. We discussed the work of Lapicque, which had intrigued me for some time. Ali Monnier was returning to Paris to set up an electrophysiological laboratory with cathode-ray oscilloscope equipment he was bringing from Saint Louis. I became determined to go to Paris to work with him and to see what there was in the idea of isochronism in the transmission of nerve impulses across synapses and at the neuromuscular junction.

Fortunately, the Rockefeller Foundation, through the National Research Council, approved of this plan. Working with Ali for two years following my doctorate in Iowa was an outstanding experience from both the scientific and personal points of view. Ali was a true biophysicist, able to handle complex mathematical theories of nerve excitability with predictive value. He gave me many practical lessons in electrophysiology, and together we learned much about the principles of nerve-membrane excitability, conduction, and rhythmicity. We proved more about "pseudochronaxies," however, than we could establish about the value of chronaxie itself, and we showed that there was a linear relationship between current and the frequency of repetitive discharge in crustacean nerve (Jasper and Monnier, 1933). It was Ali who introduced me to Adrian, Eccles, Sherrington, A. V. Hill, Bronk, Fessard; Gerard, Gasser, Erlanger, F. O. Schmitt, Rushton, and many others.

My career in neurophysiology was thus firmly established through a most pleasant personal association with Ali Monnier, though my objectives in the neurosciences had not changed from the beginning. Our paths separated when I became interested in electroencephalography in 1933, while working in the electrophysiological laboratories newly established at the Bradley Hospital of Brown University by a grant from the Rockefeller Foundation. It was here that Ali and I did our last experiment together on the artificial synapse, demonstrating in lobster nerve that impulses in one nerve could excite and generate a nerve impulse in an adjacent nerve fiber without a synapse (Jasper and Monnier, 1938). This was later called an "ephapse" by Arvanitaki. Harry Grundfest has summarized very nicely the comparative functions of ephapses and synapses in the first Neurosciences Study Program (Grundfest, 1967).

Strict axonology got left far behind for a time. Ali Monnier returned to his studies of the biophysics of nerve membranes, which were interrupted in 1939 by the beginning of the war. Herbert Gasser, who had just moved to the Rockefeller Institute from Saint Louis, became a personal friend and valued scientific adviser through my friendship with Ali Monnier. I respected but did not always follow his judgment. He felt that I was probably wasting my time by embarking on studies of the electrical activity of the brain, for he thought it unlikely that much could be learned from brain waves since they probably represented only a complex average of mass nerve action potentials whose analysis would be much too complex to be useful. But I maintained, rather naively, that there must be more than action potentials involved in the electrical activity of the brain, and at long last we could record electrical signs of brain activity in behaving alert animals and in man. The EEG was exquisitely sensitive to altered states of consciousness in sleep and waking, to altered brain metabolism, and to generalized states of excitability, with particularly dramatic changes in epileptic discharge.

Studies with implanted electrodes in freely moving animals showed that all brain areas, surface and deep, were affected by those specific sensory stimuli that had behavioral significance to the unanesthetized animal. Conditioning experiments showed that responses could also occur independently of sensory stimuli. Clinical applications such as the diagnosis and localization of epileptic processes in the human brain and the localization of brain lesions or metabolic disease were immediately apparent. I had as yet had no training in medicine, but this represented another major turning point in my career—a turning point but not a change in goal. My contact with such biophysicists as Ali Monnier, Frank Schmitt, and K. C. Cole had provided me with exciting prospects for the future, but it seemed that it would be a far distant future before our knowledge of the molecular properties of nerve tissue could be applied to an understanding of mind-brain relationships. My philosophical studies had raised questions in my mind regarding the reductionist molecular approach to this problem. Was this not a false path leading only to greater understanding of less and less?

Recordings with DC amplifiers and preliminary studies with microelectrodes had convinced me that there was much more to the electrical activity of the brain than summated nerve action potentials (Jasper, 1936a,b, 1948). Was the answer to come from more refined studies of the biophysical properties of nerve membranes, single nerve cells, and synapses, or in the microstructure and interconnections of large assemblies of neurons and their functional organization in relation to mental processes and behavior?

In collaboration with Leonard Carmichael, then director of the Department of Psychology at Brown, I was able to confirm many of Berger's findings on the electroencephalogram in man, thanks to the excellent amplifiers and recording equipment built by our electronics engineer Howard Andrews (Jasper and Carmichael, 1935; Jasper and Andrews, 1936). We were in close touch with Hal Davis and Alex Forbes in Boston, and with Adrian and Matthews in Cambridge, which added to the competitive excitement of these early days in the development of the EEG.

In 1935 I returned to Paris to defend my doctoral thesis on the electrical properties of crustacean neuromuscular systems, while presenting a second thesis on the electroencephalogram. It was the second, minor, thesis that created the greatest interest at the time. Alfred Fessard was in the audience. He had already begun important experiments on the EEG, which were to lead to its rapid development in France. Fessard, with G. Durup, was probably the first to observe the conditioning of the alpha rhythm in man, thanks to the fortuitous observation that a click on the shutter of the light stimulus became effective with repetition in blocking the alpha rhythm, even with the light turned off (Durup and Fessard, 1935). Fessard had been prepared for this observation by his early training in psychology with Henri Piéron.

While I was in Europe in 1935, I visited Hans Berger in Jena, and Tönnies and Kornmüller in Buch bei Berlin where they were actively engaged in studying the EEG in experimental animals in the laboratories of Oscar Vogt's Brain Research Institute. I was greatly impressed by the work of Oscar and Cecile Vogt, though skeptical of their overrefined cytoarchitectonic subdivisions of the brain, which Kornmüller was trying to confirm by local patterns of spontaneous electrical activity.

I returned to the United States through England, where I compared results with Adrian and Matthews in Cambridge, and with Grey Walter and Golla in London. We all had great expectations for the future of the EEG in spite of our very rudimentary understanding of its underlying physiological basis. There could be no doubt that the development of improved techniques for study of the electrical activity of the brain in unanesthetized behaving animals and in man offered exciting prospects for an improved understanding of the brain mechanisms underlying mental processes and behavior. Applications to the diagnosis and understanding of disturbances in brain function in neurological and psychiatric diseases provided added impetus to the rapid development of electroencephalography in the 1930s.

At this juncture there was another fortunate encounter. Wilder Penfield came to lecture at Brown University, and we met together in our newly constructed EEG laboratories. He was excited though skeptical about the value of the EEG in epilepsy, but he was willing to give it a try. Our common interests in the brain and in skiing brought us together immediately, for skiing in the Laurentian Mountains was a great attraction. A collaboration was established first by commuting from Providence to Montreal, then by my moving to the Montreal Neurological Institute in 1938. This collaboration proved most enjoyable and fruitful, for Penfield was truly a neuroscientist as well as a great neurosurgeon and neurophilosopher. Together we were able to study the human brain and test our ideas on experimental animals.

Then the came war. We became preoccupied with the application of electrophysiological techniques to the study of head and nerve injuries, abortive attempts to use the EEG in pilot selection, studies of "blackout" and the development of protective measures for pilots engaged in the Battle of Britain, studies of seasickness and the development of remedies to be used by our soldiers during channel crossings for the invasion of France, and studies of reactions of the brain to the newly discovered antibiotic drugs (Jasper et al., 1943). These were busy times, especially for me since I profited by the shortened three-year course at McGill to obtain a medical degree on the side, and to acquire a Canadian wife who helped make it all possible.

The opportunities provided at the Montreal Neurological Institute for studying the functional organization of the human brain, using electrical stimulation and either recording directly by means of implanted electrodes or examining the exposed brain in unanesthetized patients able to report subjective experiences, provided a wealth of data over the years. Penfield and I summarized these data in 1954, together with tentative interpretations and many unanswered questions (Penfield and Jasper, 1954).

The opportunity to combine electrophysiological and psychological observations in an experimental "animal" capable of verbal responses and descriptions of his subjective experiences was a unique situation, to be exploited in spite of the compromises necessary for neurosurgical treatment and patient welfare (Figure 23.1). With the relatively crude techniques at our disposal, we were able to determine local areas of simple sensory and motor representation in the cerebral cortex and to observe interference, distortion, and arrest of some more complex functions such as speech, perception, and memory recording. The reproduction of visual and auditory memories during electrical stimulation or epileptic discharge in temporal regions in some epileptic patients was dramatic, but it proved to be difficult of interpretation. The electrical stimulation in speech areas always served to arrest or disrupt speech, never to elicit speech. Thus, if the engrams for the memories elicited by stimulation of local regions of the temporal lobes were situated beneath the stimulating electrodes, should not they too have been blocked? Were the engrams of memory stores located elsewhere but activated by the local temporal-lobe stimulation? Memory recording could be blocked by electrical stimulation of the limbic system (amygdala and hippocampus), but it seemed

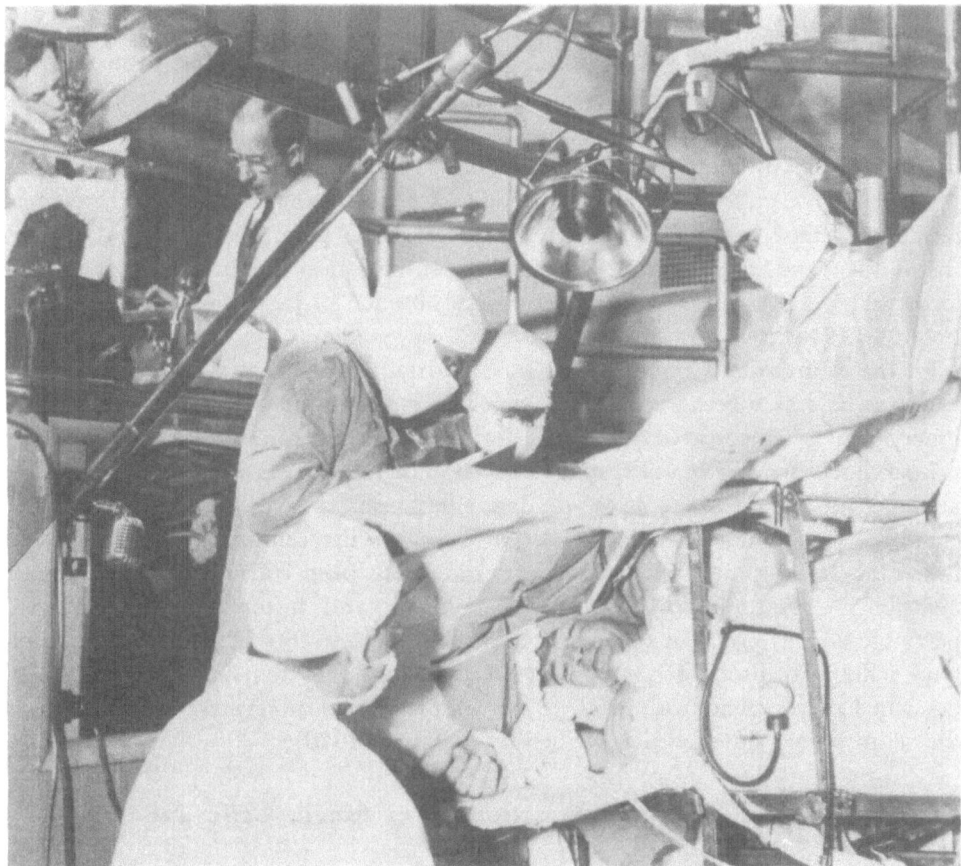

Figure 23.1    Installation for studies of the electrical activity of the cerebral cortex in man. In a glass-enclosed gallery Dr. Jasper is shown recording the electrical activity of the brain from electrodes placed on the exposed brain (hidden from view) in a patient being operated on for focal epilepsy by Dr. Penfield and associates at the Montreal Neurological Institute.

unlikely that the complex synaptic circuits underlying memory storage could be laid down solely in these relatively simple structures. Also, why could memories be induced in only a few epileptic patients, and why was only one memory elicited in a given patient?

Penfield speculated that there must be a specialized system of neurons, independent of those involved in the reception and processing of sensory information and in the direct execution of motor behavior. He called this hypothetical structure the "centrencephalic system," analogous to the "highest level of neuronal integration" of Hughlings Jackson. Conscious experience, as opposed to unconscious sensorimotor functions, was thought to require activation of the centrencephalic system concomitantly with the thalamocortical circuits involved in the transmission and processing of detailed specific sensory information and the activation of preformed patterns of motor response. This hypothesis was a real challenge to more precise neurophysiological investigation.

Moruzzi and Magoun (1949) then provided neurophysiological evidence for the reticular activating system, distinct from the principal sensory and motor pathways, which appeared to control behavioral states of sleep and waking and to exert a generalized control of the electrical activity of widespread cortical and subcortical structures, with descending projections regulating spinal reflexes. We were able to confirm these findings in experimental animals and to confirm and extend as well the previous observations of Morison and Dempsey (1942) on the regulatory and integrative functions of the intralaminar thalamus, which we proposed might be called a "thalamic reticular system." We proposed also that these reticular systems might be somehow involved, either primarily or secondarily, in the sudden and quickly reversible loss of consciousness characteristic of certain forms of epileptic attack, the "petit mal absence" (Jasper and Droogleever-Fortuyn, 1946).

At the Laurentian Conference in 1953 (Adrian, Bremer, and Jasper, 1954) the major issue was whether conscious experience, as opposed to sleep and certain other forms of unconsciousness (e.g., light barbiturate anesthesia) in which the principal sensory and motor systems seemed to be intact, was due to the integrative action of the brain as a whole or was dependent upon specialized neuronal systems capable of a general integrative action over specific sensorimotor and mental functions. During the past twenty years much has been learned by more refined anatomical, microphysiological, neurochemical, and histochemical techniques pertinent to this problem (Jasper et al., 1958; Eccles, 1966; Evans and Mulholland, 1969). The probable anatomical substrate for this system has been beautifully described by the Scheibels (1967, 1970); and thalamic integrative mechanisms at the cellular level have been described by Purpura (1970).

## The Analytic and Integral Functioning of Single Cells and Neuronal Assemblies

During the past few years, both in waking and behaving animals and in animals under light anesthesia, microelectrode studies of the firing patterns and responses of individual cells and local cell assemblies in the brain have shown that there is a remarkable functional specificity of local neuronal assemblies in the brain. This specificity may be related to a given modality and topographical location of sensory input to specific columns of cells in the cerebral cortex, as shown so beautifully for the somatic system by the work of Mountcastle (1967) and his colleagues, and for the visual system by Hubel and Wiesel (1965). In the latter studies the existence of columns of cells with varying complexities of specificity to movement and to patterns of visual stimuli has given us remarkable insight into how the brain processes information concerning visual space. The existence of relatively simple cells, as well as cells with more complex patterns of specificity, provides a model of the integrative activity of the brain that may have wide application when more hypercomplex columnar assemblies of cells have been identified. The impression gained from these studies is that there is a rigid, highly specific organization of neuronal connectivity and specialization for the analysis of sensory information in receiving areas, though some plasticity in this organization has also been demonstrated, depending upon the early experience of the developing brain.

Microelectrode recording from single units in the thalamus in conscious human subjects, as has been accomplished in collaboration with Gilles Bertrand of the Montreal Neurological Institute, has confirmed the impression of remarkable specificity of unit discharge in the human thalamus (Jasper and Bertrand, 1964, 1966a,b; see Figure 23.2). Single cells in the ventrobasal complex, for example, could be activated only by stimuli of restricted receptive fields of the skin. These cells continued to respond mechanically to a constant stimulus in an unchanging manner, regardless of the state of consciousness or attention of the patient in the

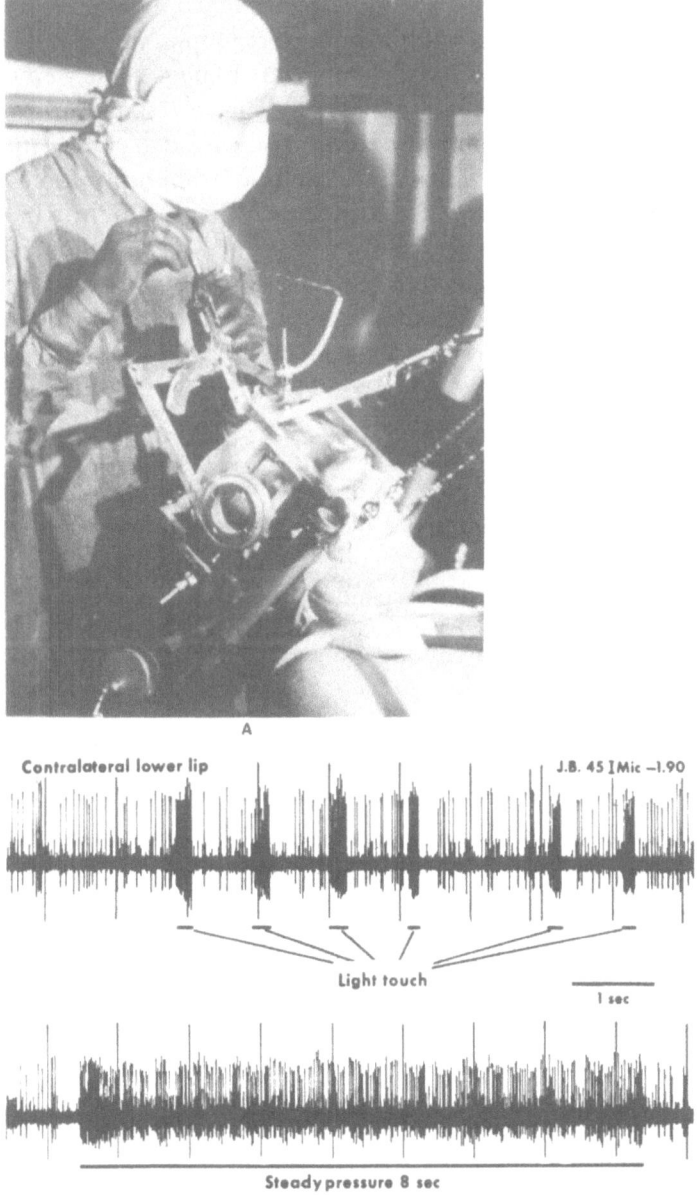

Figure 23.2   (A) Stereotaxic apparatus applied to a patient with Parkinson's disease for microelectrode recording of single cells within the human thalamus. Dr. Bertrand is shown adjusting the micrometer while viewing and listening to the single-cell discharges from the equipment located nearby in the operating theater. (B) Examples of single-unit discharges from microelectrode recording in the human thalamus. The recording point was located in the ventrobasal complex. Unit responses were recorded to light touch of the lower lip (above) and to steady pressure (below).

operating room. Other cells, apparently specifically related to joint movement or local muscle receptors, were also highly specific and uninfluenced by the general state of excitation or attention of the waking patient, or by any other form of sensory stimulation.

Cells with far more complex but still highly specific response characteristics were found in other parts of the thalamus outside these rigidly specific sensory receiving areas. In the more dorsal and mesial thalamus were found units that did not show any regular response to sensory stimuli of any kind although some would respond to a variety of stimuli and habituate rapidly, ceasing to respond on repetition of an identical stimulus. We called these cells "novelty detectors," for they did not seem to be specific to a given form or location of stimulus, but only to the novelty of the form of stimulus. There were other cells that apparently did not respond to passive movements or to reflex movements, but only to voluntary intentional movements (Phillips and Olds, 1969). There were yet others whose spontaneous firing pattern was arrested by stimuli that produced an attentive response, even at the beginning of a voluntary movement. And there were cells that, after a few repetitions of a stimulus, would respond prior to, and in anticipation of, the actual stimulus.

Thus, in the conscious waking man an extraordinarily high degree of unit specificity was apparent even in the most complex sensory or motor functions. There were many cells that we were unable to influence by any of our manipulations in the operating room and whose complex functions could not be determined. These cells continued to fire in their own pattern, independently of any applied stimuli or manipulations of the patient, and they were not influenced by any form of sensory stimulus we could devise or by any mental or emotional state in the fully conscious patient.

Specificity of complexly organized unit assemblies has also been shown in experimental animals. For example, the so-called attention units were recorded from the auditory cortex in unanesthetized cats only when the cats paid attention to the stimulus (Hubel et al., 1959). Mountcastle and his colleagues are now discovering, in the parietal cortex in awake responding monkeys, units related to complex visual-motor responses that combine coordinated movements with visual space (Lynch et al., 1973). They have also discovered specific cells that seem to be related to the initiation of directed movement in visual space, but only when these movements are voluntary or willed.

The picture of brain organization derived from these microelectrode studies suggests that even the most complex of brain functions, such as the detection of novelty, attention, voluntary movements, and the recognition of specific patterns, are encoded in specific neuronal assemblies. According to this mosaic conception, the specificity of the more complex units is dependent upon their connectivity with more simple specific cell assemblies. Is it not possible that some of these complex integrations are also dependent upon the pattern of multimodal sensory input and upon the ramifications of the widely branching axonal network that forms the reticular feltwork in the thalamus and brainstem (described by the Scheibels, 1970)?

The logical extension of this specific analytic view of cerebral organization is that there are specific assemblies of cells that must be activated when the transmission and processing of information is to be selectively transformed into conscious experience and awareness.

In contrast to this view are other observations with microelectrode recording, in behaving animals during sleep and waking and during learning processes, which suggest a much more dynamic view of brain organization. In our early experiments with microelectrode recording in unanesthetized monkeys, begun in 1956, our primary objective was to study the development of a simple, conditioned withdrawal response to a visual stimulus by multiple sampling of single cells in the frontal, sensory, motor, parietal, and occipital cortical areas (Jasper, Ricci, and

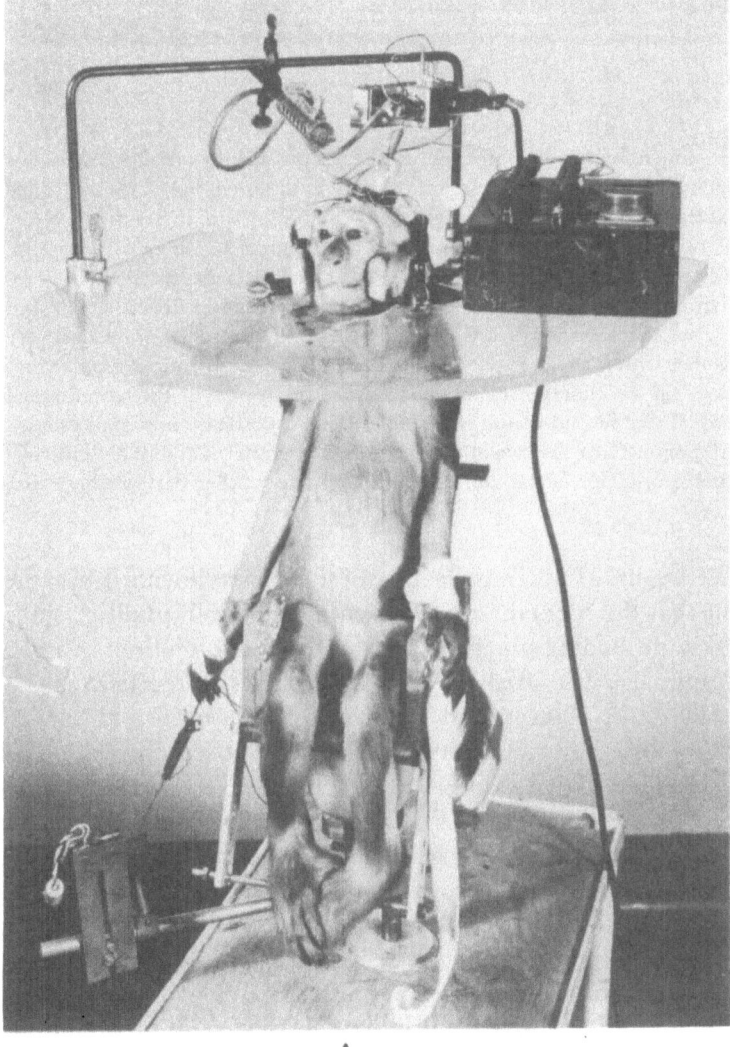

A

Figure 23.3   (A) This was one of the original installations for recording single cells from the cerebral cortex in the unanesthetized monkey during motor performance and during learning of a conditioned avoidance response (developed at the Montreal Neurological Institute in 1955 and 1956).

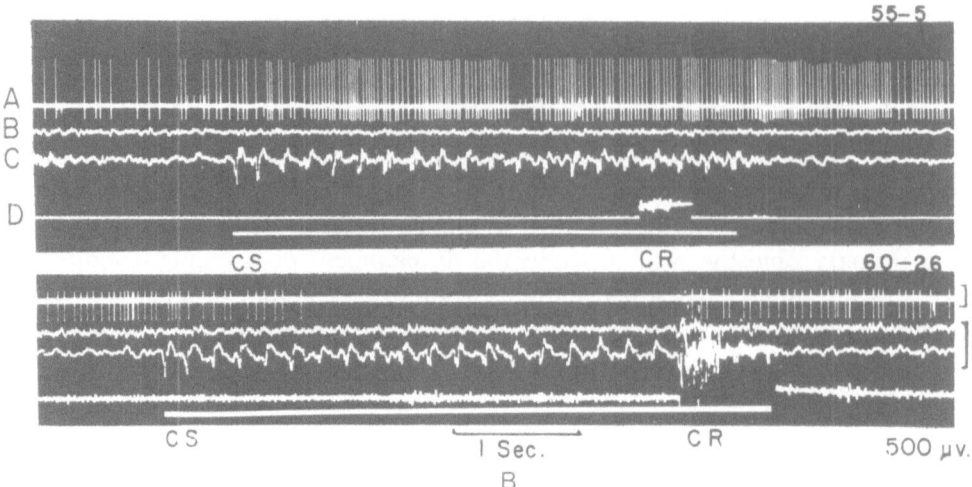

Figure 23.3  (B) Unit recording from the intact unanesthetized monkey during learning and motor behavior. The example shows conditioned response to intermittent flashes of light at a frequency of about 4.5 per sec. Light flashes begin at CS and motor response occurs with interruption of key at CR. The first line (A) indicates single-unit discharge from the motor arm area of precentral gyrus, the second (B) indicates the surface EEG from the motor cortex adjacent to the micro-electrode recording, the third (C) is from the surface of the occipital cortex recording evoked potentials to the light flashes, and the lower line (D) indicates the electromyogram from the flexors of the forearm together with a DC deflection of the base line for interruption of the avoidance response switch. Note the acceleration of motor cortical units prior to the conditioned response in the first record. In the second record inhibition of motor cortical units is also shown prior to the conditioned response. These are two of the most common patterns of unit activity seen in the conditioned avoidance behavior of the monkey, although many other patterns have also been observed.

Doane, 1958, 1960; see Figure 23.3). In the course of these experiments it became immediately apparent that the spontaneous firing patterns of cells in all of these areas varied remarkably depending upon the state of waking, attention, or sleep of the experimental animal. (The analysis of these changes in cortical firing patterns during various stages of sleep and waking has since been made more precise by Evarts, 1967.) It was obvious that the unitary responses in relation to their spontaneous background activity would have to be taken into consideration in any overall concept of how the processing of sensory information by the brain depends upon its ongoing activity and general state of excitability.

Furthermore, in the beginning of the conditioning experiments in monkeys, the alerting response to electric shock produced a generalized activation, or inhibition, of some cells in all cortical areas studied. Even an innocuous visual stimulus, given unexpectedly for the first few times, caused a change in the firing patterns of cells in frontal, parietal, and sensorimotor cortical areas, as well as in the visual cortex. As conditioning progressed, these generalized unitary responses diminished. However, even after the establishment of a highly trained conditioned motor response, units were found in frontal, parietal, visual, and sensorimotor areas whose firing patterns would change systematically depending upon whether there was a correct or incorrect conditioned response to the stimulus.

Differential responses were also found in the parietal cortex to a negative or positive visual conditioning stimulus for a given motor response. Some cells in the motor cortex, for example—presumably pyramidal-tract cells, as shown so beautifully in the study of Evarts (1973)—responded mechanically in relation to specific movements. However, other cells were not consistently related to the motor performance of the animals, but apparently responded only to significant conditioning stimuli and ceased to respond during movement. These behaved as "command cells," which were not responsible for the pattern and execution of the movement itself. Some such cells were also found in the parietal cortex, as recently described by Mountcastle and coworkers (Lynch et al., 1973).

The picture obtained from these studies is a much more dynamic one, showing that there must be an overall regulation of response characteristics in neuronal assemblies as well as a high degree of specificity in their functional organization. It is further suggested that unit assemblies subserving more highly integrative and regulatory functions may be distinct from those involved in the direct transmission and preliminary processing of sensory information, or in the more direct efferent organization of motor response. The problem of the "highest level of integration" and the mechanism for the translation of sensory information into conscious awareness and the selectivity of attention still remains unsolved. A related corollary is the problem of selective storage of experience in memory. Are these also "specific" functions of separate neuronal networks?

## The Chemical Specificity of Neuronal Systems

Rapid advances in neurochemical techniques during recent years have added a new dimension to the search for relations between "mind and molecules." Fluorescent histochemistry has identified a network of chemically specific neurons extending throughout the neuraxis independent of the principal sensory and motor pathways. The biogenic amines (noradrenaline, serotonin, and dopamine) contained within these neurons and liberated at the widespread terminals from cerebral cortex to spinal cord are substances known to have powerful effects upon mental processes and behavior. Their metabolism and mode of action have been discussed by other contributors to this volume (Bloom, Axelrod), and in the NRP Study Programs (Kety, 1967; Kravitz, 1967).

In the context of the present discussion, it is important to emphasize that the biogenic amines do not seem to be involved in the transmission and processing of information in principal sensory pathways or in primary motor systems, but act rather as modulators of these systems. It is also important to note that noradrenaline, at least, may have a metabolic link in its mode of action, acting through its effects upon adenyl cyclase and cyclic AMP, at least in the cerebellum. These neuronal systems are thought to be significantly involved in the control of sleep and waking states, and they also play an important role in mechanisms of reinforcement in certain forms of learning (Kety, 1970).

Acetylcholine has also been implicated in the neurochemical mechanisms correlated with the control of states of sleep and waking, though central transmission in sensory and motor pathways appears not to be cholinergic, as is the final common path of lower motor neurons (Krnjević, 1965; Celesia and Jasper, 1966). The cholinergic system of brainstem and diencephalon, as judged by a histochemical stain for acetylcholinesterase, has been thought to be a good candidate for an important segment of the ascending reticular activating system (Shute and Lewis, 1967). Cholinergic mechanisms have also been shown to be important in certain learning processes (for example, the amnestic syndrome produced by scopolamine).

We have also found that the pattern of free amino acids liberated from the cerebral cortex into superfusates shows striking changes during states of sleep and wakefulness and during electrical stimulation of the midbrain reticular formation (Jasper and Koyama, 1969). These changes were complex and difficult to interpret, but the most consistent change was an increase in free glutamic acid during desynchronized cortical activation. Amino acids play a role both in energy and protein metabolism, as well as being implicated as likely excitatory and inhibitory transmitter substances. Thus their role in the chemistry of the states of sleep and wakefulness is a difficult but nonetheless important problem. Indeed, an understanding of the relationship between biogenic amines, acetylcholine, and the amino acids will be most important in the attempt to construct a complete picture of the chemistry of states of consciousness. Other substances may also be involved. The chemistry of wakefulness, arousal, and attention must also be of importance in the chemistry of the learning process (Barondes and Cohen, 1968), but under certain conditions they may be dissociated as in the amnestic syndrome.

## Conclusion

I have now come to the end of my personal account of a devious path of exploration in the neurosciences, covering fifty years of research, from philosophy to physics, and from mind to molecules. I have not attempted here to make a scientific contribution to this subject, but only to trace a personal path and to speculate about its genetic and environmental determinants.

Frank Schmitt (1967), Manfred Eigen (1967), and many others taking part in NRP Work Sessions and Intensive Study Programs have developed the subject of molecular mechanisms of brain growth and function from all points of view, to the extent of present knowledge and with fascinating projections into the future. I have learned to share Frank's faith in the future of relations between molecular and neural biology without the competence to become directly involved.

It has become apparent to me while closely following the ten-year course of the NRP that the relationship between macromolecular structure and the coding of engrams for memory storage is probably no more than an analogy (Jasper and Doane, 1968). A similar conclusion has been reached by Gerald Edelman for analogies with mechanisms of molecular recognition in the immune response and in the nervous system (see Chapter 4, this volume). However, as Ted Bullock (1970) likes to remind us, we must be ready for revolutionary new ideas if we are to make real progress in the neurosciences and to find ways of bringing together new developments in molecular biophysics and protein and lipid chemistry, as well as improved conceptual approaches to systems analysis, in our search for the brain mechanisms of mind and behavior.

It is tempting at times, especially when one is near the end of the road and sees how far one is from the destination and how much work and false or unproductive paths one has followed on the way, to wonder if a true understanding of brain-mind relations will ever be achieved. Perhaps not, but the search has provided a most exciting and satisfying way of life, and has given me the greatest of pleasure, particularly through the sharing of such experiences with so many friends and colleagues of similar interest and dedication. The opportunities it has afforded to make some contribution to the betterment of those suffering from nervous and mental disease is an added reward. Excursions along the path into international relations and sociopolitical problems through the international community of neuroscientists have added much to the excitement, satisfaction, frustrations, and disappointments of my life, but they have not shaken my faith in the future, which I owe mostly to my father and my Oregon pioneer and Huguenot ancestors and to the help of many colleagues in my attempt to complete the voyage from Philosophy to Physics, and from Mind to Molecules. It remains for the future to determine whether the translation of brain mechanisms into conscious experience will be a philosophical or scientific question in the final analysis.

## References

Adrian, E. D., Bremer, F., and Jasper, H. H., eds. (1954): *Brain Mechanisms and Consciousness* (A Symposium Organized by the Council for International Organizations of Medical Sciences). Oxford: Blackwell.

Barondes, S. H., and Cohen, H. D. (1968): Arousal and the conversion of "short-term" to "long-term" memory. *Proc. Natl. Acad. Sci. USA* 61:923–929.

Berger, H. (1929): Über das Elektrenkephalogramm des Menschen. *Arch. Psychiatr.* 87:527–570.

Bullock, T. H. (1970): Operations analysis of nervous functions. *In: The Neurosciences: Second Study Program.* Schmitt, F. O., editor-in-chief. New York: Rockefeller University Press, pp. 375–383.

Celesia, G. G., and Jasper, H. H. (1966): Acetylcholine released from cerebral cortex in relation to state of activation. *Neurology* 16:1053–1063.

Durup, G., and Fessard, A. (1935): L'électroencéphalogramme de l'homme: Observations psycho-physiologique relative à l'action des stimuli visuels et auditifs. *Ann. Psychol.* 36:1–35.

Eccles, J. C., ed. (1966): *Brain and Conscious Experience* (Study Week of the Pontificia Academia Scientiarum, September 28 to October 4, 1964). New York: Springer-Verlag.

Eigen, M. (1967): Dynamic aspects of information transfer and reaction control in biomolecular systems. *In: The Neurosciences: A Study Program.* Quarton, G. C., Melnechuk, T., and Schmitt, F. O., eds. New York: Rockefeller University Press, pp. 130–142.

Evans, C. R., and Mulholland, T. B., eds. (1969): *Attention in Neurophysiology* (An International Conference). New York: Appleton-Century-Crofts.

Evarts, E. V. (1967): Unit activity in sleep and wakefulness. *In: The Neurosciences: A Study Program.* Quarton, G. C., Melnechuk, T., and Schmitt, F. O., eds. New York: Rockefeller University Press, pp. 545–556.

Evarts, E. V. (1973): Brain mechanisms in movement. *Sci. Am.* 229:96–103.

Grundfest, H. (1967): Synaptic and ephaptic transmission. *In: The Neurosciences: A Study Program.* Quarton, G. C., Melnechuk, T., and Schmitt, F. O., eds. New York: Rockefeller University Press, pp. 353–372.

Hubel, D. H., Henson, C. O., Rupert, A., and Galambos, R. (1959): "Attention" units in the auditory cortex. *Science* 129:1279–1280.

Hubel, D. H., and Wiesel, T. N. (1965): Receptive fields and functional architecture in two nonstriate visual areas (18 and 19) of the cat. *J. Neurophysiol.* 28:229–289.

Jasper, H. H. (1929): Optimism and pessimism in college environments. *Am. J. Sociol.* 34:856–873.

Jasper, H. H. (1931): Is perseveration a functional unit participating in all behavior processes? *J. Soc. Psychol.* 2:28–51.

Jasper, H. H. (1936a): Cortical excitatory state and synchronism in the control of bioelectric autonomous rhythms. *Cold Spring Harbor Symp. Quant. Biol.* 4:320–338.

Jasper, H. H. (1936b): Cortical excitatory state and variability in human brain rhythms. *Science* 83:259–260.

Jasper, H. H. (1948): Charting the sea of brain waves. *Science* 108:343–347.

Jasper, H. H., and Andrews, H. L. (1936): Human brain rhythms. I. Recording techniques and preliminary results. *J. Gen. Psychol.* 14:98–126.

Jasper, H. H., and Bertrand, G. (1964): Exploration of the human thalamus with microelectrodes. *Physiologist* 7:167 (abstract).

Jasper, H. H., and Bertrand, G. (1966a): Recording from microelectrodes in stereotaxic surgery for Parkinson's disease. *J. Neurosurg.* 24:219–221.

Jasper, H. H., and Bertrand, G. (1966b): Thalamic units involved in somatic sensation and voluntary and involuntary movements in man. *In: The Thalamus.* Purpura, D. P., and Yahr, M. D., eds. New York: Columbia University Press, pp. 365–390.

Jasper, H. H., and Carmichael, L. (1935): Electrical potentials from the intact human brain. *Science* 89:51–53.

Jasper, H. H., Cone, W., Pudenz, R., and Bennett, T. (1943): The electroencephalograms of monkeys following the application of microcrystalline sulfonamides to the brain. *Surg. Gynecol. Obstet.* 76:599–611.

Jasper, H. H., and Doane, B. (1968): Neurophysiological mechanisms in learning. *In: Progress in Physiological Psychology.* Vol. 2. Stellar, E., and Sprague, J. M., eds. New York: Academic Press, pp. 79–117.

Jasper, H. H., and Droogleever-Fortuyn, J. (1946): Experimental studies on the functional anatomy of petit mal epilepsy. *Res. Publ. Assoc. Res. Nerv. Ment. Dis.* 26:272–298.

Jasper, H. H., and Koyama, I. (1969): Rate of release of amino acids from the cerebral cortex in the cat as affected by brain stem and thalamic stimulation. *Can. J. Physiol. Pharmacol.* 47:889–905.

Jasper, H. H., and Monnier, A. M. (1933): Pseudo-chronaxies des systèmes neuro-musculaires des crustacés dues a la réponse rythmique du nerf. *C. R. Soc. Biol. (Paris)* 112:233–236.

Jasper, H. H., and Monnier, A. M. (1938): Transmission of excitation between excised non-myelinated nerves. An artificial synapse. *J. Cell. Comp. Physiol.* 11:259–277.

Jasper, H. H., Proctor, L. D., Knighton, R. S., Noshay, W. C., and Costello, R. T., eds. (1958): *Reticular Formation of the Brain* (Henry Ford Hospital International Symposium, 1957). Boston: Little, Brown.

Jasper, H. H., Ricci, G. F., and Doane, B. (1958): Patterns of cortical neuronal discharge during conditioned responses in monkeys. *In: Ciba Foundation Symposium on the Neurological Basis of Behavior.* Wolstenholme, G. E. W., and O'Connor, C. M., eds. Boston: Little, Brown, pp. 277–290.

Jasper, H. H., Ricci, G., and Doane, B. (1960): Microelectrode analysis of cortical cell discharge during avoidance conditioning in the monkey. *Electroencephalogr. Clin. Neurophysiol.* (Suppl.) 13: 137–155.

Kety, S. S. (1967): The central physiological and pharmacological effects of the biogenic amines and their correlations with behavior. *In: The Neurosciences: A Study Program.* Quarton, G. C., Melnechuk, T., and Schmitt, F. O., eds. New York: Rockefeller University Press, pp. 444–451.

Kety, S. S. (1970): The biogenic amines in the central nervous system: Their possible roles in arousal, emotion, and learning. *In: The Neurosciences: Second Study Program.* Schmitt, F. O., editor-in-chief. New York: Rockefeller University Press, pp. 324–336.

Kravitz, E. A. (1967): Acetylcholine, $\gamma$-aminobutyric acid, and glutamic acid: Physiological and chemical studies related to their roles as neurotransmitter agents. *In: The Neurosciences: A Study Program.* Quarton, G. C., Melnechuk, T., and Schmitt, F. O., eds. New York: Rockefeller University Press, pp. 433–444.

Krnjević, K. (1965): Transmitters in the cerebral cortex. *In: XXIII International Congress of Physiological Sciences* (Lectures and Symposia, Tokyo, Japan, 1–9 September 1965). International Congress Series, no. 87. Amsterdam: Excerpta Medica Foundation, pp. 435–443.

Lynch, J. C., Acuna, C., Sakata, H., Georgopoulos, A., and Mountcastle, V. B. (1973): The parietal association areas and immediate extrapersonal space. *In: Abstracts.* Third Annual Meeting, Society for Neuroscience, San Diego, p. 244.

Morison, R. S., and Dempsey, E. W. (1942): A study of thalamo-cortical relations. *Am. J. Physiol.* 135:281–292.

Moruzzi, G., and Magoun, H. W. (1949): Brain stem reticular formation and activation of the EEG. *Electroencephalogr. Clin. Neurophysiol.* 1:455–473.

Mountcastle, V. B. (1967): The problem of sensing and the neural coding of sensory events. *In: The Neurosciences: A Study Program.* Quarton, G. C., Melnechuk, T., and Schmitt, F. O., eds. New York: Rockefeller University Press, pp. 393–408.

Penfield, W., and Jasper, H. H. (1954): *Epilepsy and the Functional Anatomy of the Human Brain.* Boston: Little, Brown.

Phillips, M. I. and Olds, J. (1969): Unit activity: Motivation-dependent responses from midbrain neurons. *Science* 165:1269–1271.

Purpura, D. P. (1970): Operations and processes in thalamic and synaptically related neural subsystems. *In: The Neurosciences: Second Study Program*. Schmitt, F. O., editor-in-chief. New York: Rockefeller University Press, pp. 458–470.

Scheibel, M. E., and Scheibel, A. B. (1967): Anatomical basis of attention mechanisms in vertebrate brains. *In: The Neurosciences: A Study Program*. Quarton, G. C., Melnechuk, T., and Schmitt, F. O., eds. New York: Rockefeller University Press, pp. 577–602.

Scheibel, M. E., and Scheibel, A. B. (1970): Elementary processes in selected thalamic and cortical subsystems—the structural substrates. *In: The Neurosciences: Second Study Program*. Schmitt, F. O., editor-in-chief. New York: Rockefeller University Press, pp. 443–457.

Schmitt, F. O. (1967): Molecular neurobiology in the context of the neurosciences. *In: The Neurosciences: A Study Program*. Quarton, G. C., Melnechuk, T., and Schmitt, F. O., eds. New York: Rockefeller University Press, pp. 209–219.

Schmitt, F. O. (1970): Introduction. *In: The Neurosciences: Second Study Program*. Schmitt, F. O., editor-in-chief. New York: Rockefeller University Press, pp. v–ix.

Shute, C. C. D., and Lewis, P. R. (1967): The ascending cholinergic reticular system: neocortical, olfactory and subcortical projections. *Brain* 90:497–520.

Roger W. Sperry (b. 1913, Hartford, Connecticut) is Hixon Professor of Psychobiology at the California Institute of Technology. His researches have included split-brain studies of cerebral organization, cerebral correlates of perception, and the patterned growth of nerves in development and regeneration. In human split-brain patients he has investigated hemispheric interaction and specialization underlying various modalities and styles of cognitive processing.

# 24
## In Search of Psyche

Roger W. Sperry

To a beginner in science back in the mid-1930s, it seemed that there could be no more challenging problem at which to aim—as a long-term, ultimate goal kind of thing—than that of consciousness and the mind-brain relation, more acceptably expressed in those days as the problem of the "neural correlates of conscious experience." A naive beginner, of course, could hardly expect to approach a final solution, but it is always reassuring to feel that one's efforts are at least aimed in the general direction of something that might be of ultimate importance. Meantime, as a "brain researcher," one could find plenty of lesser but entirely respectable and more researchable corollary problems along the way, such as perception, learning, and memory.

In the 1930s it already had begun to appear that science might soon close in on the nature of the changes produced in the brain by learning and experience. With the conditioned reflex as a model, researchers had begun to draw hypothetical diagrams for the kind of new brain pathways that must be formed in conditioning to link the conditioning stimulus to the conditioned response. As time passed and further experiments eliminated one neural hypothesis after another, however, it became evident that the nature of the newly formed stimulus-response connections was much more complex than had been at first supposed. Indeed, it began to look doubtful that new nerve connections of any sort were involved in conditioning, or in any brain function. By the late 1930s the connectivity principle as a basis of central nervous integration had come under fire from many directions and was very much in question.

The theoretical impact of K. S. Lashley's brain-lesion studies and his concepts of mass action and cortical equipotentiality had at that time reached their peak. These and related findings pointing up the nonlocalizability of the engram seemed incompatible with any stimulus-response connectionist formula. So also were the Gestalt views of the 1930s that emphasized the control role of excitatory patterns as wholes and their associated metaneuronal "field" forces. The Gestalt or "figure" properties were conceived to transcend the function of individual fiber connections.

The absence of functional specificity in nerve connections seemed to have been substantially confirmed in an extended series of clinical and experimental studies, from all parts of the world, demonstrating that nerves were functionally interchangeable after surgical cross-union. The same was reported to hold with respect to both the transplantation of muscles to take over new functions and the grafting of skin flaps to new locations. It all seemed to confirm the lack of any fixed functional specificity in neural connections, and emphasized an extreme wholesale plasticity in neural integration that provided for almost unlimited readaptation capacity in the central mechanisms of the brain and spinal cord. The classic account by Stratton (1897) of his own experience in adjusting to the inverted vision produced by wearing an optical device was widely cited in this same connection. To see the world right side up is something we all had to learn, or so it seemed, and it was readily relearnable.

Additional reinforcement for the plasticity and anticonnectionist views of the 1930s came from the field of nerve growth and development, where the prevailing doctrine for more than a decade had proclaimed the outgrowth and termination of developing and regenerating nerve fibers to be diffuse and nonselective. Chemical and electrical selectivity appeared to have been ruled out, leaving only mechanical guidance in command as the primary orienting influence (Weiss and Taylor, 1944). In terms of the evidence then available, it seemed entirely impossible that the enormously intricate and precisely adjusted wiring circuits for adaptive behavior could be grown into a brain directly—that is, organized through the growth process itself without benefit of experience and learning. On these and other grounds it was widely accepted that the nerve networks for behavior could not be inherited, and the idea of "instinct" as an explanatory construct in behavioral science thus reached an all-time low in disrepute in the late 1930s. Under these conditions acceptance of the new upstart discipline called "ethology" remained quite limited, with resistance particularly strong in the United States and the Soviet Union.

Some of the strongest evidence against nerve-connection specificity came from another long series of experiments demonstrating that surgical rearrangements between nerve centers and periphery in amphibians failed to disrupt orderly coordination under conditions where relearning could be excluded. These experiments, pioneered by Paul Weiss (1936), were taken to prove that central nervous integration could not be based on selectivity in fiber connections. As an alternative to the classic connection-switchboard model of integration, a radio-broadcast model was proposed, based on resonance effects involving diffuse morphological interconnection with impulse specificity and selective neuronal and end-organ attunement. Like radio pick-up, the "resonance principle" provided selective response in the presence of diffuse nonselective synaptic connections. Meanwhile, the idea that synaptic relations within the neuropil were not morphologically selective but formed, rather, in an excessive common profusion, seemed to receive further support from C. J. Herrick's (1948) intensive anatomical analyses of the central neuropil in the brain of the tiger salamander.

These many different lines of convergent evidence, combining and mutually reinforcing each other, had built up by the end of the 1930s into quite a substantial and convincing case against the classical Sherringtonian model of central nervous integration. In many quarters it became fashionable to refer to Sherringtonian connectionism when one wished to exemplify simplistic and outmoded naivety.

Anticonnectionist thinking received a further major boost in the early 1940s when it was reported that the largest system of fiber connections in the human brain, the corpus callosum, containing over 200 million elements, could be completely transected in clinical surgery without producing any definite functional symptoms (Akelaitis, 1943). Here again the brain seemed to possess an almost mystical plasticity in its ability to achieve proper orderly function in spite of radical disruptions in its normal wiring plan. Not since the early beginnings of neuroscience had the brain looked so bafflingly obscure and resistant to physiological analysis. How could one even begin to formulate orderly laws and understanding for a mechanism that continued to operate correctly regardless of rearrangement and disruption of its interconnecting parts?

Against this background of general theoretical uncertainty, it appeared a rather poor risk in the late 1930s to invest time and research efforts directly in neural models or working hypotheses concerning such higher psychological functions as learning, memory, or consciousness, or even the conditioned reflex. What we needed first were some better answers at elemental levels, and hopefully some unifying resolution of all the divergent views and issues involved.

When we began to follow up experimentally, one at a time, various aspects of the foregoing plasticity and anticonnectionist phenomena, the results—to our initial surprise—failed to accord with previous accounts. To make a long story short, it was found that motor nerves and muscles, as well as sensory nerves, were not at all functionally interchangeable after surgical transposition, but instead persistently retained their original functions (Sperry, 1945): inverted vision produced surgically by eye rotation showed a fixed persistence, lasting indefinitely without correction by experience and training; and nerve growth in the brain and spinal centers was anything but diffuse and nonselective (Sperry, 1951a,b). To account for the kind of central nerve regeneration found in the new experiments it became necessary to reinstate the old concept of chemotaxis in an even more extreme form and to postulate a degree of cellular specificity and chemotactic guidance more extensive and refined than that previously imagined even by Ramón y Cajal.

Analysis of some of the behavioral effects of reversed vision led us to postulate the function of "corollary discharge" as a mechanism for maintaining perceptual constancy in the presence of disturbing eye, head, and body movements (Sperry, 1950). We had to conclude more generally from the nerve-growth findings that the brain's wiring diagram must, after all, be largely innate, that it is grown in with extreme precision through an enormously elaborate chemical guidance program that is under genetic control, and that it is therefore in very large part inherited. The orderly function found by Weiss to follow nerve disarrangements in amphibians was reconfirmed experimentally; but with our new findings it proved to be explainable in terms of orthodox connectionist principles, thus obviating the need to invoke resonance phenomena or *Erregungspezifität*.

Tests for the postulated electric-field forces of Gestalt theory, conducted in cats and monkeys (Sperry and Miner, 1955), gave results that pointed mainly to the absence of such influences. The data emphasized instead the remarkable capacity of the brain to preserve orderly function when confronted with gross distortions in its internal electric-field pattern and/or disruptions in its horizontal transcortical interactions.

Furthermore, studies involving surgical section of the corpus callosum showed that brain function was by no means left unimpaired. Using various measures to obtain controlled lateralized input and special tests for functional processing within each hemisphere independently, we were able to demonstrate a whole host of cross-communication deficits, first in animals (Sperry, 1961) and later, using the same principles, in human patients undergoing operations for severe, intractable epilepsy (Sperry, 1974a). This last major stronghold of anticonnectionism was shortly to be turned around into a leading bastion for the opposing views. In the evidence relating to the neocommissures, more than from any other place in the brain, we now come closest to tying higher conscious functions to specific cortical-fiber systems.

The conceptual view of the nervous system that emerged from the evidence of the late 1950s had very different properties from the view with which we had had to deal earlier, in the plasticity-equipotentiality period. Differential connection patterns now meant something in terms of functional control. The design and the operating principles of connection circuits, though enormously and perhaps overwhelmingly complex, were subject, at least in principle, to experimental analysis and to lawful formulation. We were now in a much better position to approach such problems as the nature and locus of the new neural connections established in conditioned-response learning. Curiously, the neural model for conditioning that I eventually settled on involved a rejection of connectivity in a sense. I concluded that it had been an error to search for newly formed sensory-motor connections, that we should think of the new sensory-motor linkages observed behaviorally as being effected instead by means of transient cerebral facilitating sets (that is, passing excitatory physiological states) that only temporarily open or prime the requisite stimulus-response connection paths in the conditioning situation.

The long-term "engram" changes in this model (Sperry, 1955) assumed a very different pattern and location, designed to arouse the requisite excitational facilitating set at the right time in the right context. The long-term changes, accordingly, were allocated to the realm of perceptual learning and expectancy, phenomena that involve the association systems of the cortex rather than direct sensorimotor pathways. Furthermore, the changing facilitating set was conceived to be a basic master switching system that would continually alter the functional wiring plan of the brain. By opening and closing different patterns of neural circuits for different functions, this switching system would give the brain, in effect, many different circuit design systems in one, somewhat like different computer programs and subroutines. Switching mechanisms of this sort, based on transient excitatory sets, were felt to account for a large part of the brain's readjustment capacity and its tremendous versatility.

By this time even the remote problem of consciousness had come to look at least a little less remote—mainly through a gradual process of elimination. Among the suggested interpretations of consciousness that it now seemed safe to eliminate was the one in which conscious experience was conceived to be a correlate of isomorphic electric-field forces and volume current changes in the cerebral cortex. This view, engendered in Gestalt psychology and a major contender among the-

ories of consciousness in the 1940s (Köhler and Held, 1949), appeared to be ruled
out in particular by the failure of multiple metallic and dielectric inserts (i.e.,
electric-field distorters) to produce any major disruption in visual pattern perception.

Another view that reached a peak in the 1950s (Delafresnaye, 1954) had consciousness centered in brainstem reticular and centrencephalic mechanisms. The
contention was that a person really lives, so far as conscious feeling and experience
are concerned, in these deep mesencephalic centers. The neocortex came to be
regarded as a relatively recent and superficial adjunct for enhancing and elaborat-
ing the basic qualities of conscious experience already evolved in the mesencephalon. Interpretations along these lines had to be largely abandoned in the face of
our new findings on brain bisection in which surgical separation of the cerebral
hemispheres alone, leaving the brainstem intact, proved sufficient to divide most
of the higher psychological functions in cats and primates.

The split-brain findings also helped to resolve another major dichotomy in the
theory of mind. A long-standing question in philosophy asks whether conscious
awareness is restricted to brains or is, instead, a universal inner property of all
things. Do plants, atoms, cities, ships, and molecules all have some form of inner
awareness? If one could show that consciousness is selectively localized even within brains, with some neural systems being endowed with the property of conscious
experience while others are not, this would be a strong argument against the idea
that inner conscious awareness is something universal. If consciousness is lacking
in the cerebellum and in other neural systems, if it is lacking even in the cerebrum
during dreamless sleep, in coma, or after death, why should we assume it to be
present in plants, mountains, or molecules? The added discovery that conscious
awareness could be divided into right and left realms by severing a set of forebrain
fiber systems at the neocortical level greatly strengthened the view that consciousness is a special and selectively localized property rather than something universal.
The balance of the evidence would now appear to favor a prior inference (Sperry,
1952) that consciousness is an operational derivative of activity in particular
cerebral circuit systems designed expressly to produce their own specific conscious
effects. The implication here of causal action *upon* as well as *from* neural events
was yet to be appreciated.

Section of the corpus callosum appears to divide the unified perception of the
visual field down the vertical midline, into two inner visual worlds within the
left and right hemispheres respectively (Sperry, 1968, 1970a). This and similar
split-brain phenomena begin to carry us rather close to where direct correlations
can be made between conscious mental experience and activity in specific neural
structures. Incidentally, one may now occasionally come across statements, made
with the advantage of hindsight, that this callosal syndrome had already been fully
recognized and elucidated much earlier in the writings of Maspes, Dejerine, and
the German school of neurologists, and had simply been forgotten or overlooked
in the English-language literature. Actually, the confusion during the 1940s and
early 1950s regarding the corpus callosum and its functions was worldwide. The
extensive review in French by F. Bremer and his colleagues, which appeared in

1956 in the *Archives Suisse de Neurologie et de Psychiatrie*, gives a knowledgeable and fair assessment of the world literature and of the confused picture regarding callosal function as it stood at the time.

In any case the split-brain, dielectric-plate, and related findings seemed, along with other developments, to clear the way for a modified approach to the theory of consciousness (Sperry, 1965). This was an interpretation that I had recently come to favor but had been hesitant to publish, mainly because it represented a swing toward mentalism, presenting a conceptual explanatory model for psycho-physical interaction. An alternative to psycho-physical parallelisms and psycho-physical identity theory, this modified view involved a break with long-established behaviorist-materialist doctrine, amounting almost to a full reversal of the central premise on which behaviorism had been originally founded: instead of renouncing or ignoring the subjective conscious mind, this interpretation gave full recognition to inner conscious experience as a top-level directive force or property in cerebral function.

In this view, which has held up for more than ten years now, the conscious mind is no longer set aside as a passive correlate of brain activity, but becomes instead an essential working part of the brain process and a causal determinant in cerebral action (Sperry, 1970b; 1974b). Consciousness in this scheme is not looked upon as just an inner aspect of the neural process; nor do we relegate it to some metaphysical, epiphenomenal, or other separate dualistic realm. Nor is it dismissed by semantic gymnastics as being unimportant or nonexistent, or as being identical to the neural events. Conscious mental experience in our present interpretation is conceived to be a holistic emergent of brain activity, different from and more than the neural events of which it is composed, and a real phenomenon in its own right possessing causal potency in brain function. At present a more detailed or exact description in objective terms is hardly possible, but this will presumably be achieved with further advances in brain research.

Our current interpretation can be classified as an "emergent" theory of mind, provided it is distinguished from the earlier emergent views of Gestalt psychology. In the present scheme there is no dependence on electric-field forces or volume conduction effects, nor on an isomorphic or topological correspondence between the events of perceptual experience and the corresponding events in the brain. Furthermore, the mental events are conceived to be not merely *correlates* of brain activity, but also *causes*. The causal relation involves the universal power of the whole over its parts, in this case the dynamic enveloping power and properties of conscious high-order brain processes over their constituent neurophysiological and chemical elements. As dynamic emergent properties of cerebral excitation, conscious mental phenomena are given a working role in brain function and a pragmatic reason for being and for having been evolved. I was unable to find anything quite like this interpretation expressed previously from either the mentalist or materialist side, and it seemed to offer a compromise and resolution for the two divergent approaches to the mind-brain problem.

Back in 1965, when this "mind-over-matter" model was first ventured, one had to search a long way in philosophy and especially in science to find anyone who would put into writing the view that mental forces or events are capable of causing physical changes in an organism's behavior or its neurophysiology. With rare exceptions, writings in behavioral science dealing with perception, imagery, emotion, cognition, and other mental phenomena were very cautiously phrased to conform with prevailing materialist-behaviorist doctrine. Care was taken to be sure that the subjective phenomena should not be implied to be more than passive correlates or inner aspects of brain events, and especially to avoid any implication that the mental phenomena themselves might interact causally with the physical brain process. And fifteen years earlier I would not have dreamed that I would ever accept such a concept myself.

Those few in philosophy who had earlier subscribed to psychophysical interaction had been such extreme dualists that little heed had been paid to them in behavioral science. However, once we were able to show that mental events as emergent properties could causally influence neural events in a compromise formulation without violating the principles of scientific explanation, the long-standing resistance to psychophysical interactionism began to decline. For example, the frequency of use of such terms as "mental imagery" or "visual imagery" as explanatory constructs has, after more than five decades of careful avoidance, literally exploded in the recent scientific literature dealing with perception, cognition, and other higher functions (for a critique of this trend see Pylyshyn, 1973). During this same period related philosophical positions have undergone pertinent but subtle rephrasing to encompass these changes, to a degree where it now becomes important in many instances to distinguish between "pre-1965" and "post-1965" versions of a given philosophic stance.

Among other things, the acceptance of inner mental experience as having a significant causal role in cerebral function has produced a changed picture of scientific determinism as applied to human behavior and social action. The phenomena of subjective experience, including feelings and values of all kinds, must now be recognized as positive causal factors in the brain's decision-making process. The freedom thereby introduced into the causal brain sequence leading to a volitional choice far surpasses, in both degree and kind, the notions envisaged in the more mechanistic and atomistic forms of determinism that have excluded mental events. The present interpretation may be seen to set the human brain apart in respect to free will, placing it at an apex above all other known systems in the deterministic universe of science. Our new approach in mind-brain theory thus goes far to restore to human nature some of the personal dignity, freedom, inner creativity, and responsibility of which it has long been deprived by behaviorism and by materialist science generally.

The issues involved here are basic and central to human value questions at all levels. Value problems tend today to take priority over the problem of consciousness or, indeed, any of the theoretical problems of pure science. What good will it do, one may ask, for mankind finally to crack the mind-brain problem if the whole human species is about to be blown off the globe, or starved or crowded off, or polluted out of any reasonable quality of existence? Even staunch advocates of pure science agree that the most important thing many of us can do for pure science these days is not so much to *practice* it as to try to *preserve* it along with civilization and humanity through the coming decades—or generations if one's optimism allows. It is simply a matter of first things first. The same reasoning applies, of course, to many of the aims and objectives that have held priority in biomedical and other fields. The discovery of a cure for cancer, schizophrenia, or cardio-vascular disease, for example, would have a relatively minor impact on human existence, generally, as compared to the effect of a slight shift in social values affecting world policy on abortion, birth control, species rights, conservation, and the like.

When it comes to the practicalities of the world problems that are now reshuffling priorities in science, we find, somewhat paradoxically, that the "ivory tower" problem of consciousness continues to carry a top practical rating (Sperry, 1972). World conditions and the future in general will be determined very largely by concepts and beliefs regarding the properties of conscious mind and the kinds of life goals and values that derive from these. Take, for example, the question of whether conscious mind is mortal or immortal, or reincarnate, or cosmic, or whether it is brain-bound, or universal as in panpsychism, or perhaps "supracoalescent" as suggested by Teilhard de Chardin. Clearly, each of these alternatives suggests a different system of value criteria and social priorities.

The final answer, of course, is not yet in, but advances in the mind-brain sciences during the last few decades have substantially narrowed the latitudes for realistic answers. In modern neuroscience it is no longer a question of whether conscious experience is mortal and tied to the living brain, but rather to which particular parts of the brain and which kinds of neural systems are involved. Current interpretation, strongly favoring a "this-world" concept of mind, dispenses with a large number of the "other-world" value determinants of the past. A unifying view of brain, mind, and man in nature can now be seen that provides a monistic framework for values within which science can operate. Even a science of values can be envisaged that would treat values as objective determinants in decision-making and become a basic core for behavioral and social science.

Once it is agreed that mental experience exerts directive control in brain function it follows that the world of inner experience must be given its due in the scientific description of brain action. This puts neuroscience in a position to encompass, at least in principle, all those higher subjective, humanistic aspects of man's nature that the objective approach of science has traditionally seemed to exclude. For these and related reasons (Sperry, 1972), we can no longer accept the dichotomy that has heretofore kept science and values in separate realms. When subjective values have objective consequences and are viewed as universal determinants in all social decision making, they become part of the content of science. The origins, development, and logical structure of values as powerful causal agents become important scientific concerns. More than this, science on these terms, after exclusion of various metaphysical and mystic alternatives including "other-world" mythologies, becomes man's most important means for determining ultimate value and meaning. Our current scheme (Sperry, 1972) would elevate science into a higher social role as source and arbiter of values and belief systems at the highest level. Science would become the final determinant of what is right and true, the best source and authority available to the human brain for finding ultimate axioms and guideline beliefs to live by, and for reaching an intimate understanding and rapport with the forces that control the universe and created man.

## Acknowledgment

The work of the author and his laboratory is supported by Grant No. 03372 of the National Institute of Mental Health and by the F. P. Hixon Fund of the California Institute of Technology.

## References

Akelaitis, A.J. (1943): Studies on the corpus callosum. VII: Study of language functions (tactile and visual lexia and graphia) unilaterally following section of corpus callosum. *J. Neuropathol. Exp. Neurol.* 2: 226–262.

Bremer, F., Brihaye, J., and André-Balisaux, G. (1956): Physiologie et pathologie du corps calleux. *Arch. Suisses Neurol. Psychiatr.* 78: 31–79.

Delafresnaye, J. F. (1954): *Brain Mechanisms and Consciousness*. Springfield, Ill.: C.C Thomas.

Herrick, J. C. (1948): *The Brain of the Tiger Salamander*. Chicago: University of Chicago Press.

Köhler, W., and Held, R. (1949): The cortical correlate of pattern vision. *Science* 110: 414–419.

Pylyshyn, Z. W. (1973): What the mind's eye tells the mind's brain: A critique of mental imagery. *Psychol. Bull.* 80: 1–24.

Sperry, R. W. (1945): The problem of central nervous reorganization after nerve regeneration and muscle transposition. *Quart. Rev. Biol.* 20: 311–369.

Sperry, R. W. (1950): Neural basis of the spontaneous optokinetic response produced by visual inversion. *J. Comp. Physiol. Psychol.* 43: 482–489.

Sperry, R. W. (1951a): Mechanisms of neural maturation. *In: Handbook of Experimental Psychology.* Stevens, S. S., ed. New York: John Wiley, pp. 236–280.

Sperry, R. W. (1951b): Regulative factors in the orderly growth of neural circuits. *Growth Symp.* 10: 63–87.

Sperry, R. W. (1952): Neurology and the mind-brain problem. *Am. Sci.* 40: 291–312.

Sperry, R. W. (1955): On the neural basis of the conditioned response. *Br. J. Anim. Behav.* 3: 41–44.

Sperry, R. W. (1961): Cerebral organization and behavior. *Science* 133: 1749–1757.

Sperry, R. W. (1965): Mind, brain and humanist values. *In: New Views of the Nature of Man.* Platt, J.R., ed. Chicago: University of Chicago Press, pp. 71–92. Reprinted in *Bull. Atom. Sci.* (1966) 22(7):2–6.

Sperry, R. W. (1968): Mental unity following surgical disconnection of the cerebral hemispheres. *Harvey Lect.* 62: 293–323.

Sperry, R. W. (1970a): Perception in the absence of the neocortical commissures. *Res. Publ. Assoc. Res. Nerv. Ment. Dis.* 48: 123–138.

Sperry, R. W. (1970b): An objective approach to subjective experience: Further explanation of a hypothesis. *Psychol. Rev.* 77: 585–590.

Sperry, R. W. (1972): Science and the problem of values. *Perspect. Biol. Med.* 16: 115–130. Reprinted in *Zygon* 9 (1974): 7–21.

Sperry, R. W. (1974a): Lateral specialization in the surgically separated hemisphere. *In: The Neurosciences: Third Study Program.* Schmitt, F.O., and Worden, F.G., eds. Cambridge, Mass.: The MIT Press, pp. 5–19.

Sperry, R. W. (1974b): Mental phenomena as causal determinants in brain function. *In: Mind and Brain: Philosophic and Scientific Strategies.* Globus, G., Maxwell, G., and Savodnik, I., eds. New York: Plenum (in press).

Sperry, R. W., and Miner, N. (1955): Pattern perception following insertion of mica plates into visual cortex. *J. Comp. Physiol. Psychol.* 48: 463–469.

Stratton, G. M. (1897): Vision without inversion of the retinal image. *Psychol. Rev.* 4: 341–360; 463–481.

Weiss, P. (1936): Selectivity controlling the central-peripheral relations in the nervous system *Biol. Rev.* 11: 494–531.

Weiss, P., and Taylor, A. C. (1944): Further experimental evidence against "Neurotropism" in nerve regeneration. *J. Exp. Zool.* 95: 233–257.

Wilder Penfield (b. 1891, Spokane, Washington) was the first director of the Montreal Neurological Institute and the first professor in McGill University's Department of Neurology and Neurosurgery. His researches have included studies of epileptic-seizure patterns in relation to the functional anatomy of the brain, mapping the excitable cortex in conscious man, and the brain mechanisms responsible for speech. Since his retirement in 1960, Dr. Penfield has written on subjects ranging from historical fiction and biography to neuroscience and philosophy.

# 25
## The Mind and the Brain*

## Wilder Penfield

The book of which this paper will be a part began as an address prepared for the annual general meeting of the American Philosophical Society, Philadelphia, April 21, 1973, entitled "The place of understanding." The chairman of that meeting was Francis O. Schmitt. The address was elaborated for the Hans Berger Centennial Symposium on Brain-Mind Relationships, held at the Montreal Neurological Institute, May 23, 1973.

After that I could hardly refuse the challenge, and so I have marshaled the facts as I see them and have drawn conclusions that seem to me inevitable. In the process, the book became a pilgrim's progress rather than an all-embracing analysis. I must, therefore, apologize to other workers in this field if I have not done full justice to their excellent contributions.

Karl Lashley spent thirty years of his industrious life striving to discover the nature of the "memory trace" in the animal brain, beginning with experimental investigations of the rat's brain and ending with the chimpanzee (see Lashley, 1960). He was hunting for the engram, the record, which is to say, "the structural impression that psychical experience leaves on protoplasm." He failed to find it, and he ended by laughing cynically at his own effort and by pretending to question whether it was, in fact, possible for animals or even man to learn at all.

But consciousness and the relationship of mind to brain are problems difficult to study in animals. The method of study by conditioned reflexes introduced by I. P. Pavlov (1927) can carry our understanding only a certain distance. On the other hand, clinical physicians, in their approach to man, may hope with reason to push on toward an understanding of the physiology of memory and the physical basis of the mind and of consciousness.

In June 1947 Sir Charles Sherrington wrote a foreword to his book *The Integrative Action of the Nervous System*, which was then being republished in his honor by the Physiological Society. The last paragraph of his foreword expressed his conclusion about mind-brain relationships: "That our being should consist of two fundamental elements offers, I suppose, no greater inherent improbability than that it should rest on one only."

It is a quarter of a century since Sherrington wrote these words. We have learned a good deal about man since then, and it is exciting to feel, as I do, that the time has come to look at his two hypotheses, his two "improbabilities." Either brain action explains the mind, or we must deal with two elements.

*Selections from Wilder Penfield, *The Mystery of the Mind: A Critical Study of Consciousness and the Human Brain* (Princeton University Press, 1975). Reprinted by permission of Princeton University Press.

Hippocrates, the father of scientific medicine, left behind only one discussion of the brain and the nature of consciousness. It was included in a lecture delivered to an audience of medical men on Epilepsia, the affliction that we still call epilepsy. Here, in an excerpt from this lecture, we see his amazing flash of understanding: "Some people say that the heart is the organ with which we think and that it feels pain and anxiety. But it is not so. Men ought to know that from the brain and from the brain only arise our pleasures, joys, laughters, and tears. Through it, in particular, we think, see, hear, and distinguish the ugly from the beautiful, the bad from the good, the pleasant from the unpleasant. . . . To consciousness the brain is messenger." In another part of his discussion he remarks, simply and accurately, that epilepsy comes from the brain "when it is not normal."

In retrospect, it is abundantly clear that Hippocrates came to his conclusions by listening to epileptic patients when they told him their stories, and by watching them during epileptic seizures. The reader will come to understand, in the pages that follow, that Epilepsia still has secrets to reveal. She has much to teach us if we will only listen. And today the use of the stimulating electrode as well as the electroencephalograph can help us to understand her speech.

**The Physiological Interpretation of an Epileptic Seizure**

In 1958, after I had accumulated considerable clinical experience, I reconsidered critically the physiology involved in the electrical exploration of the human brain and reported my conclusions in the Sherrington Lecture (Penfield, 1958). I realized that when an electrode passes a current into the cerebral cortex, it interferes completely with the patient's normal use of that area of gray matter. In some areas there is no evidence of any further effect. For example, as shown in Figure 25.1, an electrode on the speech cortex causes aphasia. But in other areas, as explained in Figure 25.2, stimulation gives a positive response as well. Such positive responses are produced, not by activation of the local gray matter, but by neuronal conduction along insulated axons to a distant area of gray matter beyond the interfering influence of the electrode's current.

Let me repeat: The activation is of the distant gray matter, as seen in Figure 25.1 in stimulation of motor cortex, somatic sensory cortex, or visual sensory cortex. There is always interference in the normal use of the local gray matter. On the other hand, if there is also a positive response, it is due to functional activation of distant gray matter.

Consequently, when the electrode is applied to the hand area of the motor cortex, the delicate movements of the hand that the cortex makes possible are paralyzed, but the secondary station of gray matter in the spinal cord is activated, and crude movements, such as clutching, are carried out.

In clinical epilepsy the spontaneous discharge occurs, in the great majority of cases, either in the gray matter of the cortex or in the gray nuclei of the higher brainstem. It never occurs in white matter. If it occurs in a so-called silent area of the cortex, there may be no manifestation of it unless an electroencephalogram is being taken.

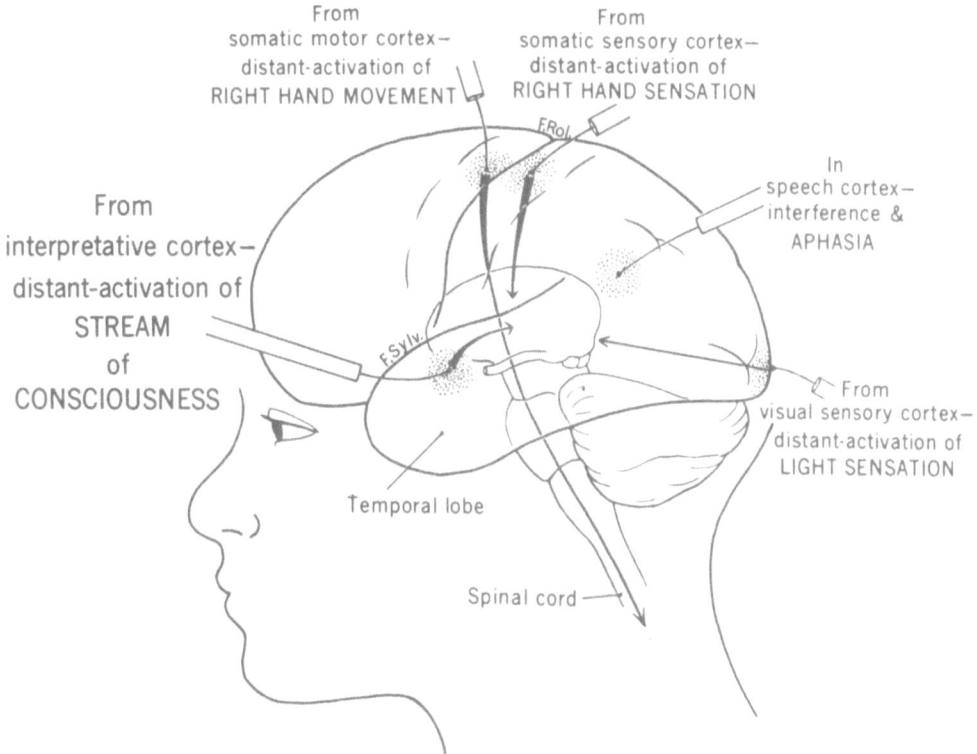

Figure 25.1  Activation of the brain's record of consciousness and some other results of stimulation. The left hemisphere of the brain is outlined, with the brainstem and spinal cord shown beneath, to illustrate the results of electrode stimulation of the cortex in motor, sensory, and what may be called psychical areas for the recall of past experience. The dotted zone about each stimulating electrode tip (unipolar) suggests the area of interference in which local cortical elaborative action is arrested by electrical interference. In addition to this interference, a positive response is described from each of these electrodes, except the one on the area where speech is localized. Stimulation of the speech cortex produces only interference aphasia. The positive responses, on the other hand, are caused by normal axonal conduction from cells near the electrode to a distant but functionally related area of gray matter. Thus the active response is a physiological activation of that distant gray matter. In the case of stimulation of the interpretive cortex, it is the sequential record of successive conscious states from the past that is activated. In the case of motor cortex the target of activation is gray matter in lower brainstem or spinal cord. In the case of sensory areas the target is in the higher brainstem. Drawing by Eleanor Sweezey.

In any fit, focal discharge begins in some local region of gray matter. If a positive manifestation occurs, it is produced, as in the case of electrical stimulation (see Figure 25.2), by axon conduction to a distance. It is due, then, to neuronal activation of some distant secondary ganglionic station.

An epileptic discharge probably continues until the discharging local neurons are exhausted. The secondary, distant response that it produces then stops. But the local paralytic interference in the primary area of discharge continues after the discharge is over, until there is recovery from the cell exhaustion. The distant response, if any, is a physiological phenomenon and stops, as I have said, as soon as axonal conduction to it stops.

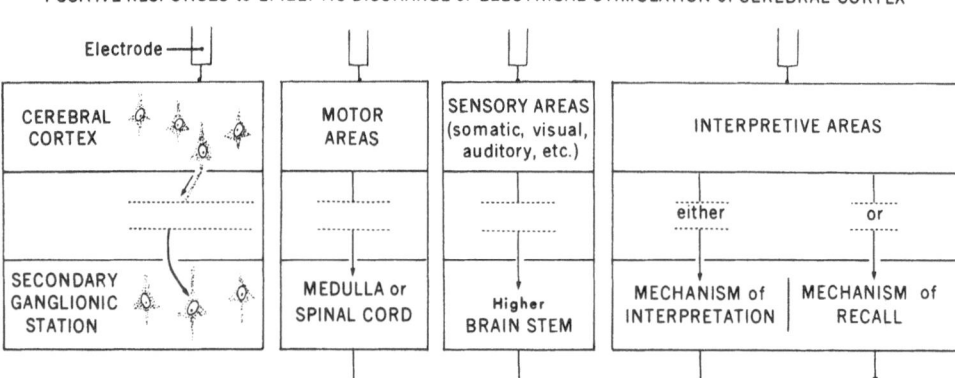

Figure 25.2   Positive responses. Electrical stimulation (or epileptic discharge) interferes with function of gray matter locally. It produces an active response only when the electrode is applied to an area of cerebral cortex from which axonal conduction along a functional tract normally activates some *distant* ganglionic station. Cortical responses are of four types: muscular movement, sensation, interpretive perception, and recall of conscious experience.

There is always a danger in exploratory electrical stimulation that the electrode may bring to the cortex a current that is too strong, causing the local gray matter to go into epileptic discharge. When the electrode is withdrawn, there is an after-discharge and a local seizure. There is also added danger then that axonal conduction from the local gray to some distant gray matter may increase enough to become a bombardment, and so produce a secondary epileptic explosion.

Spread of the local discharge in any fit may occur in one of two ways: by a "Jacksonian march" into contiguous gray matter, or at a distance (as just explained) by neuronal conduction to a functionally related area of gray matter. Spread of discharge, and thus of the epileptic fit, occurs when that conduction turns into a too-violent bombardment. The physiological activation of the distant gray matter is then replaced by discharge in that distant area. This causes a new local functional interference at a distance, instead of activation.[1]

## An Early Conception of Memory Mechanisms—And a Late Conclusion

This understanding of the physiology of electrical stimulation, and of the pattern of neuronal discharge in an epileptic seizure, led at once to a clearer understanding of what is taking place in each experiential response to electrical stimulation. It called for a reconsideration of the "flashbacks." Consequently, after the close of my own career as an operating neurosurgeon in 1960, I and my colleagues considered and published every detail of the experiential responses so others would be able to judge their meaning for themselves. I presented this reexamination of our data in the Lister Oration at the Royal College of Surgeons in 1961, and published it in full with Phanor Perot in 1963.

There were 1132 patients for us to reconsider. The brain of each had been explored under local anesthesia in the course of an operation for radical treatment

1. If this is a true statement of the physiological principles involved in epileptic seizures, and I believe it is, it calls for the thoughtful attention of clinicians and electroencephalographers.

of epilepsy. In 520 patients whose temporal lobes were exposed and explored, the experiential responses came only from the temporal lobe, never from any other part of the brain. Forty of the temporal explorations (7.7%) gave experiential responses, and 53 (10%) had complained before operation of dreamlike attacks in which past memories came to mind.

In 1951 I had proposed that certain parts of the temporal cortex should be called "memory cortex," and suggested that the neuronal record was located there in the cortex near the points at which the stimulating electrode may call forth an experiential response (Penfield, 1952b). This was a mistake, as I showed clearly in 1958 during my Sherrington Lecture (Penfield, 1958).

Although the record is not in the cortex, nevertheless the basic initial hypothesis proposed at that time is still tenable. "It is tempting to believe," I wrote, "that a synaptic facilitation is established by each original experience." If so, that permanent facilitation could guide a subsequent stream of neuronal impulses activated by the current of the electrode even years later.

Since then, as I shall point out in the next section, we have come to call the "memory cortex" by another name—the "interpretive cortex." Its boundaries and those of the speech area may be seen in Figures 25.3 and 25.4, And today we realize that stimulation of interpretive cortex activates a recording located at a distance from that cortex, in a secondary center of gray matter. Putting this together with other evidence makes it altogether likely that the activated gray matter is in the diencephalon (higher brainstem), as I shall describe below.

## The Interpretive Cortex

Let me marshal and reconsider the evidence now presented to us by epilepsy and

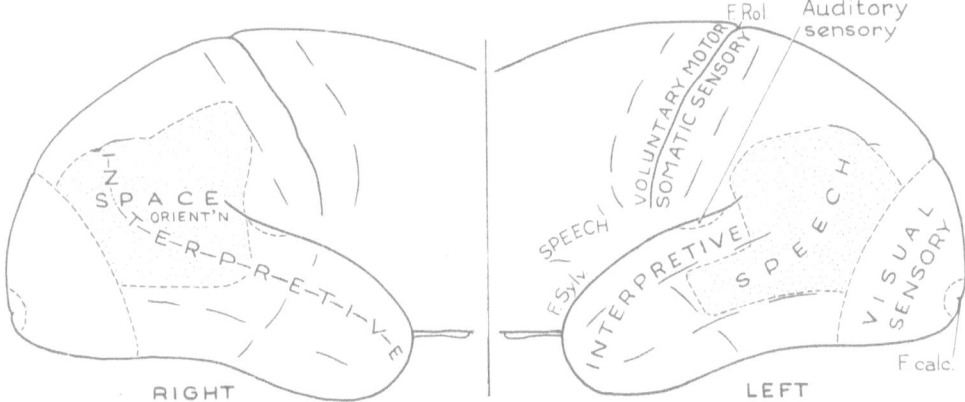

Figure 25.3   Interpretive cortex and speech cortex (lateral surfaces of the posterior parts of both hemispheres of a human adult). On the dominant, or speech side, interference aphasia is produced by stimulating in the area marked "speech." Both experiential and interpretive responses are produced by stimulating in the interpretive cortex. The area marked "space orientation" on the nondominant (right) side was outlined by study of the results of cortical excision. Complete removal of this area produced permanent spatial disorientation without aphasia. (For evidence in regard to the frontiers of the temporal speech area (Wernicke), see Penfield and Roberts (1959). For that on space orientation, see H. Hécaen et al. (1956). For the localization of the interpretive cortex, see Penfield (1959).) Drawing by Eleanor Sweezey.

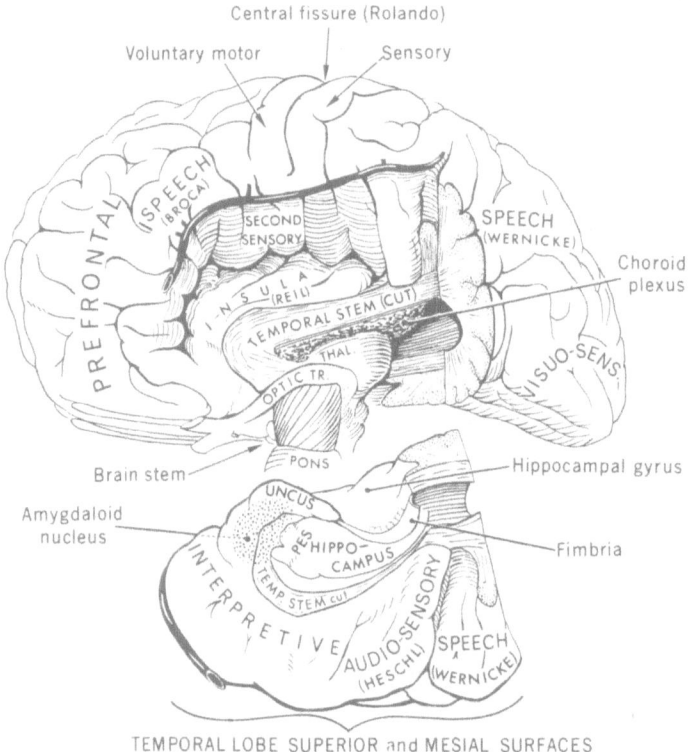

Figure 25.4  Superior and mesial surfaces of temporal lobe (left cerebral hemisphere). The temporal lobe was dissected free at autopsy, by opening the fissure of Sylvius, and was then cut across and turned down. Note that the hidden audiosensory gyrus of Heschl is seen to be bounded by speech cortex posteriorly and interpretive cortex anteriorly. Drawing by Eleanor Sweezey.

the electrode, after which we may go on to a consideration of the relationship of mind to brain.[2]

Two related mechanisms are revealed by stimulation of the interpretive cortex. First, there is a brain mechanism, the function of which is to send neuronal signals that interpret the relationship of the individual to his immediate environment. The action is automatic and subconscious, but the signal appears in consciousness. Examples of such signals are: these things are "familiar" or "frightening"; they are "coming nearer" or "going away"; and so on (Mullan and Penfield, 1959; Penfield, 1959).

Second, there is another, related brain mechanism revealed in experiential responses which is capable of bringing back a strip of past experience in complete detail without any of the fanciful elaborations that occur in a man's dreaming (Penfield, 1959; Penfield and Perot, 1963).

In ordinary life the automatic signal that informs one that present experience is familiar comes to all of us, I suppose. If it is accurate, and it usually is, one must be

2. I began to do this in a chapter of the book entitled *Basic Mechanisms of the Epilepsies* (edited by my former associates Herbert Jasper, Arthur Ward, and Alfred Pope, and published by Little, Brown and Co., Boston, 1969). It led me to a discussion of a specific mechanism for the mind. I shall here push the argument through to a conclusion.

using an automatic mechanism that can scan a record of the past, a record that has not faded but seems to remain as vivid as when the record was made.[3]

The gray matter of the interpretive cortex is part of a mechanism that presents interpretations of present experience to consciousness. In a sense, it would seem that the interpretive cortex does for perception of nonverbal concepts what the speech cortex and the speech mechanism do for speech. The localization of areas devoted to speech is reasonably clear. Although much work has yet to be done on the recognition of nonverbal concepts, I shall refer to it now as the "nonverbal-concept mechanism." These mechanisms, the one verbal and the other nonverbal, form a remarkable memory file to be opened either by a conscious call or by an automatic one (Penfield, 1958, 1959, 1968).

There is much more that could be said about the temporal lobes and memory. That mysterious doubled structure, the hippocampus, may well have much to do with memory of smell in some lower mammals, but in man it is concerned with memory of other things. It can be removed on one side with impunity when the remaining hippocampus is functioning normally. But if it is removed on both sides, the ability to reactivate voluntarily or automatically the record of the stream of consciousness is lost. The hippocampi seem to store keys-of-access to the record of the stream of consciousness. With the interpretive cortex, they make possible the scanning and the recall of experiential memory (see Penfield and Mathieson, 1974).

### An Automatic Sensorimotor Mechanism

And now there opens before us an exciting vista in which the automatic mechanisms of the brain interact with, and may be separated from, the brain's machinery-for-the-mind.

As I have pointed out, epileptic discharge may, and frequently does, confine itself selectively to one functional system, one functional mechanism, within the brain. When it does so, it paralyzes that mechanism for any normal function. If the function of gray matter is highly complicated and only partially automatic, as in the speech area of the human cerebral cortex, the epileptic discharge in it produces nothing more than paralytic silence, as in aphasia.

And so it is that the mechanism in the higher brainstem, whose action is indispensable to the very existence of consciousness, can be put out of action selectively! This converts the individual into a mindless automaton, as happens when epileptic discharge occurs in gray matter that forms an integral part of that mechanism. The tentative localization of that gray matter is shown in Figure 25.5.

If the discharge occurs primarily there, the patient's attack is called petit mal automatism. But because the temporal cortex and the prefrontal cortex have much to do with the transactions of the mind, a seizure discharge that begins locally in temporal cortex or in the anterior frontal cortex may spread by violent distant

---

3. Although the great majority of experiences thus recalled seem to be strongly visual or strongly auditory, or both, the perception of familiarity is not limited to auditory or visual experience at all, but apparently applies to all that enters consciousness. A person seen may be labeled as "seen before" (déjà vu), a bar of music as "heard before," a sequence of events as "happened before."

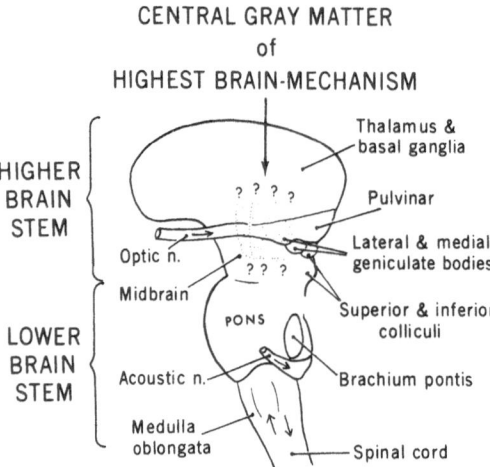

CENTRAL GRAY MATTER
of
HIGHEST BRAIN-MECHANISM

Figure 25. 5 The highest brain-mechanism. The site of the central gray matter of this brain-mechanism, the normal action of which constitutes the physical basis of the mind, is shown by the dotted lines. The question marks indicate only that the detailed anatomical circuits involved are yet to be established, not that there is any doubt about the general position of this area in which cellular inactivation may be brought about variously by pressure, trauma, hemorrhage, local epileptic discharge, and normally in sleep. Drawing by Eleanor Sweezey.

bombardment to this gray matter in the higher brainstem and thus produce an attack of automatism that differs little in character from that of petit mal.

These attacks show clearly the automatic complex performance of which man's computer is capable. In an attack of automatism the patient becomes suddenly unconscious, but since other mechanisms in the brain continue to function, he changes into an automaton. He may wander about, confused and aimless. Or he may continue to carry out whatever purpose his mind was in the act of handing on to his automatic sensorimotor mechanism when the highest brain-mechanism went out of action. Or he follows a stereotyped, habitual pattern of behavior. In every case, however, the automaton can make few, if any, decisions for which there has been no precedent. He makes no record of a stream of consciousness. Thus he will have complete amnesia for the period of epileptic discharge and during the period of cellular exhaustion that follows.

Patients are quite unable to predict when these absences of the mind will come. One patient, whom I shall call A, was a serious student of the piano and subject to automatisms of the type called petit mal. He was apt to make a slight interruption in his practicing which his mother recognized as the beginning of an "absence." Then he would continue to play for a time with considerable dexterity.

Patient B was subject to epileptic automatism that began with discharge in the temporal lobe. Sometimes the attack came on him while walking home from work. He would continue to walk and to thread his way through busy streets on his way home. He might realize later that he had had an attack because there was a blank in his memory for a part of the journey, as from Avenue X to Street Y.

If Patient C was driving a car, he would continue to drive, although he might discover later that he had driven through one or more red lights.

In general, if new decisions are to be made, the automaton cannot make them. In such a circumstance he may become completely unreasonable, uncontrollable, and even dangerous.

The behavior of these temporary automatons thus throws a brilliant light on a second mechanism, clearly distinguishable from that which serves the mind. It is the *automatic sensorimotor mechanism*. It, too, has centrally placed gray matter in the higher brainstem, where it must have a close functional interrelationship with the mechanism for the mind. The sensorimotor mechanism has its primary localization in the higher brainstem (see Figure 25.6), but the mechanism has, quite

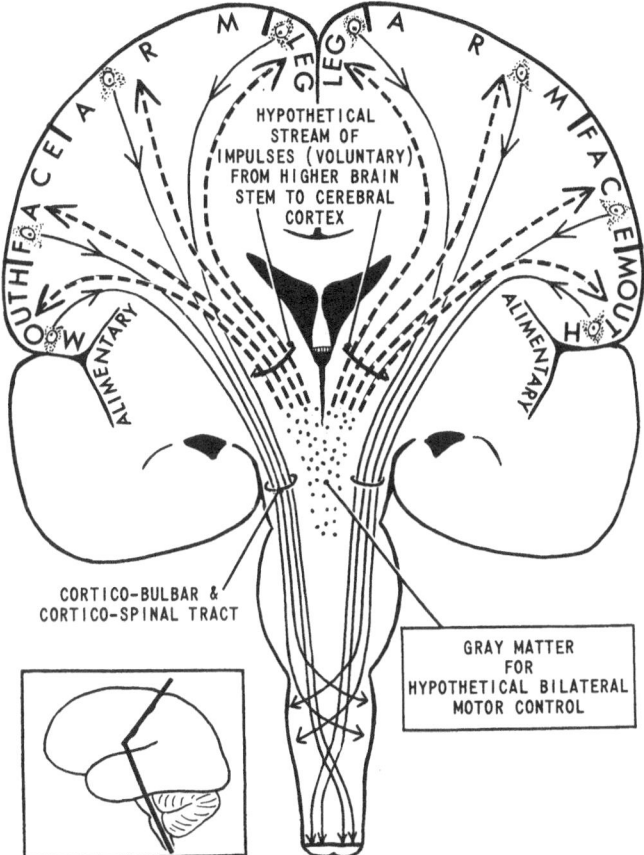

Figure 25.6  The automatic sensorimotor mechanism. This much-simplified diagram outlines only the direction of the executive, or motor, messages of the mechanism that takes bilateral control of the body either under direction of the mind or automatically. It makes use of the motor-cell stations in the precentral gyrus of both sides, as shown here from leg down to face and mouth. The entire mechanism is a portion of the centrencephalic integration and coordination system that makes effective mind-action possible. One may call it "man's computer." It makes available the many skills (including that of speech) that have been learned and recorded in the individual's past. It controls the behavior of the "human automaton" while the mind is otherwise occupied, or when the highest brain-mechanism is selectively inactivated, as in epileptic automatism. On the other hand, epileptic discharge within its central gray matter produces interference with its function and calls forth active responses from the motor centers in the cortex of both hemispheres, thus producing a generalized convulsion (grand mal). Drawing by Eleanor Sweezey.

obviously, a direct relationship to the sensory and motor portions of the cerebral cortex in both hemispheres. Thus there are two brain mechanisms that have strategically placed gray matter in the diencephalon or brainstem.

When an epileptic discharge occurs in the cerebral cortex in any of the sensory or motor areas, and if it spreads by bombardment to the higher brainstem, the result is invariably a major convulsive attack, *never*, in our experience, an attack of automatism. On the other hand, as mentioned above, a local discharge in pre-frontal or temporal cortex may develop into automatism.[4] This is a matter of considerable functional significance, and one that has been largely overlooked. We were first made aware of the differences in the manner of spread of epileptic discharge from cerebral cortex to diencephalon when Kristiansen did his study in 1951. After examining 95 cases in our clinic, he pointed out that 29 of the seizures had begun with local epileptic discharge in a motor convolution, 55 were somatic-sensory, and 11 visual-sensory. None of these patients developed automatism during the evolution of their attacks. Many, however, occasionally went directly from localized sensory or motor manifestations to generalized seizures (Penfield and Kristiansen, 1951).

William Feindel showed that automatism is frequent (78%) among patients who are subject to temporal-lobe epileptic discharges (Feindel and Penfield, 1954). He and I showed that automatism could be produced by stimulation if the electrode was passed into the temporal lobe and penetrated into or near the amygdaloid nucleus (see Figure 25.4). But this occurred only when one continued to stimulate until local epileptic discharge was produced. We assumed that this continued stimulation caused interference in the hippocampus on both sides and perhaps neuronal bombardment of gray matter in the higher brainstem that went on to epileptic discharge.

Thus, from a practical point of view, a clinician may find it useful to remember that local epileptic discharge in motor or sensory gray-matter areas of the cortex may spread by bombardment and so cause epileptic discharge in gray matter of the automatic sensorimotor mechanism in the higher brainstem. This produces a major convulsion because of its activation of all the motor areas of the cortex. The cortex of one side, being pitted against the other, causes the patient to stiffen the body and limbs rather than to turn.

The sensorimotor mechanism exerts activating control from its gray matter in the higher brainstem. This acts upon the secondary gray matter in the cerebral cortex of each hemisphere, and on the tertiary gray matter in the lower brainstem and the spinal cord. The major functional outflow of axon-conducted energy is carried to the muscles in one efferent stream. During any generalized grand mal seizure only the automatic control of breathing, which is located in the lower brainstem, escapes and continues its function.

When a local discharge occurs in prefrontal or temporal areas of the cortex, it may (or may not) spread to the highest brain-mechanism in the higher brainstem.

4. Herbert Jasper and I showed that local epileptic discharge in the prefrontal cortex, occurring either spontaneously or when we had set it off by the electrode, might spread to the diencephalon and cause an attack of automatism which was very like the automatism of petit mal (Penfield and Jasper, 1954).

When it does this, it produces automatism. This suggests that functional interaction between the mind's brain-mechanism and the sensory and motor convolutions of the cortex normally takes place through the automatic sensorimotor mechansim in the higher brainstem, rather than directly.

### Recapitulation

Mind, brain, and body make the man, and the man is capable of so much! Capable of comprehension of the universe, dedication to the good of others, planned research, happiness, despair, and eventually, perhaps, even an understanding of himself. He can hardly be subdivided. Certainly, mind and brain carry on their functions normally as a unit.

The neurophysiologist's initial undertaking should be to try to explain the behavior of this being on the basis of neuronal mechanisms alone—this biped who falls on his knees to pray to his god and rises to his feet to lead an army, or to write a poem, or to dig a ditch, or to thrill to the beauty of the sunrise, or to laugh at the absurdities of this world!

To me it seems more and more reasonable to suggest, as I did at the close of the Thayer Lectures at Johns Hopkins in 1950, that the mind may be a distinct and *different essence*. A science reporter who was present at the lectures used those words in his report as though that were my conclusion. I was not ready for that then, and I am hesitant still. But look with me for a moment at our world.

If one is to make a judgment on the basis of behavior, it is apparent that man is not alone in the possession of a mind. The ant (whose nervous system is a highly complicated structure), as well as mammals such as the beaver, dog, or chimpanzee, shows evidence of consciousness and of individual purpose. The brain, we may assume, makes consciousness possible in him too.

In all of these forms, as in man, memory is a function of the brain. Animals in particular show evidence of what may be called "racial memory." Moreover, new memories are acquired in the form of conditioned reflexes. In the case of man these reflexes preserve skills, the memory of words, and the memory of nonverbal concepts. Then there is, in man at least, a third important form of memory: experiential memory and the possibility of recalling the stream of consciousness with varying degrees of completeness. In this form of memory and in speech, the convolutions, which have appeared in the temporal lobe of man as a late evolutionary addition, are employed as "speech cortex" and "interpretive cortex."

Perhaps I should recapitulate. I realized as early as 1938 that to begin to understand the basis of consciousness one would have to wait for a clearer understanding of the neuronal mechanisms in the higher brainstem. They were obviously responsible for the neuronal integrative action of the brain that is associated with consciousness.

Since then electrical stimulation and a study of epileptic patterns in general have helped us to distinguish three integrative mechanisms. Each has a major area or nucleus of gray matter within the higher brainstem, an aggregation of nerve cells that may be activated or paralyzed.

1. *Highest Brain-Mechanism* (Figure 25.5): The function of this gray matter is to

carry out the neuronal action that corresponds with action of the mind. Proof that this is a mechanism in itself depends on the facts that injury of a circumscribed area in the higher brainstem produces invariable loss of consciousness, and that the selective epileptic discharge, which interferes with function within this gray matter, can produce the unconsciousness seen in epileptic automatism without paralysis of the automatic sensorimotor mechanism located nearby.

2. *Automatic Sensorimotor Mechanism* (Figure 25.6): The function of this gray matter is to coordinate sensorimotor activity previously programmed by the mind. This biological computer mechanism carries on automatically when the highest brain-mechanism is selectively inactivated. It causes a major convulsive attack (grand mal) by activation of cortical motor convolutions when epileptic discharge occurs in its central gray matter.

3. *The Record of Experience* (Figure 25.7): The function of the central gray

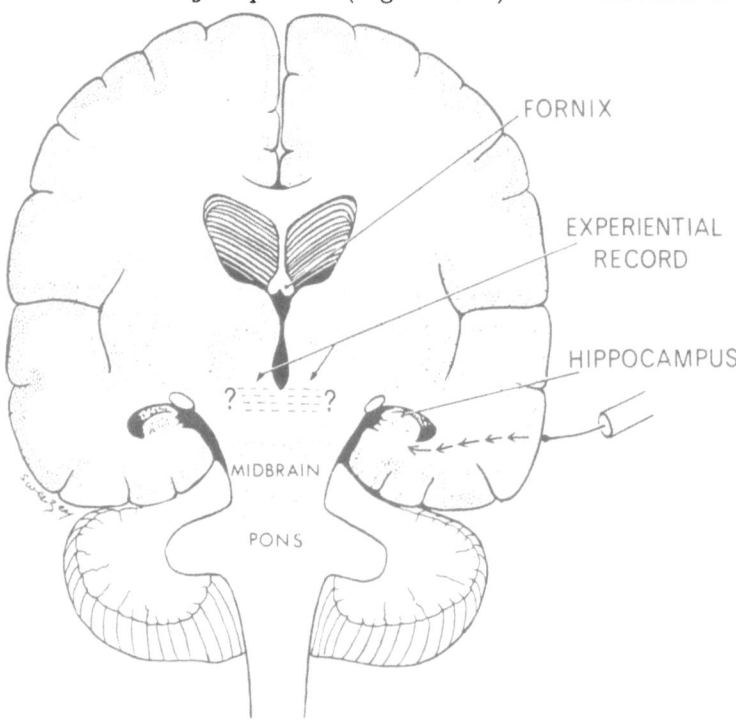

Figure 25.7   The experiential record of consciousness and its reactivation. A stimulating electrode is shown here applied to a convolution in the interpretive cortex. It recalls the past by activating a functional circuit (experiential response). The circuit has yet to be completely defined but may well include the hippocampus and fornix. The beginning of the pathway of neuronal activation is shown by the line of arrows from beneath the electrode. It may be that it passes directly through white matter into the higher brainstem. Or it may enter the hippocampus and pass along the fornix, which conducts from hippocampus to brainstem. The hippocampus evidently plays an essential role, together with the interpretive cortex, in the function of scanning the experiential record and recalling memories. Removal of one hippocampus affects memory little. Removal of both abolishes the ability to recall past experience, either voluntarily or for the purposes of automatic interpretation. In the case of a simple bilateral loss of the hippocampi, other forms of memory are retained, e.g., memory of language and of skills and of nonverbal concepts. (See Penfield and Mathieson, 1974.) Drawing by Eleanor Sweezey.

matter of this integrative mechanism, as shown by electrode activation, is to recall to a conscious individual the stream of consciousness from past time.

As I have said, the function of each of these three areas of gray matter can now be recognized. Each plays a role in the normal centrencephalic integration and coordination which is prerequisite to, and which accompanies, consciousness. As has been pointed out, this integrative activity is arrested by any form of general functional interference with neuron activity in the region outlined in Figure 25.5. The central gray matter for the automatic mechanism (1) and that of the mind's mechanism (2) are shown separately in Figure 25.5 and 25.6 because the two are capable of separate function and because I cannot picture their exact anatomical relationship.

Thanks to patients who have told us about the stream of consciousness presented to them while the surgeon's activating electrode was still in place, we can conclude with certainty what data these integrative mechanisms inhibit and thus eliminate, and what they present to the mind and thus cause the brain to record.

The imprint of memory's engram is somehow added during neuronal action. Conscious attention seems to give to that passage of neuronal impulses permanent facilitation for subsequent passage of potentials along the neuronal connections in the same pattern. Thus a recall engram is established, and this, one may suggest is the real secret of learning. It is effective in the establishment of experiential memory as well as of the memory that is based on conditioned reflexes.

If this is so, then it was correct to assume that the record of the stream of consciousness is not in some duplicated, separate memory apparatus such as is available in the hippocampus, but is, rather, in the operational apparatus of the highest brain-mechanism. The hippocampal gyrus in each hemisphere is part of the mechanism of scanning and recall. Therefore there must be a laying down in each hippocampus of some sort of duplicated key of access to the record of the stream of consciousness (Penfield and Mathieson, 1974).

### Relationship of Mind to Brain: A Case Study

Consciousness can be present whenever the highest brain-mechanism is normally active, even though adjacent parts of the brain are inactivated by some abnormal influence. A patient, whom I was called to see in Moscow under dramatic circumstances in 1962, illustrated the fact that conscious understanding may be present when motor control has been lost completely, or almost completely, and when the brain is not capable of making a permanent record of the stream of consciousness.

The patient was the brilliant physicist Lev Landau. Only intensive nursing had kept him alive during six weeks of complete unconsciousness following a head injury in an automobile crash.

On my first examination of the patient, I agreed that he was completely unconscious. I then recommended a minor diagnostic operation (ventriculogram). His limbs were paralyzed, his eyes open but apparently unseeing.

Next morning, when I entered his room to examine him again, I was accompanied by his wife. She preceded me and, sitting down at the bedside, she talked to him, telling him that I had suggested to the Soviet surgeons that he should have a brain operation.

As I stood silent, watching over her head, I became aware of a startling change in the patient. He lay unmoving still, as on the previous night. But his eyes, which had been deviated from each other then, were focused now in a normal manner. He seemed to be looking at her, and he appeared to hear and see and to understand speech! How could this be? She came to the end of her explanation and was silent. His eyes then moved upward to focus quite normally on me. I moved my head from side to side. The eyes followed me. No doubt about it! Then they swung apart again and he appeared, as he had the night before, to be unconscious.

It was clear that the man had returned to consciousness. He had been able to hear and see and understand speech, but not to speak. He could not move, except to focus and turn his eyes briefly.

Perhaps I should explain that he and his wife had been separated for a time. It was our talk of a possible operation that had led to her being summoned to Moscow and to the hospital. She was seeing him that morning for the first time since the accident.

It was thrilling to realize what had happened. He had been roused by her presence and had probably understood her message. Evidently the hemispheres above the higher brainstem, with their speech and visual and auditory mechanisms, had not been injured. The to-and-fro exchange between brainstem and cortex was free. But when he sent neuronal messages out to the peripheral motor nuclei in the lower brainstem and spinal cord, none could pass the block at the level of the hemorrhage in the midbrain. None, that is, except those to the eye-movement center, which is highest of all the peripheral centers for motor control.

If he was conscious, he must have sent down many other messages that would normally have flashed outward to the muscles. His wife was an appealingly handsome woman. His mind may well have sent a message intended to cause his hand to take hers. But his hand lay motionless.

However that may be, I went back to the other doctors and we decided that no operation was necessary. I had seen the first sure sign of recovery. He was transferred at once from the outlying hospital, in which he had been nursed so magnificently, to the Moscow Neurosurgical Institute where he would have the great advantage of supervision by Professor B. G. Egorov. Physiotherapy was begun at once, and I learned later that there was slow but continuous and progressive recovery from that day onward.[5]

For the first six weeks after the accident, Landau's fellow physicists, most of them his disciples, had joined the nurses and doctors in their gallant effort to keep the patient breathing and capable of recovery if that should prove possible. This man, who had already been awarded the Lenin Prize for his contributions to physics, was given the Nobel Prize during his convalescence. He and his wife were happy together, and she was with him on the special occasion of his acceptance of the award.

5. An account of this case was published by a science reporter in an interesting book for lay readers: Alexander Dorozinski, *The Man They Would Not Let Die* (New York: Macmillan, 1965).

I saw him nine months later, on my return from a visit to the university hospitals of China. He had been transferred to a convalescent hospital. The neurologist, Professor Propper Graschenkov, took me to visit him, in company with Landau's collaborator, the brilliant Professor Lifshitz, who had shared the Lenin Prize with his master.

I quote now from my journal note made a few days later: "Landau was sitting up in bed in a fresh white shirt looking at me anxiously. He was a handsome person with a fine head. He had an air of understanding. He was stubborn in his confusion, and I could appreciate why, several months earlier, he had been depressed and had even asked Lifshitz to bring him poison."

Landau spoke excellent English and did so eagerly. "When I asked him if he had received the Lenin Prize, he said yes, but he could not tell me when. When I asked him if someone had shared it with him, he looked around and, finding Lifshitz standing with the others, smiled and pointed to him. 'I think he and I shared it.'"

The relationship of the two men was somewhat like that of David and Jonathan. "Lifshitz feels he has lost his closest friend and his leader."

The final diagnosis was as follows: a head injury resulting in a small hemorrhage in the conducting tracts (white matter) of the higher brainstem. During the six weeks following his accident the hemorrhage had gradually been absorbed through the brain's own circulation. Eventually neuronal communication had been restored through the nerve fibers of the brainstem. There remained only a difficulty in recalling the recent past. His confusion was due to that.

It is, of course, only a matter of conjecture exactly when consciousness did return. Before he could show that he was conscious, it was necessary for him to regain control of some portion of his motor system. Evidently, control of the movement of his eyes returned before anything else.

During our first interview, while Mrs. Landau was explaining the situation to him, and when he looked at me so searchingly, I concluded that the whole explanation continued to be clear in his mind. He was remembering what she had said at the beginning as well as what she was saying at the close of her explanation. For that, the mind does not need the brain's mechanism of experiential recall. It needs only the mechanism of the stream of consciousness, which was normally active.

When I saw him at the second interview nine months later, he did not remember me or anything that had happened to him for some months after his transfer to the Moscow Neurosurgical Institute. The scanning and recall of past experience can only be carried out by a mechanism in which at least the hippocampus of one temporal lobe must play an essential role (see Figure 25.7). And that mechanism can carry out recall only for periods in which it was normal and for which it is thus able to form its own clues to the stream of consciousness, its duplicate "keys of access" (Penfield and Mathieson, 1974).

But I learned later that Landau did continue to improve in the year that followed, and he was able to tutor his son for his university entrance examinations. Great recognition came to this man whose mathematical genius has been likened to that of Einstein. His countrymen rejoiced at his recovery, but the "depression" returned to him. Perhaps he realized that his brain could no longer serve him as it had.

I have described this case in some detail because it demonstrates how it is that when consciousness is present, the highest brain-mechanism is used to activate and employ other brain-mechanisms that are capable of normal function. Beyond that, it bears out my conclusion that the mind can hold the data that has come to it during this focusing of attention and while the mechanism of the stream of consciousness is moving forward. But the mind, by itself, cannot recall past experience unless the brain's special mechanism of scanning and recall is functioning normally.

In such a case as this, in which damage brings to the brain a small area of paralyzing interference, one's knowledge of how brain-mechanisms coordinate and integrate is put to the test. But, more than that, such accidental experiments point the way toward clearer understanding of how the business of the brain is transacted. Human physiology can only wait for guidance from such accidental experiments.

Hughlings Jackson remarked in his Hunterian Oration in 1872 that, "Medical men, since they, only, witness the results of the experiments of disease on the nervous system of man, will be looked to more and more for facts bearing on the physiology of the mind."

Since he used the daring phrase "physiology of the mind," he must have thought that the brain did, or would someday, explain the mind. After a century of study by neurologists who observe the paralyzed and the epileptic, by neurosurgeons who excise and stimulate the convolutions of the brain, and by electrophysiologists who record and conduct experiments, surely the time has come to ask that question and other questions.

Has the brain explained the mind? If it has, does the brain do so by the simple performance of its neuronal mechanisms, or by the supplying of energy to the mind? Or both? Does it supply the mind with energy and at the same time provide it with basic neuronal mechanisms that are related to consciousness?

## References

Adrian, E. D. (1966): *In: Brain and Conscious Experience* (Study Week of Pontificia Academia Scientiarum, September 28–October 4, 1964). Eccles, J.C., ed. New York: Springer-Verlag, pp. 238–248.

Feindel, W., and Penfield, W. (1954): Localization of discharge in temporal lobe automatism. *Arch. Neurol. Psychiatr.* 72: 605–630.

Hebb, D. O., and Penfield, W. (1940): Human behavior after extensive bilateral removal from the frontal lobes. *Arch. Neurol. Psychiatr.* 44: 421–438.

Hécaen, H., Penfield, W., Bertrand, C., and Malmo, R. (1956): The syndrome of apractognosia due to lesions of the minor cerebral hemisphere. *Arch. Neurol. Psychiatr.* 75: 400–434.

Hippocrates (c. 460–377 B.C.): The sacred disease. *In: Hippocrates, Medical Works.* Vol. 2. Jones, W.H.S., trans. Cambridge, Mass.: Harvard University Press (1923), pp. 127–183.

Jackson, J. H. (1873): On the anatomical, physiological and pathological investigation of the epilepsies. *West Riding Lunatic Asylum Med. Rep.* 3: 315–339.

Jackson, J. H. (1931): *Selected Writings of John Hughlings Jackson.* Vol. 1. *On Epilepsy and Epileptiform Convulsions.* Taylor, J., ed. London: Hodder and Stoughton.

James, W. (1905): *The Principles of Psychology,* Vols. I and II. New York: Holt, Rinehart and Winston.

Lashley, K. S. (1960): *The Neuropsychology of Lashley. Selected Papers of K. S. Lashley.* Beach, F. A., Hebb, D. O., Morgan, C. T., and Nissen, H. W., eds. New York: McGraw-Hill.

Mullan, S., and Penfield, W. (1959): Illusions of comparative interpretation and emotion. *Arch. Neurol. Psychiatr.* 81: 269–284.

Pavlov, I. P. (1927): *Conditioned Reflexes. An Investigation of the Physiological Activity of the Cerebral Cortex.* Anrep, G.V., trans. and ed. London: Oxford University Press.

Penfield, W. (1938): The cerebral cortex in man. I. The cerebral cortex and consciousness. *Arch. Neurol. Psychiatr.* 40: 417–442.

Penfield, W. (1952a): Epileptic automatism and the centrencephalic integrating system. *Res. Publ. Assoc. Res. Nerv. Ment. Dis.* 30: 513–528.

Penfield, W. (1952b): Memory mechanisms. *Arch. Neurol. Psychiatr.* 67: 178–191.

Penfield, W. (1955): The permanent record of the stream of consciousness. *Acta Psychol.* 11: 47–69.

Penfield, W. (1958): *The Excitable Cortex in Conscious Man* (The Sherrington Lectures, V). Springfield, Ill.: C. C Thomas.

Penfield, W. (1959): The interpretive cortex. *Science* 129:1719–1725.

Penfield, W. (1968): Engrams in the human brain. Mechanisms of memory. *Proc. R. Soc. Med.* 61: 831–840.

Penfield, W. (1969): Consciousness, memory and man's conditioned reflexes. *In: On the Biology of Learning.* Pribram, K. H., ed. New York: Harcourt, Brace & World, pp. 129–168.

Penfield, W. (1969): Epilepsy, neurophysiology, and some brain mechanisms related to consciousness. *In: Basic Mechanisms of the Epilepsies.* Jasper, H.H., Ward, A.A., Jr., and Pope, A., eds. Boston: Little, Brown, pp. 791–805.

Penfield, W., and Jasper, H. H. (1954): *Epilepsy and the Functional Anatomy of the Human Brain.* Boston: Little, Brown.

Penfield, W., and Kristiansen, K. (1951): *Epileptic Seizure Patterns. A Study of the Localizing Value of Initial Phenomena in Focal Cortical Seizures.* Springfield, Ill.: C. C Thomas.

Penfield, W., and Mathieson, D. (1974): Memory autopsy: Findings and comments on the role of hippocampus in experiential recall. *Arch. Neurol.* 31.

Penfield, W., and Perot, P. (1963): The brain's record of auditory and visual experience. A final summary and discussion. *Brain* 86: 595–696.

Penfield, W., and Rasmussen, T. (1950): *The Cerebral Cortex of Man. A Clinical Study of Localization of Function.* New York: Crowell-Collier and Macmillan.

Penfield, W., and Roberts, L. (1959): *Speech and Brain-Mechanisms.* Princeton, N.J.: Princeton University Press.

Sherrington, C. (1941): *Man on His Nature* (The Gifford Lectures, Edinburgh, 1937–38). New York: Crowell-Collier and Macmillan.

Sherrington, C. (1947): *The Integrative Action of the Nervous System.* New Haven: Yale University Press.

# The Neuroscience Community:
# People and Ideas

Ralph W. Gerard (b. 1900, Harvey, Illinois; d. 1974) was one of the world's foremost physiologists, a pioneer in the study of the chemical and electrical activity of nerve and brain. He developed the modern capillary micro-electrode, and his interests ranged from its use in the study of individual neurons to the biological bases of normal and disturbed human behavior. This article was perhaps his last completed work.

# 26
## The Minute Experiment and the Large Picture

Ralph W. Gerard

My scientific career must surely begin with my father, a brilliant man with great intellectual curiosity who was a born, but frustrated, teacher. Coming to this country from Central Europe, after a stop in England, to obtain a degree in engineering, he made a career in industry but clearly should have been a university don. As an only son, and being apparently bright enough to ignite his hopes and ambitions that I would achieve what had been denied him, I was the beneficiary of his pent-up devotion to science and to learning. He had had a trying life and was not an easy man to live with, but above all else there came through clearly a passionate devotion to matters of the intellect and to communicating to his young son some of the golden understanding reverberating in his own mind. He was a great admirer of Ralph Waldo Emerson, for whom I am named, and of Thomas Huxley, whose lay lectures in science were a model for his own teaching. Sunday morning walks always provided an opportunity for a Socratic examination of some phenomenon of nature; I recall vividly his picking up a rounded pebble on the beach and leading me to formulate the action of waves and other natural forces in producing this shape. Surely my continued movement into a generalist role, my interest in teaching and use of the Socratic method, and my devotion to reason as a way of life trace back to these very early experiences with a gifted teacher.

He loved mathematics and managed to teach me some rudiments of algebra. When I encountered this subject in my first year at high school, I told the teacher that the first day's lesson was entirely familiar. She was a wise person and said, "There will be an examination in three weeks. Why don't you see if you can pass it and go on to geometry?" On looking over our text I soon realized that my knowledge of algebra corresponded to about half of the first chapter, but such an opportunity was not to be shrugged off. With some help from my father, I mastered the book in the time available and did in fact pass the examination and move up to geometry. This was heady stuff, and I did more or less the same thing in geometry, learning the subject in a few weeks of intensive study at home. When I then launched into solid geometry, I discovered that I had good space visualization and sometimes had to convince my father of the correctness of the solution to a theorem. Partly because of my accelerated progress through high school, I entered university very young, but still with parental guidance.

I have grave doubts about the wisdom of this rapid progress, not only on a psychosocial basis, but even in terms of my learning of mathematics. I whipped through trigonometry and calculus in great shape (at that time regarded as very adequate preparation for chemistry of medicine, to which I was inclined), but I have all my life regretted the lack of a really firm foundation in this universal tool for thinking. At the University of Chicago, certainly guided by my father's attitudes as well as my own tastes, I managed to take at least one course in every science offered and in a great many of the nonsciences as well.

My "game plan" was to obtain a thorough grounding in chemistry, then take medicine, and end up doing biological research. I did, indeed, take a great deal of chemistry as an undergraduate and in my earlier graduate work, and I was greatly influenced by Julius Stieglitz, who taught organic chemistry. His lectures were well-polished jewels, rhetorically barren but intellectually as satisfying as any I have ever heard. Incidentally, he lectured with no notes—a trait that I promptly adopted as my own hallmark. But some discouragement at assistantship arrangements in chemistry and the positive lure of Ajax Carlson in physiology soon led me to shift my base for a Ph.D. program from chemistry to physiology.

Neurophysiology constituted the last third of the year's medical course in physiology and was normally taught by Carlson. He had been overseas in World War I, but he returned just as I was about to take this course and I was in his discussion section. Ajax had quite a reputation for bullying the students, and I decided to meet this situation head on. Early in the course a laboratory experiment with a frog sciatic-gastrocnemius preparation was intended to prove that nerve is nonfatigable. An ether block was applied to the nerve near the muscle, and the nerve tetanized central to the block continuously for an hour or two; then the block was removed while tetanization continued, and the muscle soon went into tetanus. In discussion section Carlson asked another student how this proved that nerve didn't fatigue, and the young man tried to say that it didn't really prove it since the block might stop all impulses from traveling. He was not too clear about this idea, though, and Carlson started to make mincemeat of him; I decided that the time had come and called out, "You're all wrong, Dr. Carlson." This large man came slowly to my desk and towered over me, "That's interesting, why am I wrong?" I put my colleague's idea more clearly—if you turn a stopcock in a tube, no fluid flows above it as well as below. "Yes," he said, "but action currents go down the nerve all this time." I didn't know yet about action currents, but managed to salvage my position by answering, "Yes, but that isn't part of this experiment." Carlson respected my courage and my reasoning and we soon became great friends.

Although I was, even then, interested in neurophysiology, I was assigned a research problem with Lester Dragstedt because of my chemical background: to search for histamine in the fluid in an obstructed intestine. Appropriate extracts were assayed for this substance, not only chemically but pharmacologically, by contraction of the guinea-pig intestine strip. I had noticed that the kymograph regularly showed a slight movement of the lever, indicating gut contraction, just as I was about to introduce the test solution. Finally I asked Carlson to help with this mystery, and it was soon solved. As he approached the apparatus, there was a considerably larger movement than I had observed, and it was apparent that the floor was yielding a little under the weight of the experimenter.

Ralph Lillie, also a member of the physiology department, helped me develop a lifelong interest in what was then called general physiology during several summers at Woods Hole.

One other experience during my formal college years deserves note. I somehow managed to register for a course in histology without having had the normal prerequisite introduction to a microscope in beginning zoology or botany. But, after floundering for a few days, I learned to deal with the instrument and with sections and completed the course very creditably. Ever since, I have been a profound disbeliever in formal prerequisites and lockstep learning. More important, our professor, George Bartelmez, was a kind and permissive man. In the late teens there was still great argument among histologists as to how much of what one saw under the microscope was present in the living cell or tissue and how much was fixation artifact. (This even included myofibrils.) I suggested to Professor Bartelmez that if a quartz needle was moved steadily across a living muscle fiber, the tip would move smoothly if the protoplasm was homogeneous but in a sort of cogwheel fashion if viscous fibrils were imbedded in fluid sarcoplasm, and this could be followed by reflecting a beam of light from a mirror attached to the needle. He was enthusiastic, and unearthed from a storage shelf and presented to me the original micromanipulator that had been developed in the department by Kite. Protoplasm proved to be vastly more viscous than I had dreamed, and this particular experiment did not work. But I discovered a number of other interesting things about cells and their surfaces, remained permanently interested in micromanipulation, and became forever hospitable to callow students with novel ideas.

At the end of the summer of my second graduate year, when my thesis research had been completed, the dean of the Medical School of the University of South Dakota turned up with an emergency problem. His professor of physiology had accepted a position elsewhere, and he needed an immediate replacement. No one else was about and, although I had not yet earned my Ph.D. (I returned that Christmas to take the examination) and was only 21, I was offered the position and went off to be head of the Departments of Physiology, Biochemistry, and Pharmacology for an extremely educational year in my life. Though tempted to stay longer, I returned to Chicago at the end of the school year and married Margaret Wilson, who had just completed her Ph.D. in neuroanatomy with Ranson at Northwestern, and we both embarked upon our clinical years in medicine at Rush Medical College. She then interned at Children's Hospital in Chicago while I went to Los Angeles General Hospital for a strenuous and wonderful year of rotating internship.

I soon became a close friend of Emil Bogen, a first-year resident, who later developed the universally used breath test to determine degree of drunkenness. The two of us, with various other interns or residents, would play twenty questions during a Sunday hike, where the entity to be identified was usually some rare disease. Emil and I and three other members of the medical staff came into close contact with a case of pneumonic plague that had been admitted to the hospital as a case of pneumonia and was finally diagnosed from sputum I sent to the pathology laboratory. All of us who were exposed lived for a solid week in the firm conviction that we would be dead by the end of it.

At the end of the internship I was faced with the major career decision of my life; within a few days I was offered both the head residency in medicine and a two-year National Research Fellowship in Europe, for which I had been recommended by Dr. Carlson. I chose the latter, and more than once I have wondered whether this was the right decision.

Another major bifurcation occurred on my return to the University of Chicago after the years in London and Berlin. Carlson offered me an assistant professorship in physiology, and Dallas Phemister, just creating the outstanding Department of Surgery in the nascent Medical School on the campus, offered me a like position in surgery at twice the salary. I chose physiology since I could not satisfy myself that there would be a full opportunity for research in the clinical department. But this was certainly a mistake, for surgery from the start offered much better research support, in funds and equipment and assistants, than physiology ever did, and would also have given me the opportunity to keep my hand in with clinical medicine, which I thoroughly enjoyed.

Boy and man I remained at Chicago from the fall of 1915 until that of 1952, partly because my wife was practicing child psychiatry (learned in Vienna from Anna Freud and other psychoanalysts). No offer that came to me seemed sufficiently important to disrupt her professional work until, in 1952, I was asked to develop and head the research laboratories of the Neuropsychiatric Institute of the University of Illinois, just across town.

During the next three years flourishing research developed in electrophysiology, neurochemistry, tissue culture, and animal behavior; and at the same time I was given an appointment as professor of behavioral science at the University of Chicago and participated one afternoon a week in a vigorous multidisciplinary discussion group. At that time my wife died, and I, footloose, spent a year as a Fellow in the pioneer cohort at the Center for Advanced Studies in Behavioral Science above Stanford. I also married my present wife, whose background was in government and administration.

Following this, I joined James Miller, until then head of psychology at the University of Chicago, in setting up the Mental Health Research Institute at the University of Michigan—a major angling of my career line. We were concerned with behavioral science and developed strong research activities across all organizational levels: large group, small group, individual psychology and behavior, organismic neurophysiology, and the other activities that had been left behind at Illinois. There was also a strong program in computer science, which stood me in good stead when I eventually moved on to help create the new campus of the University of California at Irvine. This last move took me away entirely from active research, since I was dean of the Graduate School and responsible for fostering a vigorous program in the use of computers and other technical aids in higher education. I was 63 at that time and becoming concerned with an appropriate place to be at retirement; I was also feeling increasingly pressed to remain current with the exploding advances in neuroscience and science in general, and to keep up with the vastly sharper mathematical and instrumental sophistication of young scientists.

The research laboratories at the University of Illinois occupied a large basement area connecting twin towers for psychiatry and neurology, and I accepted the self-imposed task of bringing these disparate groups together—even introducing to each other two men who had worked in their separate towers for over a decade. The institute at Michigan was also concerned with bringing together people of far-flung interests, and I was led into the thick of this by the problem of our name. This name had brought us generous support but had also, despite initial and complete disclaimers of a practical bent, produced a continuous though subtle pressure to show some results in mental health. I had become heavily involved with the National Institute of Mental Health, in connection with the advent of psychoactive drugs, and with considerable support from that institute set up a research program on schizophrenia and psychopharmacology at a nearby mental hospital in Ypsilanti. The main thrust of the study was to characterize schizophrenics in general, and more particularly as meaningful clinical subgroups, by subjecting an adequate number of individuals to a wide-ranging battery of measurements, ranging from anthropometric through chemical, biological, and psychological, to psychiatric and sociological.

A research position at a state mental hospital was not highly alluring during this period of great personnel shortage and begging university positions, but an able group was finally recruited and forged into a cooperating team. I knew that we had arrived when, at a weekly staff conference to decide on further procedures, the head of our biochemistry section objected to the suggestion that spinal fluid be drawn for chemical study on the grounds that this would disturb the entire psychological situation of the patients. Support was discontinued after all the data were in but before the monograph had been written, and the various section chiefs began to drift away before the contribution of each was completed. At about this time I also moved (to California), and the still incomplete monograph weighs heavily on my conscience.

This story has jumped ahead too rapidly, and I should return to earlier formative influences that shaped my scientific career. Several books that I read, probably during my graduate student years, and that influenced me strongly come to mind. Bayliss's scholarly *Principles of General Physiology* opened thought vistas in all directions, and his evidence that a nerve did not increase its metabolism when actively carrying impulses left me disturbed until, years later, my work with A. V. Hill proved this incorrect. I remember especially my delight at the elegant simplicity of an experiment by Engelmann, illustrated by a picture in the book, to determine the action spectrum of photosynthesis. Under the microscope a spectrum was projected onto the length of a single filament of a green algae, and the clustering of oxygen-avid bacteria along the filament was observed. In a relatively short time the bacteria had distributed along the spectrum in a density curve that directly recorded the action spectrum of chlorophyll. Other books that I read with excitement and growth were: Raymond Pearl's *The Biology of Death*, Lawrence Henderson's *The Fitness of the Environment*, Keith Lucas's *Conduction of the Nerve Impulse*, d'Arcy Thompson's *On Growth and Form*, and of course Sherrington's *The Integrative Action of the Nervous System*. The companion volumes of two of my teachers at Chicago, C. Judson Herrick in neurology and Charles Manning Child in zoology, dealing with the evolution of the nervous system and of behavior, influenced my thinking, especially in terms of metabolic and behavioral gradients; and later in life Cannon's *The Wisdom of the Body* brought up to date vividly Claude Bernard's understanding of the constancy of the internal environment.

The two years in Europe as a National Research Fellow constituted my real entry into professional physiology. A. V. Hill, then a Royal Society Research Professor at University College of the University of London, to my great fortune was about to reundertake the measurement of nerve heat, having failed some years earlier to obtain any positive results with his myothermic technique that had been so successful in muscle. He and his indispensable instrument man, Downing, were developing particularly sensitive thermopiles, with hundreds of junctions to contact with stretched-out frog nerves, and a marvelously sensitive moving-magnet galvanometer. This was still not sufficient, and I was commissioned to bring back from Delft, Holland, a sensitive moving-coil galvanometer and a thermal relay from Kipp and Zonen. The final precarious structure was the moving-coil galvanometer on a plate of glass, floating on mercury in a dish on a rubber suspension that was carried on a heavy concrete column built into the ground without touching any of the building structure. An intense light beam, reflected from the galvanometer mirror, played upon a thermojunction, which activated the moving-magnet galvanometer. Six or eight frog sciatic nerves were dissected and mounted on the thermopile during the afternoon, and the whole container was placed in a thermos flask in a sawdust-filled box and allowed to come to temperature equilibrium until late evening, when I began the all-night measurements. Only at night was the vibration of London traffic, above and underground, sufficiently subdued to permit the necessary stability for these measurements.

It was pretty cold working in an unheated London basement in the early morning hours throughout a winter, making a galvanometer reading almost every five seconds for hours on end, but it was an excellent introduction to the drudgery and excitement of serious research. During a few days when the laboratories were essentially deserted for the Easter holidays, I discovered the delayed heat of nerve. When A. V. returned late at night from his holidays to bring something to the laboratory and I was able then and there to demonstrate the phenomenon to him, I won many "Brownie points," and I was soon invited to his weekend cottage on the North Sea.

The following fall (1926) at the International Physiological Congress at Stockholm, he presented the basic methodology and the primary result of successfully measuring the heat of a nerve impulse, and he invited me to present the fuller details of nerve heat production, including the existence of a delayed-heat component. These reports were very well received, and I had emerged on the international scene. (To set the record straight, Franklin, in his *History of Physiology*, stated that the reports on nerve heat were given by Hill, an oversight picked up by Wallace Fenn in writing the history of the IUPS.)

The congress was a truly memorable occasion, held at the beautiful City Hall of Stockholm, where even the flagstone paths in the garden made modern designs in two shades of brown. On this trip I first met Herbert Gasser and was naive enough to argue with him that heat measurements would now replace electrical ones in the study of nerve—certainly one of my most outstandingly wrong guesses.

A few other scientific matters stand out in my mind from this meeting. Kato offered his proof that conduction with a decrement, presumably established by Adrian, was not the case for nerve; the apparent decremental conduction through pools of depressant, with block in one of nine millimeters but not in two separated ones of six, was the unlucky consequence of an unfortunate choice of actual dimensions and diffusion along the nerve. I also recall Ivy's reporting with some triumph that the myenteric reflex previously studied by Starling was an isolation artifact and could be profoundly modified by the extrinsic nerves to the gut. I was greatly impressed by Starling's comment to me later that it was only by eliminating these extrinsic complicating influences that it had been possible to reveal in clarity the major intrinsic mechanism. The importance of paring down an experimental system to the simplest possible situation was deeply impressed upon me.

I do not recall whether it was at the 1926 Congress in Stockholm or the 1929 one in Boston that Rudolph Hess demonstrated his experiments on a sleep center in the basal brain. Electrodes were buried into this structure in a cat, and when stimulated, the cat promptly made a few circling movements in seeking a bed and lay down to sleep. Hess's ingenious and completely convincing proof that this was truly sleep, and not some uncontrollable coma, was simply to place a wet towel on the bottom of the cage. The cat then circled, but not finding a satisfactory bedding-down place, it leaned against the wall of the cage and went to sleep in that position. I am certain that it was in 1929 at Boston that Gasser and Erlanger presented their fabulous demonstration of the nerve action potential on the cathode-ray tube and their proof that conduction velocity is a function of nerve fiber diameter, with different groups of fibers carrying functionally different messages.

In 1922 Hill and Meyerhof had shared the Nobel Prize for their thermal and chemical work on muscle, and my plan was to spend a year with Meyerhof in Berlin following my stay in London. The transition was particularly easy, for the Hills invited my wife and me to join them, the Meyerhofs, and some others in a small auto cavalcade from London to Berlin to Stockholm for the Congress, and we found ourselves warmly welcomed by the Meyerhofs. As a concrete example of the inseminating effect of visits, the proof that Warburg's *"Atmungsferment"* contained a heme and acted much like hemoglobin resulted from this visit to Berlin. At Cambridge Anson and Mirsky had just demonstrated that light broke up carbon monoxyhemoglobin, and Hill mentioned this at dinner. Warburg, who was in the midst of experiments with yeast, soon excused himself from the party, and the next morning he sent in to *Naturwissenschaften* his classical note showing that light reverses the carbon monoxide inhibition of yeast respiration.

England is a small country and people see a good deal of one another; so I had met most of the important English physiologists and biochemists before I went on to Germany. From London to Oxford or Cambridge and back is an easy day's trip, and the monthly meetings of the Physiological Society took me to many other important centers as well. E. H. Starling and Lovett Evans were active in the same building in which I worked at University College, a bust of John Stuart Mill presided over the central corridor of the main building, and many other great or future great informed the premises with their presence. The fine neurological hospital at Queen's Square, where Hughlings Jackson had presided, was within walking distance, and since my wife was working there with F. M. R. Walshe and Gordon Holmes, I soon established comfortable relations there as well.

Hill was quite shy in personal contacts, and at first I seemed to make my way around pretty much on my own; but when he became convinced that I had some merit, he obviously spread kind words about me. Not only had he been pleased by my discovery of delayed nerve heat, but he had given me the assignment when I first arrived of calibrating a decade capacitance box—obviously as a means of introducing me to some physical apparatus and procedures—and my results checked precisely with the official calibration that had been obtained earlier. I also recognized the virtue of Hill's strategy of writing a paper before doing the experiments, thus assuring that one has reasoned the situation through fully and that when the experiments are actually done there will be no loopholes—ordinarily one has merely to insert the actual findings in the already written paper and it is ready to publish. In any event, when I attended a meeting of the Physiological Society at Cambridge, I found Sherrington, Adrian, Dale, and other great men listening to my paper and pointedly being kind to me during my stay there.

Sherrington urged me to visit Oxford soon and to let him know when I was coming; but I was hesitant to impose upon his time and I arranged with Derek Denny-Brown to show me around when I did journey to Oxford a few weeks later. "Sherry" encountered us in the hall, promptly took me in tow, carried me off to luncheon and to dinner (I have often shuddered at what must have gone undone for this spur of the moment hospitality), and won my undying admiration by his response to a problem I put to him at lunch. At this time, in 1926, he had just formulated his CES (Central Excitatory State) and CIS (Central Inhibitory State) interpretation of integration at central neurons, and I had read this paper and some others carefully on the train to Oxford. Since each of these effects was postulated to sum for the entire neuron, yet local increments were released at synapses all over the cell surface, both the excitation and the inhibition must somehow manifest themselves at a distance. Yet, if this action was achieved by propagation in the usual manner, two kinds of influences must propagate—which would violate the all-or-none law. When I pointed out this problem, he thought for several moments and then smiled warmly and said, "You know, I never thought of that." I then timidly suggested the idea of integration by actions upon an overall cell potential (what I later called the somatic potential between axonal and dendritic poles of the soma) as the integrating mechanism, and he was very encouraging to this approach. It was over a decade later, when Ben Libet was doing his Ph.D. work with me on the isolated frog brain, that I returned seriously to the problem of somatic potentials and their functioning in neuron masses as well as at individual cell elements.

I believe it was on this same trip to Oxford that I first met Jack Eccles, just launching his own brilliant career, and I still remember my feeling of awe and admiration when he assured me that every day, no matter what experiments or other commitments he had, he read 300 pages of physiology. Our paths have crossed happily many times during subsequent decades, and our encounters have always been stimulating. Perhaps most memorable was an almost daylong argument we had lying on our backs in the garden of his home in Dunedin, New Zealand, Jack supporting the existence of free will and I denying it. The same debate resurfaced when the two of us, I fear, rather monopolized a panel discussion on a much broader topic at one of the famous public series at the San Francisco Medical School. Despite these philosophic clashes (or perhaps partly because of them) we have remained close friends.

At the end of my stay in Berlin I spent two or three months in Adrian's laboratory at Cambridge; and since my brain showed a good alpha rhythm, I was one of his favorite subjects at a time when he was entering this area of research. This was following the extraordinarily fruitful years in which he, with Bronk and others, demonstrated the trains of impulses in a sensory nerve when a receptor is stimulated and determined the relation of frequency to intensity. I recall that my alpha rhythm would continue even when a light was shined in my eyes, if I consciously avoided paying attention to it. Conversely, when I was told I was showing a splendid alpha rhythm, it promptly broke up. (Years later I showed that the rhythm breaks up in hypnotized subjects when it is suggested that a light has been turned on.)

Adrian let me play about a little with the capillary electrometer that had been used since Keith Lucas's time and that he had recently discarded in favor of the new Matthews oscilloscope, which I also explored. Adrian himself was experimenting with Brian Matthews on the latency of visual optic responses, and I was directly involved only in some of the interpretations of the results. I had the pleasure of having Adrian as my guest in subsequent years, in a summer home in Vermont, in my home at Chicago, and fairly recently in California.

We were also invited to stay with the Adrians on later visits to Cambridge, in the Master's quarters of Trinity College. Here was the quintessence of British science, with mementos of her great men and with a splendid informal picture of Newton reminding one of his spirit hovering over the quadrangle where his great work was done. Nor were the rituals and appurtenances of the Fellows' dining hall less impressive than they had been over the centuries.

William Rushton was also working in the physiology department at Cambridge at this time, having recently returned from a stay with Det Bronk at the Johnson Foundation in Philadelphia, where he had played havoc with Lapicque's interpretations of chronaxie, and I treasure a memorable night which began with his explaining to me what hyperbolic trigonometric functions were and continued until the morning light while we discussed innumerable aspects of nerve conduction as each of us in turn would walk the other home, intending to get some sleep but never quite managing to tear ourselves apart. Keilin was also a joy to talk with, and I still treasure his wise response when I mentioned that Szent-Györgyi had been proved in error in his formula for ascorbic acid: "I would rather have one of Szent-Györgyi's errors than the correct result of most scientists." Dale and

Barcroft and "Hoppy" also became admired friends; indeed, by the time I had spent a total of four years in England in dribs and drabs over the following decades, I had come to treasure many of her scientists as friends as well as colleagues.

To return to my time in Berlin, I found that Meyerhof had prepared for my arrival some special small-sized chambers for the Warburg manometer, with electrodes sealed in, so that almost immediately I could measure the oxygen consumption of frog nerve, at rest and on stimulation. It was a happy arrangement to be able to make comparable experiments under comparable conditions and show that the respiration corresponded quantitatively to the heat production. This led to measurements of lactic-acid changes, of ammonia production, and a start on phosphate turnover. Creatine phosphate had just recently been discovered by the Eggletons in England, and pyrophosphate was isolated by Lohmann in Meyerhof's laboratory while I was there. For a period Fritz Lipmann shared my laboratory and Hans Krebs was working with Warburg on the floor above. Otto Meyerhof was very kind to me and tolerated my informal American ways (I would sometimes perch on his desk while talking with him in his office), and since his English was much better than my German, at first we had many exciting discussions in my language rather than in his. In time, however, I did manage to give a seminar report in German that was tolerably intelligible.

When I returned to the United States from Cambridge, England, I spent some time in Cambridge, Massachusetts, with Alex Forbes. He introduced me to sophisticated electrophysiological techniques and to sailing, and contributed richly to my store of anecdotes. Rosenblueth was working with Cannon in the same building and was reported to have stated that he had started a research one day and sent the completed paper to press 24 hours later, which led to some unkind comments; but sometime later I did in fact have just such an experience. I visited A. V.'s laboratory on a Saturday afternoon and found McKeen Cattell busily stimulating crab nerve at varied high frequencies. Responses soon stopped, and I suggested this might be due to cumulating polarization—which could be checked for by introducing a commutator to alternate the direction of the shocks. We did several experiments that afternoon, I wrote up the results that evening, and the finished paper was handed to A. V. Sunday morning at his home.

Hal Davis was also working in Forbes's laboratory (he and Dave Brunswick were busy on an electrical model of nerve), and I have always treasured his classic remark in praise of the advantages of recording from single axons rather than a mixed trunk: "It is hard to deduce the properties of an egg from studying an omelette." Phil Bard was also around the laboratory, working with Cannon and doing his important localization of the sham rage release to the hypothalamus. The studies of Cannon and of Bard seemed to give a splendid physiological basis for understanding Freud's superego and id, and strongly influenced my own later thinking on the physiology of psychoanalytic processes. Davis and Bard, and soon Fenn and Bronk, became close and lasting friends. These men, along with Gasser, Erlanger, Bishop, and Schmitt of the Washington University group, and Forbes, and later Lorente de Nó, Kacy Cole, Grayson McCouch, Dave Lloyd, and perhaps another man or two, became the famous Axonologist group—which sprang into existence at my home in Chicago, where I had invited everyone working on nerve who was present at the meeting of the American Physiological Society that year.

When I returned to Chicago and set up my own laboratory, I received a munificent research grant of $3000 from the university. This amount hardly permitted me to push ahead with elaborate instrumentation for heat studies, but my students and I continued to press on with nerve-metabolism experiments. About 1927 Bill Amberson took over the physiology course at Woods Hole and invited Bard, Davis, and myself to join him in teaching it, and Keffer Hartline and Kacy Cole moved strongly into my horizon.

A few years later, when Hans Winternitz challenged the increased respiration of stimulated nerve as being an artifact of artificial impulses, Hartline and I agreed to test this out on the *Limulus* optic nerve, isolated along with the attached eye. The first attempt, using small Warburg vessels, was clearly far below the required sensitivity; but the problem was solved that same night by threading the optic nerve into a capillary through a vaseline seal, the eye being outside and the far end being closed with a measuring drop. Two such capillaries in a large closed test tube in a thermostat were arranged so that light could be shined on the eye of either nerve, and each one thus constituted a control for the other. The movement of the index drop was followed with an ocular micrometer minute by minute. The oxygen consumption when "natural" nerve impulses were carried was established, and a valuable microrespirometer became available. Since our time commitments were such that we had less than a week to work together, experiments were continued day and night and neither of us was out of his clothes for the entire period.

Also at Woods Hole, some years later, John Young, who had come on a Rockefeller fellowship to work with me at Chicago, discovered the giant fiber of the squid. He, Bronk, and I attempted in the two days we had together to measure its action potential, but we were not then successful.

At Chicago, Karl Lashley and Paul Weiss became new colleagues and stimulating friends, and each in his own way made it clear that the central nervous system was much more than a simple summation of reflexes. Paul and I quarreled vigorously over his resonance theory, but we collaborated on a bit of research attempting to record the development of brain waves in the chick embryo. The isolated brain was placed in a small silver spoon as one electrode and a silver wire touched the upper surface of the second electrode, the whole preparation being immersed in a thermos flask in air above an appropriate salt solution. We recorded marvelous sharp, regular waves of a bewildering variety of shapes and frequencies, which did not look biological but did vary with temperature in a reasonable manner and did stop promptly when a drop of ether was put into the flask. Nonetheless, they turned out to be artifacts, due to condensation or evaporation from the electrodes, as became all too evident when we lifted out the preparation after a long and vigorous series of waves had been recorded, and found that the silver wire had been displaced into the air and was not in contact with the brain.

I was led into the study of action potentials and electrophysiology in general by a felt need for a direct indicator of function to correlate with metabolic studies.

But, although I built an amplifier from scratch one summer, I never felt really on top of the electronic equipment, and I leaned heavily on the contributions of a series of assistant-students, such as Wade Marshall, Frank Offner, and Michael David, for help at this level. Slow integrating galvanometers served many purposes and were, of course, necessary for finding the very long delayed potentials, after nerve tetanization, that matched the delayed heat production and oxygen consumption of nerve.

For all its allure, the central nervous system seemed a morass of different cell types and of highly individualized regions, and it seemed to present an unknown mixture of excitation and inhibition when manipulated. But when Leon Saul moved from Davis's laboratory at Harvard to the Institute for Psychoanalysis in Chicago and sought my collaboration, I was finally tempted to make the plunge, and, with Saul and Marshall, I explored the electrical activity of the cat's brain in considerable detail with oscilloscope and loudspeaker. Stephen Ranson at Northwestern kindly made available his stereotaxic landmarks, and a local machinist made for us a modernized version of the Horsley-Clarke instrument (and developed quite a worldwide business supplying this instrument for some years). I was thus launched on the great ocean of brain waves. Although our first major paper was not published until some years later, we followed evoked potentials—which we named—from sight, sound, touch, and proprioception into all sorts of regions where these sensory impulses were not supposed to go, including Ammon's Horn and the cerebellum. We were also intrigued by the richness and variety of the "spontaneous" rhythms picked up with concentric electrodes, which were heard especially well and which changed abruptly with small movements of the electrode or with changes in the physiological state of the brain.

In the continuing search for simplified preparations, Ben Libet and I turned to the isolated frog brain, which is small enough and has a sufficiently low metabolism to remain viable for hours, and we found standing and traveling waves that proved of great interest. In particular, under the action of caffeine large regular rhythms spread from the olfactory bulb over the entire hemisphere, and even crossed a completed anatomical section if the cut surfaces were well reapposed. This, in turn, led me back to the somatic potential and to steady potential influences in general.

In the brain studies, even the Adrian-Bronk concentric electrodes seemed too large, and I kept searching for a true microelectrode, again finding the way from my earlier adventure with striated muscle as a histology student. With Judith Graham, I developed a salt-filled capillary with a tip small enough (up to five microns) that a muscle fiber could be impaled without excessive damage. Gilbert Ling soon picked up these studies, and the electrode was pushed down to a few tenths of a micron and extensive measurements were made on membrane potentials, primarily of muscle (including cardiac muscle). I reported some of our findings at a meeting of the American Physiological Society at which Alan Hodgkin happened to be present. He asked for full details, carried the technique back to Cambridge, improved it by introducing a concentrated salt solution into the capillary and having it feed into a cathode follower before routine amplification, and soon thereafter carried out his classic studies on the membrane of the squid giant fiber. Again, we were rather slow to publish, and Alan graciously wrote that he was about to release a paper and invited me to get mine into press before his.

I was timid about impaling neurons in the brain in vivo without seeing what I was doing; but Eccles reasoned soundly that if moving the electrode tip produced a sudden large negative potential, it must be inside a cell, and if this cell behaved properly physiologically, one could safely assume that it was a neuron. Andrew Huxley also used the microelectrode on muscle in his actinomycin studies, and these three men shared the Nobel Prize some years later. I was nominated for developing the microelectrode, and this seems to be my best-known contribution, although I personally have been more excited about some discoveries and interpretations than about methodological contributions.

In the mid-thirties Alan Gregg, of the Rockefeller Foundation, asked me to spend a year traveling around Europe to investigate research on the nervous system and to recommend projects and individuals for future support. This was an intensely educational venture, and I came in contact with nearly all workers on the nervous system, in laboratories or clinics; I carried ideas, like pollen, from one laboratory to another and enjoyed seeing the fruits of this cross-fertilization in many papers that appeared during the next two or three years. My report to the Rockefeller Foundation occupied three volumes of typescript (during one long day in Copenhagen I had forty separate talks arranged by August Krogh). It would not be in order here to attempt any kind of summary of the many exciting experiences, but two incidents require mention. At Cambridge I was invited by Adrian to attend a student tutorial with Alan Hodgkin and one or two others. Alan had made the observation that when a nerve impulse is stopped at a region of block, the electrical threshold just beyond the block is lowered. This mystified everyone, but Alan found a sound explanation in terms of eddy currents advancing beyond the block but not strong enough to activate a propagated response.

At the other end of the spectrum from this bright omen of the future, I visited Pavlov not long before his death and he told me that his experiments on conditioned neuroses had been stimulated by reading some of Freud's work. A week later I had moved from Leningrad to Vienna and mentioned this observation to Freud. He snorted and exclaimed, "It would have helped me enormously if he had said that a few decades earlier."

Pavlov was the focus of another memorable experience. Following the 1956 International Congress in Brussels, Ernst Guttmann had organized a conference in Prague on trophic phenomena. This was the first scientific meeting in a decade to which Western scientists had been invited behind the Iron Curtain, and the long, narrow meeting room was arranged as if the Iron Curtain ran up the central aisle: communist scientists were seated to the right, others to the left. Moreover, the leader of the Russian delegation, Kostoyanz, was placed in the front row center on one side, and I, cast in the role of doyen for the visitors, was seated across the aisle. Our host presented a very creditable paper on trophic nerves to muscle, beginning by attributing the discovery of trophic action to Pavlov but refuting Pavlov's theory on how this worked. Kostoyanz immediately opened the discussion with a rather impassioned defense of Pavlov, and I placed a note on his desk while he was at the rostrum, to the effect that all physiologists recognized Pavlov's enormous contributions but I was puzzled as to why the Russians felt it necessary to insist on Pavlov's godlike infallibility. I added that this would pose a serious dilemma to the Soviet regime, since Pavlov had told me that he had no use for communism; I asked him to tell me whether Pavlov had been right or wrong.

When he read the note, he returned to the platform and asked me if I did not want to make some comments, but when I suggested that he read the message I had given him, he said simply that it wasn't so that the Russians insisted on Pavlov's infallibility. I suggested we discuss it further, and at lunch the next day, where leaders of physiology in six Iron Curtain countries chanced to be sitting at a table with us, he simply repeated his denial of my statement of fact. No one else at the table commented at the time, but in the following half hour, while returning to the meeting room, each of them managed to get me aside for a moment to say he agreed with the point I had made.

At the end of the conference there was a sort of plenary session (part of which I missed while renewing my acquaintance with Prague under the guidance of Josef Charvat, a professor of medicine whom I had come to know on an earlier medical mission to Czechoslovakia), and I was hurried back to the meeting room in time to hear Kostoyanz say that physiology depended on the work of many great men from all countries, mentioning Cannon in the United States. The audience exhibited a startle as if a cannon had been fired, and I was told later that that was the first time a Western scientist had been mentioned in a public meeting in ten years. Then it was my turn to speak. I said I preferred to emphasize the future and congratulated the country on the improvement in the work of its young scientists since my previous visit, but I had to confess that much of the work reported still seemed to lack careful controls. I said I would like to pass on the advice given me by a great scientist, Ludwig Hektoen, who had received it from his teacher: "Young man, take care lest you find what you are looking for." Jerzy Konorski, who was present through all this, told me a year or two later that there was inscribed over the front of his lecture hall the Polish translation of this statement. Medical or scientific missions and conferences took me to all parts of the world over the course of two or three decades, and by the mid-fifties I had certainly been exposed to a vast amount of neuroscience.

I have lingered, perhaps too fondly, over many of the formal and informal experiences that helped form and direct my interest in neurophysiology. In later decades I continued to receive stimulation from graduate and postdoctoral students. Some have already been mentioned, but since only a small part of our work can be covered here, I am eager not to overlook a handful of excellent students who made important contributions in my laboratory and went on to distinguished careers of their own. Norris Brookens was in the first course I taught at the University of Chicago, a sort of catchall undergraduate introduction to physiology that I turned into an adventure in undergraduate research. Norris arrived as a freshman expecting to make a career in classics, but he took a Ph.D. in physiology and went on in medicine as a successful practitioner and educator. Robert Cohen, another in that small class of 18 guinea pigs, also obtained his Ph.D. with me, working on brain metabolism, and then an M.D., and became director of research in the National Institute of Mental Health and a distinguished psychoanalyst. T. P. Feng joined me as a postdoctoral student shortly after my arrival at Chicago and soon discovered the influence of the nerve sheath in greatly retarding diffusion into the nerve trunk. He then spent some time with Hill and returned to China, where he is now a leading figure in the science establishment. Julian Tobias came from Michigan to do his graduate work with me, mainly on nerve respiration, and he remained as a colleague in the department, where he broke new paths in studying physical changes in nerve during conduction; he was making a powerful attack on the properties of nerve membranes at the time of his death. Bob Tschirgi, who was captured for physiology by the lectures he heard in the undergraduate biological science course, worked on central nervous metabolism, and later, at UCLA, on the blood-brain barrier and on problems concerning the evolution of symmetries in organisms.

During the war years Albert Potts, now a distinguished investigator in ophthalmological chemistry, and Harvey Patt, who contributed importantly to the study of radiation sickness, joined me in a team that was concerned with phosgene poisoning and other problems in gas warfare. In the early 1950s the returning veterans brought a new surge of excitement into the department, and Bob Doty went on to an especially productive career studying the neural mechanisms of conditioning. Leo Abood joined me after obtaining his degree in pharmacology and was my right-hand man in building up a strong research program at the Illinois Neuropsychiatric Institute in the early 1950s. Sid Ochs has become a leader in axon-flow research. When I moved to Michigan in the mid-fifties, Bernie Agranoff joined me to head up the laboratory program and developed his exciting studies on memory phenomena in the goldfish. Arthur Yuwiler similarly became the mainstay of the schizophrenia research program at Ypsilanti Hospital. The story is told that Harvard is such a repository of knowledge because freshmen come in knowing so much and seniors leave knowing so little; in a more serious vein, I am greatly indebted to my many students, including any number unnamed here, for bringing their ideas and capacities to my laboratory and enriching my own development.

One line of investigation that early intrigued me, partly because of the challenge of the failure to obtain regeneration in the central nervous system and partly to answer the theoretical question of why a nerve fiber separated from the cell should degenerate, was the field of nerve degeneration and regeneration. Several students and colleagues worked with me on this, and the final and most definitive study of central regeneration was done with Oscar Sugar, who is now, appropriately enough, a leading neurosurgeon. It was first proven that degeneration of the cut nerve fiber was not due to the absence of conducted impulses, and we were then able to show that substances moved down the nerve fiber, presumably enzymes formed in the cell nucleus or body that were necessary to maintain the metabolic integrity of the fiber.

Steve Kuffler came from Australia in the late 1940s. He brought his splendid dissection techniques for studying a single nerve-muscle unit and demonstrated the special small-nerve motor system in frogs and later in mammals. I had convinced myself that the small nerve fibers innervated the same skeletal units as did the large motor fibers, but Steve stubbornly insisted on additional experiments and finally proved that the two were entirely separate systems.

At the University of California at Irvine I succeeded in giving a strong initial push to the field of computer-aided learning and to the use of other technical aids in the educational process. These interests, in turn, led back to studies of the nervous system in relation to learning and to information processing. Other research byways have led me entirely outside the nervous system, to studies on the respiration of sea-urchin eggs and other invertebrate organisms, to a study of aging processes and whether they are indeed inevitable, and to the examination of groups of organisms and societies as epiorganisms with many basic similarities to systems at the cellular and organismic levels. This, in turn, led to considerable interest in the general theory and properties of systems, explored extensively with Miller and Rappaport and others at the Mental Health Research Institute.

Still another research offshoot, the ability of thyroid to increase the respiration of the brain and of tissues in general, suggested that the thyroid itself must somehow be depressed by thyroid hormone, since otherwise an explosive situation would result. This in fact proved to be the case, and a negative feedback system was established, although, at the time, the loop through the pituitary was not recognized.

To return to our main line of research, extensive studies on anoxia were made as an approach to quantitative metabolic differences in the central nervous system, measured by the survival of electrical activity, the recovery time on readmission of oxygen, and the time of anoxia before irreversible damage. The findings brought out clearly the existence of a general metabolic gradient from the rostral to the caudal portions of the nervous system—in line with Hughlings Jackson's clinical findings and, more generally, with Child's studies of cephalocaudad gradients in invertebrate organisms. An offshoot of this work was the study of local vascular changes in brain regions in relation to state of activity; with Herman Serota I demonstrated, by the flow of heat from a warmed thermocouple, that marked dilatation occurred in the occipital cortex upon illumination of the eyes. A clear temperature gradient normally exists between the cooler cortex and the warmer deep structures.

As earlier we had used electrical studies to monitor our metabolic ones, so later we used behavioral studies to monitor the electrical findings in the central nervous system. By the use of hibernation and cold in chipmunks, and later of electroshock in rats, it proved possible to stop all central electrical activity without abolishing recent learning, proving that some morphological change had occurred; and it was further found that prior learning did disappear under these conditions if some minutes had not elapsed between the completion of the learning experience and the application of the cold or electroshock. This interval, which we called fixation time, presumably was a measure of the amount of impulse reverberation necessary to produce enduring material changes in the involved circuits. It was later possible to show that an asymmetric discharge of impulses from the cerebellum or vestibular system to the lower spinal motor neurons would produce asymmetries in posture which, if allowed to persist some 45 minutes, endured after section of the upper spinal cord. This fixation time was lengthened by drugs that interfered with neuron metabolism and was significantly shortened by one that enhanced RNA formation.

I have already mentioned our studies on schizophrenia. Along with these there was considerable interest in the psychotomimetic and other psychoactive drugs and in the physiological mechanisms underlying mental illness.

A last thread in my professional life that I would like to touch upon is an interest in general biological principles and the effort to relate phenomena at one level to those at others, as in the early recognition of certain basic likenesses between immune phenomena and memory stated in my first book, *Unresting Cells* (1939), a presentation of biology for the layman. It was from my experiences in teaching a comprehensive biology survey course to beginning students at the University of Chicago that I became interested in the evolution of division of labor and differentiation in organisms, in the recognition of major levels of organization (from molecule through cell, organ, organism, small group, large group, and total biota), and also in the identification of the three major groups of system attributes—those essentially constant in time (being, morphology), those changing reversibly in time (behaving, functioning), and those changing irreversibly in time (becoming, evolution, development, learning, social history). An exciting matrix was developed on these three dimensions, and it was recognized that cause and effect relations spiraled upwards, so that structure at the molecular level determined function at the cellular level (oriented actinomycin fibers and directional shortening of striated muscle; directed tension of muscular contraction and the development of bony trabeculae).

I was also led to interesting excursions into more general ethical and philosophical problems touching strongly on biology. A recognition of the need for bringing valid science (an understanding of its method and procedures more than of its findings) to the layman led me early and continuously into extensive lecturing and writing and, of course, contributed to my lifelong concern with education in general.

In bringing to a close this report on my scientific life, I should make a concluding statement on my research concerns and on my personal approach. Certainly a major concept in my neuroscience thinking was that of an active nervous system—modulated by input from the outside, from other portions of the nervous system, and from its bathing fluids—rather than an inactive organ awaiting arousal by particular stimuli. Even in peripheral nerve the emphasis was on processes enduring in time; after the electrochemical burst associated with conduction were the continuing processes of afterpotentials and delayed heat and metabolism, and the still longer changes of "becoming"—enzyme increases and synaptic enhancement and actual growth of new processes as a result of activity. Brain waves to me have always been the manifestation of groups of neurons beating in unison rather than envelopes of spikelike actions distributed in time over the wave form.

In terms of my mode of thought, it must by now be abundantly evident that, as

Coleridge said of himself, "My illustrations swallow up my thesis. I feel too intensely the omnipresence of all in each, platonically speaking; or, psychologically, my brain-fibers, or the spiritual light which abides in the brain-marrow, as visible light appears to do in sundry rotten mackerel and other smashy matters, is of too general an affinity with all things, and though it perceives the difference of things, yet is eternally pursuing the likenesses, or, rather, that which is common." I am, indeed, a generalist (different from a generalizer), but have earned the right to this broad approach by meticulous devotion to many particular areas of research, and by the exploration in simplified systems and with micromethods of the precise mechanisms involved in their operation. Thus, at successive periods I helped found "The Axonologists," *The Journal of Neurophysiology, Behavioral Science*, and the Society for Neuroscience. I am probably best known for the microelectrode, on the one hand, and for synthetic summaries of wide-ranging symposia, on the other. I treasure the terse introduction when closing a weeklong meeting, "Ralph Gerard, citizen of science." I have, indeed, been committed to the minute experiment and the large picture.

Richard Jung (b. 1911, Frankenthal, Germany) is professor of neurology and clinical neurophysiology at the Albert-Ludwigs-Universität, Freiburg i. Br. His research work has encompassed such topics as the electrophysiology of the brain, clinical neurophysiology, experimental epilepsy, neuronal mechanisms of the visual cortex, visuovestibular coordination, and visual perception.

# 27

## Some European Neuroscientists: A Personal Tribute

## Richard Jung

**Summary**

1. Personal recollections of prominent neuroscientists whom the author met during the years 1930–1950 are given with special reference to neurophysiology. Some traits of Paul Hoffmann, Walter R. Hess, Hans Berger, Jan F. Tönnies, Erich von Holst, and the philosopher Nicolai Hartmann are sketched.

2. In 1930–1932 Paul Hoffmann gave the first suggestions about neurophysiology to the young medical student. A visit to Vogt's Brain Research Institute in Berlin in 1934 demonstrated the wide extent of the brain sciences and brought the first contact with Jan Tönnies. There followed neuroanatomical work under H. Spatz in Munich and a short clinical intermezzo with K. Beringer in Freiburg. The award of a Rockefeller Fellowship in 1936 for work in London, Zürich, and Berlin provided an introduction to the actual mainstream of neurophysiology created by Adrian, Berger, Hess, Dale, and Eccles during the early 1930s. This resulted in the author's choice of the electrophysiology of the brain as a research field.

3. The fact-oriented inductive British physiology and the function-oriented integrative physiology of W. R. Hess based upon systematic theories showed a way to combine experiment and theory in brain research. In addition Nicolai Hartmann's philosophical concepts offered an ordered synthesis.

4. For neurophysiology in man Hoffmann's reflex recordings, Berger's and Adrian's EEG findings, and Tönnies's cooperation were important prerequisites in building up in 1938–1939 a neurophysiological laboratory in Freiburg to apply neurophysiological methods to human neurology.

5. The importance of combining multiple traditions in a new research field is stressed. Some general remarks are made on motivation and methods of research, and on qualitative versus quantitative proceedings. The value of pilot studies and of hypotheses as research guides is stressed.

6. Two failures of research are discussed: one due to lack of a new theory to integrate various facts, which were wrongly interpreted by conventional concepts; the other caused by a too early quantification with neglect of broad-scale, qualitative pilot studies.

### Introduction

This paper describes the influence of my teachers on my work and on the development of the neurosciences in Europe. During the early 1930s, when I became interested in neurophysiology, two main research fields had been opened up: (1) the electrophysiology of the brain; (2) objective sensory physiology with recordings from single nerve fibers. It was a favorable time to be entering these new fields of research, as one had an opportunity for personal contact with the pioneers in the brain sciences. At this early period communication between different centers was lively, despite difficult travel conditions. Personal contact could be maintained by exchanges of results and letters since only a few laboratories in the world were engaged in brain research. One could know the work of others with an intimacy that would be impossible today, given the current mass production of publications.

I received important stimuli for my own research from many scientists. These influences came about not only through close cooperation in the laboratory with such great physiologists as Hoffmann and Hess, or with such technical innovators as Tönnies, but also through brief encounters with such prominent scientists as Adrian and Eccles, and remote communication with Hans Berger. Furthermore, contacts with such great philosophers as Nicolai Hartmann stimulated thought and encouraged systematization.

The German universities in the early 1930s were at the height of their scientific achievement and offered many opportunities for broadening the cultural and philosophical background of one's research. There was less specialization, and many workers aimed to achieve a synthesis of science and the humanities. It is the purpose of this paper to present a few highlights to illustrate this notable period.

In these years research in neurophysiology was more adventurous than today, and chance discoveries were common. Nearly every electrode inserted into an unexplored region of the cat's brain revealed something new, and each good experiment pointed the way toward further goals. Related fields, such as neuroanatomy, psychophysics, and the comparative physiology of lower forms, showed parallels and gave one stimuli to ask the right question at the right time. Relations with other sciences and philosophy helped to distinguish the important from the trivial problems.

Each of the personalities with whom one comes in contact while acquiring experience has, of course, his own individual style of teaching and research. As a young worker, one adapts what one regards as the best aspects of each in forming one's own research style.

I was equally interested in the experimental, theoretical, and practical aspects of brain research. But it was not an easy task to combine animal experimental work in the physiological laboratory with clinical neurology in man, and failures were bound to occur. Toward the end of this paper I would like to show how research failure may result from lack of an adequate hypothesis or from neglecting qualitative exploration of a new field in favor of a too-early quantification.

Science cannot be separated from man, and investigation of the human brain is an important aim of neurophysiology. I learned to explore human spinal-cord reflexes from Paul Hoffmann and the human cerebral cortex from Jan Tönnies and Hans Berger. Meanwhile, over three decades clinical neurology remained a daily task supplementing the laboratory experiments in animals. Clinical responsibilities often hinder experimental work, but they also provide valuable pointers that help one to avoid wrong directions and blind alleys.

In this paper I wish to evoke the memory of a few of the outstanding scientists who were among my many teachers and friends. The neurophysiologists Paul Hoffmann, Walter R. Hess, Jan F. Tönnies, Alois E. Kornmüller, Edgar D. Adrian, John C. Eccles, Yngve Zotterman, H. Keffer Hartline, and Ragnar

Granit influenced my work, and this paper will pay special tribute to the first three as well as to the neuroanatomist Hugo Spatz, the neuropsychiatrist Hans Berger, the biologist Erich von Holst, and the philosopher Nicolai Hartmann. Many other teachers should also be mentioned, but this will have to be left for another occasion.

The names of some clinicians who guided my early path from 1934 to 1940 should also stand here: Kurt Schneider, who did pioneer work in psychopathology; Kurt Beringer, who made the first systematic studies of experimental psychoses with mescalin; E. Arnold Carmichael, who taught me to apply simple methods in clinical neurophysiology; and Karl Kleist, the promotor of cortical localization and of a subcortical regulation of consciousness. The neuroanatomists Hugo Spatz and Oskar Vogt gave me the morphological basis. From the continuous succession of coworkers who have helped build the Freiburg research group in the last three decades, Günther Baumgartner, Rudolf von Baumgarten, Otto Creutzfeldt, Otto-Joachim Grüsser, and Hans Kornhuber have now founded their own schools in other places.

Although I will treat mainly recollections about a historical period, I think it might not be out of place to add some points of my own biographical background and a few general remarks on research approaches in the neurosciences.

## Neurophysiology in Man: Paul Hoffmann and Hans Berger

### Paul Hoffmann's influence

From my first teacher in physiology, Paul Hoffmann in Freiburg, I learned two lessons about electrophysiology in man. The first was that one can obtain exact data about timing and delays in the human central nervous system by electrical stimulation of reflexes. (Thus Hoffmann had developed his concept of the two-neuron reflex by very simple experiments, mostly done on himself.) The second was that electrical recordings in man may help to answer difficult questions in sensorimotor regulation since the discharge of motoneurons may be recorded indirectly from human muscles. This may seem rather obvious today, but it appeared new in 1929 when Adrian and Bronk had just begun to pick up single-motor-unit discharges with a concentric electrode in humans.

Hoffmann's concept of the two-neuron *Eigenreflex* had been developed systematically since 1910 by electrical recordings from human subjects with skin electrodes. In his first paper in 1910 he described the reflex response, now called the H-reflex in the electromyogram (see Figure 27.1), and later, in 1922, he proved its monosynaptic nature.

After publishing these early papers on human reflexes and electromyography from Piper's institute in Berlin, Hoffmann continued his reflex studies as von Frey's assistant in Würzburg. During the First World War (1915), he described the mechanical hypersensitivity of regenerating fibers in peripheral nerve injuries, later known as the Hoffmann-Tinel sign. In 1924 he became von Kries's successor to the Freiburg chair of physiology, where I heard his lectures in 1929–1931. Hoffmann followed the traditional academic career at German universities, but he

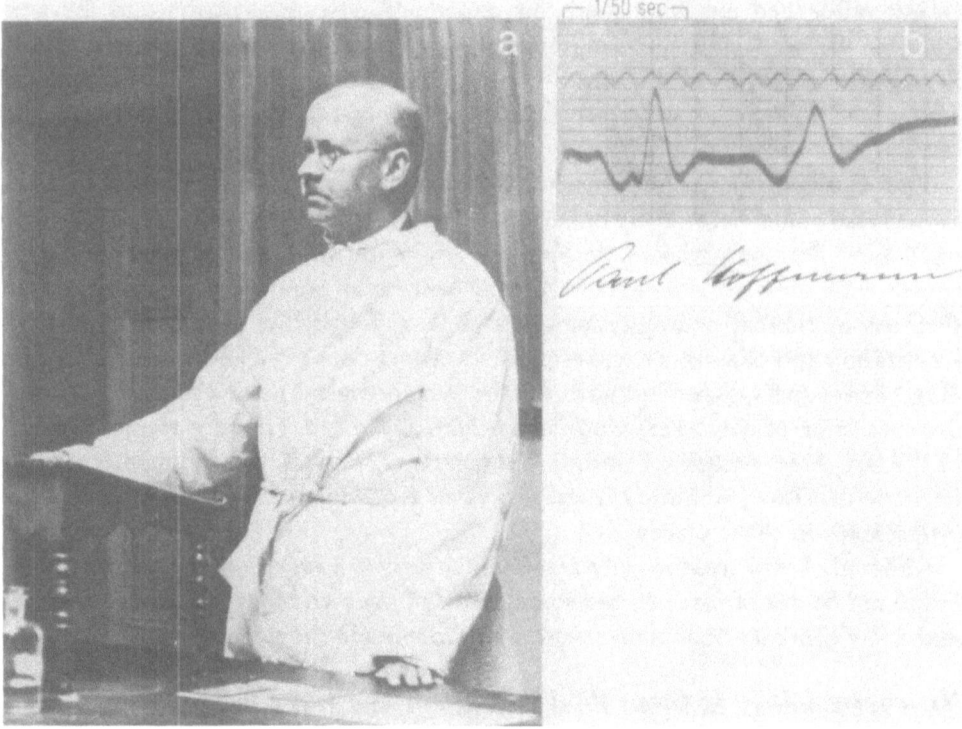

Figure 27.1   Paul Hoffmann (1884–1962). (a) This picture was taken four years after his call to the Freiburg chair of physiology as successor of von Kries. Hoffmann's teaching was clear and simple, accentuating essential facts with historical comments. His manner was modest and even shy, he avoided rhetorical brilliance, hated scientific propaganda, rarely mentioned his own work, and expressed criticism by making dry asides. In the institute, Hoffmann wore a brown, rather shabby coat. The white coat seen in the picture was used only for lectures. (b) A record from Hoffmann's first paper on the direct and reflex response elicited by electrical stimulation of motor and proprioceptive nerve fibers in man. An electrical stimulus of the tibial nerve (stimulus artifact) is first followed by the direct responses (7 msec latency), then by a lower amplitude response of the monosynaptic reflex (28 msec latency). The upper tuning-fork record marks 4-msec intervals. From Hoffmann (1910).

differed from the older type of professor in his modesty, informality, and self-criticism. His sometimes caustic jokes about scientists with exaggerated self-esteem and his humorous remarks about himself when he was mistaken as the *"Labor-diener"* were reported in many anecdotes (Jung, 1969).

Hoffmann's experiments on the *Eigenreflex* convinced me of the importance of electrophysiology in man. When I returned to Freiburg in 1935, I worked in electromyography in Hoffmann's institute with W. Eichler, who in 1937 first recorded human nerve potentials in situ through the skin. I was never an official member of his institute, but the close relation with Hoffmann continued from 1935 for over two decades and throughout his retirement until his death in 1962.

## Tönnies's direct EEG recordings from the human brain

A decisive early experience in brain research was my first meeting with Jan Tönnies in February 1934 in Berlin. I volunteered as a student at the neurosurgical service of Professor Heymann, who had succeeded Fedor Krause, the founder of neurosurgery in Germany. Both had been general surgeons but were primarily interested in neurosurgery. At the beginning of the century, long before Foerster, Krause had already used electrophysiological methods during his brain operations by stimulating the human sensorimotor cortex. Later, in 1924, he also performed the first electrical stimulation of the human visual cortex.

During the winter term of 1933–1934 J. F. Tönnies had arranged with Heymann to record brain potentials from the exposed human cortex. His inkwriter electro-encephalograph was installed in the operating theater, and I assisted in these direct EEG recordings during operations on cerebral tumors. (Until then Berger had only obtained records from the cortex and white matter in one case.) At about the same time Adrian and Foerster also began electrocorticographic recordings in brain operations. Of course, Tönnies's first records did not succeed as well as those obtained in animal experiments: they were disturbed by artifacts and alternating current, although Tönnies's new differential amplifier attenuated the latter considerably. This exciting experience of recording cerebral potentials directly from the human brain gave my plans for scientific work a new direction toward neurophysiology.

**Vogt's Brain Research Institute**
In order to learn more about the brain sciences I visited Kornmüller and Tönnies's laboratory in Vogt's *Hirnforschungsinstitut* at Berlin-Buch during March 1934. There I assisted in a demonstration by Oskar Vogt of the program of his Brain Research Institute: Vogt surveyed the work of his groups, which covered genetics, neuroanatomy, neuropathology, neurophysiology, and neurochemistry and included a clinic of sixty beds for neurological patients. Figure 27.2 shows the impressive view of this large institute.

Although primarily interested in anatomy and architectonics, Oskar Vogt was more than a neuroanatomist. He conceived brain research as a coordinated application of many different methods fifty years before the multidisciplinary concept became popular. The term "neurophysiology," often attributed to J. F. Fulton with his founding of the *Journal of Neurophysiology* in 1938, was first coined by Oskar Vogt in 1902. In his programmatic paper on the methods of brain research introducing the first volume of the *Journal für Neurologie und Psychologie*, Vogt mentioned *"Neurophysiologie"* and *"Neurochemie"* as important disciplines supplementing neuroanatomy, which dominated brain research at the beginning of this century. In 1930 Vogt also recognized the importance of *technology* for the neurosciences when only a few neurophysiologists such as Erlanger, Gasser, Adrian, and F. O. Schmitt foresaw the possibilities of modern techniques in research. Thus Vogt promoted the young Jan Tönnies to be head of an autonomous technical department of his Brain Research Institute in 1931.

My personal relations with C. and O. Vogt began much later, when both had retired from Berlin and built up a new and smaller institute in Neustadt near Freiburg in 1937. During the visit in 1934 I was particularly impressed by Jan Tönnies's neurophysiology and the possibilities that modern electronics offered

Figure 27.2 Oskar Vogt's Brain Research Institute in Berlin-Buch, built in 1929–1931. Located within the large complex of the Kaiser-Wilhelm-Institut für Hirnforschung, the main building in the center contains the laboratories of neuroanatomy, neurophysiology, neurochemistry, genetics, and technology, along with the animal house. The building on the left is the clinical department for neurological patients. Among the trees, behind the main building, Oskar Vogt lived with his wife Cécile and his daughters Marthe and Margarete, who also were engaged in research. The building at right contains the apartments for scientific and technical workers. Vogt's institute was the largest among the many research centers of the Kaiser-Wilhelm-Gesellschaft, a foundation started in 1911, now called the Max-Planck-Gesellschaft.

for electrophysiological work (see Tönnies, 1932, 1934). With his five-channel electroencephalograph, the *Polyneurograph* built in 1933, Tönnies recorded without arc distortion simultaneously from five different brain regions. In comparison with Paul Hoffmann's single string galvanometer, used without amplifiers until 1934, the Berlin-Buch equipment was of superb technology and was probably the most advanced in the world at that time. This technical superiority of the Berlin group produced a "we-know-it-all-better" attitude toward Berger.

Just two weeks after Tönnies's recordings from the human brain, I saw W. R. Hess demonstrating his film of subcortical stimulation in freely moving cats, and I heard his discussion with Bethe at a congress in Wiesbaden in March 1934. Hess's demonstration induced the idea of combining his technique of stimulation and coagulation with the recording of brain potentials from deep cerebral structures.

In spring 1934 I began to work in neuroanatomy and neuropathology with Spatz in Munich. Three years later, when Spatz was called to Berlin as successor to Vogt and I had finished my neurophysiological training with Hess, I entered the Brain Research Institute to work with Kornmüller after Tönnies had gone to New York in 1936.

### Neuroanatomy with Hugo Spatz

In 1934, during the fifth year of my medical studies which took me to the universities of Vienna, Freiburg, Paris, Berlin, and Munich, I worked in the neuroanatomical laboratory of Hugo Spatz in Munich and completed my doctoral thesis on cerebellar angioblastomas. My earlier introduction to neurophysiology as a student of Paul Hoffmann, my assistance with Tönnies's recordings from the human brain, and Hess's demonstration of the effects of intracerebral stimulation in cats, had all influenced my decision to become a neurophysiologist, but I followed the advice of most experts and began with neuroanatomy as a preparation for physiology. Spatz had made some fundamental discoveries in the histochemistry of the brain during the early 1920s (see Spatz, 1922) and had just finished his studies on the blood-brain barrier when I entered his group in the spring of 1934. Figure 27.3 shows Spatz in the early 1930s, when he was at the height of his research activity.

I learned much from Spatz about cerebral morphology and about the intimate relation of structure and function in brain research. Spatz was a true morphologist. During the whole of his long life, he maintained his youthful enthusiasm for the structural wonders of the brain and for its phylogenetic and ontogenetic development. As professor emeritus he continued working in the Frankfurt Brain Research Institute, which had been built after the institute in Berlin-Buch ceased to exist in 1945.

My change from neuroanatomy to neurophysiology was preceded by a visit of Dr. Alan Gregg and Dr. O'Brien from the Rockefeller Foundation to Spatz's

Figure 27.3    The neuroanatomist Hugo Spatz (1888–1969) in Munich at the age of 45. Trained by Nissl in 1913–1914, Spatz made his main discoveries on brain histochemistry and the extrapyramidal system in 1919–1923 at the neuropathology laboratory of the psychiatric clinic in Munich. He concentrated his teaching on a small group of young coworkers, which I joined in 1934. This picture was taken in Munich, four years before Spatz received the call to the Berlin Brain Research Institute as successor of Oskar Vogt in 1937.

laboratory, while they were touring the European centers on a program to promote the neurological sciences. I told Gregg about my idea to combine Hess's method with the recording of potentials from deep regions of the brain and said that I hoped to learn the method in Zürich if I could get a stipend. In the following year a Rockefeller Fellowship was awarded to me for work in London and Zürich. In the meantime I had returned to Freiburg in 1935 for training in neuropsychiatry in Beringer's clinic and for neurophysiological work in Hoffmann's laboratory with Eichler.

Spatz was an editor of the *Archiv für Psychiatrie*, in which Berger's papers on the EEG appeared from 1929 onwards (Berger, 1929, 1931, 1933). The page proofs of these papers were given by Spatz to his coworkers, and I eagerly collected all of Berger's papers that I could find among the proofs. Although Spatz tried to keep me in neuroanatomy, his proof distribution led to the opposite result: it became a second stimulus for my choice to work on the electrophysiology of the brain (Tönnies's recording from the human cortex in Berlin had been the first). The third stimulus came two years later when I saw Adrian demonstrating the spread of activity in the cat's cortex at a meeting of the Physiological Society in Cambridge. This definitely fixed my decision for brain physiology as my working field in the summer of 1936.

## Hans Berger and his EEG

Berger worked in the neuropsychiatric clinic of Jena for four decades, and in 1924 he inaugurated the recording of electrical brain waves in man. His preliminary experiments in dogs from 1902 to 1910 had yielded doubtful records, but stimulation of the precentral cortex in man through skull defects, begun in 1923, led to a new approach. Berger called the small rhythmic waves, barely detectable in his first string-galvanometer curves, *Elektrenkephalogramm* (EEG), now spelled electroencephalogram in English. Berger was born in 1873 and spent his life as a lonely and little-recognized scientist at the University of Jena until the discovery of the EEG brought him world fame in the late 1930s.

Berger's motivation for his assiduous attempts from 1902 to 1931 to overcome the immense methodological difficulties of recording electrical phenomena of the human brain was his concept of "psychic energy": he believed that the chemical energy of brain metabolism was transferred into heat, electrical, and psychic energy, and he hoped to extrapolate the latter by measuring the heat production and electrical activity of the brain.

In 1929 Berger published the results of five years' pioneer work on the EEG in man. (I have described details of his discovery with a biography and excerpts of Berger's diaries in an earlier paper.) Berger's priority in the discoveries of the human EEG remains undoubted, although his simple methods were considered "amateurish" by the workers in Berlin-Buch who had the best modern technological refinements at their disposal. Berger's work was not in the British style of apparent amateur workmanship with its understated methodical refinement; rather it was a more naive, German enthusiastic dilettantism with *Gründlichkeit*.

Berger's work was also criticized severely in his own faculty by the famous electrophysiologist Biedermann, who held the chair of physiology. Biedermann was an old bachelor with queer, uncompromising obstinacy, and he used to ride every afternoon for hours in the woods around Jena. He set high standards in electrophysiology and said bluntly that Berger's idea of recording the heat production and electrical activity of the brain was a hopeless task and an amateurish illusion. He was the main opponent in the faculty when Berger's promotion to a full professorship had to be decided in 1919. In spite of this objection, Berger became the successor of Binswanger as director of the neuropsychiatric clinic. Somewhat reluctantly, Biedermann allowed an assistant to help Berger rearrange his string galvanometer, which had remained unused since 1910, when Berger had discontinued his first trials of brain-wave recordings in dogs. Besides Biedermann, many other electrophysiologists questioned the value of Berger's records until his EEG was brought to public notice and international fame by Adrian's demonstration in 1934.

Between 1924 and 1931 Berger recorded the main features of the EEG. Figure 27.4 shows some unpublished examples from his pioneer experiments. Figure 27.4a is probably the first historical record of alpha-wave arrest during mental activity, obtained in 1926. Berger had expected the opposite, activation of cortical rhythms, and only later recognized the alpha-blocking, as described in his second EEG paper in 1930. Further decisive advances were his direct proof of EEG origin in the human cortex by comparison of cortical and subcortical leads in 1930 (Figure 27.4b) and his records of specific epileptic discharges in 1931 (Figure 27.4c). Berger's single or double short photographic EEG records appeared less attractive than the long multiple inkwriter curves of later authors, but they showed all the characteristics of normal and pathological EEGs. Last year, when looking through our Berger-Archiv in Freiburg, I found out that Berger had already recorded typical epileptic spike-wave potentials of the EEG in 1930–1931 (Figure 27.4c). However, Berger's overconscientious attitude prevented the publication of these beautiful curves because he noted myoclonic twitches in his subject during a petit mal attack and feared mechanical artifacts. In 1930 Berger had also recorded a complete petit mal seizure, but he did not publish it until 1933 when his doubt about artifacts was relieved by Fischer and Kornmüller's description of similar high-amplitude convulsive potentials in strychnized animals. Two years later Lennox, Gibbs, and Davis in Boston showed that epileptic discharges could often be recorded from the human brain if one concentrated on petit mal and psychomotor attacks instead of grand mal, which Berger had studied initially without success.

## Some early EEG groups outside Germany

From 1937 onwards scientific relations with Anglo-American neurophysiologists, especially the Boston groups, developed out of our common interest in EEG. Hallowell Davis had initiated American EEG recordings with Pauline Davis in Cannon's institute (see Chapter 18, this volume), and Gibbs and Lennox extended EEG research to clinical diagnosis in epilepsy and brain lesions. During their European travel I met the Gibbses and Lennox after I had read their systematic

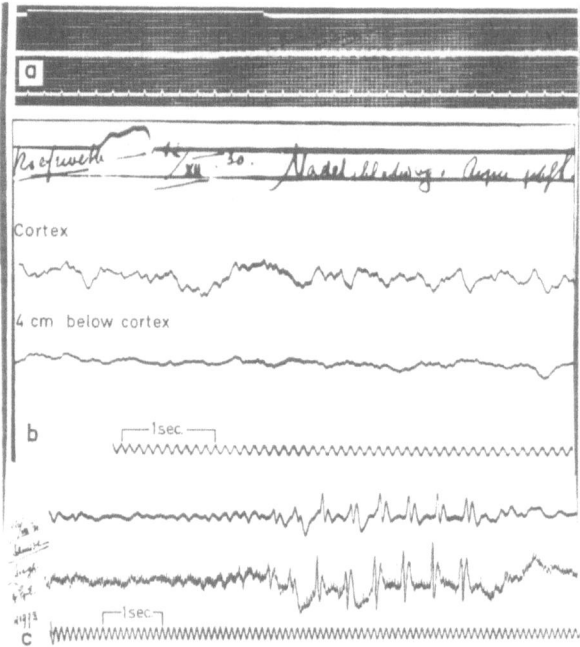

**Figure 27.4**  Samples of Berger's early EEG records from 1926, 1930, and 1931 (unpublished string- and coil-galvanometer records from the Freiburg Berger-Archiv). (a) A string galvanometer curve taken on September 6, 1926, from the intact skull convexity by two coated silver needles in frontal and occipital position (female with subtentorial decompression). This was Berger's 58th EEG experiment after he started recording from a trephined patient in July 1924. In these early years Berger investigated mainly the influence of mental activity such as calculation or explanation of proverbs (marked on top). The string record (middle) shows scarcely visible alphalike wobbles of 8 per second appearing after the mental task has been performed, as marked by a noise (Berger's writing "Hammer niederwerfen" above middle record). Time marker: 0.2 sec. Amplitude: about 1 mm for 0.1 mV. Record no 58/2. (b) First direct recordings from the human cerebral cortex and the subjacent white matter, December 17, 1930. In a 20-year-old trephined patient with brain tumor, Berger used a diagnostic *Hirnpunktion* to record simultaneously from two pairs of coated silver needles inserted in the cortex and 4 cm below. The different amplitudes demonstrate that the EEG waves originated in the cortex and that the white matter (lower record) had less electrical activity. Berger published only a small fragment of these records in 1931. In this Siemens double-coil galvanometer without amplifier, 0.1-mV amplitudes correspond to about 2 mm in the upper, and 1.8 mm in the lower record. Time marker: 0.1 sec. After name and date, Berger wrote on top: "Nadelableitung. Augen geschlossen" (needle recording, eyes shut). Record no. K 1765. (c) An early EEG record of epileptic discharges taken by a Siemens coil galvanometer and mirror oscillograph on August 26, 1931. In 1931 Berger had just acquired one amplifier with a Siemens electromagnetic mirror oscillograph (lower curve), which he combined with his old coil galvanometer (upper record, no amplification). Berger was uncertain whether the spikes and waves were epileptic brain potentials or artifacts because he observed myoclonic twitches in the face during the seizures and wrote on the record "Lidzuckungen." Record no. K 1993.

study of the EEG in epilepsy, which enlarged so effectively on Berger's first attempts. Lindsley's studies of EEG development in children, also from Harvard, and the night-sleep recordings of A. L. Loomis, made in his villa in Tuxedo Park after consulting Davis and Tönnies, showed exciting new EEG applications. Good communications developed with these EEG laboratories and others, including H. Jasper's in Brown and McGill University and W. G. Walter's in Bristol. My correspondence continued with Lindsley even after the war had begun.

## Berger's last years

My contact with Hans Berger from 1939 until his sudden death in 1941 was restricted to an exchange of letters and reprints, although I tried several times to meet him personally. After his abrupt dismissal from the clinic in Jena, Berger retreated to a small sanatorium in Blankenburg, and all EEG work in Jena was stopped by his successor in 1938. The relations between Berger and the EEG group in Berlin-Buch were hampered by personal tensions, and my return to Freiburg from Berlin in 1938 enabled me to maintain a neutral position between Jena and Berlin. Thus Berger regarded the Freiburg laboratory as the place where his EEG work continued in Germany, even though I had never worked in his clinic. Berger's former coworker Hilpert, from whom he had expected further EEG research, died, and Berger wrote in a letter of June 28, 1939: "My hope that my faithful coworker and scholar Hilpert would continue the EEG was dashed by his tragic death. . . . The fate of my EEG appears remarkable after my discovery of 1924 was made known in Germany by my friend Adrian [in] Cambridge in 1935." And in a letter of April 19, 1941: "I thank you that you always defend me so bravely. . . . I am glad to know that my EEG-child is being so well cared for and developed in your hands." ("Ich freue mich, dass ich mein EEG-Kind bei Ihnen in so guter Hut und Aufzucht weiss.")

Prewar difficulties and the fact that I was drafted into the army in May 1940 prevented personal contact. Berger was interested to hear that we used the EEG for recording from brain-injured soldiers in Kleist's clinic in Frankfurt a.M., where Vogt's inkwriter EEG machine had found some practical application. In Frankfurt I had to evaluate both the clinical EEGs from Freiburg and those from the Army Hospital in Frankfurt. I informed Berger that our EEG work begun in Freiburg with two coworkers was being continued as far as possible. In May 1941 I received a letter from Berger in which he complained about his health, but I did not recognize that this was a sign of a serious depression until three weeks later the sad news of his death by suicide arrived. This letter appears to be the last he wrote. It is reproduced in Figure 27.5 as a personal document of this remarkable scientist who founded the electrophysiology of the human brain and continued his keen interest in this field until the last days of his life. Figure 27.6 shows Berger's death mask taken by the University of Jena in 1941.

## Experimental Brain Research and Three Neurophysiologists: Walter Hess, Jan Tönnies, and Erich von Holst

In 1936 I obtained a Rockefeller Fellowship to work first in England and then with Hess in Zürich. During my stay in London at the National Hospital, Queen's Square, with E. A. Carmichael, I had ample occasion to visit the main neurophys-

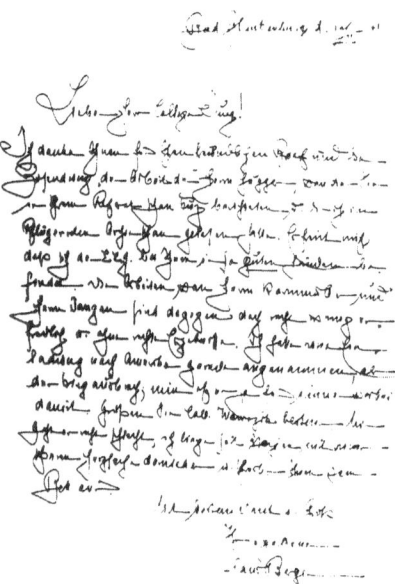

Figure 27.5   The last letter of Hans Berger referring to EEG (May 12, 1941). It was written in a state of severe depression seventeen days before his death on June 1, 1941. Despite his depressed mood and his isolation after retirement, Berger regularly followed the EEG literature. This can be seen from the letter, in the first lines of which he expresses thanks for a reprint of an EEG paper "which I had already read in *Pflügers Archiv*." He then mentions his concern about the development of German EEG research: "I am glad that my EEG is in so good hands with you." After critical remarks on other papers, he comments on his current state of health: "I will now never be able to take up the American invitation which I accepted before the war. My condition is bad. I have been lying with severe heart trouble for weeks and am writing in bed."

iological centers in Cambridge, Oxford, and London. There I met E. D. Adrian, B. H. C. Matthews, J. C. Eccles, D. Denny-Brown, G. L. Brown, and saw H. H. Dale, A. V. Hill, and other leading scientists at the meetings of the Physiological Society.

### Hess's integrative physiology

When I moved from London to Zürich, the change from fact-oriented British physiology to the systems-oriented physiology of W. R. Hess (Figure 27.7) made a lasting impact on my work. In the early 1920s Hess had developed his general concept of autonomic functions (Hess, 1925). It remained the basic theory upon which he planned his experiments. After some preliminary studies he inaugurated in 1926–1934 the method of brainstem stimulation of the unrestrained cat by implanted electrodes (Hess, 1932). He pursued these investigations systematically for twenty years to clarify the cerebral correlates of autonomic functions, under which he included *sleep* as an intracerebral "vegetative" regulation, essential for recovery.

Hess always thought in terms of biological relations and systematic integration for his physiological concepts. He considered single facts only in their context

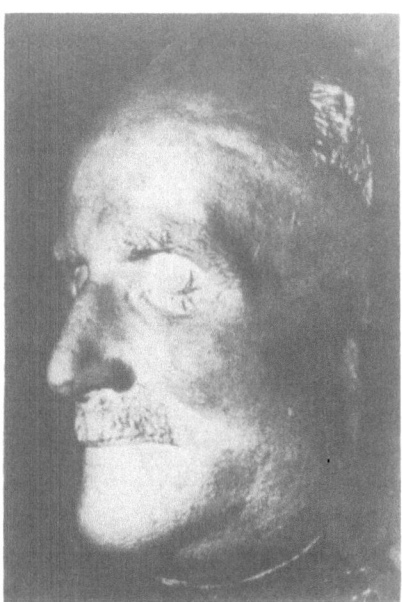

Figure 27.6 . Hans Berger's death mask, taken at the University Clinic of Jena on the day of his death (June 1, 1941). Berger began his research work in the Neuropsychiatric Clinic of Jena University as house physician in 1897 and continued it during four decades as *Privatdozent* (1901) and full professor of neurology and psychiatry (1919) until he became emeritus professor in 1938. His main scientific publications included the effects of visual deprivation on the occipital cortex in dogs and cats (1900), studies of cerebral circulation in man (1901, 1904, and 1907), temperature measurements of the brain (1910), and fifteen papers on the electroencephalogram in man (1929–1938). He began his experiments on the brain potentials of dogs in 1902 and his recordings from human skull defects in 1924, and he published the first paper on the EEG in 1929.

with functional systems or in their significance for the organism. Indeed, his very first paper, conceived as a student, typified his method of research. A variation of leg vessels that was only a descriptive curiosity for anatomists had a functional meaning for Hess: from the vascular configuration he concluded that there were mechanically induced laws for the structural development of the vascular system, and he wrote about this to Wilhelm Roux, the founder of experimental research on animal development *(Entwicklungsmechanik)*. Roux's interest in his hypothesis gave an early and lasting stimulus to Hess's further research.

Hess's concept of interaction between somatic and autonomic systems distinguishes two categories of integrative vegetative functions that are in reciprocal relation, the ergotropic and the trophotropic. The *ergotropic subsystem* prepares the *readiness* of the organism and facilitates somatic functions, whereas the *trophotropic subsystem* inhibits somatic action and regulates *recovery* of somatic function in various tissues (Hess, 1933, 1949). It also induces a restoration of the brain and the entire organism by sleep. In the periphery, the ergotropic system is related to, but not identical with, the anatomically defined sympathetic nervous system, and the trophotropic to the parasympathetic. Cannon's view of autonomic functions was somewhat similar, but relied more upon anatomical distinctions. Hess, in contrast

Figure 27.7   Walter R. Hess (1881–1973). (a) A photograph taken in 1936. During this year, when I began to work in Hess's physiological institute at Zürich University, he was finishing his monograph on the cerebral mechanisms of respiration and circulation (Hess, 1938) and was beginning to organize the Zürich International Congress of Physiology, which would be held in 1938. Hess spoke clearly and slowly in his lectures, but his discussions were very lively, with sharp and well-aimed arguments. (b) A conversation with Hess in 1948. Hess looked skeptical when I told him about my plans to record with microelectrodes from the brain by applying Hartline's and Granit's methods to the cat's visual cortex. (This photograph was taken on a visit to Kleist's home in Frankfurt by his daughter during a meeting of the German Physiological Society in September 1948.)

to Cannon, was less interested in peripheral autonomic innervations and hormones than in the *central coordination* of the two vegetative subsystems in the brain.

In Zürich I worked with Weisschedel on stimulation and coagulation of extrapyramidal motor structures, the substantia nigra and the caudate nucleus, whereas Hess concentrated on the hypothalamic regulation of respiration and circulation (Hess, 1938). I saw only a few of Hess's experiments on sleep. I admired the systematic organization but wondered about various electrode localizations and the long and variable latencies of the sleep effects. However, I was impressed by Hess's concept of sleep as an active central vegetative regulation of the brain with inhibition of the cortex from below and inactivation of body activity (Hess, 1933). This remained a challenge for neurophysiological experiments for several decades.

Figure 27.8 illustrates the laboratory milieu in Hess's institute, in which he did his well-prepared experiments in later years. In the thirties the personnel was restricted to only two: the secretary who did the protocols and the able technician, Jenni, who assisted Hess in conducting the experiment and filming the cat.

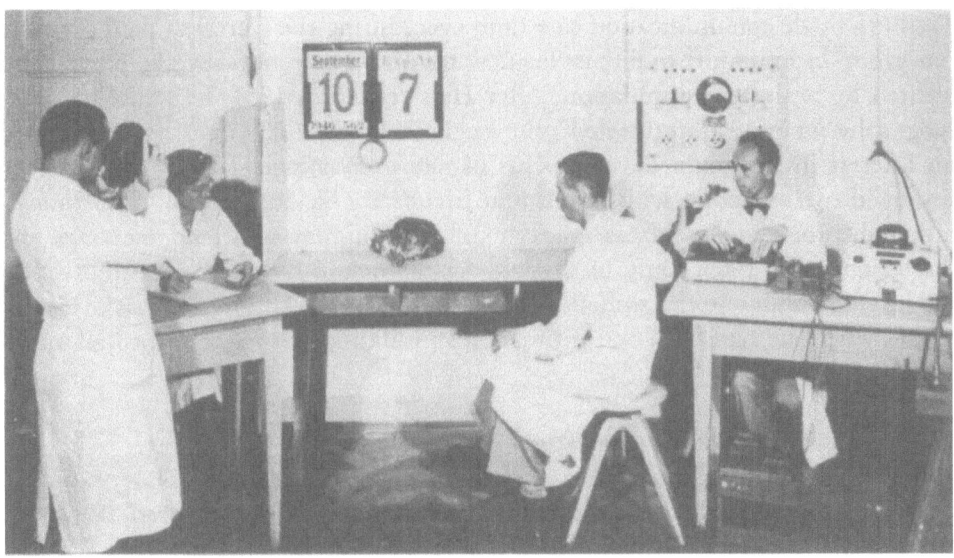

Figure 27.8   Hess's experimental setup in 1946 for subcortical stimulation of the unrestrained cat. In the years 1928–1939 Hess conducted these experiments himself with the camera in his hands and with the help of his technician, Jenni, who is sthown on the right of this picture. After 1946 the group became larger: Jenni managed the stimulation apparatus, two assistants filmed and observed the cat, and the secretary (the former nurse of Hess's children) wrote the experimental protocols. Hess performed all his experiments from 1928 to 1950 with a standard arrangement having the clock, date, and experiment number in the background, which appears on the cat's film.

Hess's admonition that single facts mean nothing for physiology if they are not related to functional systems or biological goals agreed with the teaching of Nicolai Hartmann, whose philosophy of science is discussed below. Hess's research principles impressed me as a contrast to the often heard assertion that physiology is a science of facts. Hess continued to explain to his pupils that physiology is *leistungs-bezogen* and demands a functional synopsis besides an analysis of the various mechanisms: to understand the functional order of the organism one must consider the *context* of experimental facts, the *aim* of the animal's action, and the biological *goals* of the behavior. It is not sufficient to record single events by optimal techniques. Although Sherrington (1906) had similarly considered functional aims for the integrative action of the nervous system, Hess thought that physiology in Anglo-American countries was neglecting this aspect. In comparing British fact-oriented inductive and Hess's systems-oriented integrative physiological research in the years 1936 and 1937, I could see the advantages and disadvantages of the two conceptions. My discussions with Hess moderated my overestimation of British physiology, which had been induced by Hoffmann, but also heightened my admiration for those physiologists working in England who had developed general concepts from their experimental findings, such as Adrian, Dale, and Eccles.

When I expanded the Zürich experiments in Berlin-Buch, and in Freiburg after my return there in 1938, Hess often gave me advice, and we maintained our personal and scientific relationship for over 35 years. The proximity of Zürich and

Freiburg made communication easy, and even during the war Hess paid a visit to our group in Frankfurt to discuss his new results on direction-specific movements elicited by brainstem stimulation. After Hess retired in 1951, he wrote an autobiography which was translated into English (Hess, 1963), and he maintained an interest in the main developments of neurophysiological research. He even arranged two symposia with pupils and friends for his 70th and 80th birthdays, although progressive deafness made it difficult for him to attend meetings and discussions after 1965. Only in the last decade before his death in August 1973 did he stop coming to the annual meetings of the Deutsche Physiologische Gesellschaft, where he had been one of the most stimulating participants for half a century.

### Work in Berlin with Kornmüller

Following my stay in Hess's laboratory, I was eager to realize my plan of electrophysiological exploration of subcortical brain regions when I entered the Brain Research Institute in Berlin in 1937. Since Tönnies had left for New York, I worked in the laboratory of A. E. Kornmüller, who had done basic work on local brain potentials of cytoarchitectonic areas (Kornmüller, 1937). With Hess's electrodes we recorded the theta waves from the rhinencephalon after sensory arousal and the relations of brain potentials in striatum, motor cortex, and thalamus (Jung and Kornmüller, 1938/39). In addition to making brain potential recordings with Kornmüller, I began local coagulations of the anterior commissure, the caudate nucleus, and the mesencephalic tegmentum of cats using Hess's electrodes.

Major changes had occurred at the Brain Research Institute between my 1934 visit and my arrival in May 1937: Vogt had just handed over the direction to Spatz and left for Neustadt, being persona non grata to the National Socialist government; Tönnies had gone to New York and had been succeeded by J. A. Schaeder. During my work with Kornmüller, who had previously worked with M. H. Fischer in Prague on optokinetic and vestibular nystagmus, I became acquainted with some of the traditions of the Prague schools of visual physiology which had flourished from 1820 to 1930 under Purkinje, Mach, Hering, and Tschermak. In the Berlin institute the research group under the Russian geneticist Timofejev-Ressovsky, working with *Drosophila*, had little relation with neurobiology, but we liked to discuss various themes of brain research with the universally interested Timofejev. I remember a nightlong discussion on brain localization that was provoked when I uttered a word in defense of Lashley and dared to say that the 200 cortical areas of Vogt must be interrelated by physiological coordination for behavior and thought. When I expressed my belief that functional cooperation would be a more serious problem to attack than the anatomical delimitation of many cortical areas, I was surprised that Timofejev stoutly defended area localizations and their *haarscharfe Grenzen* as described by C. and O. Vogt (1919). Thus I got an impression of the strong influence that Vogt had had on his coworkers.

**Friendship and work with Jan Tönnies**

Jan F. Tönnies (Figure 27.9) was born in 1902, the son of the founder of German sociology, Ferdinand Tönnies. His father was a friend of O. Vogt, both coming from the same town, Husum in Holstein. So Vogt was able to follow Jan Tönnies's early studies in technology. In 1932 Tönnies became head of the department in Vogt's Brain Research Institute, where he built the first inkwriter electroencephalograph (1932) and made basic contributions to EEG research (1933, 1934).

Tönnies was eight years older than me. This age difference and his extensive knowledge of technology and neurophysiology put him in the position of experienced scientist and admired "elder brother" until I had acquired sufficient neurophysiological experience myself. Following the short Berlin encounter in 1934, Tönnies and I were separated from 1936 until 1939 by his appointment to Gasser's department at the Rockefeller Institute. We worked independently in different fields, and when I consulted him about amplifier and experimental problems by letter, his answers returned promptly by transatlantic sea mail within two weeks. Personal meetings were restricted to Tönnies's brief European visits to Berlin in 1937 and to the Zürich International Physiology Congress in 1938.

In 1938 Jan Tönnies constructed the first "imperative" cathode-follower input to record from high-resistance electrodes for Alan Hodgkin in New York (see

*Jan Friedrich Tönnies*

Figure 27.9  Jan F. Tönnies (1902–1970) in his electronic laboratory, 1947. From 1929 to 1931 Tönnies organized the building of Vogt's Brain Research Institute. There he constructed the first inkwriter electroencephalograph in 1931 and the differential amplifier in 1934; in New York he built the first cathode follower in 1938. After moving back to Berlin in 1939 and from Berlin to Freiburg in 1942, Tönnies built his workshop in his home. In the years after the war he reconstructed the Freiburg neurophysiological laboratories and contributed to many new developments in brain research and technology. The wall picture shows his parents.

Hodgkin, 1938). These amplifiers became the methodological basis of modern microphysiology after microelectrodes of less than 1 $\mu$ tip were introduced in 1951 by Ling and Gerard (1949). In 1939 Tönnies himself began to make a few multiple recordings from the spinal cord with rather simple microelectrodes (sewing needles). These records demonstrated a dipole field of slow potentials in the ventral horn, superimposed on rapid small discharges. Tönnies thought that the latter were groups of synaptic potentials, but they were later known as Renshaw-cell discharges. When Tönnies showed these records to me after his return to Germany, I got rather excited and planned to do further experiments in Hoffmann's laboratory to elucidate spinal reflex circuits and their relations to the slow root potentials that Barron and Matthews (1938) had discovered. The war cut short all these plans, and Tönnies's records were published only in one figure of his postwar paper (1949) after we had resumed our spinal-cord experiments in 1946.

In June 1939 Tönnies returned to Germany and married, but war conditions prevented collaborative research work. However, I used my army EEG service in Frankfurt to arrange a few personal meetings with Tönnies during the first years of the war. The Frankfurt apparatus was the only EEG machine with inkwriter in Germany outside Berlin-Buch, since we had only optical oscillograph recordings for our EEG in Freiburg before and during the war. When I returned from the Russian campaign in 1943, Tönnies had moved with his family to Freiburg from bomb-damaged Berlin. From that year onwards we remained together for the nearly three decades until his death in 1970. Our relation with Oskar Vogt's institute, set up in 1937 in Neustadt near Freiburg, remained more personal than scientific, since Vogt limited his work to neuroanatomy and neuropathology after his neurophysiological assistant von Ledebour was killed in the war.

In our thirty years of friendship Jan Tönnies and I exchanged scientific ideas freely and happily and stimulated each other with new ideas and methods to test them in experiments. We never knew who was the first to find a new approach or hypothesis, but I think many original ideas in our work came from Tönnies when he applied his engineer's knowledge of communication and control to the nervous system. For four years after the war I planned all my animal experiments with Tönnies (Tönnies and Jung, 1948; Jung and Tönnies, 1950), and we also collaborated on vestibular research in man. Tönnies was both scientist *and* technologist. His strength lay in this combination and in the originality of his thinking, and his successes were many: the creation of the multichannel electroencephalograph, the differential and the cathode-follower amplifiers, the physical analysis of EEG distribution through skull and scalp structures (Tönnies, 1933), the dorsal-root reflexes (Tönnies, 1938), the theory of synaptic welding for memory (Tönnies, 1949), the flexor-reflex feedback model (Tönnies and Jung, 1948), the aperiodic EEG-spectrum-analyzer EISA (Tönnies, 1969). Tönnies's weakness was anatomy, and this sometimes resulted in unconventional ideas and findings. But the story, which Herbert Gasser used to tell, that Tönnies discovered the dorsal-root reflexes of the spinal cord because he mistook the dorsal for the ventral roots was, of course, exaggerated. We checked localization and the spread of excitation in the brain anatomically, and experiments on local brain stimulation led us to postulate a restraining control *(Bremsung)* of the cerebral machine to prevent convulsive discharges or to limit their spread (Jung and Tönnies, 1950).

In his last years Tönnies developed a new method of aperiodic-interval spectrum analysis (EISA) of the human EEG that clearly shows the different periods of a whole night's sleep condensed onto a record of about one meter in length (Tönnies, 1969). After receiving an honorary degree from Freiburg medical faculty in 1968 and while involved in the development of a new optical version of his EISA apparatus, this creative scientist died unexpectedly of a heart attack on Christmas Eve 1970.

### Erich von Holst and his concepts

Erich von Holst (Figure 27.10), who stimulated my work on sensorimotor mechanisms, was born in 1908, of a German-Baltic family in Riga, the son of a neurologist. He was one of the most original German zoologists and a friend of K. Lorenz and O. Koehler.

Von Holst was trained in physiology by Bethe and called himself a "*Systemphysiologe.*" He preferred to experiment on unrestrained animals and to relate the analysis of functional systems to behavior. In 1935 von Holst began his work on the relative coordination of the fin movements of spinal fish, in Naples, and three years later I found similar phenomena in human tremor. Most of von Holst's concepts, developed from experiments in fish and lower forms, can also be applied to man and are thus of general importance.

Figure 27.10   Erich von Holst (1908–1962). When this picture was taken (1958), he had just moved from Wilhelmshaven to the new Max Planck Institute for Behavioral Research in Seewiesen. There he continued his brainstem stimulations for eliciting instinctive behavior in chickens. Von Holst's personality was dynamic. His lively gesticulations and the sharp and surprising turns of his arguments made his brilliant lectures and discussions fascinating.

This is the story of our acquaintance: In 1938 I had made multiple recordings of parkinsonian tremor which showed certain periodicities and mutual influences of the rhythm in arms and legs (Jung, 1941). One day, looking through *Pflügers Archiv*, I experienced a déjà vu feeling when I came across the same periodic irregularities in von Holst's tracings of fin movements as were in my tremor recordings. Von Holst had developed his principles of relative coordination from experiments in the spinal fish. I sent him my human tremor records, and we agreed that these showed the same characteristics of "relative coordination" as his fish. In 1939 von Holst reproduced these human recordings in a synopsis of his work, and I also found similar principles for the interhemispheric coordination of the alpha rhythm of the human EEG (Jung, 1939a). I concluded that von Holst's relative coordination guided various central rhythms, including locomotion and brain waves. The neurophysiological mechanisms of interaction between central rhythms remained obscure, and a plan to investigate this by electrical recordings in the fish's spinal cord with von Holst in Naples was cut short by the war. After the war von Holst moved from Göttingen to Heidelberg and in 1948 to Wilhelmshaven. In 1958 he and K. Lorenz founded the Max Planck Institute for Behavioral Research in Seewiesen near Munich.

Although von Holst and I used very different animal species in our experiments, we remained in research contact until his untimely death in 1962. Our common interests were autorhythmicity and feedback in the central nervous system and the regulation of eye movements and vestibular functions. In 1954 von Holst sent his coworker Lore Schoen to our laboratory, and she succeeded in finding neuronal correlates of otolith functions in the fish's vestibular nuclei (Schoen, 1957).

I had tried in vain to find neurophysiological explanations of the stability of the visual world, being unsatisfied by general concepts such as Hering's "shift of attention." Then, in 1949, von Holst told us about his concept of reafference. He discussed with us whether his *Efferenzkopie* could be related to Tönnies and my concept (1948) of neuronal feedback *(Rückmeldung)* over the dendritic tree and to ocular motoneurons, but we had to disappoint him. Even today a neurophysiological correlate of von Holst's *Efferenzkopie* is still lacking, and the negative feedback of the Renshaw cells has been clarified by Eccles (1957) as a different, special mechanism, while we had assumed dendrodendritic interaction. (Tönnies and Jung, 1948).

Von Holst had an excellent capacity for quickly recognizing the essential points in a multitude of facts and observations and then proceeding immediately to new methods of experimental systems analysis of specific functions. This is shown by his experiments on otolith functions in fish (Holst, 1950), using the intact animal. His lively speech and intellectual brilliance fascinated every audience, but they also provoked reactions of either total acceptance or emotional opposition, and his changes of mood were often difficult to understand. During his periods of depression von Holst retreated from the laboratory to his side interest of building model violas. In his last years this interest developed into systematic investigations on the physical and physiological basis of the playing of stringed instruments. (His notes on this *Geigenkunde* were printed posthumously in a private edition.)

A touch of genius can be seen in the way von Holst dealt with research problems such as relative coordination (1939), flying models of birds (1969), vestibular analysis in the intact fish (1950), the reafference principle (1951), and brainstem stimulation (Holst and Saint Paul, 1960). The idea and the approach were in every case new and entirely unconventional. Jan Tönnies had a similar ability, but in other fields. However, both Tönnies and von Holst had the same weakness of neglecting anatomical connections in their functional syntheses. When I asked von Holst in 1939 whether the brainstem cut of his spinal fish left a connection with the vestibular nuclei, he said he was not interested in such details, and in his last years he neglected anatomical controls of his stimulation points in the chicken diencephalon. He preferred systems concepts to neuronal circuits.

**Work in Freiburg**

From 1938 to 1940 my research was centered primarily on brain potentials and polygraphy (Jung, 1939a,b,c). Studies on eye movements and vision, begun in 1938, were continued for over two decades. Research on sleep and alterations of consciousness also started in 1938, following the lines indicated by Hess, but had to be interrupted during the war and early postwar years.

After 1946 the work with Tönnies on spinal-cord reflexes and experimental epilepsy in animals and on vestibular and optokinetic nystagmus in man was followed by neuronal recordings from the brain. From 1950 onwards the Freiburg research groups concentrated on the neuronal physiology of vision and visual-vestibular coordination. Eye movements were investigated first in man by nystagmography, and then in animals by neuronal recordings. Our main concept was to make use of the phenomena of visual perception in man in planning neuronal experiments in animals, and to coordinate psychophysics and neurophysiology. Animal experiments also provided stimulation for work on human perception. The concept of a visual receptive field was used for psychophysical measurements of perceptive fields in man. Collaboration with Tönnies and personal contacts with Swedish, British, and American groups amplified the effect of the stimuli received from Hoffmann, Hess, and von Holst before the war and gave the work a new impetus. Many young coworkers joined the Freiburg laboratory between 1948 and 1960.

All the publications of the Freiburg groups in the years 1939–1971 are listed in a bibliography (Schriftenverzeichnis Richard Jung und Mitarbeiter, 1971), and I will refrain from mentioning any particular work. The main promotor of neuronal recordings in the visual system was G. Baumgartner. Visual-vestibular coordination and multisensory coordination were investigated by Grüsser and Kornhuber and their groups. Creutzfeldt extended his work in neuronal physiology to the sensorimotor cortex, Spehlmann to the electrophoresis of synaptic transmitters.

**Nicolai Hartmann's Philosophy of Science**

During my early student years I was mainly attracted by experimental sciences such as physiology; in the humanities, art history interested me more than philosophy. When I came to Freiburg in 1929, Heidegger had just been appointed to succeed Husserl as professor of philosophy. Compared to objective research, Husserl's phenomenological "reduction," which deliberately excluded the facts of science, appeared to me artificial, and the verbal appeal of Heidegger's lectures, which expressed deep skepticism toward modern science, could not deter me from pragmatic research. Hence, in spite of some genuine interest in philosophy, I regarded phenomenology and existentialism as too remote from science and acceptable only to psychiatrists. In consequence, I at first adopted the naive approach to research that disregards its basic philosophical relations. Hearing a lecture of E. Cassirer, although it was impressive in its formal elegance, did not move me to change my attitude of reserve.

To my great surprise I found in 1932, at Berlin University, an entirely different thinker who taught that modern philosophy, including metaphysics, should not construct systems but should, rather, build its problem-centered research within the ordered frame of present-day science. This was Nicolai Hartmann (Figure 27.11), who warned esoteric philosophers that their "play is lost when they disregard the results of modern science" (Hartmann, 1946). In this philosophical attitude "Diesseits von Idealismus und Realismus" (Hartmann, 1924), I found the way I was seeking.

### The Kantian basis and Hartmann's order of sciences
After reading Kant's "Kritik der Urtheilskraft," which analyzes the basic concepts of biology, and the "Kritik der reinen Vernunft," which postulates that philosophy should use experimental methods and imitate natural science, I was

Figure 27.11  The philosopher Nicolai Hartmann (1882–1950). This picture was taken three years before his death, after he had moved from Berlin to Göttingen in 1946 and had finished his last books on the philosophy of nature (Hartmann, 1950) and teleological thinking (Hartmann, 1951), both of which appeared posthumously.

prepared to understand Hartmann's philosophy with its aim of incorporating the general results of the natural sciences into philosophical research. My admiration for Kant was not diminished by his dependence on the zeitgeist of the eighteenth century. It was a *Kantseminar* of Nicolai Hartmann, which I attended in 1937 after returning to Berlin, that made me unwilling to accept the flat positivistic and pragmatic attitude of most natural scientists. I adopted Hartmann's clear position "at this side of idealism and realism." Since then I have retained a weakness for the relation between science and philosophy in spite of my continuing fascination with new techniques, experimental research, and clinical applications of the neurosciences. To obtain an orderly overall view of the complex relations of the various sciences that explore the world and to systematize the achievements of human culture, I have constructed a scheme based upon Hartmann's concepts (Figure 27.12).

Figure 27.12 shows the stratified structure of the world in relation to the mental and cultural activity of man. This is a simplified diagram derived from Hartmann's writings (1940, 1946) to demonstrate the hierarchy of sciences and the relationship between brain research and psychiatry. The lowest level, with the inorganic processes, is investigated by physics and chemistry. The second level, comprising the vital functions, is explored by physicochemical methods, and more specifically by biology, which includes the neurosciences for investigation of the nervous system. According to Hartmann's laws of level dependency (1940), the higher mental and social levels can only exist on the basis of accurate functioning of the lower inorganic and biological strata. All lower levels, however, are independent of the higher levels. Thus mental functions become possible only on the basis of the subjacent physicochemical and biological processes. Species peculiarity, individualization, and social grouping in biological and mental levels are indicated by dotted and interrupted lines.

Nicolai Hartmann, born in 1882 in Riga of a Baltic-German family, maintained contact with natural science until his death in 1950. He was in opposition to both

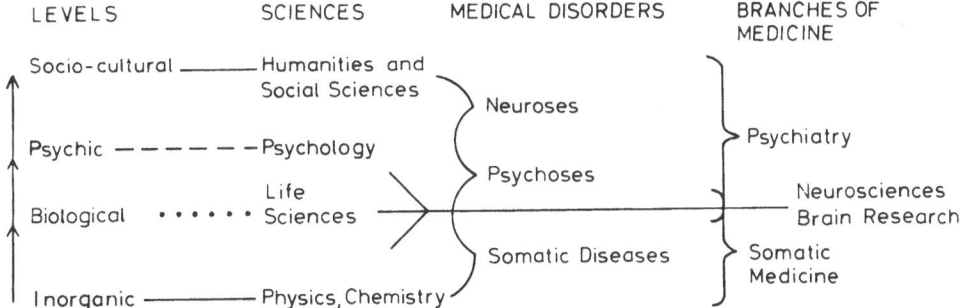

Figure 27.12  Nicolai Hartmann's level concept of the world and its relation to the hierarchy of sciences. This diagram demonstrates the structural stratification in four levels (inorganic, vital, psychic, and cultural), and shows the related order of sciences, including neurobiology and medicine. Each stratum depends on the basic functions of all subjacent levels but has its specific new laws. These special laws increase in complexity and differentiation for the higher levels, although the natural laws of the lower are also valid in all upper levels (as indicated by the arrows). Similar rules apply to health disorders and their treatment in medicine. Adapted from Jung (1967).

the system-constructing and the phenomenological schools of German philosophy. He began to study medicine in Dorpat and then changed to philosophy and classical philology. In 1905 he joined the Marburg Neo-Kantian school of Cohen and Natorp. After becoming Natorp's successor to the chair of philosophy in 1922, he was called to Köln in 1925 and to Berlin University in 1931, where I first heard his lectures in 1932.

Interested in the sciences since early youth, Hartmann tried to integrate their results into his problem-oriented philosophy beginning with his 1912 book on the philosophical basis of biology. In his writings, however, he conserved the classical philosophical style. Although he used drawings occasionally for elucidation in a seminar, Hartmann avoided schematized tables, fearing popularized simplifications.

During my Berlin semesters I heard parts of Hartmann's systematic lectures in 1932 and 1933 and participated in his *Kantseminar* in 1937 when I worked at the Brain Research Institute. Hartmann published the first draft of his level concept in 1936 *(Kategoriale Gesetze)*, and the zoologist Max Hartmann incorporated it into his philosophy of science (1937). Traditional philosophers, however, remained skeptical and told me that a scheme like that of Figure 27.12 would be a crude simplification and a "biologistic" distortion of his thinking. In the following years many pressing problems left little time for systematic thought and philosophy. However, during some quiet weeks in 1941, before the Russian campaign started, I was able to read Hartmann's 1940 monograph on the structure of the real world. In this third volume of his ontology I found a confirmation of my scheme, which I have used since for several lectures and published in a general synopsis of neurophysiology and psychiatry in 1967.

Hartmann was more impressive as man and teacher than as the author of long books in which he proved his points with detailed explanation and close reasoning. He talked slowly in his clear voice and sharply pronounced Baltic accent, and he fascinated even large audiences. In his seminars he liked sharp discussions, which he then led to a conclusion; with a few guiding words he helped his pupils to find the right formulation. The best synopsis of his concepts can be found in his book on the basis of ontology (1946), which is written in a beautifully concise style.

Hartmann's world concept allows a complete determinism only at the lowest causal levels. His philosophy gives no place to God. For him, free will and final decisions are possible at the psychic and mental level by selecting the means among different causal sequences. Human choice demands some knowledge of causal laws at lower levels. The freedom of selection at the higher levels must be consistent with the causal laws at the lower and evidently cannot act against them.

Hartmann's ethics recognized moral values as a challenge to human decision. His ethical atheism left room for the choice between good and bad or between different causal conditions. When some physicists attempted to use Heisenberg's principle of uncertainty in support of Christian belief, Hartmann disapproved, as is shown in his correspondence with Kurt Schneider.

There was a fundamental disagreement between Hartmann and the existentialists, represented by Heidegger and Jaspers, who overrated the individual thought of man and devaluated objective science. I admired Jaspers, whom I had met in Basel, more for his early work in psychopathology than for his philosophy. I lost my Hartmannian reserve against *Existenzphilosophie* rather late, when I attended Heidegger's seminar on Aristotle and discussed with him his mistrust of science and technology. I found Heidegger more open to science and more natural in his discussion than in his stylized lectures. He often impressed more by his personality than by his philosophy, which could be obscured by his verbal mannerisms. During a discussion he could integrate a complex situation, often after a long silence, into one clear sentence or characterize a man in two or three succinct words.

Nicolai Hartmann spent the last four years of his life at the University of Göttingen, where he found a large interdisciplinary audience. In spite of the war and postwar difficulties, he finished his monograph on the philosophy of nature (1950), the fourth and final volume of his ontology, and rewrote his *Teleologisches Denken*, conceived in 1945 in Berlin. He died in November 1950 after suffering a stroke while cycling to his seminar. Hartmann was one of the rare universal thinkers who synthesized the classical German philosophical tradition with an extensive knowledge of natural sciences.

## Hartmann's categorical dependence and the neurosciences

Nicolai Hartmann's *Schichtenlehre*, the hierarchic concept of stratified levels of the world that reach from the inorganic to the highest mental and social strata, demonstrated both the necessary dependence of the higher levels upon the lower and the order of the scientific approach to these levels. When I put this idea into a simple schematic diagram (Figure 27.12), I suddenly understood both the interrelations of the various relevant sciences and their research limitations.

In applying this hierarchic concept to the neurosciences, I became aware of the need to use different approaches in brain research. One science can only answer the specific questions that are accessible to its methods. Therefore, different methods must be used for each of the various functions and structures of the brain, and each method will yield a different aspect that will complement the others. This concept facilitated my later attempts to combine neurophysiological and psychophysical approaches in visual research (Jung, 1961).

It also became clear that single facts are more important for history or the humanities than for the natural sciences, and that isolated facts are worthless in neurobiology if they do not lead to a general rule or are not seen in context with many other facts. When I read Carl Stumpf's 1906 article on the order of sciences in 1933, I came to understand the importance of theories and natural laws for science, and later related it to Popper's (1935) and to Hartmann's (1940) postulates. During 1936 and 1937, when I was acquiring some practice in animal experiments, I tried to combine experiment and theory to find out how a scientist should plan his work, order his experimental facts according to principles, and test their significance by new experiments. In spite of Hartmann's skepticism about direct access to biological processes, I had become convinced that the neuroscientist can force the living brain by appropriate methods to answer his questions. Further experience dampened this optimism and showed that failures in research may occur because our faculty of combining theory and experimental methods is limited.

## Discussion

In the preceding review I have tried to revive the memory of teachers and friends who influenced my research work in the two decades after 1930, the year when, as a student of Paul Hoffmann in Freiburg, I had my first glimpse of neurophysiology. The example set by these scientists shaped my course in research, even when I did not always take their advice. In this discussion I will add some general remarks on research and on research failures which may illustrate the role of theory and experiment in the neurosciences. All this should be viewed against the historical background of neurophysiology in the 1930s, when individual research was given more prominence than teamwork.

As a young neurophysiologist, I learned much from brief conversations with distinguished scientists. I remember how stimulating it was to find that even somewhat naive questions were taken seriously by leaders of international renown and that they were criticized constructively. The founders of neurophysiology, such as Adrian and Hess, even demonstrated their results to newcomers. So I tried later to encourage young coworkers by showing interest in their thoughts and adding critical remarks, if necessary, to restrain too much speculation. Sometimes, original workers disproved my skeptical prognosis of difficult research programs by successful experiments. On the other hand, some promising research failed, and it may be useful to discuss such failures.

### Two research failures

A recognition of the reasons for past failures can provide a valuable lesson to the scientist and an aid in research planning. Two examples, from 1937 and 1957, will now be described to demonstrate some general rules for research. In the first we missed the ascending reticular activation system because we lacked a theory to integrate our facts and adhered too closely to conventional anatomical concepts. In the second our group did not recognize the orientation-specific neurons of the visual cortex, being misled by methodological limitations, premature quantification, and a neglect of qualitative pilot studies.

*The reticular-formation experiments and their misinterpretation.* Working in Hess's laboratory in 1936–1937, Weisschedel and I extended our experiments from the substantia nigra to the midbrain tegmentum and caudate nucleus. We continued these stimulations and coagulations with Hess's method in the Berlin Brain Research Institute during my recordings of subcortical potentials with Kornmüller in 1937. By the time I returned to Freiburg in March 1938, we had collected a considerable number of observations. These were as follows: (1) stimulation of the midbrain tegmentum in the nonanesthetized cat elicited behavioral and acoustic arousal, with coordinated head, ear, and eye movements besides pupillary dilatation: (2) bilateral coagulation of the same regions, mainly in the midbrain reticular formation, caused a sleeplike state of one or two weeks' duration: (3) intense sensory stimulation of all tested modalities elicited rhythmic waves of 5–6 per sec in the hippocampus and other allocortical regions, with simultaneous flattening in all isocortical areas (Jung and Kornmüller, 1938/39). We sometimes saw similar EEG responses after electrical midbrain stimulation, but we interpreted these as stimulus-spread to sensory pathways and did not pay attention to them.

We thought that the sleep state following lesion of the midbrain tegmentum, found while we were investigating very different problems of oculomotor coordination, was accidental. Our midbrain coagulation was aimed at the brainstem substrate of optokinetic nystagmus, since Lorente de Nó (1931) had demonstrated that nystagmus and eye movements were coordinated with vestibular afference in the pontomesencephalic reticular formation. The sleep state was unexpected, and not until twelve years later, after Moruzzi and Magoun's work (1949), did we understand the significance of these observations for brainstem regulation of wakefulness and attention. In interpreting all available data (i.e., stimulations, lesions, and EEG recordings) we might have discovered what was later called the "ascending activation system" (Magoun, 1963) *if* we had postulated ascending pathways, from the midbrain to the diencephalon and from there to the allocortex, eliciting the theta waves, and to the isocortex, causing EEG flattening (later called "arousal EEG"). Berger had already (in 1930) described the latter phenomenon in man after sensory stimuli and mental activity, and twenty years later Magoun, Lindsley, and their coworkers related it to the reticular formation (Magoun, 1963). Although Magoun's concept oversimplified and overgeneralized the reticular functions, which also include special coordination of eye movements (Lorente de Nó, 1931), it supplemented Hess's theory of trophotropic regulation by brainstem centers during sleep. Only in 1939, after I had developed polygraphy in man (Jung, 1939a), did I apply Hess's concepts of ergotropic and trophotropic regulation to the human EEG: in normals Berger's "active EEG" was compared to ergotropic brain activation, and the "passive" and synchronized sleep EEG was explained by trophotropic inactivation (Jung, 1939c); the blocking of epileptic slow waves of petit mal by sensory arousing stimuli was also conceived as ergotropic (Jung, 1939b).

In 1947 both Weisschedel and I thought that Hess did not pay enough attention to anatomical connections. We therefore introduced the Marchi method for tracing degenerating pathways, and we found that the main fiber degenerations after coagulation of the midbrain tegmentum were descending, except when well-known specific, long, ascending pathways to the thalamus were also damaged. Thus Weisschedel (1938) explained the ear movements with acoustic arousal following midbrain stimulation in the cat by descending pathways to the rhombencephalic auditory nuclei. Although this had also appeared as a new hypothesis in 1938 and was verified much later, it misled us into rejecting an ascending system, which could have been conceived of from Hess's theories.

One may learn from this experience that in pursuing special problems there is a risk of losing sight of their wider context, and that unexpected experimental results should lead to a revision of research programs through an infusion of new hypotheses. In 1938 we had a multitude of experimental facts but were not able to integrate them into a synoptic theory because we were too dependent upon conventional anatomical data.

*Missing the orientation specificity in the visual cortex.* A second, quite different research failure occurred in 1957–1959; we missed the receptive-field axis orientation of cortical neurons, which Hubel and Wiesel described so clearly in 1959. This can be explained by a premature quantification and a too rigid methodological restriction when Baumgartner began his receptive-field measurements by contrast-contour shifts. In 1957, after Hartline had told me that he used to search for the frog's "off" neurons by moving a stick through a diffused light, I proposed this simple "shadow search" to Baumgartner. He said, however, that he did not want to do sloppy experiments, but planned to construct a rather complicated machine that would move a vertical contrast border exactly through the receptive field in small steps. After more than two years this machine was built, and he began his receptive-field experiments in 1958 with Hakas. But only concentric or vertical fields could be measured by this method since the contrast border could not be tilted. When I was asked later why we missed the orientation specificity during five years' work on cortical neurons, I used to tell this story and remark that we might have found them in one experiment if we had used the stick with its easy movements in all orientations instead of the quantifying machine.

Other explanations of this failure may be that in 1956–1958, before Hubel and Wiesel began their experiments, our attention was concentrated too heavily on nonspecific and multisensory afferences, and that, owing to administrative duties, I had to reduce my own laboratory work. During this time Creutzfeldt was experimenting on the thalamoreticular system, and Grüsser and Kornhuber were investigating multisensory responses with visuovestibular interaction. I had had to cut short all experiments after accepting election as dean of the faculty in 1954.

It may be learned from this experience that one should not leave the laboratory during a productive phase of research and that a field should first be explored by simple qualitative pilot studies to find out the most rewarding approach, before experiments are reduced to quantification with a single method.

## Some remarks on research procedures in the neurosciences

*Methodology, theory, and intuition.* Experimental results obtained by new methods should be fitted into a concept. It may be less exciting to find that they correspond to an old conception than to formulate a new theory, but one should not make new hypotheses unnecessarily; the principle of economy is as valuable in science as in trade. In the course of scientific discussions, usually the most simple and widely applicable theories survive; special hypotheses are often ad hoc and short-lived. However, our experience with brainstem stimulation in 1938 demonstrates that one may miss a discovery if one remains within old pathways of thinking. The outcome might have been more successful if we had enlarged Hess's concepts of the interrelations between diencephalon and cortex for an ascending system.

Even when the motivation of curiosity remains operative, the theoretical and methodological aspects of modern science demand many other incentives for creative work in the neurosciences. The development of new methods and of experiments involving their first application may also become a motive in itself. This was Jan Tönnies's attitude, and, until 1951, also mine, as we applied novel methods to attack hitherto unsolved questions but left the systematized development to others. In 1931 Tönnies began with purely technical developments in Vogt's institute and then gradually planned his own neurophysiological experiments. After he had seen what the animal experiment needed, he constructed an adequate apparatus and gave it to the physiologist for trial. He always assisted in the pioneer experiments with genuine joy, and then left them to his friends for further utilization.

The role of intuition is often underrated in the natural sciences. Many people recognize only inductive methods and experimental results as scientific and declare deductive postulates from general theories to be inadmissible. However, the history of science shows that the main discoveries have been made by intuition, which can link together apparently nonconnected observations through a new hypothesis. Of course, loose generalizations and oversimplifications are of little use, and general theories can only be accepted after many experimental verifications and with some specifically defined exceptions under special conditions. However, even an erroneous theory may be used as a guide to research until a better theory is developed. Chance observations may also lead to important discoveries if their general implications are recognized.

*Chance, curiosity, and motivation.* Most basic discoveries in neurophysiology have been made by qualitative or semiquantitative methods of exploration. In the early period when the electrophysiology of the brain and of single-fiber recording began, each experiment was a scientific adventure and chance observations were frequent. Success depended not only on chance, but also on the intuitive feeling of what could be a new scientific approach. Our drive for research was just curiosity and not theory. The fitting of the new results into concepts came later. New fields could be opened up by chance observations made possible by the newly acquired tools of electrobiology. For this period, I agree with Bronk's statement (1951) that "the primary and potent motive of the scientist is curiosity." I think, however, that curiosity as a scientific impetus is only sufficient for individual starts in new fields. Planned advancement with a working hypothesis is a second and necessary step. This may be followed by a third step which comprises systematized teamwork and *Arbeitsteilung.* Popper's more rigid principles about scientific theories seem to leave no place for chance observations. However, every scientist knows how such accidental findings may stimulate research. Clearly, after a primary observation has been made and recognized as important, efficient research needs theories. It cannot depend on new methods or chance findings alone; it must use ideas. I agree with Eccles (1970) that in the neurosciences creative imagination is more important than technology and that a good scientist should quickly recognize the significance of unexpected results. Quantification should come later.

For successful research a good idea is more important than a good on-line computer—though of course it is nice to have both. But theory should precede measurements. Even when the idea is wrong, as in Berger's case, it may become a useful motivation to overcome the technical setbacks and disappointments that often hamper research work. The wish to verify a pet idea may enhance the personal assiduity and focus the curiosity of the experimenting scientist.

*International relations in science and the Rockefeller Foundation.* Another prerequisite for good research projects during the rapid advance of modern science is freedom of communication with scientists of other countries working in related fields. During the 1930s and again after the Second World War, scientific relations were hampered by political tensions and a shortage of money for research. Although personal contact on an international level was as necessary for successful research as it is today, there were no grants available to enable scientists to go abroad, except for the fellowships of the Rockefeller Foundation. I had the good fortune to receive one of these fellowships at the age of 25 for work in England and Switzerland. Again in 1952, the Foundation gave me a traveling grant as a former Fellow which enabled me to make my first transatlantic trip and to see what had been done in Anglo-American research centers during and after the war. In the United States I met old friends again and contacted new groups of scientists.

All this was organized by Alan Gregg of the Rockefeller Foundation, whose biography was written so well by Penfield (1967). Fourteen years after our first encounter in 1935, I met him again when he twice visited Freiburg in the postwar years. Gregg maintained that international contact in science helps to break down political barriers. This was a consolation in the politically difficult period after 1933, and I soon realized how important these relations were for preserving intellectual freedom. As the political grip of Hitler's government gradually tightened on the universities, German science appeared in imminent danger of isolation. Grant organizations such as the Rockefeller Foundation opened up closed frontiers that could not be crossed on our own initiative. I felt that this was a unique opportunity for a young scientist which would have to be used reasonably. So on the way to England in 1936, I visited several Dutch laboratories, with introductions from Spatz and the Foundation, and on my return to the continent I stopped at many French and Belgian research laboratories. It seems remarkable how many scientists of my generation owe their first international contacts to a Rockefeller fellowship. These fellowships made it possible to integrate multiple scientific traditions of several countries for successful research.

*Deduction and teleology.* Both induction and deduction have their place in science. I have learned to appreciate both fact-oriented inductive and systems-oriented deductive research. Living organisms and physiological systems act *zielgerichtet* and are success-oriented in the sense of Hess (1949). Biological order contains finalistic tendencies, although, according to Kant (1790), we should try to proceed with mechanistic analyses as far as possible. But Hartmann (1940, 1951) made me conscious of the essential differences between the inorganic, vital, and psychic levels and their laws of dependency. Whether biological finality can be explained by natural selection is a matter of surmise, and even Hartmann (1950) had no other explanation for this *Nexus organicus*. However, the rule remains that purposeful action is dependent on knowledge of causal relations. Furthermore, the biologist must accept an a priori structural order of the sensory and neural systems that may be compared to Kant's apriorisms (M. Hartmann, 1937; Lorenz, 1943).

Many leading neurophysiologists have held a similar view and have accepted teleological thinking in biology (see Sherrington, 1906; Granit, 1972; Eccles, 1970). Popper's statement (1935) that the empirical sciences are systems of theories may be somewhat exaggerated, but it contains an important truth. To students who adore facts and neglect theories I used to say: "In natural science single facts are worthless if they are not reproducible or do not follow general laws or are not systematized according to experience."

I know that our old-fashioned neurophysiology of the 1930s may be called "antediluvian" by the younger generation of computer-oriented researchers. But, considering my own research failures, I still maintain that one will inevitably miss essential discoveries if one simply feeds experimental results into a computer before making a good hypothesis or exploring the field by pilot experiments. To combine inductive-experimental and deductive-theoretical approaches appears to me the *via regia* in the neurosciences.

## Conclusions

I learned more from my teachers and friends than scientific techniques. Their styles of research were individual, but they shared a respect for ethical rules and for the personal decisions of their coworkers. From them I learned that teamwork should combine objective criticism with individual freedom. Research guidance means helping gifted workers in their first steps, promoting first-rate young scientists in their scientific careers, and also discouraging bad or mediocre people from taking up research work. Original scientists can and should find their own way. In founding my own research school, I learned not to impress my personal style of work on others, but rather confine myself to giving initial advice and providing optimal research facilities to good workers.

In the modern world, developing toward rationalization and technology, personal relationships must compensate for the impersonal trends of civilization. Respect for individual values and a balance of friendship and reserve facilitate research achievements. Productive science develops best against a background of general culture with a knowledge of history and philosophy. The old German tradition of *Bildung* includes a broad basic teaching in the sciences and humanities, and this produces an independent attitude to research.

## Acknowledgment

My hearty thanks are due to many old friends and to the relatives of deceased scientists for providing pictures and for refreshing or correcting my memory. I trust they will accept this collective expression of gratitude in place of a long list of names that would never be complete. Special thanks for writing and correcting the manuscript are due to Frau H. Kremer and Frau S. Brinkmann.

## References

Adrian, E.D., and Bronk, D.W. (1929): The discharge of impulses in motor nerve fibres. Part II. The frequency of discharge in reflex and voluntary contractions. *J. Physiol.* 67: 119–151.

Barron, D.H., and Matthews, B.H.C. (1938): The interpretation of potential changes in the spinal cord. *J. Physiol.* 92: 176–321.

Baumgartner, G., and Hakas, P. (1959): Reaktionen einzelner Opticusneurone und corticaler Nervenzellen der Katze im Hell-Dunkel-Grenzfeld (Simultankontrast). *Pflügers Arch. Gesamte Physiol.* 270: 29.

Berger, H. (1929): Über das Elektrenkephalogramm des Menschen. *Arch. Psychiatr. Nervenkr.* 87: 527–570.

Berger, H. (1930): Über das Elektrenkephalogramm des Menschen. 2. Mitteilung. *J. Psychol. Neurol.* (Leipzig) 40: 160–179.

Berger, H. (1931): Über das Elektrenkephalogramm des Menschen. 3. Mitteilung. *Arch. Psychiatr. Nervenkr.* 94: 16–60.

Berger, H. (1933): Über das Elektrenkephalogramm des Menschen. 7. Mitteilung. *Arch. Psychiatr. Nervenkr.* 100: 301–320.

Bronk, D. W. (1951): The unity of the sciences and the humanities. *Wiley Bull.* 34: 1–8.

Eccles, J. C. (1957): *The Physiology of Nerve Cells.* Baltimore: The Johns Hopkins Press.

Eccles, J. C. (1970): *Facing Reality. Philosophical Adventures by a Brain Scientist.* New York: Springer-Verlag.

Eichler, W. (1937): Über die Ableitung der Aktionspotentiale vom menschlichen Nerven in situ. *Z. Biol.* 98: 182–214.

Granit, R. (1972): In defence of teleology. *In: Brain and Human Behavior.* Karczmar, A. G., and Eccles, J. C., eds. New York: Springer-Verlag, pp. 400–408.

Hartmann, M. (1937): *Philosophie der Naturwissenschaften.* Berlin: J. Springer.

Hartmann, N. (1912): *Philosophische Grundlagen der Biologie.* Göttingen: Vandenhoeck & Ruprecht.

Hartmann, N. (1924): Diesseits von Idealismus und Realismus. Ein Beitrag zur Scheidung des Geschichtlichen und Übergeschichtlichen in der Kantischen Philosophie. *Kant Studien* 29: 106–206.

Hartmann, N. (1940): *Der Aufbau der realen Welt. Grundriss der allgemeinen Kategorienlehre.* Berlin: W. de Gruyter.

Hartmann, N. (1946): *Neue Wege der Ontologie.* Stuttgart: W. Kohlhammer.

Hartmann, N. (1950): *Philosophie der Natur; Abriss der speziellen Kategorienlehre.* Berlin: W. de Gruyter.

Hartmann, N. (1951): *Teleologisches Denken.* Berlin: W. de Gruyter.

Hess, W. R. (1925): *Über die Wechselbeziehungen zwischen psychischen und vegetativen Funktionen.* Zürich, Leipzig, Berlin: O. Füssli.

Hess, W. R. (1932): *Die Methodik der lokalisierten Reizung und Ausschaltung subcorticaler Hirnabschnitte.* Leipzig: G. Thieme.

Hess, W. R.: (1933): Der Schlaf. *Klin. Wochenschr.* 12: 129–134.

Hess, W. R. (1938): *Beiträge zur Physiologie des Hirnstamms.* II. *Das Zwischenhirn und die Regulation von Kreislauf und Atmung.* Leipzig: G. Thieme.

Hess, W. R. (1949): *Das Zwischenhirn. Syndrome, Lokalisationen, Funktionen.* Basel: Benno Schwabe.

Hess, W. R. (1963): From medical practice to theoretical medicine: An autobiographic sketch. *Perspect. Biol. Med.* 6: 400–423.

Hodgkin, A. L. (1938): The subthreshold potentials in a crustacean nerve fibre. *Proc. R. Soc. Lond. B.* 126: 87–121.

Hoffmann, P. (1910): Beiträge zur Kenntnis der menschlichen Reflexe mit besonderer Berücksichtigung der elektrischen Erscheinungen. *Arch. Anat. Physiol./Physiol. Abt.* 1910: 223–246.

Hoffmann, P. (1922): *Die Eigenreflexe (Sehnenreflexe) menschlicher Muskeln.* Berlin: Springer-Verlag.

Holst, E. von: (1939): Die relative Koordination als Phänomen und als Methode zentralnervöser Funktionsanalyse. *Ergeb. Physiol.* 42: 228–306.

Holst, E. von (1950): Die Tätigkeit des Statolithenapparates im Wirbeltierlabyrinth. *Naturwissenschaften* 37: 265–272.

Holst, E. von (1951): Zentralnervensystem und Peripherie in ihrem gegenseitigen Verhältnis. *Klin. Wochenschr.* 29: 97–105.

Holst, E. von (1969): *Zur Verhaltensphysiologie bei Tieren und Menschen. Gesammelte Abhandlungen.* 2 volumes. Munich: R. Piper.

Holst, E. von, and Saint Paul, U. von (1960): Vom Wirkungsgefüge der Triebe. *Naturwissenschaften* 47: 409–422.

Hubel, D. H., and Wiesel, T. N. (1959): Receptive fields of single neurones in the cat's striate cortex. *J. Physiol.* 148: 574–591.

Jung, R. (1939a): Ein Apparat zur mehrfachen Registrierung von Tätigkeit und Funktionen des animalen und vegetativen Nervensystems (Elektrencephalogramm, Elektrokardiogramm, Muskelaktionsströme, Augenbewegungen, galvanischer Hautreflex, Plethysmogramm, Liquordruck und Atmung). *Z. Gesamte Neurol.* 165: 374–397.

Jung, R. (1939b): Über vegetative Reaktionen und Hemmungswirkung von Sinnesreizen im kleinen epileptischen Anfall. *Nervenarzt* 12: 169–185.

Jung, R. (1939c): Das Elektrencephalogramm und seine klinische Anwendung. I. Methodik der Ableitung, Registrierung und Deutung des EEG. *Nervenarzt* 12: 569–591.

Jung, R. (1941): Physiologische Untersuchungen über den Parkinsontremor und andere Zitterformen beim Menschen. *Z. Gesamte Neurol. Psychiatr.* 173: 263–332.

Jung, R. (1961): Neuronal integration in the visual cortex and its significance for visual information. *In: Sensory Communication.* Rosenblith, W. A., ed. Cambridge, Mass.: The MIT Press.

Jung, R. (1963): Hans Berger und die Entdeckung des EEG nach seinen Tagebüchern und Protokollen. *In: Jenenser EEG-Symposion. 30 Jahre Elektroenzephalographie.* Werner, R., ed. Berlin: VEB Volk und Gesundheit, pp. 20–53.

Jung, R. (1967): Neurophysiologie und Psychiatrie. *In: Psychiatrie der Gegenwart. Forschung und Praxis.* Band I/1A. *Grundlagenforschung zur Psychiatrie.* Gruhle, H.W., Jung, R., Mayer-Gross, W., and Müller, M., eds. Berlin: Springer-Verlag, pp. 325–928.

Jung, R. (1969): Paul Hoffmann 1884–1962. *Ergeb. Physiol.* 61: 1–17.

Jung, R., Baumgarten, R. von, Baumgartner, G. (1952): Mikroableitungen von einzelnen Nervenzellen im optischen Cortex der Katze: Die lichtaktivierten B-Neurone. *Arch. Psychiatr. Z. Neurol.* 189: 521–539.

Jung, R., and Kornmüller, A.E. (1938/39): Eine Methodik der Ableitung lokalisierter Potentialschwankungen aus subcortikalen Hirngebieten. *Arch. Psychiatr. Nervenkr.* 109: 1–30.

Jung, R., and Tönnies, J. F. (1948): Die Registrierung und Auswertung des Drehnystagmus beim Menschen. *Klin. Wochenschr.* 26: 513–521.

Jung, R., and Tönnies, J. F. (1950): Hirnelektrische Untersuchungen über Entstehung und Erhaltung von Krampfentladungen: Die Vorgänge am Reizort und die Bremsfähigkeit des Gehirns. *Arch. Psychiatr. Z. Neurol.* 185: 701–735.

Kant, I. (1790): *Critik der Urtheilskraft.* Berlin, Libau: Lagarde u. Friederich.

Kornmüller, A. E. (1937): Die bioelektrischen Erscheinungen der Hirnrindenfelder. Leipzig: G. Thieme.

Ling, R., and Gerard, R.W. (1949): The normal membrane potential of frog sartorius fibres. *J. Cell. Comp. Physiol.* 34: 382–396.

Lorente de Nó, R. (1931): Ausgewählte Kapitel aus der vergleichenden Physiologie des Labyrinthes. Die Augenmuskelreflexe beim Kaninchen und ihre Grundlagen. *Ergeb. Physiol.* 32: 73–242.

Lorenz, K. (1943): Die angeborenen Formen möglicher Erfahrung. *Z. Tierpsychol.* 5: 235–409.

Magoun, H. W. (1963): *The Waking Brain.* 2nd ed. Springfield, Ill.: C. C Thomas.

Moruzzi, G., and Magoun, H. W. (1949): Brain stem reticular formation and activation of the EEG. *Electroencephalogr. Clin. Neurophysiol.* 1: 455–473.

Penfield, W. (1967): *The Difficult Art of Giving. The Epic of Alan Gregg.* Boston: Little, Brown.

Popper, K. R. (1935): *Logik der Forschung.* Vienna: Springer-Verlag. (*The Logic of Scientific Discovery,* New York: Basic Books, 1959).

Schoen, L. (1957): Mikroableitungen einzelner zentraler Vestibularisneurone von Knochenfischen bei Statolithenreizen. *Z. Vergl. Physiol.* 39: 399–417.

Schriftenverzeichnis Richard Jung und Mitarbeiter. (1971): Herausgegeben anlässlich des 60. Geburtstages von Richard Jung. Berlin, Heidelberg, New York: Springer-Verlag.

Sherrington, C. S. (1906): *The Integrative Action of the Nervous System*. New Haven: Yale University Press.

Spatz, H. (1922): Über den Eisennachweis im Gehirn, besonders in Zentren des extrapyramidal-motorischen Systems. I. *Z. Gesamte Neurol. Psychiatr.* 77: 261–390.

Stumpf, C. (1906): Zur Einteilung der Wissenschaften. *Abh. Preuss. Akad. Wiss. Phil. Klasse:* 1–93.

Tönnies, J. F. (1932): Der Neurograph, ein Apparat zur Aufzeichnung bioelektrischer Vorgänge unter Ausschaltung der photographischen Kurvendarstellung. *Naturwissenschaften* 20: 381–384.

Tönnies, J. F. (1933): Die Ableitung bioelektrischer Effekte vom uneröffneten Schädel. Physikalische Behandlung des Problems. *J. Psychol. Neurol.* 45: 154–171.

Tönnies, J. F. (1934): Die unipolare Ableitung elektrischer Spannungen vom menschlichen Gehirn. *Naturwissenschaften* 22: 411–422.

Tönnies, J. F. (1938): Reflex discharge from the spinal cord over the dorsal roots. *J. Neurophysiol.* 1: 378–390.

Tönnies, J. F. (1949): Die Erregungssteuerung im Zentralnervensystem. Erregungsfokus der Synapse und Rückmeldung als Funktionsprinzipien. *Arch. Psychiatr. Nervenkr.* 182: 478–535.

Tönnies, J. F.: (1969): Automatische EEG-Invervall-Spektrumanalyse (EISA) zur Langzeitdarstellung der Schlafperiodik und Narkose. *Arch. Psychiatr. Nervenkr.* 212: 423–445.

Tönnies, J. F., and Jung, R. (1948): Über rasch wiederholte Entladungen der Motoneurone und die Hemmungsphase des Beugereflexes. *Pflügers Arch. Gesamte Physiol.* 250: 667–693.

Vogt, C., and Vogt, O. (1919): Allgemeine Ergebnisse unserer Hirnforschung. 1–4. *J. Psychol. Neurol. (Leipzig)* 25: 277–462.

Vogt. O. (1902/03): Psychologie, Neurophysiologie und Neuroanatomie. *J. Psychol. Neurol. (Leipzig)* 1: 1–3.

Weisschedel, E. (1938): Zur Methodik experimenteller Untersuchungen am Hirnstamm. *In: Proceedings of the 16th International Physiological Congress, Zurich, August 1938*. Vol. 2, pp. 354–356.

# The Neuroscience Community: Some Organizational Developments

Horace W. Magoun (b. 1907, Philadelphia, Pennsylvania) is emeritus professor of psychiatry at the Medical School of the University of California at Los Angeles. His research focused on the influences of the brainstem upon spinal and cerebral functions, especially the role of the reticular system in sleep and wakefulness. His career has been marked by a long succession of creative collaborations with graduate students, young postdoctorals, and more senior colleagues. He has received a number of awards for his work, including the Jacoby Award of the American Neurological Association and the Lashley Prize of the American Philosophical Society.

# 28
# The Role of Research Institutes in the Advancement of Neuroscience: Ranson's Institute of Neurology, 1928-1942

## Horace W. Magoun

The focus of this symposium upon "Paths of Discovery" is certainly appropriate to the Boston community, where, a generation ago and at a neighboring institution, an earlier neuroscientist, Walter Cannon, summed up aspects of his research career in a modest publication entitled *The Way of an Investigator* (Cannon, 1945). I can still recall the impact of this volume upon me as a young man, and, on rereading it recently, I found a number of its themes as fresh and pertinent today as when it was published.

The theme most commonly recalled concerned the role of unconscious processes and serendipity in discovery, a fact to which, I presume, many of us can join in bearing witness. Another theme that Cannon emphasized was the importance of the individual and independent investigator. No one will wish to debate this importance, but I would here like to say something supportive of organized research activity, which has also contributed a great deal to the development of the neurosciences. In this latter direction, Cannon himself stressed the importance of collaboration in research and education, and he commented sensitively upon mentor-apprentice relationships and upon the genealogical aspects of a succession of productive investigators in the sciences. Cannon had begun his own scientific career under Bowditch, whom he later succeeded in the chair of physiology at Harvard. Thus, Cannon suggested, Bowditch might be considered his scientific father. Bowditch, in turn, had earlier gained his start in physiology as a student of Karl Ludwig in Germany, and so, Cannon added, he might claim Ludwig as his scientific grandfather.

### The Rise of Research Institutes in the Nineteenth Century

Over much of the latter half of the nineteenth century, as the unpretentious professor and director of the Institute of Physiology at the University of Leipzig, Karl Ludwig was engaged in supervising the investigative development of some 250 students—typically young postdoctorals, as we would call them today. Another American who trained with Ludwig in this period was Warren Lombard, later professor of physiology at the University of Michigan. On his arrival at Ludwig's institute, the young man told the professor he was interested in fatigue but had no idea of how to study it. Ludwig defined a problem, assembled apparatus, set him to work, and provided much help during the course of the study. At its conclusion Lombard summarized the project's results and submitted a jointly authored paper. It was shortly returned, almost entirely rewritten, with only Lombard's name at the top. The young man went to the professor and protested: "You set the problem, showed me how to use the apparatus, solved my difficulties, and rewrote the paper—your name should appear with mine!" "Not so," Ludwig replied. "You did the work and should receive the credit. But if you never do anything more, people will think that *I* did it!" (Cannon, 1945).

Some research institutes, such as Ludwig's, had been established by the German universities to attract and hold outstanding professors, but from 1850 on, steps were also being taken to adapt the institutional form to deal with the practical problems of German agriculture. By 1880 the number of these agricultural experiment stations in Germany had increased to fifty. Similarly, in this country, the Morrill Act of 1862, which established a system of publicly supported land-grant colleges and state universities, was followed in the 1880s by the establishment of agricultural experiment stations at each of these institutions.

A second development leading to the establishment of research institutes in the nineteenth century stemmed from advances in microbiology and their application to immunotherapy against infectious disease. In Germany Koch's achievements were recognized in 1880 by the establishment in Berlin of an Institute for Infectious Diseases under his direction. In France Pasteur's discoveries led the French Academy to open an international subscription for establishment of the Pasteur Institute, inaugurated in Paris in 1888. By the turn of the century other such centers had been established in Saint Petersburg, London, Tokyo, and New York.

## Research Institutes in the Neural Sciences

These developments in agriculture and the fields of immunology and infectious disease were followed by the provision of organized research units for study of the brain and nervous system. At an inaugural assembly of the International Association of Academies, held in Paris in 1901, the Royal Academy of Sciences of Saxony, in the person of Wilhelm His, professor of anatomy at Leipzig, formally moved to encourage increased research upon the brain. The association appointed an International Brain Research Commission, dedicated to establishing central research institutes in each country where special collections of investigative material or other resources and facilities could be made available for study. Within a decade, eight existing or newly established units had been made associates of the International Brain Research Commission and had become designated as cenrtal or national institutes: at Vienna, Leipzig, Zurich, Madrid, Frankfurt, Saint Petersburg, Amsterdam, and Philadelphia (Magoun, 1962).

The manner in which the participation of the United States developed in the commission's program is of considerable interest. One of Wilhelm His's most distinguished students, Franklin Mall, professor of anatomy at Johns Hopkins, had been appointed a member of the Brain Commission. In 1905 Mall was consulted by Piersol, professor of anatomy at the University of Pennsylvania, about plans for reorganization of the Wistar Institute for Anatomy and Biology in Philadelphia. This institute had originated from activities of Caspar Wistar, who had occupied the chair of anatomy at the University of Pennsylvania in the nineteenth century. Following Wistar's death many of his anatomical preparations were presented to the university, where they ultimately gained the status of a museum. In 1892 the university had succeeded in interesting a descendant, General Isaac Wistar, in the incorporation of this museum as the Wistar Institute.

Mall recognized that the institute's reorganization provided an opportunity to advance the goals of the International Brain Commission, which had already requested the National Academy of Sciences to designate an institute for study of the brain in this country. He wrote, "It seems to me that it would be natural and proper that the Wistar Institute should be so designated. Through it, all work in neurology in America could be enlarged and correlated with activities abroad. It would be a brilliant opportunity and almost criminal if the Wistar Institute did not take part." The proposed introduction of a program of research was also plainly supported by the general, who wrote, "I fully agree that the Wistar Institute should be designed for the use of investigators, rather than a mere gaping public!" Consequently, the Wistar Institute became the U.S. facility in this international program, and, on Mall's recommendation, Henry H. Donaldson, professor of anatomy at the University of Chicago, was appointed its director. At the Wistar, Donaldson gathered a large quantity of data on the growth of the nervous system; his interest in the establishment of biological norms also found expression in standardization of the Wistar Albino Rat as a common laboratory animal.

### Ranson's Institute of Neurology at Northwestern

Among the last of Donaldson's doctoral students at the University of Chicago was a reserved and hardworking young man named Stephen Walter Ranson. Figure 28.1 shows Ranson in his customary laboratory garb about 1940, at the peak of his career as director of his own research institute. Though he had also received an M.D., Ranson much preferred an academic career and soon became professor of anatomy at Northwestern University's Medical School in Chicago. Here his penetrating analysis of the unmyelinated fibers in peripheral nerves and their role in pain conduction led Washington University in Saint Louis to invite him to its faculty. Following the earlier pattern of the German universities, Northwestern subsequently countered by establishing an Institute of Neurology, in 1928, and brought Ranson back to Chicago as its founding director (Sabin, 1945). Among his first steps was the appointment of a young associate, Walter R. Ingram, shown in Figure 28.2 in his maturity. Young Dr. Ingram was assigned responsibility for the experimental part of the institute's program, including instruction of entering students in the laboratory's instrumentation and techniques. I was most fortunate to be among the small initial group, which swelled rapidly as graduate students, postdoctorals, and visiting scientists were annually assimilated into a growing number of investigative programs. Although Rex Ingram was my instrumental uncle, to paraphrase Cannon, Ranson was my scientific father. I am very pleased to acknowledge this paternal relationship. However, although I was born in Phil-

Figure 28.1    Stephen Walter Ranson, 1880–1942.

Figure 28.2    Walter R. Ingram, professor of anatomy, University of Iowa.

adelphia and have a naturally pale complexion, I want to make it perfectly clear that the notion that an Albino Wistar Rat was my grandfather—though often suspected—has never been confirmed!

## Revival of the Horsley-Clarke Instrument

On initiating a program of studies on the brainstem in his newly acquired research facility, Ranson revived use of the Horsley-Clarke stereotaxic instrument to gain experimental access to this deep-lying part of the brain. The instrument had been developed early in the twentieth century by Robert Clarke (seen in riding garb in Fig. 28.3), a physician/physiologist who had collaborated amicably for a number of years with Sir Victor Horsley, the pioneer English neurosurgeon, first at the Brown Institution and then at University College, London, in the use of the instrument to stimulate the cerebellum of the animal brain. As time passed, however, Clarke's relations with his mentor and associate became increasingly strained. Horsley, seen in Fig. 28.4 as a young man, was a crusading teetotaler and also objected strongly to smoking. He came by these zealous attitudes naturally, for his father, a highly regarded portrait painter of the Victorian era, was so vigorously opposed to the painting of the female nude that he became widely known as clothes-Horsley. Clarke, on the other hand, enjoyed riding to hounds, drinking toddy, and smoking cigars. More substantively, however, he became increasingly concerned that his role in developing the stereotaxic instrument had been inadequately recognized, all accolades and references being directed principally to Horsley, whose professional status far exceeded his own (Davis, 1964; Tepperman, 1970).

Ranson had first seen Clarke's instrument on an earlier visit to Horsley's laboratory at University College, London, a cramped space beneath a sloping amphitheater where it was possible to stand upright only at one end of the room (Fig. 28.5).

Figure 28.3  Robert Henry Clarke, 1850–1926. Reproduced from an original photograph through the courtesy of the Wellcome Historical Medical Museum, London.

Figure 28.4   Sir Victor Horsley, 1857–1916. Reproduced from an original photograph through the courtesy of Drs. William H. Feindel and Wilder Penfield, Montreal Neurological Institute.

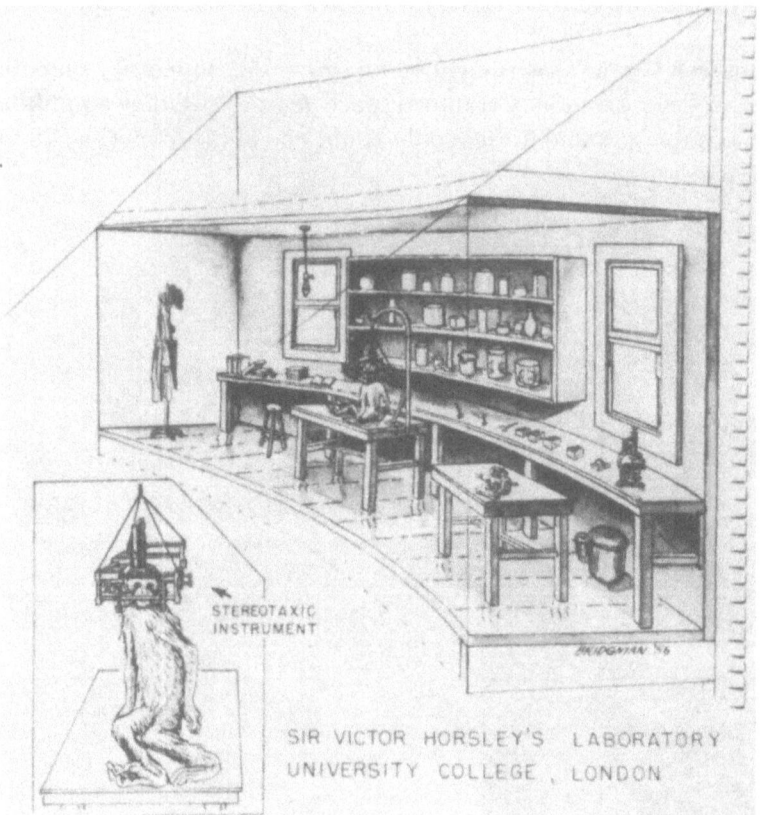

STEREOTAXIC
INSTRUMENT

SIR VICTOR HORSLEY'S   LABORATORY
UNIVERSITY  COLLEGE , LONDON

Figure 28.5   A sketch of Horsley's laboratory at University College, London, illustrating early use of Clarke's stereotaxic instrument. Reconstructed, from an original photograph, by and reproduced through the courtesy of Dr. Charles F. Bridgeman, director, National Medical Audiovisual Center, National Library of Medicine.

Ranson had also seen the instrument at Washington University, where Ernest Sachs, professor of neurosurgery, following a period of training with Horsley, had brought a precise replica of the original frame, prepared by the same instrument maker in London (Hinsey, 1961). A photograph of Sachs's instrument, which he most graciously presented me a number of years ago after his retirement, is shown in Figure 28.6. No award could have been more appreciated, for my initial experience with the instrument had almost ended my career. For those unfamiliar with it, the frame was fastened to an animal's head by bars that fitted into plugs inserted into each external auditory canal. In the awkwardness and tension of my student effort, one of the plugs slipped out of the animal's ear, and all efforts to retrieve it were futile. Everything in the operating room was searched, and I went over the floor on my hands and knees, but without finding any trace of the plug. Uncertain whether to slit my throat then and there or wait until morning, I went despairingly home, where, on disrobing for bed, the plug dropped out of the cuff of my trouser leg. Needless to say, the blissful relief of its recovery far exceeded that following any other trauma of my research career.

### The Achievements of Ranson's Institute

With the use of this instrument, through the 1930s and until Ranson's death in 1942, there flowed from his institute a profusion of contributions on the role of the hypothalamus and lower brainstem in visceral integration, emotional expression, the regulation of feeding, fighting, mating, and other vital functions that contri-

Figure 28.6   Lateral view of a precise replica of Clarke's original stereotaxic instrument, presented to the author by the late Dr. Ernest Sachs.

bute to the stability or well-being of our internal environment. The succession of young investigators participating in these studies contributed in a major way to the progress of the research. At the same time, they benefited greatly from experience with the instrumentation and resources available, as well as from the association with and the sophisticated counsel and guidance provided by Ranson and Ingram. Most of all, perhaps, the sense of satisfaction and accomplishment that these young scientists gained from mastering an investigative approach and using it to extend the range of their knowledge and insight into their subjects, reinforced them and helped to shape their subsequent careers.

Happily, Ranson's own achievements were increasingly recognized. For example, he was elected president of the American Association of Anatomists in 1938 and, relatedly, would have enjoyed knowing that three of the young men who had begun or extended their investigative careers in his institute subsequently became presidents of the American Physiological Society. In the late thirties Ranson was also invited to deliver a number of prestigious lectures (the Weir Mitchell Oration, a Harvey Lecture, the Dunham Lectures, and the Hughlings Jackson Lecture), and a volume of contributions entitled *The Hypothalamus and Central Levels of Autonomic Integration*, presented at the annual meeting of the Association of Research in Nervous and Mental Disease, was dedicated to him in 1939. The program also had its lighter sides. Carried away by the gargantuan polyuria of little monkeys with experimental diabetes insipidus, following supraoptic lesions and atrophy of the neurohypophysis, John Fulton pointed out that the hypothalamus was located beneath the thalamus and, if *thalamus* was Latin for bed or chamber, it appeared that the hypothalamus lay not only under the bed, but also under the chamber!

Following Ranson's death in 1942, I was invited to move from his well-appointed institute, in the tower of the Medical School, down among the bare bones of the anatomy department, and I abruptly became aware of the expensive nature of the kind of research that had been going on. If anything further was to be undertaken, I needed first to raise funds for its support. Research grants for young scientists were unheard of in the 1940s, but the annual epidemics of poliomyelitis in that era had led to the establishment of the National Foundation for Infantile Paralysis and its March of Dimes. A substantial number of the Chicago cases were of the bulbar type, and, with trepidation, I applied to the foundation for some of those dimes to support a study of the brainstem pathology in fatal cases, with related animal experiments on the function of the injured parts. To my astonishment and gratitude, the project was supported. In contrast to the loss of anterior horn cells in spinal poliomyelitis, the analogous nuclei of the cranial motor nerves were relatively uninjured in the bulbar disease, in which the central reticular core of the brainstem was the principal seat of inflammation and neuronal destruction (Barnhart et al., 1948). In related animal studies, stimulation of this reticular formation did not itself elicit motor responses of either the face or limbs. If, however, a background of motor activity was initiated either reflexively or by exciting the motor cortex, concomitant reticular stimulation facilitated or inhibited responses markedly, depending upon the site of stimulation.

### Percival Bailey and the Illinois Neuropsychiatric Institute

In following up these descending influences of the brainstem reticular formation, which appeared to mediate many of the extrapyramidal motor functions of the cerebrum and cerebellum, I enjoyed participating in some of the activities of a second research unit that had by then become established in Chicago. Percival Bailey, professor of neurosurgery at the University of Chicago, had earlier attempted to develop a comprehensive program in the neurosciences there, but without success. Invited to move to a new neuropsychiatric institute under construction at the University of Illinois Medical Center in Chicago, Bailey spent an interim period at Dusser de Barenne's laboratory at Yale. On the latter's death, Bailey transported the entire laboratory—staff, stock, and barrel—to the Illinois institute. In this feat, like something out of the *Arabian Nights*, the research collaborators—Bailey, Warren McCulloch, and Gerhardt von Bonin—hardly missed an experiment in their exploration of the functional interrelations of areas of the cerebral cortex, as determined by strychnine neuronography. To a young man brought up in the rather formal and reserved behavior patterns of Ranson's tower institute at Northwestern, the informal atmosphere of high-spirited excitement characteristic of the basement laboratory at the Illinois Neuropsychiatric Institute proved a novel and stimulating experience.

Among the many young postdoctorals at the INI at that time was Jack French (Figure 28.7), who had completed his neurosurgical training at the University

Figure 28.7   John D. French, director, Brain Research Institute, University of California, Los Angeles.

of Rochester and was engaged in research under Bailey's general supervision. A part of his work, undertaken with Oscar Sugar from the University of Chicago, explored the consequences of thalamic lesions on pain conduction in the monkey. Following operation, their initial animal was secured, after considerable struggle, in an examining chair where, despite vigorous painful stimulation, the monkey continued to gaze imperturbably into a corner. Jubilant at the outcome, Sugar shouted, "Get McCulloch!" Warren arrived, repeated the examination with the same result, began to jump up and down, and shouted, "You've found the seat of pain; get Bailey!" Bailey arrived and, as seeming analgesia was demonstrated once more, shouted only, "Get a normal monkey!" Following another struggle, this animal was fastened in the chair. To the chagrin of all but Percy, when marked painful stimulation was again applied, the normal monkey also continued to gaze imperturbably into a corner. Though failing to determine the cerebral focus of pain perception, these experiments vividly illustrated the capability of intense emotional excitement to mask or block reactions to painful stimulation.

## EEG Studies at the INI and Northwestern

To one not previously familiar with the newly acquired instrumentation, some of the most valuable experience to be gained at the INI was with the multichanneled, inkwriting electroencephalograph, used there most productively (as earlier at Yale) to record spikes conducted from a strychnized focus of the cortex to other cortical areas or to the brainstem. Shortly thereafter Don Lindsley (Figure 28.8) moved to Northwestern from the psychology department at Brown University. He was one of the early participants in the development of EEG studies at Alex Forbes's laboratory of neurophysiology in Cannon's department at Harvard,

Figure 28.8   Donald B. Lindsley, professor of psychology and physiology and chairman of advisory committee, Brain Research Institute, University of California, Los Angeles. Reproduced through the courtesy of Jeannette (Mrs. H. W.) Magoun, a stalwart contributor to the neurosciences, to whom the photo is inscribed.

along with Hallowell Davis, Fred Gibbs, Bill Derbyshire, Albert Grass, and others. A decade before, Herbert Jasper and Leonard Carmichael had undertaken their pioneering recordings of the electroencephalogram in man at Brown and, curiously, Lindsley and Jasper had still earlier been graduate students together under Lee Edward Travis at the University of Iowa, where initial electromyographic studies of reflex time were later followed up by other early studies of the EEG (Lindsley, 1969). In addition to his research at Northwestern's Evanston campus, Don established recording facilities in the Department of Anatomy at the Medical School, where collaborative research was undertaken on the effect of arousing stimulation upon the EEG following experimental lesions of the central reticular formation or more lateral afferent pathways in the brainstem.

As a culminating and related association in Chicago, Giuseppe Moruzzi (Figure 28.9) spent a visiting period at Northwestern. Moruzzi had earlier pursued postdoctoral research with Adrian and with Bremer on electrocortical activity, so it was natural for him to investigate ascending brainstem influences upon the EEG. In initial experiments, the large electrocortical slow waves of lightly anesthetized animals were recorded at relatively low amplification, and during reticular stimulation, the record became absolutely flat. After ruling out a variety of possible artifacts, it appeared that the experiments had stumbled upon some perplexing type of ascending inhibition, perhaps related to the suppressor areas of the cortex or to spreading depression, both of which were then in vogue. Only after some delay, and quite by chance, was the gain finally turned up, and it was then possible to see the large slow waves give way during reticular stimulation to the low voltage, fast activity of EEG arousal characteristic of attention and alert wakefulness. Evocation of this generalized electrocortical alteration has since become commonplace, but it is still possible to recall the arousal evoked in the investigators by its initial display as a response to ascending reticular stimulation.

Figure 28.9    Giuseppe Moruzzi, director, Institute of Physiology, University of Pisa, Italy.

## The Offspring of Ranson's and Bailey's Institutes

Although Ranson's Institute was dismantled at the end of the 1940s, its influence had been felt and had spread to other settings. The advantages of its pattern of organized research in advancing progress in the neurosciences so impressed a number of the students, postdoctorals, and senior scientists participating in the program that a number of them subsequently joined or became involved in developing similar units in this country or abroad. Teizo Ogawa returned to Japan to become director of the Brain Research Institute at the University of Tokyo. Giuseppe Moruzzi returned to Italy to become director of the Institute of Physiology at the University of Pisa. Bill Windle, who had succeeded Ranson as director of the Northwestern institute, became, first, chief of the Laboratory of Neuroanatomical Sciences at the National Institutes of Health, Bethesda, then chief of the Laboratory of Perinatal Physiology at the NIH in San Juan, Puerto Rico, and finally director of research at the Institute of Rehabilitative Medicine at New York University. John Brobeck and Dick Davis joined, and Jim Sprague returned to, the University of Pennsylvania, where they contributed importantly to the establishment and success of its Institute of Neurological Sciences (Sprague is presently its director). Ray Snider joined the University of Rochester, where he became the founding director of its productive Center for Brain Research. Jack French, Don Lindsley, and I heeded the advice of Horace Greeley and went West to join the developing campus and new School of Medicine of the University of California, Los Angeles. With the major participation of Tom Sawyer, Bob Livingston, Ross Adey, Fred Worden, and many others, a Brain Research Institute subsequently became established there and has since continued to flourish, with Jack French its director.

On balance, therefore, Ranson at the Institute of Neurology at Northwestern's Medical School, and Bailey at the Illinois Neuropsychiatric Institute, contributed importantly to the growth of the neural sciences in this country through the 1930s and 1940s. Fascinating as neuroscience may sometimes be when considered conceptually and impersonally, its pathways of discovery plainly are generated by, converge at, or branch from such outstanding individual investigators as Ranson and Bailey. When a program of research can become organized around such unusually catalytic persons, the range and progress of discovery is almost always accelerated and enhanced. For me, the particular attraction of this symposium lies in its celebration of the birthday of a third such remarkable individual who, over the past decade, in a feat undreamt of in the *Arabian Nights*, has established an invisible organized research unit—the Neurosciences Research Program—at the Massachusetts Institute of Technology!

## References

Barnhart, M., Rhines, R., McCarter, J. C., and Magoun, H. W. (1948): Distribution of lesions of the brain stem in poliomyelitis. *Arch. Neurol. Psychiatr.* 59:368–377.

Cannon, W. B. (1945): *The Way of an Investigator. A Scientist's Experiences in Medical Research.* New York: W. W. Norton.

Davis, R. A. (1964): Victorian physician-scholar and pioneer physiologist. *Surg. Gynecol. Obstet.* 119: 1333–1340.

Hinsey, J. C. (1961): Ingredients in medicine research—The story of a method. *Pharos* 24: 13–23.

Lindsley, D. B. (1969): Average evoked potentials—Achievements, failures, and prospects. *In: Average Evoked Potentials. Methods, Results, and Evaluations* (Proceedings of a conference sponsored by NASA, San Francisco, California, 10–12 September 1968).

Magoun, H. W. (1962): Development of brain research institutes. *In: Frontiers in Brain Research.* French, J. D., ed. New York: Columbia University Press, pp. 1–40.

Sabin, F. R. (1945): Biographical memoir of Stephen Walter Ranson, 1880–1942. *In: National Academy of Sciences. Biographical Memoirs.* Vol. 23. Memoir 14. Washington, D. C.: National Academy of Sciences, pp. 365–397.

Tepperman, J. (1970): Horsley and Clarke: A biographical medallion. *Perspec. Biol. Med.* 13: 295–308.

Judith P. Swazey (b. 1939, Bronxville, New York) is an associate professor in the Department of Socio-Medical Sciences at Boston University Medical School. and in the graduate faculty, Department of History, Boston University. Her research interests have centered on the history of the neurosciences, the nature of therapeutic innovation and clinical investigation, and the social implications of contemporary biomedical developments.

# 29

# Forging a Neuroscience Community: A Brief History of the Neurosciences Research Program

## Judith P. Swazey

Less than fifteen years ago the word neuroscience, and the multidisciplinary research effort that it denotes, did not exist. If any one man can be credited with the concept and genesis of neuroscience, a field uniting historically disparate disciplines concerned with brain structure and function and behavior, it is Francis O. Schmitt. His vision of neuroscience took shape and substance in 1962 with the founding of the Neurosciences Research Program (NRP), a new type of scientific organization that has sought to implement Schmitt's convictions about how the complex problem of understanding central-nervous-system functions, particularly conscious cognitive behavior, might be most fruitfully approached. The history of the NRP thus is a core chapter in the brief history of neuroscience (as distinct from the long history of work on mind, brain, and behavior), a chapter that relates how a scientific community was forged and that offers a picture of that community's structure and the dynamics of its scientific endeavors.

### "Mental Biophysics"

During the 1950s Schmitt became convinced that significant strides in understanding the neural bases of behavior and of the functions of the human mind were urgently important, and demanded a new type of interdisciplinary conceptual and experimental approach. His long-range vision of what this new approach might yield was stated as follows at the first anniversary convocation of the NRP in February 1963:

There is urgency in effectuating [a] quantum step in an understanding of the mind; not only as an academic exercise of scientific research; not only to understand and alleviate mental disease, the most crippling and statistically significant of all diseases; not only to create an entirely new type of science through vastly improved intercommunication between minds and hence to survive this present world crisis and advance to a new quantum jump . . . in human evolution; but perhaps through an understanding of the mind to learn more about the nature of our own being.

For Schmitt, who "grew up [as a scientist] with the idea of applying biophysics to neural phenomena," the great strides of molecular biology during the 1950s "in the age-old search for the physical basis of life" provided a model for the sorts of accomplishments that he felt could be attained by establishing a "new science of the mind and of human communication." "Those of us who have lived through the period in which molecular biology grew up and have had an active part in it," he said at the 1963 convocation, "and some of us who not only took an active part in molecular biology and macromolecular chemistry but had a very strong and early interest in nerves, realized that here was an enormous opportunity for a new synthesis." As Schmitt wrote in a 1963 progress report on the NRP's work, this "new synthesis" was an

approach to understanding the mechanisms and phenomena of the human mind that applies and adapts the revolutionary advances in molecular biology achieved during the postwar period. The breakthrough to precise knowledge in molecular genetics and immunology—"breaking the molecular code"—resulted from the productive interaction of physical and chemical sciences with the life sciences. It now seems possible to achieve similar revolutionary advances in understanding the human mind. A wealth of research literature on the mind stems from the classical approaches of physiology and behavioral sciences. By making full use of these approaches and by coupling them with the conceptual and technical strengths of physics, chemistry, and molecular biology, great advances are foreseeable.

Prompted by advances in molecular biology, ideas about a multidisciplinary attack on brain research that focused on biochemical and biophysical aspects had begun to germinate in Schmitt's mind by the fall of 1958. His thoughts had been further stimulated by his activities as organizer and chairman of a new NIH Study Section on Biophysics and Biophysical Chemistry. To Schmitt, the work of the study section, particularly its monthlong Biophysical Sciences Study Conference at Boulder, Colorado (see below), demonstrated his conviction about the need to "get the dry and wet approaches together," and about the worth of "bringing disparate fields together into a new biological discipline."

A three-month trip to European research institutes in the summer of 1959, to investigate possible roles of solid-state physics and fast reactions in nerve and brain function, gave Schmitt additional insights and impetus for developing a new approach to brain-mind research. Prior to his trip, one stumbling block to his interest in a biophysical and chemical approach to the brain was the fact that, to his knowledge, chemical reactions were not fast enough to account for information retrieval processes in the brain. At the Max Planck Institute in Göttingen, however, he met physical chemist Manfred Eigen, and from him heard about "the type of fast reaction I was looking for, giving a basis to go on and do what I did in terms of the NRP."

Meeting Eigen in August, I was excited about his paper with DeMaeyer on "Self-dissociation and protonic charge transport in water and ice" (*Proc. R. Soc. A* 247: 505–533, 1958). This particularly stimulated me to a speculation, which Eigen thought possible, that neurofilaments, which course the entire length of nerve axons, might be the substratum of information processing by fast transport of elementary-charge carriers. The still very vague idea was entertained that to account for processes like retrieval of memory and indeed the mind in general would, of necessity, require fast reactions.

In addition to discussing ideas about fast reactions in central-nervous-system functions, Schmitt seized the opportunity to sound out leading European scientists about the idea of a multidisciplinary research organization, one that would bridge the barriers between traditional research disciplines concerned with biological and behavioral phenomena. Two men with whom he shared his embryonic plans were Manfred Eigen and physicist Werner Heisenberg at the Max Planck Institute in Munich. "Eigen," Schmitt recalled in a recent letter to Walter Rosenblith, "felt enthusiastic about organizing a multidisciplinary group. I remember driving to Cornell, where he was Visiting Lecturer in 1961, to spend some hours planning the possible composition of the group to be organized. I owe much to him for reinforcing my conviction of the potential value of such an organization." Heisenberg also endorsed Schmitt's idea, but "he advised me that it might be good, before such formal organization, to convince myself through personal investigation of the correctness of the hypothesis being made concerning the phenomena."

Accepting Heisenberg's advice to investigate further the fruitfulness of pursuing biophysical and biochemical hypotheses about brain processes, Schmitt organized two seminars at MIT in the spring terms of 1960 and 1961. The topic for the first seminar grew out of his discussions with Eigen during his 1959 trip to Europe. Following those discussions, Schmitt wrote:

Eigen's International Colloquium on Fast Reactions in Solutions, at Hahnenklee, September 14–17, 1959, further galvanized my interest in such processes to the point that in the spring semester of 1960 at MIT, I organized a series of seminar lectures "on the possible physiological role of fast transfer reactions in eliciting specific interactions in ordered macromolecular structures." Because of my growing interest in the biophysical investigation of brain processes, I organized a similar seminar series in the spring term of 1961. Abstracts of these papers were published in a book entitled *Macromolecular Specificity and Biological Memory* (MIT Press, 1962).

The information gained from these seminars, and the interest that they generated among the participants, fueled Schmitt's beliefs about the need for and potential yield of a more unified interdisciplinary approach to neural and behavioral phenomena. The fact that he had the time to begin acting upon those beliefs, Schmitt notes, owes much to his having been made an Institute Professor at MIT in 1955, freeing him from many of the demands imposed by having been head of the biology department since 1942. "With the freedom to seek challenging directions in which to move," Schmitt and a small group of collaborators had, by the summer of 1961, begun to lay concrete plans for what would become the NRP. In this endeavor Schmitt skillfully combined several roles: the researcher with a comprehensive grasp of the state of the art; the theoretician framing major questions to be asked and mapping possible routes to their answers; and the "scientific impresario" who, by virtue of his wide range of professional contacts with scientists, federal agencies, and foundations, was able to command the collaboration and support requisite to the success of his project.

Two central matters to which Schmitt directed his attention during 1961 were the sort of interdisciplinary synthesis that was needed, and the type of organizational or institutional vehicle that would most effectively help to create this synthesis and maintain it on a long-term basis. Another question, of no small moment, was what to call his project—what word or phrase would best conceptualize and thus help to channel the type of research endeavor that Schmitt envisaged?

As we have noted, the first area of disciplinary merger that Schmitt wanted to effect was among biophysical approaches to the central nervous system, bringing into a closer and more fruitful liaison what he termed "wet, moist, and dry" biophysics. A specific aim of his research plan, as he later wrote in a 1962 grant application, was

to investigate the "wet and dry" biophysics of central nervous system function, i.e., to study the physical basis of long-term memory, learning, and other components of conscious, cognitive behavior, by effective utilization of the biophysical and biochemical sciences, from the physical chemistry of neuronal and glial constituents (wet biophysics) through bioelectric studies (moist biophysics) to studies of fast transfer of elementary charged particles, organized microfields, stochastic models, and application of computer science (dry biophysics).

This orientation is reflected in the titles that Schmitt first gave his project: "mental biophysics," and "biophysics of the mind." In August 1961 he wrote to one of his principal collaborators in the project, Dr. Humberto Fernández-Morán, professor of biophysics at the University of Chicago, about developments "which may put meat on the skeleton of plans we have been dreaming about for the development of the biophysics of the mind." "Is there any possibility you could curtail your trip [to Switzerland] to be present at [a] conference [during the week of September 11]?," Schmitt asked Fernández-Morán. "The iron is getting hot and must be shaped soon into the facility needed or it will cool off, i.e., the human components in the self-organizing system will have diffused sufficiently to have lost critical interaction potential. Also, world events are moving with a pace that makes dalliance doubly dangerous."

On September 12 Schmitt met at MIT with Fernández-Morán and a second collaborator, physicist John B. Goodenough from MIT's Lincoln Laboratory, to discuss the type of institutional organization that should underly the mental-biophysics project. Schmitt came to the meeting with some specific ideas about the basic disciplines that should be involved, and the general structure, location, and possible costs of a research center. In a handwritten memo dated September 1961, he listed nine "basic disciplines" for mental biophysics: solid-state physics, quantum chemistry, chemical physics, biochemistry, ultrastructure (electron microscopy and x-ray diffraction), molecular electronics, computer science, biomathematics, and literature research.

When Schmitt began considering the creation of a mental-biophysics research center, in 1960, his thoughts were oriented toward combining disciplines in a laboratory investigative program. Such a program, as he later wrote in 1966, would have involved establishing

an institute or complex of laboratories in which world leaders in each of the cognate sciences pertaining to molecular neuropsychology might be given superb laboratories, thus accomplishing the double function of establishing a *community* of research scholars interested in the brain and of establishing the roots of each scholar firmly in the soil of the institute in which his laboratories would be located.

By the fall of 1961, however, Schmitt had concluded that a research-laboratory complex was not, after all, the type of institute that would fulfill the needs he saw for a new approach to work on the brain and mind. "Even if fortunate," he wrote in 1966,

such a scheme would eventually produce but another research laboratory which, though multidisciplinary (like any university), would not necessarily introduce a novel scheme of scientific communication and interaction. Moreover, since the conceptual mandate (understanding man's brain) is a large one, many disciplinary subjects would have to be cultivated. Where time-consuming laboratory work is involved, such broadening inevitably leads to diversification and thus to dilution of the zeal per investigator, the determination not to be deflected from the main task: brain research. The bricks and mortar laboratory institute therefore lacks the necessary ingredients that would provide a new, more effective method of detecting and processing scientific information and of fast publication of the results for worldwide utilization.

A more visible and fruitful type of multidisciplinary research center, Schmitt decided, would be one oriented toward probing and evaluating the state of the art, seeking to map new research and conceptual directions, and providing new modes of education for the communication among a worldwide network of scientists. In his September 1961 memo he recorded his current thoughts about this type of research center and how it might be funded. He listed nine "desiderata" for a center, and dwelt, at greater length, on the need for a "self-organized" project.

Close to high concentration of research; conditions desirable to students and postdocs; excellent library facilities; good transportation facilities—near airports; endowed posts for key scientists; no control by government, universities, or industry; no pressure and/or prohibition to publish; publish a house archival journal or proceedings; avoid project-type of applications for funds except for major equipment and facilities . . . .
Because one of the cardinal features of the idea is that the Research Center be independent of universities, government, and industry, these cannot be used as references or bases of operations. Therefore the solid core of the project must be "self-organized," consisting of convinced scientists and lay leaders in finance and administration . . . . Since the effectiveness and attractiveness of the concept depends absolutely on the excellence of the personnel, a substantial fraction of the key men (ca. 6) must organize themselves and be prepared to operate as a group in seeking major funds to endow the project.

In addition to the organization and funding of a research center, possible formats for nonlaboratory investigative programs were also taking shape in Schmitt's mind, drawing upon the example of three such programs in which he had been involved. The first program, providing a model for what would become the NRP's Intensive Study Programs, was the monthlong 1958 Boulder Study Conference on Biophysical Science. The conference, attended by over 100 scientists and their families, had been planned for more than a year by the NIH Study Section on Biophysics and Biophysical Chemistry. The three aims of that conference (summarized by Schmitt in a 1963 progress report on the NRP's work) were: (1) to help characterize the field of biophysics, then beginning to emerge as a unified branch of the life sciences; (2) to expose a group of carefully picked young investigators, as well as a dozen mature physicists and engineers, to lectures by the most brilliant scientists who could be brought to his program; and (3) to harvest the lectures in a book which could be published rapidly and which, through special subsidy, could be sold at a price which would aid in rapid, worldwide dissemination.

More recently, the two seminars that Schmitt had conducted at MIT in 1960 and 1961 provided models for the future NRP Work Sessions—carefully planned two- to three-day meetings in which leading experts gather to review the state of the art of selected topics and to explore new conceptual developments. Although "there is literally no previous experience with this project," Schmitt wrote in his first grant application for the NRP, he had conducted a weekly seminar series which provided

convincing evidence of the value of the more ambitious program proposed in this application. Eminent authorities from the United States and Europe lectured before a broad representation of the Greater Boston academic community in a series devoted in 1960 to a study of "Fast Fundamental Transfer Processes in Aqueous Biomolecular Systems" and in 1961 on "Macromolecular Specificity and Biological Memory." Following the weekly two-hour formal lectures and informal discussion in the afternoon, there was a dinner for about a dozen scientists (including half a dozen core scientists who regularly attended the dinners with other specially invited guests) at which the subject of the lecture and its bearing on the general topic of the series were further discussed. Many valuable suggestions for new experiments sprang from these discussions.

By September 1961, then, as summarized in the record of his meeting at MIT with Fernández-Morán and Goodenough, Schmitt had formulated a plan that contained many of the major organizational features of what would become the NRP:

Instead of attempting to found a new institute comparable to the Institute for Advanced Study at Princeton, involving great expense in providing the facility and for the running expenses, it is suggested [by Schmitt] that as a preliminary stage the organization be concerned primarily with regular and coordinated consultation of a core group on problems of the physics of the mind. According to such a scheme, the first step would be the selection of six to ten individuals in various of the cognate fields who would serve as Charter Members. These Members would then come to a consensus regarding the rules by which they wish to proceed. One of the first steps would be the incorporation of the group under a Board of Trustees or Corporation. This group would then be in a position to receive moneys and to sponsor the program. For certain of the desired purposes, private financial support would be essential, though for the remaining large fraction of the expense, funds from government agencies and other foundations would be satisfactory.
A central facility would be established which would house the permanent staff, library, conference rooms, and possibly some laboratories.
The members would be selected with a view to covering the various fields bearing on investigation of the biophysics and biochemistry of the mind and

with due regard for personal characteristics of commitment to the program, cooperation, etc.

Regular meetings and consultations of the entire group would take place with sufficient length and continuity to permit meaningful interaction, thorough discussion of the literature, and possible directions of research, and a determination of investigative programs as the group may determine. It is visualized that something of the order of two to four days per month on the average would be devoted to such conferences.

## The Mens Project

As Schmitt and his collaborators continued to develop their plans during 1961, they began to broaden the scope of the interdisciplinary synthesis that they sought to effect in the research-center concept described above. On October 4, 1961, Schmitt, Fernández-Morán, and Michael Kasha, director of the Institute of Molecular Biophysics at Florida State University, prepared a new list of 25 possible fields to be represented in the project, greatly expanding the nine basic "mental-biophysics" disciplines listed by Schmitt in September.[1] This broadening multidisciplinary scope was reflected by a change in the project's title: instead of the more limited phrase "biophysics of the mind," Schmitt's group began to discuss "the Mens Project," from the Latin word for mind. Through the Mens Project, Schmitt hoped to bring about a synthesis of intellectual effort not only among "wet, moist, and dry" biophysicists, but among "the whole array of disciplines" concerned with elucidating the nature of mental processes. "If biophysical (wet and dry) and biochemical sciences are to be brought effectively to bear in the investigation of mental processes," he wrote in 1962,

adequate support must be given not only for research of the types already well established and highly productive, e.g., neurophysiology, neuroanatomy, neuropathology, etc., but also to facilitate maximum participation of scientists from mathematics, physics, biophysics, chemistry, biochemistry, bioengineering, etc. In few cases are physical and engineering scientists trained also in neurology, psychology, or other life sciences. Even the most versatile and eminent representatives of these individual disciplines, relatively discrete and noncontiguous with the other disciplines, are seldom acquainted with more than the rudiments of any of the other disciplines in their own general areas of science, e.g., solid-state physics, quantum chemistry, crystallography, molecular electronics, computer science, etc. Collaboration between representatives of the whole array of disciplines, including molecular biology and the neurological disciplines, is likely to be ineffective and sporadic unless a substantial intellectual overlap is achieved, figuratively as pi-orbital overlap facilitates electron transfer through molecular aggregates.

1. The 25 fields were: molecular biology; electron microscopy; crystallography (?); quantum chemistry; semiconductor solid-state physics; physical polymer chemistry; biochemistry—RNA, DNA; biochemistry—enzymal, protein; neurophysiology—electrophysiology; neuroanatomy; neurology; general and comparative neurology; psychology; psychiatry (?); molecular electronics; instrumentation; computer technology; biomathematics (?), biometrics; information theory; developmental neurology; ultraminiaturization; human genetics; comparative neuropaleontology; radiation biology (not high energy); radiation biology (high-energy).

By the beginning of 1962 Schmitt and his associates felt that their plans for the Mens Project were concrete enough to begin actively implementing them. Thus, on January 22, 1962, Schmitt held an "exploratory meeting" in Washington, D.C., with representatives of various federal agencies that might lend support to the project, discussing with them the nature and estimated costs of the sort of research center he wanted to establish. Within the next week he also held conferences with officials at MIT, to inform them of the status of his plans and to consider MIT's possible sponsorship role for a research center.

Then, on February 1, Schmitt convened an organizational meeting for the Mens Project in New York City, at the apartment of close friends, Mr. and Mrs. Louis E. Marron. Schmitt invited eleven researchers to the meeting. This then was the founding group of NRP Associates:

Leroy G. Augenstein
Professor of Biophysics
Biology Research Center
Michigan State University

Michael Kasha
Director, Institute of Molecular
 Biophysics
Florida State University

Leo DeMaeyer
Max Planck Institute
Göttingen, Germany

Heinrich Klüver
Sewell L. Avery Distinguished
 Service Professor of Biological
 Psychology
University of Chicago

Manfred Eigen
Max Planck Institute
Göttingen, Germany

Albert L. Lehninger
Professor of Physiological Chemistry
Johns Hopkins School of Medicine

Humberto Fernández-Morán
Professor of Biophysics
University of Chicago

Severo Ochoa
Professor of Biochemistry
New York University Medical School

John B. Goodenough
Lincoln Laboratory
Massachusetts Institute of Technology

Francis O. Schmitt
Institute Professor
Massachusetts Institute of Technology

Holger V. Hydén
Director, Institute of Neurobiology
University of Göteborg, Sweden

William H. Sweet
Chief, Neurological Service
Massachusetts General Hospital

At the meeting Schmitt briefed the participants on the scientific background of the project, reviewing the status of the biophysics and biochemistry of mental processes, and the need that he saw for a "new type of productive interaction" among scientists from various disciplines. He then detailed the nature of the Mens Project, revolving around the establishment of a new type of research center that would have both an "internal program" for a core "study-section council" and "external programming" such as a "Boulder-type" intensive study program.

The next topic was the matter of finances—how the costs of a center, personnel, and research projects would be funded. Schmitt reviewed his January 22 meeting in Washington with federal-agency representatives, and discussed the idea of preparing a pilot NIH grant application for a March 11 meeting of the National Advisory Health Council (NAHC). In addition to government foundation support, Schmitt also related plans to incorporate a private foundation "to insure independence and continuity" for the center. The final item on the agenda was a discussion of candidates for "additional members of the central group." (At that first meeting, as today, election to the small, collegial group of NRP Associates was an informal process.)

With the endorsement of the Associates, Schmitt moved swiftly after the February 1 meeting to secure funding for the project and to acquire a physical base of operations—a center where a staff could begin working and the Associates could convene. On February 6 the Grants Applications Division of NIH was notified that an application would be submitted, and on February 14 Schmitt conferred with the NIH's associate directors about the grant for the Mens Project. A five-year grant application, for a "Neurophysical Sciences Study Program," was submitted on March 1, with MIT as the sponsoring institution. Five days later an NIH Site Visit Committee came to MIT, and on March 14, just two weeks after the application had been submitted, it was approved by the NAHC.

**The NRP is Launched**

With the awarding of the five-year NIH grant, Schmitt's project was ready to be launched. The program still had various designations in the spring of 1962—the Mens Project, the Neurophysical Sciences Study Program, and, emerging in a grant application to NASA, the now-familiar title, the Neurosciences Research Program.

Two immediate objectives were to obtain additional support, beyond the NIH grant, and to house the NRP in a research center. Schmitt and his associates moved in several directions to secure funding for the NRP. In keeping with the desideratum of maintaining an independent center, not tightly bound to government, university, or industry support, bylaws for a nonprofit corporation were drafted at the end of February 1962, and on May 25, 1963, the Neurosciences Research Foundation was incorporated by the Commonwealth of Massachusetts. The foundation subsequently received tax-free status in 1964, and was certified as a private operating foundation under the 1971 tax law. The foundation's purposes have been to provide for long-range financial planning and stability, to help defray program expenses for which federal funds are not appropriate, to make available a "financial buffer" for rapid action on new ideas pending action on grant applications, or to temporarily support projects in case of the withdrawal of a grant.

In addition to receiving the NIH award and establishing the foundation, Schmitt and his colleagues also broadened and secured the NRP's financial base by seeking research grants from other federal agencies and from private foundations. Beyond costs covered by federal funds, the meetings held by the NRP from 1963 to 1967 were made possible by an unrestricted five-year award from the Rogosin Foundation of New York.

The question of where to house the center had been a topic of discussion among Schmitt and his collaborators since they first began planning the mental-biophysics project. When Fernández-Morán, Goodenough, and Schmitt met at MIT in September 1961, they agreed that only two locales "were favorable"—in the San Francisco area, possibly near Palo Alto, or in the New England area, close to Boston. Once the project was formally organized in February 1962, with Schmitt as chairman and principal investigator for research grants to be awarded through MIT, a location near Boston became the obvious choice. One possible site that occurred to Schmitt was the spacious headquarters of the American Academy of Arts and Sciences in Brookline—a location that would offer proximity to Boston and Cambridge and at the same time be remote enough from those university centers to insure intensive, uninterrupted work by the NRP Associates. Schmitt found that Hudson Hoagland, then president of the academy, was most receptive to the idea of having an enterprise like the NRP housed with them, and on March

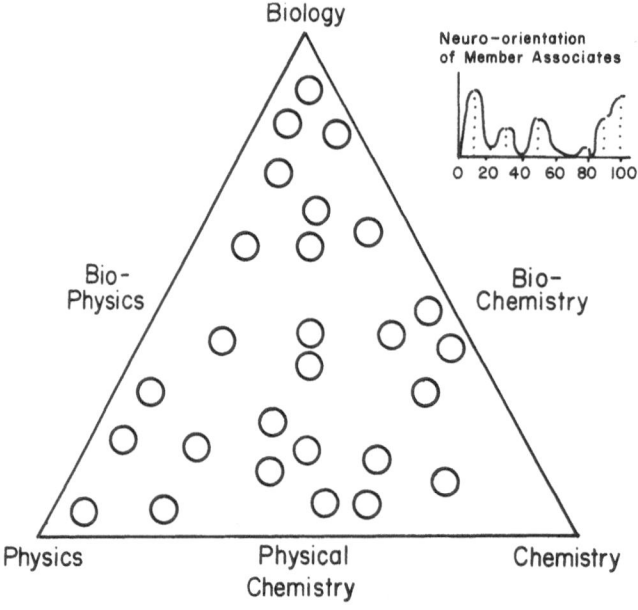

Figure 29.1  An organizational chart of the Neurosciences Research Program. Drawn from a 1963 Progress Report to the National Institutes of Health.

15, 1962, the day after the NIH funds were awarded, Schmitt met with William Allis of the academy to arrange the terms of a lease.

While the academy's third floor was being refurbished for the NRP, the program's initial three-person staff temporarily began working at MIT (Schmitt; Miss Harriet Schwenk, his assistant; and H. K. Gayer, executive officer). On August 20–24, the NRP Associates convened at the academy house in Brookline for their third meeting, and on October 15 the NRP staff moved into their new quarters.

The NRP celebrated its first birthday on February 1, 1963. The program's organizational structure at this time, and the scientific background of its Associates are illustrated in Figures 29.1 and 29.2, drawn from a 1963 Progress Report to the NIH. In its first year the number of Associates had increased from 12 to 27, the staff had grown from 3 to 10, the program's first resident scientist, R. G. Ojemann, was working at the center, and the NRP's first four Work Sessions, scheduled for June, July, and September, were being planned. These and other signs of growth and early success were celebrated at the center on February 1–2 by a

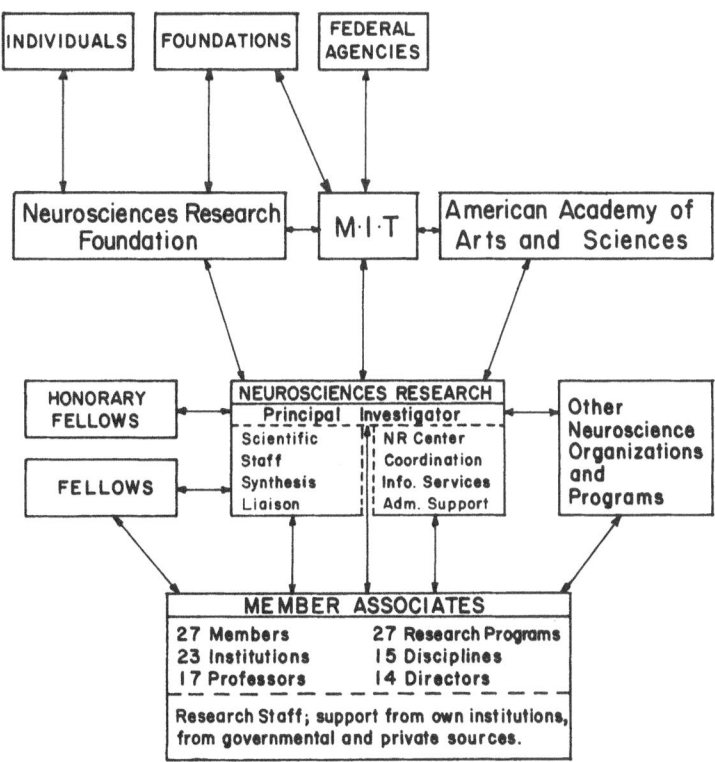

Figure 29.2  A chart showing the distribution of base fields of Associates of the Neurosciences Research Program. Drawn from a 1963 Progress Report to the National Institutes of Health.

First Anniversary Convocation, attended by the Associates and by some 100 leading scientists and representatives of granting agencies and foundations. During the first day of the convocation, nine of the Associates lectured on "Mind, Brain, and Molecules," dealing with the spectrum of molecular, neural, and behavioral sciences to which the NRP addressed its work; the second day was devoted to discussions of the NRP's practical and fiscal planning.

The lead speaker at the convocation's scientific program, appropriately, was the NRP's founder and chairman. In his introduction to the program and in his lecture, Schmitt addressed the types of issues that confronted neuroscience, and some of the developments that he hoped this new multidisciplinary synthesis might help to effect. Drawing from the molecular-biology model that had so strongly influenced the genesis of the NRP, Schmitt designated the core field of neuroscience, one touching on all other disciplines concerned with work on the brain and mind, as "molecular neurology"—the application of molecular biological principles to brain function in an attempt to understand the mechanisms of the mind, memory, and so on. He spoke of the NRP's "wish to develop for this field a solid science," but also emphasized that the new synthesis embodied in the NRP "does not involve merely molecular neurology. Rather it involves an attempt to bring together the disciplines involved in molecular neurology with the behavioral sciences in a fusion without which we will never understand the mind."

In his lecture Schmitt elaborated on how such a disciplinary fusion might, vis à vis an understanding of the brain and mind, overcome what he called the "system-component dilemma":

> There are perhaps three ways of studying complex problems in the life sciences. One is behavioral-holistic—a systems type of method. This is usually descriptive; it may be taxonomic . . . and [is] frequently very intuitive. . . . There is a second way which might be called analytical or reductionist—I would like to call it molecular, because in this way one attempts to get at the componentry, the individual components which play a dominant role in determining the phenomenon. This method . . . attempts to make incomprehensible and complex phenomena understandable in terms of something at least that is comprehensible. I think . . . that in the past, and probably in the future, we shall go on oversimplifying and realizing later that it has been an oversimplification and start over again. There is a third method, and that is to try to do both simultaneously; that is to say, to attempt to keep in mind the molecular approach, the isolation and identification of the molecules or components of great importance, whilst at the same time keeping in mind the systems properties. This is extremely difficult.

Historically, Schmitt pointed out, the development of increasingly sophisticated techniques enabling life scientists to begin probing the physical basis of life has too often involved a failure to distinguish between the structure and function of an organism's components (e.g., cells, molecules) and the nature of the system formed by those components. "[As soon as] our technical developments make it possible to look closer into the questions of components, we assume and clothe those components with properties which really inhere in the system." To Schmitt, the state of the art of molecular genetics in 1963 exemplified this system-component dilemma and underscored "the more difficult problem of understanding the nature of the mind by the development of . . . molecular neurology."

Now this present preoccupation with the sophisticated macromolecular coding theory tends to obscure the fact that the cell is not—and here I quote Paul Weiss, another one of our members—an inert playground for a few almighty masterminding molecules, but is a system—a hierarchically ordered system—of independent molecules, molecular assemblies, and organelles. Life depends on the ordering of their interaction, on systems properties, in other words . . . how does this preformed DNA code know when to tell particular cells to do things like when to make a muscle, when to make a nerve. They are all the same code, and every cell has the same chromosomes. Something has to tell the cells to do *this* now; it is appropriate that you should make a leg, or you should make an eye in an embryo. What tells the chromosomes to be read out like that? This is the difficult systems property that the physicists, the biophysicists, the biochemists, and the molecular biologists are just getting around to trying to explain.

"Great success has been achieved in molecular biology in dealing with this cell-molecule problem in molecular genetics," Schmitt stated. Comparably, he declared, "through intensive and dedicated interdisciplinary effort [we can] find out how we can discover something new about [the] mind-brain-neuron-molecule dilemma. Eventually such new knowledge will be coupled with the concepts and techniques of systems analysis and behavioral science to forge a new science of the mind and of human communication."

### The NRP and Neuroscience: A Not-So-Invisible College

It is beyond the scope of this essay to trace in detail the development of the NRP and to relate its development to that of the neurosciences. Rather, I will indicate briefly in this final section some of the methods the NRP has evolved in its attempt to fulfill the stated aims of "investigation, synthesis, and education-communication," and will consider the NRP's creation and functioning in relation to the development of a neuroscience community.

As a historian and sociologist of science, I would like at some time to study more closely the types of scientific societies that have developed over the past three centuries, and to reexamine the NRP's creation and functions within such a broader analytic context. In general terms, it may be said that the scientific societies of the seventeenth century established patterns of tradition and continuity, and formats for the rapid communication of new ideas and discoveries, that served to generate new canons of universality for the pursuit of science. With particular respect to their communication function, Stillman Drake has noted that

it would be hard to imagine modern science without scientific societies. The progress of scientific ideas is heavily dependent upon communication; hence the need for a particular kind of organization and a special class of publication. Scientific societies, by the selection of persons with highly specialized interests, greatly reduce the hazard that avenues of useful communication will be cluttered up with rubbish or damaged by false or misleading announcements in their field. They also provide an important means for the organized defense of the interests of their members against interference with free research and communication and other disturbances from outside which occur from time to time. (Drake, 1966, p. 1194)

The first "modern" organized scientific society, historians generally agree, was the Accademia del Cimento, founded in Florence in 1657, whose creation had been influenced by two earlier seventeenth-century Italian societies, the Accademia Curiosorum Hominum and the Accademia dei Lincei. In marked contrast to the NRP, speculation was prohibited at meetings of the Accademia del Cimento during its ten years of existence. Its members' activities included experimentation, the creation of instruments, the establishment of standards of measurement, and the devising of exact methods of research; the book of their meetings and work, the *Saggi*, has been described as the "laboratory manual" for the eighteenth century.

The 1650s and 1660s saw the establishment of many other scientific societies in England and continental Europe, among them the Royal Academy of London and the Académie des Sciences in Paris. Schmitt and his associates have often referred to the NRP as a "kind of invisible college," likening their organization to that formed in 1645 by Robert Boyle and his fellow "virtuosi," whose weekly meetings to study and discuss the "new philosophy" (science) formed the soil out of which grew the Royal Society.

The activities of Schmitt and his "circle" of internationally based colleagues may indeed have constituted a kind of invisible college in the late 1950s and early 1960s, when they began meeting to discuss the ideas that developed into the NRP. But once the NRP was formally launched in 1962, its activities soon transformed the "invisible college" into an organization with an increasingly high degree of visibility and influence among mind-brain-behavior researchers (see Crane, 1972, on modern invisible colleges).

The NRP's growing sphere of influence over the past decade seems ascribable, in large measure, to three major features of its structure and operation which, taken together, distinguish it from more traditional types of scientific institutions or professional bodies: the relationships between and the work of the Associates; the larger worldwide network of scientists that the NRP has engaged in its work; and the nature of the organization's programs.

Both because of their small number and because of their long-term and regular participation together in the NRP's activities, its Associates have developed an unusually cohesive working relationship. Although representing diverse disciplines, with specialized languages, research strategies, and concepts of their own, the Associates have learned to communicate and work together as neuroscientists. Their "reeducation" has, in turn, enabled the Associates to begin transmitting new, interdisciplinary ways of thought and work to colleagues and students at their own institutions, helping to propagate a new genealogical tree of neuroscientists that did not exist a generation ago. One example of this feedback effect is provided by the comments of NRP Associate Robert Livingston on the establishment of a neurosciences department at the University of California, San Diego, School of Medicine. "As you know," Livingston wrote to Schmitt in November 1970,

the NRP has been the principal inspiration for whatever we have succeeded in achieving here at this new University of California campus at La Jolla. We were invited to bring together the neurological disciplines, basic and clinical, in the establishment of the first department in a new medical school. The nervous system is recognized as an integrative system in education as well as in life. We took inspiration from the NRP by identifying this group as the Neurosciences Department. Our teaching program, the Neurosciences Study Plan, is greatly influenced in its design by the frame of orientation in this field established by the NRP. If imitation is flattering, you are being greatly flattered by many institutions. The Neurosciences Study Plan has also been greatly aided by people like Ted Melnechuk, Gardner Quarton, and yourself, as well as by Ted Bullock, Bob Galambos, and another Associate, Mac Edds, who was with us on sabbatical leave during a formative year. We think of ourselves as being a Pacific outpost of the NRP, with a year-round marine laboratory and other neurosciences research and teaching resources available to NRP personnel and program ventures. We find NRP meetings immensely useful for our own education and NRP publications indispensible for our graduate training programs.

The NRP's core personnel consist of the Associates and the professional staff, including a small number of resident and staff scientists invited to work at the center, both on NRP activities and their own projects, for periods of several months to one year or longer. Over the years the NRP's initially small professional staff has grown both in size and in the scope of its activities. To help implement one of the programs' major purposes, the rapid dissemination of information, Theodore Melnechuk joined the staff as communications director in September 1963 and, in Schmitt's words, "was largely responsible for keeping the shop going in those early days." Three more key staff positions were created in 1964 with the appointments of George Adelman as managing editor and librarian, Katheryn Cusick as administrative officer, and L. Everett Johnson as business manager. Another new position, that of program director, was established in January 1965 to aid Schmitt in the increasingly demanding task of overseeing the NRP's burgeoning activities. Gardner C. Quarton, a research psychiatrist, served as the organization's first program director until 1967, when he was succeeded by Frederic G. Worden, a research psychiatrist and neurophysiologist who was appointed executive director.

Beyond its core group, the NRP over the years has built up a worldwide communications network among scientists in many fields who have worked as NRP consultants or attended its various programs. In its first decade over 1000 scientists were engaged in the NRP's endeavors, providing a large resource pool from whom the staff and Associates can garner knowledge and who can, in turn, avail themselves of the NRP's resources.

This neuroscience network has been woven since 1962 by the NRP's programmatic activities—meetings, publications, and informal and formal teaching endeavors. Although the initial emphasis of the NRP was on harvesting and disseminating the ideas that might be generated by bringing together the expertise of the Associates, it became clear during 1962 that the small size of the NRP staff and Associates and the volume of subjects to be examined and reported demanded new program formats. "Our first Associates' meeting," Schmitt remembers, "lasted for five days, and we were tutoring each other day and night. We then had several other three-day meetings that first year, and saw that this schedule was too frequent and too intensive. So we decided to hold two Stated Meetings of the

Associates each year, and to develop another type of fact-finding meeting." In response to this need, the staff, spearheaded by Ted Melnechuk, evolved the Work Session format in 1963, and began worldwide dissemination of the results of these sessions in the *NRP Bulletin*. Like other types of NRP meetings, the major purpose of the six to eight Work Sessions held each year is not only to report new research data, but to provide a forum for developing new ideas and concepts from which may flow new research leads and theoretical formulations. In this emphasis, and in the long and intensive planning for and follow-ups to the Work Sessions, they differ significantly from most other types of scientific meetings and symposia.

The second category of NRP meetings is the one-day conference, which, over the years, has evolved into four distinct types. "Information-seeking" conferences on specialized subjects are held, often seriatim, as part of the process of determining Work Session topics. "Core-consultant" conferences bring together experts retained by the NRP as long-term consultants to explore new subjects and examine selected problem areas. A third, less frequent, type of conference is one to "nucleate" a particular research finding:

Occasionally NRP learns of a new laboratory discovery which, if coupled with other types of research, would have its value amplified or would be speeded in its application to neuroscience. An example of this type of activity concerned the rapid investigation and exploitation of the brain-specific protein S-100. In an NRP one-day conference the discoverer, Dr. B. W. Moore of Washington University's Department of Psychiatry, provided purified protein to another member of the conference, Dr. Lawrence Levine of Brandeis University's biochemistry department, an immunochemist who subsequently made an antiserum against the poorly immunogenic, acidic protein S-100. This antiserum was generously sent by Dr. Levine to workers around the world, and in only a year it was shown that the S-100 protein is primarily glial and may be involved in certain physiological processes.

In 1964 the NRP staff introduced a fourth type of conference, "Whither Neuroscience," in which small groups of experts periodically convene, at locales around the world, to attempt to anticipate important developments in the neurosciences.

The "Whither Neuroscience" conferences were designed to supplement other NRP procedures for assessing and anticipating important developments in the field of the neurosciences, broadly conceived. Whereas Work Session participants are selected to provide knowledge relevant to a particular research problem or topic, "whither" participants are selected to provide expertise spanning diverse topical areas and levels of organization ranging from physics and chemistry to neurology and psychology. Half a dozen such authorities are selected not only for their knowledge, but also because their personalities and interests favor meaningful group interaction and foster a mood of creativity.

The NRP's other major category of meetings, the origins of which have been described earlier in this paper, are its triennial Intensive Study Programs (ISPs). Like the other types of NRP meetings, the three ISPs held to date, in 1966, 1969, and 1972, have gradually evolved in their format to meet the changing shape and needs of the neurosciences. Thus the 1966 ISP sought to define a newly emerging field, neuroscience, the 1969 ISP was organized around three major themes, and, in 1972, the ISP participants dealt with twelve topics "chosen to represent the most promising and catalytic new developments in neuroscience research."

Another new type of NRP program was inaugurated in August 1973, combining elements of the Work Sessions, one-day conferences, and ISP's. The program, a five-day conference entitled "Functional Linkage in Biomolecular Systems," examined the role of functional linkage in terms of theoretical principles and applications in various biological systems. This type of conference, substantially longer than the Work Session format, departed from the ISP model "in that it considered a single broad physicochemical principle and its possible application to various aspects of neuroscience" (Schneider, 1974).

The NRP's Work Sessions, conferences, and ISPs obviously all have a large educative component, serving as they do to bring together in common endeavors experts from many disciplines, and to disseminate their work to a larger community of scientists through the NRP's publications. Other types of educational activities undertaken by the NRP have included an MIT graduate-level seminar, begun in 1970, which offers a small number of students the opportunity to work intensively for a semester with the NRP staff and Associates. Another teaching enterprise is the NRP's tutorial programs, designed to introduce "novitiates" from fields such as mathematics, physics, and engineering—leading experts in their own fields—to the complexities of the brain sciences. The tutorial programs began in 1965 with a two-week lecture-laboratory course on the fundamentals of human brain structure and function. As geneticist Seymour Benzer recalled at the "Paths of Discovery" symposium, his "first formal instruction in neurosciences" took place when Schmitt invited him to participate in the 1965 tutorial program.

It was the epitome of a crash course, consisting of dawn-thru-dusk sessions by Walle Nauta speaking to a motley crew of physicists, crystallographers, biochemists, and other types. I learned some qualitative things about the nervous system and some quantitative things. One of the qualitative things was that the easiest way to remember the wiring of the brain is to assume that everything is connected to everything else unless otherwise specified. And I would not be surprised if in the future human brain anatomy is taught in reverse that way—it might be easier to remember. One of the quantitative things I remember is that all the physicists insisted on knowing how many neurons there were in the brain, and Walle said $10^{10}$; later on he was talking about the cerebellum, and when he was asked how many neurons there are in the cerebellum and he said $10^{10}$, everyone got very upset. Walle settled the problem by announcing that the total number of neurons in the brain is $10^{10}$—not counting the $10^{10}$ in the cerebellum!

By far the most important thing I learned in this course . . . took place at the great moment in the laboratory when we were handed human brains by the dozen to tear apart with our fingers. The thing that was very striking to me was that the brains differed from each other as much as the faces and the behavior of the people in the course, and one must wonder how much of the structure of the brain is determined by the genes of the individuals and how much by their environment.

To this observer of the NRP, perhaps the most significant overall influence of the organization to date has been its catalytic role in creating and promoting the growth of a neuroscience community. As we have seen, the term and concept of "neuroscience," of a new type of integrated multidisciplinary and multilevel approach to the study of the brain and behavior, came into being with the NRP. If queried about their scientific specialty, most of those working in this new discipline might still identify themselves as molecular biologists, neuroanatomists, biophysicists, psychologists, and so on. But, at the same time, these specialists also recognize that they now belong to a broader scientific community, that they are neuroscientists, sharing certain common models, research strategies, and goals that cut across more traditional and narrow specialty lines.

In the second edition of his book *The Structure of Scientific Revolutions*, Thomas Kuhn recognizes that his earlier implicit "one-to-one identification of scientific communities with scientific subject matters . . . will not . . . usually withstand examination, as my colleagues in history have repeatedly pointed out."

There was, for example, no physics community before the mid-nineteenth century, and it was then formed by the merger of parts of two previously separate communities, mathematics and natural philosophy *(physique expérimentale)*. What is today the subject matter for a single broad community has been variously distributed among diverse communities in the past. (Kuhn 1970, p. 179)

In much the same way as a physics community developed in the mid-nineteenth century, I would maintain, a neuroscience community has begun to emerge since the early 1960s, formed by the joining together of specialists from a number of previously separate communities in the shared pursuit of penetrating the complexities of the human brain and behavior. As an institution created expressly to effect, channel, and propagate such a merger, the NRP has been a principal agent in forging this neuroscience community.

## References

Crane, D. (1972): *Invisible Colleges. Diffusion of Knowledge in Scientific Communities*. Chicago: The University of Chicago Press.

Drake, S. (1966): The Accademia dei Lincei. *Science* 151: 1194–1200.

Kuhn, T. (1970): *The Structure of Scientific Revolutions*. 2nd ed. Chicago: The University of Chicago Press.

Oncley, J. L., editor-in-chief (1959): *Biophysical Science—A Study Program*. New York: John Wiley.

Quarton, G. C., Melnechuk, T., and Schmitt, F. O., eds. (1967): *The Neurosciences—A Study Program*. New York: Rockefeller University Press.

Schmitt, F. O., editor-in-chief (1970): *The Neurosciences—Second Study Program*. New York: Rockefeller University Press.

Schmitt, F. O., and Worden, F. G., editors-in-chief (1974): *The Neurosciences—Third Study Program*. Cambridge, Mass.: The MIT Press.

Schmitt, F. O., et al., eds. (1966–1975): *Neurosciences Research Symposium Summaries*. Vols. 1–8. Cambridge, Mass.: The MIT Press.

Schneider, D. M. (1974): The Neurosciences Research Program. *Fed. Proc.* 33: 6–8.

# On the Nature of Research in Neuroscience

Paul C. Dell (b. 1915, Alsace, France) is director of the Neurobiological Unit of the Institut National de la Santé et de la Recherche Médicale (INSERM), Marseilles. His work has included studies of humoral and vegetative control of the brainstem reticular formation, the induction of slow-wave and paradoxical sleep by vagoaortic baroceptive messages, and various methods of reticular deactivation.

# 30

## Creative Dialogues: Discovery, Invention, and Understanding in Sleep-Wakefulness Research

## Paul C. Dell

"Seit ein Gespräch wir sind . . . "
Friedrich Hölderlin

"L'être, l'objet dans le final mental, ne sont plus narcissiques: l'altercation ininterrompue avec le réel, celui que nous dégageons et celui qui s'oppose à nous, ne l'autorise pas. Et ce n'est pas l'espace gris et grandiose où l'homme depuis peu bourlingue qui ajoutera à cet évènement."
René Char

The scientific dialogue that is pursued by a researcher between his curiosity and the objects of his research is in reality the culmination of two earlier ones which may be called the evolutionary and the cultural dialogue. Born as a human being into a given culture, his mind has been shaped by his social and cultural milieu. And as *Homo sapiens* he is one of the links in the dialogue of the species with the outside world.

At first sight these three dialogues appear to have much in common. Under many changing and versatile faces, each one hides an inner core of unity and continuity bolstered by an inherent logic. New arguments are constantly being introduced; some of these arguments are minor in importance and serve only to sustain the dialogue. Others are, by their freshness or their boldness, surprising and provocative. It is these arguments that bring the dialogue to its occasional crises, moments at which a choice must be made. And it shall be made because one of the possible combinations of the new argument and the ones that have kept the dialogue going until then will appear to be more interesting to and more productive for the fruitful continuation of the dialogue, as well as leading to a more coherent and intelligible organization of the material.

The three dialogues are, for this reason, creative and inventive and by the same token irreversible. As in commedia dell'arte, where each character invents his part as the play goes along and there is no totally preestablished plan for the comedy, these eternally unfinished dialogues are open dialogues; the uncertainty of the future confers upon them both the charm of the unexpected and the anxiety of the unknown. It is as if some inner forces were pushing forward this succession of creative acts, spurring on the capacity of imagination and invention to overpower the world's eternal resistance ("investir l'éternelle résistance du monde" (Duvignaud, 1973)).

Furthermore, we, as human beings, believe that we are capable of detecting the intrinsic motives and ends of our choices, actions, and creations. For a long time we were led by our anthropocentrism to attribute the changes in natural phenomena, physical or biological, to motives similar to our own. We have had to abandon this simplistic view of the world as we have left the surface of things to delve into the inner reaches. Evolution, which appears as a "total creative freedom" ("totale liberté créatrice"), would seem to illustrate perfectly this sentence of Democritus which Monod (1970) used as an epigraph to his book: "All that exists in the universe is the result of chance and of necessity." The invention of a new form results from an accidental and aleatory alteration in the genetic text (a chance happening in the microscopic realm) whose consequences at the level of the organism are accepted or rejected depending on their survival value for the organism in a given surrounding (the organism and its environment making up the macroscopic universe).

The dialogue of human beings with the physical world and the sociocultural environment is obviously a continuation of the one they started during the perinatal maturation of the nervous system. It is reasonable, therefore, to wonder whether the same principles are involved. We know the extent to which recent investigations of such varied topics as the plasticity of the nervous system, behavioral patterns studied with the ethological approach, language acquisition studied according to Chomsky (1968) and his disciples, and the raison d'être of sleep are now beginning to alter our ideas of the respective roles of the "innate" and the "acquired" in adult personality development. Finally, with regard to the scientific dialogue, the researcher's motives, to be sure, are a source of creation and invention; as will be seen, however, he is not carrying on this dialogue with a passive and inert object, but rather with a "scientific problem" that haunts the mind and stimulates the desire to know.

Prudence requires us to hold off our judgment on whether the multiple analogies just drawn between the three dialogues are superficial and fortuitous or whether they embody a profound interconnection. We will therefore confine ourselves here to an analysis of the essential features of the scientific dialogue exclusively. Nonetheless, each dialogue does provide the foundation for the one to follow. Our capacity to utilize symbols, which is acquired during our education, and our capacity to "explain," which results from the scientific dialogue, bestow upon us "distance receptors in time" (Craik, 1943) whose survival value for the individual is obvious. By word or by pen the acquired knowledge is transferred and expanded from generation to generation; in this way the chances of survival of the human species are increased.

### The Structure of the Scientific Dialogue

Like the main character in a play (the "protagonist" in Greek drama), it is a "scientific problem" that instigates and animates the action in the scientific dialogue. It is the nucleus around which the dialogue is organized. In choosing the phenomenon he wishes to investigate, explore, and explain, the scientist believes he is taking a slice of reality, yet the chosen phenomenon never comes before him in its primordial originality. In fact, what the researcher is confronting is a mental representation of the problem that has been defined and shaped for him by his predecessors.

Frequently the problem embraces disparate elements that have been artificially put together and constitute a chimera in relation to reality. The history of science offers many such cases in which the arbitrary boundaries placed on reality by a problem's definition do not respect the natural ones. The scientist's mental representation of a problem is therefore never a faithful image of reality. The emphasis put upon certain aspects casts shadows on others, or else the exaggeration of the relative importance of one of the components of the problem creates distortions in the arrangement of the overall pattern, which as a result bears little resemblance to the real one. The mental pictures created by the scientific genius are deformed just as are those created by the artistic genius; they are therefore more similar to a Picasso painting than a photograph.

Unfortunately, such distortions are unavoidable. The zeitgeist prevailing in a given field as well as the zeitgeist of the epoch in general impose on us forms of thought and even an entire conceptual context that we usually accept unconsciously. Incidental factors as well interfere with the way we see the problem. Assistance provided by a new discovery in a rapidly expanding adjacent field, the fresh reasoning of a newcomer to the field who has put an accent on some previously neglected part of the problem, or the invention of some new, easier-to-use technique, can cause one aspect of the problem to be overemphasized at the expense of others which might be more important in reality.

Thus, instead of having before him a fragment of reality to explore, the research worker is confronted by a "protagonist" fashioned by his predecessors. This creation possesses a structure and a logic of its own, acquired during its formation. Indeed, the scientific problem is a living being who creates and inspires questions.

**Mental tools of the scientific dialogue**
Now that we are acquainted with the participants, it is necessary to describe the means of communication used by them in order to conduct their dialogue. Three such means can be identified: discovery, invention, and understanding. These mental tools complement one another and are generally used jointly.

*Discovery and invention.* Discovery and invention are always linked. In English the world "discovery," by itself, embraces both concepts; French, however, makes the distinction between the two complementary actions (see, for example, Hadamard, 1954). "To discover" connotes the revealing of the existence of something that has always existed but has been hidden by appearances. "To invent," on the other hand, connotes the imagining or conceiving of something completely new. Columbus *discovered* America since the continent had always existed; Franklin *invented* the lightning rod; mathematicians *invent* new objects or new mathematical theories.

In the experimental sciences, sometimes it is the invention that comes first and sometimes it is the discovery; but if progress is to be made, they must occur together. A casual observation might go undeveloped unless it is followed up with an invention, that is to say, unless it is placed in a context that permits it to be properly appreciated. Conversely, invention, which generally consists in formulating a hypothesis, will have no consequences unless it is supported by experimental results that lead to discovery. One example of an invention would be the application of a new technique, elaborated for the circumstances or borrowed from a related field of study; if important new results are obtained in this way, there will be a discovery.

Momentous inventions are nearly always unexpected and above all unforeseen; they often take the form of an extended synthesis reordering material until then considered to be extraneous or conflicting. Terms such as "illumination" (Henri Poincaré) or "happy idea" (Helmholtz) aptly describe how their creators experience them. On the other hand, the examples that we will borrow from the field of sleep-wakefulness physiology are instances of the first alternative, i.e., a casual observation becoming through an invention a discovery. In order for this to happen, three conditions must be fulfilled: (1) The researcher must be capable of mentally concretizing information to which the phenomenon observed by him can be linked; this information is the phenomenon's "true" context, which is able to clarify its meaning. It is in this sense that "discovery smiles on well-prepared minds." (2) The researcher must understand the importance of an observation, grasping its multiple theoretical and experimental consequences. (3) The resulting scientific explanation must bring with it a new arrangement of information, a new order of things, which everyone recognizes as closer to reality. Medawar (1967) expressed it as follows: "The regulation and control of hypotheses is more usefully described as a *cybernetic* than as a logical process: the adjustment and reformulation of hypotheses through an examination of their deductive consequences is simply another setting for the ubiquitous phenomenon of negative feedback. The purely logical element in scientific discovery is a comparatively small one, and the idea of a *logic* of scientific discovery is acceptable only in an older and wider use of 'logic' than is current among formal logicians today."

*Understanding.* As Granit (1972) recently pointed out in an article aptly titled "Discovery and understanding," Sherrington had to his credit no spectacular discovery of striking originality or innovation, and yet he was the dominant personality of his time in neurophysiology. His lifelong efforts produced year after year new insights and experiments invented in service of a precise and well-defined line of research. A scientific problem's gradual maturation is the result of consecutive forays that include both successes and failures. "Moreover, while wrestling with [the problem, the scientist] will not merely learn to understand the problem, but . . . will actually change it. A change of emphasis may make all the difference —not only to our understanding, but to the problem itself, to its fertility and significance, and to the prospects of an interesting solution" (Popper, 1963). Without going into an epistemological discussion of the criteria of testability and falsifiability of hypotheses (Popper, 1959, 1969), it should be said that hypotheses that are disproved experimentally are just as important as those giving positive results to the maturing process of a scientific problem. Our colleague

Eccles (1963) felicitously expressed this idea when he said: "I can now rejoice even in the falsification of a cherished theory, because even this is a scientific success."[1]

## How the scientific dialogue proceeds

Guided by its inherent logic, highlighted by moments of discovery and invention, shaped through understanding, constantly influenced by the scientific progress in adjacent fields of study and by the general zeitgeist, the scientific dialogue proceeds not in a steady flow but rather by stages, somewhat as a play unfolds act by act. When a new stage is reached and the elements of the problem are established in a new order, a more coherent explanation, seemingly acceptable to the majority of the participants in the dialogue, is created. With regard to the behavioral sciences, a field from which we shall be taking our examples in examining the evolution of the sleep-wakefulness problem, a coherent scientific explanation ought to include three things: (1) The sequence of the physiological mechanisms that engender the behavior and its outward manifestations (established biophysically, biochemically, etc.) must be clarified. (2) The position of the given behavior within the general organization, namely, its interrelations with other functions and behaviors, as well as with the physiological subsystems which are integrated in it, must be precisely stated. (3) An acquaintance with its phylogenetic and ontogenetic development should give some hints into its raison d'être, that is, its survival value for the individual and the species. After all, an explanation made within the framework of the behavioral sciences is only a particular case of a biological explanation, and biology is a science in which "to describe a living creature is to consider the logic of its organization as well as that of its evolution. Biology today is concerned with the algorithms of the living world" (Jacob, 1970).

Since it is unusual for a discovery or an effort at understanding to touch all three of these aspects, they develop unevenly. As a matter of fact, it is often difficult to comprehend the reasons that have made an explanation seem acceptable to a majority of the participants. As Craik (1943) said: "It is possible that the meaning of 'explanation' is different for different people; it may be one of those things which no one really understands, but which every scientist, or anyone else in a mood of curiosity, feels he desires. . . . Explanations are not purely subjec-

---

1. Or again in more detail when he wrote:

Whereas according to Popper, falsification in whole or in part is the anticipated fate of all hypotheses, and we should even rejoice in the falsification of an hypothesis that we have cherished as our brain-child. One is thereby relieved from fears and remorse, and science becomes an exhilarating adventure where imagination and vision lead to conceptual developments transcending in generality and range the experimental evidence. The precise formulation of these imaginative insights into hypotheses opens the way to the most rigorous testing by experiment, it being always anticipated that the hypothesis may be falsified and that it will be replaced in whole or in part by another hypothesis of greater explanatory power. In this way conceptual developments lead to experimental testing, which is always designed in relation to hypotheses. The status of a scientific hypothesis is given by the effectiveness with which it challenges rigorous experimental testing and by its explanatory power, which should be far beyond the existent knowledge, i.e. its status may be measured by its predictive scope. (Eccles, 1970)

tive things; they win general approval, or have to be withdrawn in the face of evidence and criticism; and the man who can explain a phenomenon understands it, in the sense that he can predict it, and utilize it more than other men."

The scientific problem's successive stages have, however, a precarious existence. Certain scientists will soon, in a confused manner at first, perceive the internal inconsistencies of the proposed solution. Much as in an embryo, where the conditions for accepting new inputs are fulfilled at each successive epigenetic phase, a newly proposed explanation exposes facets of the problem that were formerly less discernible; these facets, in turn, provide new points of impact for both new ideas and new techniques.

In this way a scientific dialogue is continuously creating new situations and new questions. It is open, and yet we don't understand its wandering and its detours very well. Its most interesting aspects are often lost forever because the authors of a discovery have rationally reordered their data and interpretations before presenting it to their colleagues. Few of us ever recount the intellectual adventure that a discovery or invention has been for us. Once we have become influential members in our fields, we write syntheses in which we logically assemble our new views with the results and theories of our predecessors and contemporaries; and posterity most often comes to know a scientific era through books and articles of this type.

### Discovery and Invention in the Field of Sleep-Wakefulness Cycles

I shall now describe two "discovery-inventions" whose consequences have been of major importance to the advancement of the study of sleep-wakefulness cycles. Subsequently I shall try to analyze the way in which our understanding of the problem has changed as the result of some recently acquired notions in various branches of neurobiology.

In my opinion, an opinion obviously biased by personal preferences, the major discoveries in the field have been, first, that of the ascending reticular activating system by Moruzzi and Magoun in 1949, and, second, that of the REM sleep mechanisms and the role of the central monoaminergic systems in the waking and sleeping mechanisms, clarified by Jouvet in the years between 1960 and 1963. In both instances the authors' inventive minds transformed a casual observation into a discovery.

### The ascending reticular activating system

"Stimulation of the reticular formation of the brain stem evokes changes in the EEG, consisting of abolition of synchronized discharge and introduction of low voltage fast activity in its place, which are not mediated by any of the known ascending or descending paths that traverse the brain stem. The alteration is a generalized one. . . . The reticular response and the arousal reaction to natural stimuli . . . appear identical. . . . The possibility is considered that a background of maintained activity within this ascending brain stem activating system may account for wakefulness, while reduction of its activity either naturally, by barbiturates, or by experimental injury and disease, may respectively precipitate normal sleep, contribute to anesthesia or produce pathological somnolence."

These citations taken from the conclusion of the article by Moruzzi and Magoun (1949) entitled "Brain stem reticular formation and activation of the EEG" summarize a discovery due to a casual observation; the authors, as far as we know, have never told how the idea of studying the mechanisms underlying the waking state came into their minds. (It should be noted in passing that "wakefulness" was not even a physiological problem being considered at the time.) Nevertheless, with the aid of some hindsight, one can reconstruct the story by noting certain aspects of Moruzzi and Magoun's article that appear somewhat odd and by knowing the scientific interests that they were expressing at the time in other publications.

First of all, we are surprised to find in the article's first illustration experiments using chloralose anesthesia. This seems a very poor choice of experimental condition for someone who has the intention of studying arousal reactions. Nor can we see the need for exploring the tectocerebellar nuclei or for citing contemporary research on the inhibiting pathway from the cerebellum to the bulbar reticular formation. If, however, these two oddities are taken as vestiges left by the original experiments, and if we reread the lectures given by Moruzzi on cerebellar physiology upon his arrival at Chicago in the fall of 1948 (Moruzzi, 1950), we can distinguish the aim of the original experiments. In fact, by using chloralose's well-known ability to produce a hyperexcitation of the pyramidal tract, the two scientists were trying to analyze paleocerebellar inhibition of motor-cortex discharges and to determine their pathways. They decided, therefore, to place stimulating electrodes at the level of the cerebellum and the bulbar reticular formation, which they thought to be a possible relay in the pathway; they provoked discharges from the motor cortex by means of strychnine (a method then common), and for keeping track of the occurrence of these cortical strychnine discharges they used EEG recordings. Being skilled experimenters, they started by testing the effects of a bulbar stimulation on the cortex before strychnine application and thus observed the abolishment of the chloralose cortical wave.

It would be very gratifying indeed to know what was said. The dialogue provoked by this observation must have been animated, and it certainly contained the very seeds of invention. The observers surely agreed that the abolishment of the chloralose cortical waves by the stimulation at the brainstem level (which had surely been noticed by others before them) was akin to the observations made by Berger (1929), Adrian (1936), Rheinberger and Jasper (1937), and Bremer (1935, 1937, 1938) of the alpha-wave blockage and cortical desynchronization produced by sensory stimuli. They knew, of course, that they were not stimulating a classic afferent pathway, and they were thus forced to admit the possibility of some unknown pathway that could activate the whole cortical mantle rather than a specific cortical area. They must have also understood the signification and the importance of their observation and, as a result, had sufficient reasons to abandon the research project that had originally brought Moruzzi to Chicago. In addition, they must also have envisioned experiments that would be suitable for proving and developing the newly suggested idea.

Magoun and Moruzzi were especially well prepared for the task. Magoun had been working with Ranson at the Institute of Neurology at Chicago's Northwestern University Medical School since 1932. Together they had written a long paper on the states of hypokinesis, drowsiness, and sleep, which Ranson had observed to follow bilateral injuries to the lateral portion of the posterior hypothalamus (Ranson and Magoun, 1939). They emphasized the existence of a "waking center" in this region of the brain and explained sleep as a passive phenomenon resulting from the suppression of the activity of this structure, a hypothesis proposed by Purkinje as early as 1846 when he wrote about an "Organ des Wachens" (a "wakefulness organ"). On the other hand, on countless occasions Magoun had used the Horsley-Clarke apparatus, which Ranson had reintroduced into neurophysiology around 1930. He was thus well aware that large *areas* of the brainstem, rather than restricted and well-defined structures, command vegetative responses as well as the facilitation or inhibition of large sectors of the musculature. At the time these areas of the brain were still terra incognita and were labeled as "reticular" by anatomists.

Moruzzi, for his part, was a former student of Adrian and of Bremer. He was well acquainted with brain waves (discovered by Berger in 1929 and studied in animals by Adrian) and knew that they were objective signs of the various states of waking and sleeping. Furthermore, he knew how to use the *cerveau isolé* and the *encéphale isolé* preparations in cats, which Bremer had introduced as early as 1935 (see Chapter 15, this volume, for further discussion of the effects of this discovery). These techniques, which are indispensable in this branch of physiology, enable the research worker to record the sleeping and waking electrocorticograms of an immobile preparation susceptible to exploration using stereotaxic methods. It was thanks to these preparations that Bremer had obtained the first experimental proof of the hypothesis describing sleep as the consequence of a sensory deafferentation of the diencephalon and cortex. In one form or another this hypothesis had been coming up in "scientific" writing since Lucretius (see Moruzzi, 1964). As a result of Bremer's work, sleep research took on some new dimensions. He saw sleep as the result of a "withdrawal" from a considerable onrush of afferent messages; the nature of the techniques he used and the importance he attributed to ascending encephalic projections of sensory afferences put an accent on "brain sleep" at a time when all other experiments were turned toward "body sleep." Hess, who at the time was using mesodiencephalic stimulation to study the characteristics of the sleep-inducing mechanism in cats, esteemed that "sleep belongs within the functional scope of the vegetative nervous system, and, indeed, is the endophylactic restitutive component, of which the parasympathetic system is the extracentral mediator. In the context of the entire coordinated effort, the resulting resting state of the animal system is plainly but a means to an end" (translated from Hess, 1948: p. 137).

In the interpretation of his results, Bremer, permeated with the zeitgeist of that time in neurophysiology (see below), utilized terms and concepts distinctly Sherringtonian. The deafferentation of the cortex produced by severing the brainstem diminishes the "cortical tonus" in the same way that severing the dorsal roots suppresses the decerebration rigidity in limbs; hence "the conception of sleep as the consequence of a fall in reflex tonus of the diencephalon and the telencephalon and thus a deafferentation in the wide sense of the term" (Bremer, 1954).

Thus it is clear that the two research workers were technically and intellectually "made-to-order" (so to speak) for easily grasping the implications of their discovery and for immediately performing the experiments suggested by their invention. It should be added that at the same time Lindsley, Bowden, and Magoun (1949), using lesion techniques, judiciously complemented Moruzzi and Magoun's stimulation experiments; the results appeared in the same issue of the journal *Electroencephalography and Clinical Neurophysiology*.

The essential conclusions drawn by Moruzzi and Magoun from their experiments are well known: (1) The arousal reaction to reticular stimulation is identical to the one provoked by sensory stimuli. (2) On their way up the brainstem, sensory messages reach the reticular formation through collateral pathways stemming from the major sensory pathways. (3) An arousal reaction can be obtained from several points along the brainstem, all of these points being situated in the reticular formation as described by anatomists. This set of points constitutes the "ascending reticular activating system." (4) The maintenance of the waking state is dependent upon the integrity and the tonic activity of this system, which is situated in parallel to the known sensory pathways. (5) Sleep is not due to a deafferentation of higher centers by the severing of sensory pathways, but rather to the elimination of the waking influences of the ascending reticular system.

To be able to appreciate the novelty of this discovery and to measure its impact, we need to recall the zeitgeist in neurophysiology at the time of its occurrence. We should then see how it contradicted certain widely held ideas and in how many ways it has changed our conceptions of the functional organization of the central nervous system.

The prevailing conceptions in neurophysiology on the eve of the Second World War can be summed up in the few statements that follow. The quotations accompanying these statements were selected during a rereading of works by great neurophysiologists influential at the turn of the century, of Sherrington's books, of Fulton's *Physiology of the Nervous System*, which was published in 1938, and also of historical texts that discuss those works (Soury, 1899; Granit, 1966; Swazey, 1969; Young, 1970).

1. "In retrospect we can clearly see that the two leading physiologists at Cambridge, Gaskell and Langley, also had in common a major interest in physiological function as reflected in anatomical organization. This is the typical problem of the biologically minded, how organization or form expresses itself in function. . . . He [Sherrington] pinned his hopes to the rapidly advancing science of histology and wanted to perform experiments which made function interpretable in histological terms" (Granit, 1966). Ramón y Cajal and Golgi shared the Nobel Prize in 1906, the same year in which *The Integrative Action of the Nervous System* was published.

2. "The physiologist tends to reduce this vast organ [the central nervous system as a whole] to its ultimate anatomical units; or to such groups of units as are involved in the conventional reflex or in certain levels of functional activity. Having subdivided the nervous system in this manner, he should be in a better position to build it up again into an integrated whole than those who have approached the problem more philosophically in the normal intact animal." The Sherringtonian method, summarized in these few lines by Fulton (1938), actually has two features: (1) the determination of the elementary mechanisms of neuron-to-neuron transmissions in the reflex arc (a field in which extraordinary progress has been made thanks mainly to research into synaptic transmission and the neuronal networks using electrophysiological methods at the single-neuron level); (2) the integrative perspective that seeks to determine how these elementary mechanisms operate within definite contexts (as exemplified by the statement that "biologically the importance of a reflex is as an item of behaviour; hence biological study presents for each reflex the issue of its meaning as an animal act" (Creed et al., 1932).

3. The notion of the localization of functions held that specific and well-defined structures at the central-nervous-system level corresponded to each function. The organization of the sensory and motor systems, as outlined first in anatomy and later in electrophysiology by thalamic and cortical charting, and the localizations attributed to vegetative centers are all examples of this conception, which implied that the whole system was more or less a juxtaposition of all the specific systems.

4. The notion of a hierarchical organization within the nervous system is illustrated by the idea of levels of function: "[the] evolutionary principle of levels of function . . . implies that headward segments of the brain have become dominant over caudal, and that when higher parts are removed, many activities of lower segments are, after a time, 'released' and can be more readily analysed. . . . [One] should remember however that in man, owing to gradual 'encephalization,' a greater number of functions have been taken over by the central cortex than in the mammals" (Fulton, 1938). This notion of a hierarchical organization of organisms, and in particular of the nervous system, is obviously derived from a general hierarchical concept deeply rooted in the human mind, and it appears to have both an introspective and a social origin. Descartes's description of the organization of the nervous system was, of course, in accordance with his distinction between animal movements and human behavior; the nervous system was made up of a machinery ("the arrangement of the organs") that was capable of producing all the movements observed in an animal; in man, however, this machinery was dominated by a soul (or a spirit) capable of commanding it in a "reasonable" and "critical" (with discrimination) way.[2] Moreover, one has only to read the papers of Goltz (Sherrington's first undertaking was to study the degenerations

---

2. "The movements which we do experience as not depending on our mind [reflexes] should not be attributed to the soul, but only to the disposition [arrangement] of the organs; and that even the movements, which are described as *voluntary*, result mainly from this disposition of the organs, without which they could not take place, whatever our will of them, *even though it is the soul which decides them*" (translated from Descartes, 1664; italics ours).

following the telencephalic excisions performed by Goltz) in order to realize that his conception of the organization of nervous functions closely copied the organization of Prussia under Bismarck. The notions of "levels of function" and of "levels of integration" are, if in spirit only, notions of Spencer's era, even though they have not been described in these terms (see Needham, 1943).

These conceptions of the functional organization of the central nervous system have certainly been modified by Moruzzi and Magoun's discovery and by the extensive experimental research that followed (see Magoun, 1950, 1963).

1. In addition to the main sensory and motor systems, which provide the basis for our relationship with the outside world, we now know that the nervous system includes certain internal systems. One highly developed such system is located at the level of the brainstem reticular formation. Unlike the sensory and motor systems, which are equipped with a very precise organization and are responsible for receiving and sending specific and well-defined signals, the brainstem reticular system triggers no one certain effect in the somatic sphere. Rather, it facilitates or inhibits various synaptic relays within the specific systems; it plays the role of an "activator" responsible for the excitatory state in the sensory and motor systems, and in a more general sense, it is a control system.

2. Sensory messages of all sorts (exteroceptive and interoceptive), messages already elaborated by the cerebral cortex and messages sent out by the motor command structures, all converge upon this reticular system. The information provided by this barrage of messages received at various times is sorted out and integrated while passing through the reticular network, which acts as a transactional system (Livingston, Haugen, and Brookhart, 1954). At the same time these messages sustain the excitatory state of the reticular system and produce the different levels of reticular activity, which can be measured by their effects on the specific systems.

3. The reticular effects, both tonic and phasic, are simultaneously transmitted by ascending and descending projections in the direction of the thalamus and the cortex as well as the spinal cord. Histology shows us that most reticular-neuron axons, being bifurcated, have an oral and a caudal branch (Scheibel and Scheibel, 1958). This arrangement explains how all somatic activities (sensory and motor) and all vegetative activities energized by the reticular formation can be simultaneously mobilized; the coordination and fitness in action of an organism are grounded on this internal system of the nervous machinery.

4. The organization of the nervous system in successive horizontal stages, described as the levels of integration in the classical conception, is complemented by a vertical organization that is probably even more important to the complete integration of the organism. The reticular brainstem system and its ascending or descending projections are the essential elements in this vertical organization.

5. These new views in the field of neurophysiology soon crossed over into psychophysiology. Lindsley (1951) proposed an "activation theory of emotion," and later Schlosberg (1954) looked upon the series of states leading from sleep to the emotional state as the result of a continuous series of reticular activity levels. It was Hebb (1955) who finally expanded this conception to include the sensory

functions: the effectiveness of the cue functions that govern behavior depends on the organism's level of vigilance.

### The pontine pacemaker of REM sleep and the role of monoaminergic systems in the sleep-wakefulness cycle

A posteriori, the discovery of the mechanisms of REM sleep and that of the role of monoaminergic systems in the sleep-wakefulness cycle would each seem to be the logical consequence of the other. And yet the former originated from a casual observation and the latter from a hypothesis formulated while reading an anatomical document that appeared in the literature.

Jouvet himself (1973a) has told us of the unexpected observation that put him on the track of the REM sleep mechanisms. Back in Lyon in 1957 after his stay at the Brain Research Institute that Magoun had founded at UCLA, he was keenly aware of the antinomy that existed between Pavlovian conceptions, in which the cerebral cortex played a vital role in every aspect of conditioning, and Western conceptions, Magoun's in particular, which emphasized the integrative role of the reticular formations of the brainstem. In order to localize the structures that are indispensable for the habituation of Pavlov's orientation reaction to occur, Jouvet and Michel undertook a study of chronic pontine and mesencephalic preparations. To their surprise, they observed, every 30 to 40 minutes, the sudden occurrence of 5- to 6-minute periods of complete muscular atony, without intervention of any external stimuli. These periods were characterized by repeated large-scale electrical discharges in the area of the pontine tegmentum (at that time these discharges were described as "spindlelike," but later they came to be known as the ponto-geniculo-occipital (PGO) discharges of REM sleep) and by jerking movements of the cat's whiskers.

Others had certainly already observed the same phases of muscular atony since this type of preparation had been frequently used before the war; they had, however, not been inventive enough to recognize their significance. Jouvet hypothesized that these phases in preparations with severed brainstem were synonymous with the REM sleep phases described a few years before in humans by Dement and Kleitman (1957) and in cats by Dement (1958). Two experiments proved his hypothesis to be exact: (1) In cats with an intact encephalon, the phase of muscular atony and the pontine discharges coincided with the cortical desynchronization, the jerky eye movements, and the twitching movements of the limbs, whiskers, and ears that Dement had pointed out as characteristic of REM sleep. (2) After undergoing lesions localized to the dorsolateral part of the pontine tegmentum (the location of the pontine discharges as described above), the cat continued to present all the signs of slow-wave sleep, but the periods of muscular atony, cortical activation, PGO, and saccadic movements of the eyes completely disappeared. Thus Jouvet had discovered the pontine pacemaker, which, flashing on and off at set intervals during sleep, initiates the special phases, known as REM sleep, during which dreaming occurs.

Jouvet's second discovery, involving the role of central monoaminergic systems in the mechanisms of the sleep-wakefulness cycle, emerged from a hypothesis.

Certain Japanese laboratories had perfected histochemical techniques for the detection of monoamine oxidase (MAO); in one of their publications (Hashimoto et al., 1962), the locus coeruleus was indicated as a structure particularly rich in MAO. This localization almost coincided with Jouvet's localization for the pontine pacemaker of REM sleep. He immediately understood the potential of this coincidence because his mind had been fully prepared by his past experience to envisage the possible role of monoaminergic mechanisms in sleep. He had been astonished when he had noted the unusually long length of time necessary for the recuperation of the pontine pacemaker after each phase of activity; he had noticed an increase in REM sleep after REM-sleep deprivation; finally he had studied the effects of various drugs (eserine, atropine, reserpine) on this mechanism. In 1960 his report at a Ciba Foundation Colloquium concluded: "Finally, the REM-sleep phase can be triggered off in animals by triggering the lower part of the brainstem, and it is suggested that this phase depends upon a neurohumoral mechanism" (Jouvet, 1961).

In the light of some additional information uncovered at that time, Jouvet was quickly able to verify his hypothesis. Dahlström and Fuxe (1964, 1965) had just published the cartography of the central monoaminergic systems that they had obtained using the histofluorescence method developed by Falck and Hillarp (Falck et al., 1962). Moreover, pharmacological studies had produced a series of drugs that could be used to block selectively the various stages in the synthesis and the destruction of monoamines. In a long series of experiments, Jouvet and his collaborators established, first, the intervention of the serotoninergic systems in the mechanisms of slow-wave sleep; second, the importance of the noradrenergic systems and especially the locus coeruleus in setting off the pontine pacemaker and in initiating the different signs of REM sleep; and third, the part played by the noradrenergic system in sustaining the waking state (Jouvet, 1967, 1972).

These kinds of experiments in the field of the sleep-wakefulness cycle are good examples of the significant changes that have taken place in the state of mind of the scientists involved in brain research. Ever since the 1920s the field had been dominated by the electrophysiological techniques introduced by Erlanger, Gasser, and Adrian. The effectiveness of these methods has been increased by more recent techniques of unit and intracellular recordings. The renewal of neuro-anatomical techniques (electron-microscopic, histochemical, and autoradiographic) and advances in the fields of neurochemistry and neuropharmacology have considerably enriched our perspective of the intimate arrangement of the elements of the nervous system and have brought us to the very doorstep of molecular biology.

**Understanding and Explaining the Sleep-Waking Cycle**

It would exceed the limits of the present essay to retrace the very long dialogue of our endeavors to understand the sleep-waking cycle or to identify its different stages. Several histories of "sleep theories" have been written (Piéron, 1913;

Kleitman, 1963; Moruzzi, 1964, 1972). On the other hand, a brief summary of the present-day explanations is in order and will permit us to point out that, probably for the first time, possibly satisfactory answers are being provided for the three main questions that necessarily arise when one is attempting to explain a behavioral process (see above). (1) Notable progress has been made in the comprehension of the mechanisms that produce the sleep-waking cycle. The interested reader can consult recent general reviews on the subject (Moruzzi, 1972; Jouvet, 1972). (2) Pertinent arguments have been offered to demonstrate how sleep is integrated within the complex of instinctive behavioral processes. (3) Some very original hypotheses on the raison d'être of sleep and especially of REM sleep and dreams have just been proposed.

Through incorporation of notions recently acquired in fields as varied as molecular biology and ethology, the phenomenon of sleep, traditionally considered as a physiological function, has been relocated within the much larger framework of neurobiology. We are now beginning to perceive its primordial role for the survival of the higher organisms.

## Sleep and instinctive behavior

By combining notions borrowed from ethology, reticular physiology, and psychophysiology, Moruzzi (1969, 1972) has recently suggested a series of arguments that permit the integration of sleep into the context of instinctive behavioral processes:

1. Sleep can be described along the lines of a classical ethological model. It is preceded by an "appetitive phase" characterized by various preparations for sleep. In certain cases there is a search for a suitable place (the "sleep trees" of birds, the sleep platforms built by orang-utans, etc.); in cats a dozing position is adopted, while humans lose interest in the outside world. The "consummatory phase" of deep sleep, which occurs only after an intermediary stage of dozing off and light sleep, corresponds to the stages of deep slow-wave sleep and REM sleep.

2. The various instinctive behavior patterns observed in animals during the waking state require a certain amount of reticular activity which facilitates the functioning of the sensory, motor, and sympathetic systems supporting the instinctive behavior. It can be said, therefore, that these behaviors depend on the interplay of a reticular activation and specific elements of this behavior, in other words, on "the accumulation of endogenous activity within a specific center (the motivation) combined with a proper environmental situation." In making this statement, Moruzzi is obviously drawing upon notions borrowed from the theories developed by Lindsley (1951), Schlosberg (1954), and Hebb (1955) on the "activation continuum" and from psychophysiological findings.

3. Just as the different types of instinctive behavior demand a higher or lower level of reticular activity, sleep behavior corresponds to a low but critical level of reticular activation. The attainment of this level by either an active or passive reticular deactivation (Moruzzi and Magoun, 1949; Bremer, 1954; Dell, Bonvallet, and Hugelin, 1961; Dell, 1963) is a necessary condition for sleep. Once attained, a state of "nonwakefulness" appears; this state is accompanied by a loss of consciousness and a release of the thalamocortical synchronization system. The

reticular-formation deactivation is a necessary preliminary condition for the hypnogenetic (neurogenetic and monoaminergic) mechanisms to come into play.

It is evident that Moruzzi's proposals do not solve the entire problem. As is done in regard to all instinctive behavioral processes, one should ask what type of mechanism governs the regular occurrence of sleep. At present we must be satisfied with hypothesis. It could involve, for example, the accumulation of metabolites in the monoaminergic systems and their subsequent destruction (Jouvet, 1972); or, perhaps, a "timing device" (biological clock) for the waking-sleeping cycle that is activity-modulated by all sorts of external and internal factors (Berlucchi, 1970). It could also be a timing device that is "absolute" and entirely free of feedbacks, having only one-way relationships with the mechanisms initiating the various primary needs, including wakefulness and sleep (Richter, 1967).

**The raison d'être of sleep**

Recent very interesting speculations, supported by experimental evidence, suggest that REM sleep plays a vital role in proper brain functioning. Two particular features of REM sleep have guided some researchers toward these new lines of investigation. First, REM sleep is unquestionably a state in which a high amount of brain activity occurs. This is confirmed by the cortical desynchronization (which is almost the same as that occurring during attentive wakefulness) and by the repetitive high-amplitude discharges labeled the ponto-geniculo-occipital (PGO) spikes. Second, it has been observed that infants spend the greatest part of their first weeks of postnatal life in REM sleep (which, as a matter of fact, takes place already in the fetus), and that thereafter the amount of REM sleep steadily decreases until adulthood.

From these points one derives the hypothesis that, during the fetal and postnatal development of the central nervous system, periods in which few external stimuli touch the brain, the PGO discharges have a role similar to that of external stimuli for the myelinization, maturation, and differentiation of the higher nervous centers. In fact, two different versions of this hypothesis have been proposed. (1) Roffwarg, Muzio, and Dement (1966) suggest that the PGO spikes (which are not, in fact, limited to the visual system) "assist in structural maturation and differentiation of key sensory and motor areas within the CNS, partially preparing them to handle the enormous rush of stimulation provided by the postnatal milieu." (2) Following suggestions originally made by Jacobson (1969), Jouvet (1972, 1973b) remarks that two types of neurons appear to exist within the central nervous system: on the one hand, neurons that develop according to a strict genetic program, such as those on certain neuronal pathways in the visual system which subsequently require epigenetic stimulation from the outside in order to function; on the other hand, neurons whose development occurs later during maturation, and whose specifications and connections can be modified. It is on this second type of neuron that the PGO discharges triggered by the pontine pacemaker act during REM sleep and dreams.

Dreams would thus represent, in the fetus or during sleep, a rehearsal of the numerous motor and integrative mechanisms that will sustain the innate behavior patterns of the waking state. . . . We interpret REM sleep as the organizer and programmer- of the innate behavior patterns that occur at each stage of the individual's development: for example, postnatal behavior (the search for food), the singing repertoire of certain birds, aggressive behavior, the marking out and defending of a territory, sexual behavior, and finally the whole set of hereditary character traits that constitute an individual's personality. In this hypothesis, REM sleep prepares, organizes, and programs the motor sequences inherent to each stage of development so that they will be ready when the conditions of the external and internal milieu are suitable. In short, REM sleep is to the motor organization of the innate behavior patterns what epigenetic events are to the maturation of the sensory systems. (Translated from Jouvet, 1973b)

Moreover, although REM sleep occurs during only a fraction of the time that an individual consecrates to sleep, it is a genuine need. After deprivation of REM sleep, the accumulated debt is compensated for by a rebound of REM sleep; the duration of REM sleep also depends on the duration of the waking period that precedes sleep. One of the most attractive hypotheses provoked by the discovery of REM sleep considers this stage of sleep as being complementary to the waking state and indispensable for the proper functioning of certain brain processes taking place during wakefulness (see, for example, Feinberg and Evarts, 1969; Hennevin and Leconte, 1971). While in the waking state, the brain is bombarded by an endless variety of stimuli, many of which are unimportant, while a few are meaningful and significant. They are all stored away and, during REM sleep, a sorting out and a restructuring of this material takes place: consolidation of engrams, the selective forgetting of nonmeaningful material, a reorganization of the stored data, and a combination of the new elements with similar ones already stored in the memory. Saint-John Perse expressed the same idea in a more poetic way when he wrote: "Aux idées pures du matin que savons-nous du songe, notre ainesse?."

The same hypothesis was formulated by Jouvet (1972), but under a different form, when he suggested that during REM sleep and dreams the phenotypic (epigenetic) information stored during the conscious waking period in the cerebral cortex is combined with the pontine pacemaker's genotypically coded discharges of REM sleep. From this combination results a new code that serves to organize acquired phenotypic schemes of activity.

The concept shared by these hypotheses is that of an inner brain system that only performs during sleep. Its role is to act upon the structures sustaining the most elaborate processes of the waking state in such a way that only that part of the "life experience" that is essential for confronting forthcoming experiences is retained. The importance of such a system to the survival of the individual is obvious.

Finally, it must be noted that it is thanks to F. O. Schmitt that these recent investigations on wakefulness and sleep were begun, for it was he who convinced the scientists interested in brain research that if they wished to accomplish their task, they should seek aid from all related fields of study, notably from genetics, immunology, and molecular biology. By establishing the Neurosciences Research Program in 1962, he assumed the leadership of this movement, and the *Ortgeist* that arose at MIT became the *Zeitgeist*. And thus one day we shall be able to describe and explain the evolutionary dialogue from which the human brain emerged, the dialogue of its maturation and of the development of its faculties through contact with the physical and sociocultural environment, and, finally, the dialogue which we as neurobiologists pursue in order to decipher the enigma of our own brains; and on that day we will be able to write "the natural history of the human mind."

## Acknowledgments

I give warmest appreciation to my friend Professor Walter Rosenblith, who unstintingly gave both of his time and his energy during an entire weekend to aid me in translating and clarifying the text of my oral presentation. I also wish to thank Lawrence Lockwood and Andy Corsini, who undertook the translation of the present text.

## References

Adrian, E. D. (1936): The spread of activity in the cerebrate cortex. *J. Physiol.* 88:127–161.

Berger, H. (1929): Über das Elektrenkephalogramm des Menschen. *Arch. Psychiatr. Nervenkr.* 87: 527–570.

Berlucchi, G. (1970): Mechanismen von Schlafen und Wachen. *In: Ermüdung, Schlaf und Traum.* Baust, W., ed. Stuttgart: Wissenschaftliche Verlags GMBH, pp. 145–203.

Bremer, F. (1935): Cerveau isolé et physiologie du sommeil. *C. R. Soc. Biol. (Paris)* 118:1235–1242.

Bremer, F. (1937): L'activité cérébrale au cours du sommeil et de la narcose. Contribution à l'étude du mécanisme du sommeil. *Bull. Acad. Med. Belg.* 4: 68–86.

Bremer, F. (1938): L'activité électrique de l'écorce cérébrale et le problème physiologique du sommeil. *Boll. Soc. Ital. Biol. Sper.* 13: 271–290.

Bremer, F. (1954): The neurophysiological problem of sleep. *In: Brain Mechanisms and Consciousness.* Adrian, E. D., Bremer, F., and Jasper, H. H., eds. Oxford: Blackwell, pp. 137–162.

Chomsky, N. (1968): *Language and Mind.* New York: Harcourt, Brace & World.

Craik, K. J. W. (1943): *The Nature of Explanation.* Cambridge, Eng.: Cambridge University Press.

Creed, R. S., Denny-Brown, D., Eccles, J. C., Liddell, E. G. T., and Sherrington, C. S. (1932): *Reflex Activity of the Spinal Cord.* London: Oxford University Press.

Dahlström, A., and Fuxe, K. (1964): Evidence for the existence of monoamine-containing neurons in the central nervous system. I. Demonstration of monoamines in the cell bodies of brain stem neurons. *Acta Physiol. Scand.* (Suppl.)232: 1–55.

Dahlström, A., and Fuxe, K. (1965): Evidence for the existence of monoamine-containing neurons in the central nervous system. II. Experimentally induced changes in the intraneuronal amine levels of bulbospinal neuron systems. *Acta Physiol. Scand.* (Suppl.) 247: 1–36.

Dell, P. (1963): Reticular homeostasis and critical reactivity. *In: Brain Mechanisms. Progress in Brain Research.* Vol. 1. Moruzzi, G., Fessard, A., and Jasper H. H., eds. Amsterdam: Elsevier, pp. 82–103.

Dell, P., Bonvallet, M., and Hugelin, A. (1961): Mechanism of reticular deactivation. *In: The Nature of Sleep* (A Ciba Foundation symposium). Wolstenholme, G. E. W., and O'Connor, C. M., eds. London: Churchill, pp. 86–107.

Dement, W. (1958): The occurrence of low voltage, fast, electroencephalogram patterns during behavioral sleep in the cat. *Electroencephalogr. Clin. Neurophysiol.* 10: 291–296.

Dement, W., and Kleitman, N. (1957): Cyclic variations in EEG during sleep and their relation to eye movements, body motility and dreaming. *Electroencephalogr. Clin. Neurophysiol.* 9: 673–690.

Descartes, R. (1664): *Description du Corps Humain. In: Oeuvres Complètes* (édition Adam & Tannery). Vol. XI. Paris: Cerf (1897–1910), p. 225.

Duvignaud, J. (1973): *Fêtes et Civilisations.* Paris: Librairie Weber.

Eccles, J. C. (1963): Biographical note (written on the occasion of receiving the Nobel Prize). Quoted from Popper (1969).

Eccles, J. C. (1970): *Facing Reality.* New York: Springer-Verlag.

Falck, B., Hillarp, N. A., Thieme, G., and Torp, A. (1962): Fluorescence of catecholamines and related compounds condensed with formaldehyde. *J. Histochem. Cytochem.* 10: 348–354.

Feinberg, I., and Evarts, E. V. (1969): Changing concepts of the function of sleep: Discovery of intense brain activity during sleep calls for revision of hypotheses as to its function. *Biol. Psychiatry* 1: 331–348.

Fulton, J. F. (1938): *Physiology of the Nervous System.* London: Oxford University Press.

Granit, R. (1966): *Charles Scott Sherrington. An Appraisal.* London: Thomas Nelson and Sons.

Granit, R. (1972): Discovery and understanding. *Annu. Rev. Physiol.* 34: 1–12.

Hadamard, J. (1954): *The Psychology of Invention in the Mathematical Field.* New York: Dover.

Hashimoto, P. H., Maeda, T., Toru, K., and Shimizu, N. (1962): Histochemical demonstration of autonomic regions in the central nervous system of the rabbit by means of a monoamine oxydase staining. *Med. J. Osaka Univ.* 12: 425–464.

Hebb, D. O. (1955): Drives and the C. N. S. (conceptual nervous system). *Psychol. Rev.* 62: 243–254.

Hennevin, E., and Leconte, P. (1971): La fonction du sommeil paradoxal: Faits et hypothèses. *Ann. Psychol.* 2: 489–519.

Hess, W. R. (1948): *Die funktionnelle Organisation des vegetativen Nervensystems.* Basel: Benno Schwabe.

Jacob, F. (1970): *La Logique du Vivant.* Paris: Gallimard.

Jacobson, M. (1969): Development of specific neuronal connections. *Science* 163: 543–547.

Jouvet, M. (1961): Telencephalic and rhombencephalic sleep in the cat. *In: The Nature of Sleep* (A Ciba Foundation symposium). Wolstenholme, G. E. W., and O'Connor, C. M., eds. London: Churchill, pp. 188–208.

Jouvet, M. (1967): Mechanisms of the states of sleep: A neuropharmacological approach. *In: Sleep and Altered States of Consciousness.* Kety, S. S., Evarts, E. V., and Williams, H. L., eds. Baltimore: Williams & Wilkins, pp. 86–126.

Jouvet, M. (1972): The role of monoamines and acetylcholine-containing neurons in the regulation of the sleep-waking cycle. *Ergeb. Physiol.* 64: 166–307.

Jouvet, M. (1973a): Telencephalic and rhombencephalic sleep in the cat. *In: Sleep and Active Process.* Webb, W. B., ed. Glenview, Ill.: Scott, Foresmen, pp. 12–32.

Jouvet, M. (1973b): Essai sur le rêve. *Arch. Ital. Biol.* 111: 564–576.

Kleitman, N. (1963): *Sleep and Wakefulness.* 2nd ed. Chicago: The University of Chicago Press.

Lindsley, D. B. (1951): Emotion. *In: Handbook of Experimental Psychology.* Stevens, S. S., ed. New York: John Wiley, pp. 473–516.

Lindsley, D. B., Bowden, J. W., and Magoun, H. W. (1949): Effect upon the EEG of acute injury to the brain stem activating system. *Electroencephalogr. Clin. Neurophysiol.* 1: 475–486.

Livingston, W. K., Haugen, F. P., and Brookhart, J. M. (1954): Functional organization of the central nervous system. *Neurology* 4: 485–496.

Magoun, H. W. (1950): Caudal and cephalic influences of the brain stem reticular formation. *Physiol. Rev.* 30: 459–474.

Magoun, H. W. (1963): *The Waking Brain.* 2nd ed. Springfield, Ill.: C. C Thomas.

Medawar, P. B. (1967): *The Art of the Soluble.* London: Methuen.

Monod, J. (1970): *Le Hasard et la Nécessité. Essai sur la Philosophie Naturelle de la Biologie Moderne.* Paris: Editions du Seuil.

Moruzzi, G. (1950): *Problems in Cerebellar Physiology.* Springfield, Ill.: C. C Thomas.

Moruzzi, G. (1964): The historical development of the deafferentation hypothesis of sleep. *Proc. Am. Phil. Soc.* 108: 19–28.

Moruzzi, G. (1969): Sleep and instinctive behavior. *Arch. Ital. Biol.* 107: 175–216.

Moruzzi, G. (1972): The sleep-waking cycle. *Ergeb. Physiol.* 64: 1–165.

Moruzzi, G., and Magoun, H. W. (1949): Brain stem reticular formation and activation of the EEG. *Electroencephalogr. Clin. Neurophysiol.* 1: 455–473.

Needham, J. (1943): *Time: The Refreshing River.* London: George Allen & Unwin.

Piéron, H. (1913): *Le Problème Physiologique du Sommeil.* Paris: Masson & Cie.

Popper, K. R. (1959): *The Logic of Scientific Discovery.* London: Hutchinson.

Popper, K. R. (1963): Science: Problems, aims, responsibilities. *Fed. Proc.* 22: 961–972.

Popper, K. R. (1969): *Conjectures and Refutations. The Growth of Scientific Knowledge.* 3rd ed. London: Routledge and Kegan Paul.

Purkinje, J. E. (1846): Wachen, Schlaf, Traum und verwandte Zustände. *In: Handwörterbuch der Physiologie.* Section 9, part 3, volume 2. Wagner, R., ed. Braunschweig: Viewag & Sohn, pp. 412–480.

Ranson, S. W., and Magoun, H. W. (1939): The hypothalamus. *Ergeb. Physiol.* 41: 56–163.

Rheinberger, M. B., and Jasper, H. H. (1937): Electrical activity of the cerebral cortex in the unanesthetized cat. *Am. J. Physiol.* 119: 186–196.

Richter, C. P. (1967): Sleep and activity: Their relation to the 24-hour clock. *In: Sleep and Altered States of Consciousness.* Kety, S. S., Evarts, E. V., and Williams, H. L., eds. Baltimore: Williams & Wilkins, pp. 8–29.

Roffwarg, H. P., Muzio, J. N., and Dement, W. C. (1966): Ontogenetic development of the human sleep-dream cycle. The prime role of "dreaming sleep" in early life may be in the development of the central nervous system. *Science* 152: 604–619.

Scheibel, M. E., and Scheibel, A. B. (1958): Structural substrates for Integrative patterns in the brain stem reticular core. *In: Reticular Formation of the Brain* (Henry Ford Hospital International Symposium). Jasper, H. H., et al., eds. Boston: Little, Brown, pp. 31–55.

Schlosberg, H. (1954): Three dimensions of emotion. *Psychol. Rev.* 61: 81–88.

Soury, J. (1899): *Le Système Nerveux Central. Histoire Critique des Théories et des Doctrines.* Carré, G., and Naud, C., eds. Paris.

Swazey, J. P. (1969): *Reflexes and Motor Integration. Sherrington's Concept of Integrative Action.* Cambridge, Mass.: Harvard University Press.

Young, R. M. (1970): *Mind, Brain and Adaptation in the Nineteenth Century.* Oxford: Clarendon Press.

# 31
# On the Nature of Research in Neuroscience

Judith P. Swazey and Frederic G. Worden

## Introduction

What is the nature of the creative process in scientific research, and how does it engender discovery and the invention of new scientific understanding? This book provides an opportunity to clarify these questions in the descriptions it presents of the complex transactional processes that affect scientists as persons, as participants in social and organizational groups, and as thinkers whose intellectual and emotional energies are heavily invested in responding to new observations, technological advances, and new conceptual models of reality.

Many different approaches could be taken in analyzing the material in this book. We have chosen to see how well the participants' accounts of research and discovery fit the concepts Thomas Kuhn has developed about the nature of science. This endeavor appeals not just because Kuhn's concepts have been both highly influential and controversial (see Lakatos and Musgrave, 1970, and Scheffler, 1967, for examples of the controversy over Kuhn's work), but also because they were developed largely from studies of the physical sciences. As the most challenging and enigmatic of the life sciences, neuroscience poses an extreme test of whether Kuhn's ideas can be usefully generalized from physical to biological science.

## Traditional and Kuhnian Views of Science

For readers unfamiliar with Kuhn's theses, a brief survey will be useful for the purpose of this essay. For a detailed explication, see *The Structure of Scientific Revolutions* (Kuhn, 1970).

A central feature of Kuhn's model is the distinction he draws between "normal science," which articulates and adjusts the paradigm shared by its scientists, and "revolutionary science," which invents new paradigms the acceptance of which overthrows the older paradigm. This proposition is controversial in part because it contradicts the traditional view that science progresses incrementally, by adding each discovery, like a new "brick," to the preceding bricks in the slowly rising edifice of "truth."

As Ragnar Granit has noted, this traditional image of science, in which major discoveries play starring roles, is mirrored in one of the most prestigious symbols of scientific achievement, the Nobel Prize:

The young scientist often seems to share with the layman the view that scientific progress can be looked upon as one long string of pearls made up of bright discoveries. This standpoint is reflected in the will of Alfred Nobel, whose mind was that of an inventor, always loaded with good ideas for application. His great Awards in science presuppose definable discoveries. (Granit, 1972: p. 3)

In traditional formulations, science is an "objective" endeavor that, as historian of science C. C. Gillispie (1960) argues, uses the cutting "edge of objectivity" to separate truth from error. Traditional accounts of the nature of science (for example, those perpetuated in textbooks) also place strong emphasis on the experimental character of science, as well as on the scientist's skepticism about dogma: "There is only one established dogma in science—that scientists do not blindly accept established dogma" (Brush, 1974).

Tied closely to the canons of eschewing dogma and utilizing experiment, a third emphasis in the traditional view of science is on the importance of the hypotheticodeductive approach. S. G. Brush summarizes this tenet as follows:

Daniel Bell writes that "as part of a general education all students should be aware of the nature of . . . hypothetical-deductive thought" as it has been developed in science. Bell, a sociologist, quotes biophysicist John Platt's statement that some fields in science move more rapidly than others, in part because "a particular method of doing scientific research is systematically used and taught." This method relies on "devising alternative hypotheses for any problem, devising crucial experiments, each of which would, as nearly as possible, exclude one or more hypotheses." (Brush, 1974: p. 1165)

Views of science contravening this traditional image have long been proffered by philosophers, historians, and by scientists themselves. The following four examples of such "unorthodox admissions" by scientists suggest why Kuhn felt the need to develop a detailed alternative interpretation of the nature of science.

1. *Albert Einstein* (replying in 1926 to Heisenberg's statement that only observable magnitudes must be used in formulating a theory like that of relativity): "Possibly I did use this kind of reasoning, but it is nonsense all the same. Perhaps I could put it more diplomatically by saying that it may be heuristically useful to keep in mind what one has observed. But on principle, it is quite wrong to try founding a theory on observable magnitudes alone. In reality the very opposite happens. It is the theory which decides what we can observe" (quoted in Brush, 1974: p. 1167).

2. *Thomas Henry Huxley:* "It is a popular delusion that the scientific enquirer is under an obligation not to go beyond generalization of observed facts . . . but anyone who is practically acquainted with scientific work is aware that those who refuse to go beyond the facts, rarely get far" (quoted in Beveridge, 1957: p. 200).

3. *John Szentágothai:* "I still suffer from the misconception that it is the interpretation that counts and not whether the finding is obvious and beautifully demonstrated, or is barely recognizable" (this volume, p. 110).

4. *John C. Eccles:* "I experienced a great liberation in escaping from the rigid conventions that are generally held with respect to scientific research. Until 1944 I held the following conventional ideas . . . . : First, that hypotheses grow out of the careful and methodical collection of experimental data. . . . Second, that the excellence of a scientist can be judged by the reliability of his developed

hypotheses. . . . Finally . . . that it is in the highest degree regrettable and a sign of failure if a scientist espouses an hypothesis that is falsified by new data so that it has to be scrapped altogether. When one is liberated from these restrictive dogmas, scientific investigation becomes an exciting adventure opening up new visions" (this volume, p. 162).

Kuhn's major theses are presented schematically in Figure 31.1. They will be discussed further in the following sections.

## Paradigm and Disciplinary Matrix

Kuhn's central concept, "paradigm," is one that has troubled many of his readers, in part because they have difficulty distinguishing the differences between paradigms and more familiar constructs such as concepts, laws, or theories. Additional confusion arises because, in his original text, Kuhn used the term paradigm in several different senses, some of which were indeed equivalent to law, concept, or theory.

One of the major senses in which Kuhn originally used paradigm was sociological, to denote "the entire constellation of beliefs, values, techniques and so on shared by the members of a given [scientific] community." Second, a paradigm denotes "one sort of element in that constellation, the concrete puzzle-solutions which, employed as models or examples, can replace explicit rules as a basis for the solution of the remaining puzzles of normal science" (Kuhn, 1970: p. 175). For Kuhn, this second usage is "philosophically . . . the deeper of the two, and the claims I have made in its name are the main sources for the controversies and misunderstandings that the book has evoked" (p. 175).

In discussing paradigms, then, Kuhn is talking principally about "exemplars of past achievements," which he defines as

the concrete problem-solutions that students encounter from the start of their scientific education, whether in laboratories, on examinations, or at the ends of chapters in science texts. To these shared examples should, however, be added at least some of the technical problem-solutions found in the periodical literature that scientists encounter during their post-educational research careers and that also show them by example how their job is to be done. More than other sorts of components of the disciplinary matrix, differences between sets of exemplars provide the community fine-structure of science. (p. 187)

Kuhn further seeks to clarify his use of "paradigm" by asking, "What do [members of a scientific community] share that accounts for the relative fulness of their professional communications and the relative unanimity of their professional judgments?" (p. 182). His original text, in its sociological use of the term noted above, gave the answer "a paradigm or set of paradigms." But, more narrowly defined as exemplars, paradigm becomes an "inappropriate" answer. "Scientists," Kuhn continues, "would say they share a theory or set of theories, [but] as currently used in philosophy of science, . . . 'theory' connotes a structure far more limited in nature and scope than the one required here" (p. 182). Thus Kuhn proposes that members of a scientific community share a "disciplinary matrix," and he identifies four major components of such a matrix: symbolic generaliza-

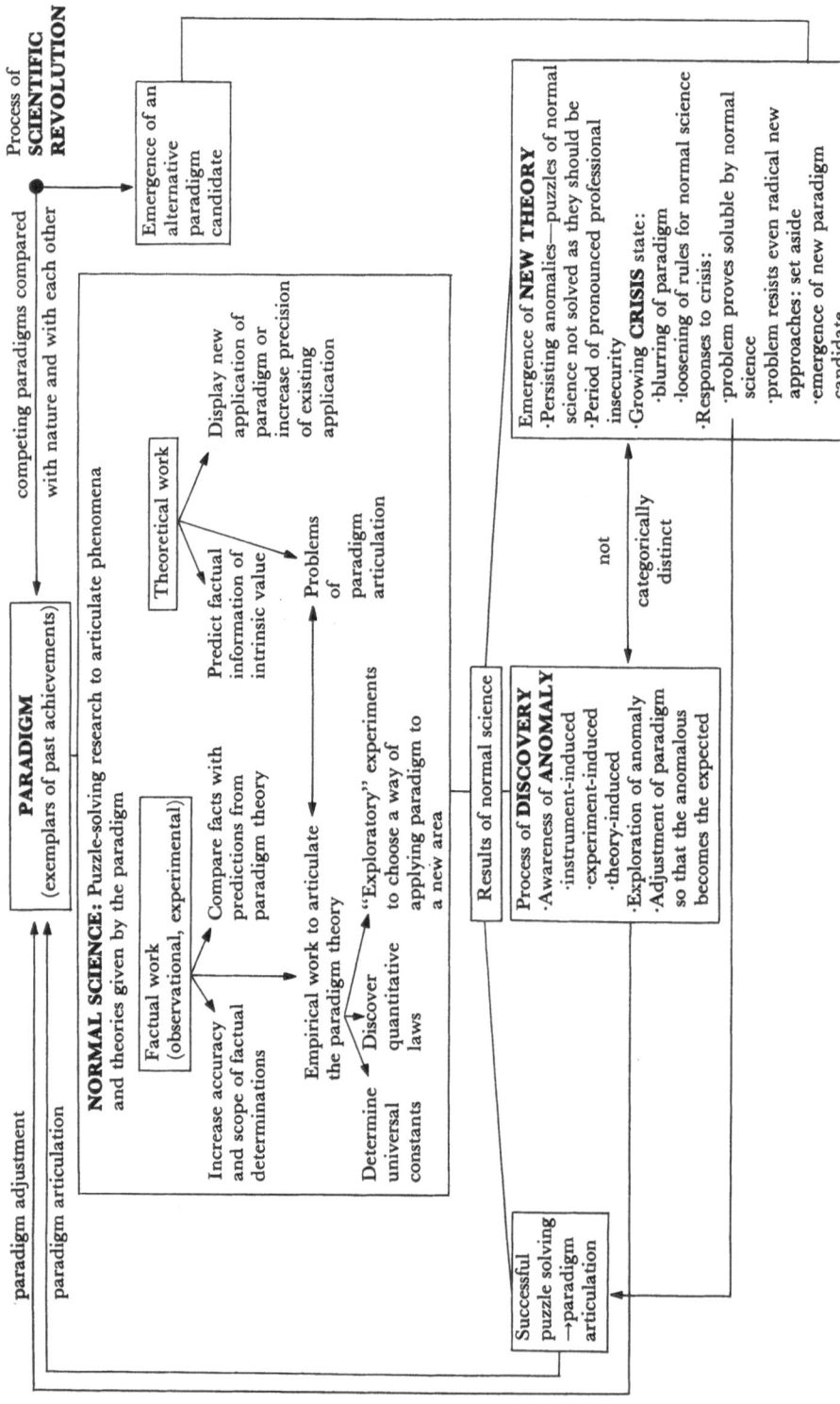

Figure 31.1  A schematization of Thomas Kuhn's theory: Normal science, discovery, crisis, and revolution.

tions, beliefs in particular models, values, and exemplars or paradigms (pp. 172–191).

## Normal Science as Puzzle Solving

For Kuhn, "normal science" is "research firmly based upon one or more past scientific achievements . . . that some particular scientific community acknowledges for a time as supplying the foundation for its further practice" (p. 10). That is, normal science is research conducted within a paradigm, or within the framework provided by exemplars of past achievements that include "law, theory, application, and instrumentation together" (p. 10).

In examining the history of science, Kuhn has found that "perhaps the most striking feature of . . . normal research problems . . . is how little they aim to produce major novelties, conceptual or phenomenal" (p. 35). Thus he views normal science as *puzzle-solving research* directed toward articulating those phenomena and theories supplied by the paradigm and thus adding to "the scope and precision with which the paradigm can be applied." Part of the fascination of scientific research, Kuhn contends, is that

bringing a normal research problem to a conclusion is achieving the anticipated in a new way, and it requires the solution of all sorts of complex instrumental, conceptual, and mathematical puzzles. The man who succeeds proves himself an expert puzzle-solver, and the challenge of the puzzle is an important part of what usually drives him on. (p. 36)

Insofar as this may be a correct objective description of research, it is worth noting that it conflicts sharply with the subjective views that many if not most scientists hold of their own research. The hope, secret or overt, to make "breakthrough" discoveries of great novelty and revolutionary force animates most scientists until that mature stage of their lives when reality forces them to abandon it in favor of a view more consonant with Kuhn's description of normal science.

As illustrated in Figure 31.1, Kuhn finds that the literature of normal science, both empirical (observational and experimental) and theoretical, falls into three classes of problems: determination of significant fact, matching of fact with theory, and articulation of theory. Scientific literature, he points out, also deals with "extraordinary problems," but these "are not to be had for the asking."

They emerge only on special occasions prepared by the advance of normal research. Inevitably, therefore, the overwhelming majority of the problems undertaken by even the very best scientists usually fall into one of the three categories outlined above. Work under the paradigm can be conducted in no other way, and to desert the paradigm is to cease practicing the science it defines. (p. 34)

If normal science "does not aim at novelties of fact or theory and, when successful, finds none," Kuhn then asks, how do new discoveries and theories emerge? Research under a paradigm, he suggests, "must be a particularly effective way of inducing paradigm change" (p. 52).

As shown in Figure 31.1, the process of scientific discovery commences with the recognition "that nature has somehow violated the paradigm-induced expectations that govern normal science" (pp. 52–53). The awareness of anomaly can be induced by a new instrument, an experiment, or a theory. For example, Kuhn cites Roentgen's discovery of x rays as an instrument-induced discovery.

"Consciously or not, the decision to employ a particular piece of apparatus and to use it in a particular way carries an assumption that only certain sorts of circumstances will arise" (p. 59). Roentgen's discovery began when, during a normal investigation of cathode rays, his barium platinocyanide screen "glowed when it should not." But, Kuhn emphasizes, "the perception that something had gone wrong was only the prelude to the discovery—x rays did not emerge without a further process of experimentation and assimilation" (p. 57).

After an initial recognition of anomaly, Kuhn maintains that the process of discovery

then continues with a more or less extended exploration of the area of anomaly. And it closes only when the paradigm theory has been adjusted so that the anomalous has become the expected. Assimilating a new sort of fact demands a more than additive adjustment of theory, and until that adjustment is completed—until the scientist has learned to see nature in a different way—the new fact is not quite a scientific fact at all. (p. 53)

### Crisis and New Theories

For analytic purposes Kuhn distinguishes between scientific discoveries ("novelties of fact") and inventions ("novelties of theory"), although he recognizes that this distinction "immediately proves to be exceedingly artificial." Thus a new theory emerges, or is "invented," by processes much like those of discovery, but usually they result in a far larger "destructive-constructive" change or shift of paradigm.

A new theory, according to Kuhn, usually emerges during a "crisis" period caused by a persistent and pronounced failure of normal problem-solving activities that in turn generates a "period of pronounced professional insecurity" (pp. 67–75).

In discussing how scientists respond to such a crisis state, Kuhn advances a thesis that departs sharply from the traditional view that theories stand or fall solely on the basis of objective experimental evidence. "When confronted by even severe and prolonged anomalies," Kuhn asserts, scientists "may begin to lose faith and then to consider alternatives [but] they do not renounce the paradigm that has led them into crisis."

They do not, that is, treat anomalies as counter-instances, though in the vocabulary of philosophy of science that is what they are. Once it has achieved the status of paradigm, a scientific theory is declared invalid only if an alternate candidate is available to take its place. No process yet disclosed by the historical study of scientific development at all resembles the methodological stereotype of falsification by direct comparison with nature. The act of judgment that leads scientists to reject a previously accepted theory is always based upon more than a comparison of that theory with the world. The decision to reject one paradigm is always simultaneously the decision to accept another, and the judgment leading to that decision involves the comparison of both paradigms with nature and with each other. (p. 77)

Crises, Kuhn concludes, seem to have two universal effects, neither of which depend upon recognition that a breakdown of normal science has occurred. First, "all crises begin with the blurring of a paradigm and the consequent loosening of the rules for normal research" (p. 84). Second, all crises end in one of three ways.

Sometimes normal science ultimately proves able to handle the crisis-provoking problem despite the despair of those who have seen it as the end of an existing paradigm. On other occasions the problem resists even apparently radical new approaches. Then scientists may conclude that no solution will be forthcoming in the present state of their field. The problem is labelled and set aside for a future generation with more developed tools. Or, finally, the case that will most concern us here, a crisis may end with the emergence of a new candidate for paradigm and with the ensuing battle over its acceptance. (p. 84)

This last "mode of closure," the emergence of an alternative paradigm candidate and the battle over its acceptance, is the process of scientific revolution.

## On the Nature of Neuroscience Research

How well do Kuhn's concepts, based largely on studies of the physical sciences, fit the process of neuroscience research as described in this symposium? Although our examination of this question is limited by the symposium's relatively small sample of the field, we hope that it nevertheless provides at least a first approach to a conceptual model of neuroscience research.

To start with, we would assert that neuroscience, as described in the participants' papers, is generally more congruent with Kuhn's analysis of the nature of science than with traditional conceptions.

Those familiar with neuroscience may immediately object that his central concept of "paradigm" does not seem applicable. Kuhn writes about sciences that use paradigms such as those we historically term Ptolemaic and Copernican astronomy, Aristotelian and Newtonian dynamics, corpuscular and wave optics, and so on.

Clearly there is as yet no paradigm with the scope and generality to encompass neuroscience, broadly conceived as including all avenues of research on the nervous system. But as Kuhn points out, science is made up of many communities: the global community of all natural scientists; main professional groups such as physicists, chemists, and zoologists; and subgroups such as solid-state physicists, organic chemists, and neurophysiologists. Each community, in turn, possesses a disciplinary matrix containing paradigms or exemplars that can be far more specialized than a Copernican-type paradigm.

Recognizing that neuroscientists come from different specialties, and that each operates within paradigms from his own field, one can apply to neuroscience Kuhn's concepts of paradigm, normal research as puzzle solving, anomaly and discovery, and crisis and theory invention.

This helps to explicate neuroscience research activities, but it also highlights the question of whether "neuroscience" is an entity waiting to be recognized and encompassed by some powerful new paradigm, or whether it is intrinsically a collection of sciences identifiable only on the basis of a common interest in the nervous system.

As noted in the quotes from Szentágothai and Eccles, some of our authors explicitly recognize that their work as neuroscientists does not fit with traditional accounts of "doing science." Others, such as Kennedy, are uncomfortable with any formal analysis of "how science works," whether traditional or Kuhnian. Kennedy suspects that he is one of many "closet violators . . . who . . . develop sweaty palms when the word *paradigm* is mentioned," for, he feels, "scientific progress occasionally resembles the blind staggers more than the measured tread of rationality." Yet, though it may discomfort Kennedy, his account of how the "comparative zoologist's basic belief system" produces investigative styles that may differ usefully from medical physiology in the study of neurobiology can be formulated as an account of how his disciplinary matrix, with its own exemplars or paradigms, has structured his puzzle-solving research.

Why then is Kennedy uncomfortable about paradigms? Could it be that the paradigms available to comparative zoologists impose less constraint and direction than do those available to physical scientists? That is, the life scientist confronts systems much more complex and open-ended than physical systems, and his research options are thus defined less by paradigmatic expectations than by the exploratory opportunities offered in the myriad forms and hierarchical levels of life phenomena. Thus, although Kennedy recognizes that the basic belief systems of comparative zoology shape an investigative style, apparently he feels that the choice of what to investigate in what kind of a creature is still a mixture of blind staggers and luck.

A detailed analysis of how each participant's paper accords with Kuhn's view of science is not feasible here, but representative examples from a few of the papers will be used as a basis for discussion of three aspects of Kuhn's view: normal science as puzzle solving, competition between paradigms, and the processes of discovery and theory invention.

**Puzzle-Solving Research**

Reading the neuroscientists' accounts of their work reveals that much of their research does not fit the concepts of "discovery," "theory invention," and "revolutionary science," as defined by Kuhn. Rather, their work can largely be characterized as normal puzzle-solving research that seeks to determine significant fact, match fact with theory, or articulate theory. Considering the symposium papers in terms of Kuhn's analysis, one finds that most of the research discussed falls into his "factual" (observational or experimental) category. That is, reflecting the state of the art of neuroscience more generally, the volume contains little discussion of theoretical activity. Moreover, many of the research activities recounted by the participants can be placed in Kuhn's "exploratory-experiment" category of "empirical work undertaken to articulate the paradigm theory," a type of experimentation that is "particularly prevalent in those periods and sciences that deal more with the qualitative than the quantitative aspects of nature's regularity" (see Figure 31.1 and Kuhn, 1970: pp. 27–29).

For Kuhn, the metaphor of normal science as puzzle solving and the scientist as aspiring puzzle solver reveals much about the "rules" governing normal science and the network of conceptual, theoretical, instrumental, and methodological "commitments that scientists derive from their paradigms." For example, he states that

explicit statements of scientific law and about scientific concepts and theories . . . help to set puzzles and to limit acceptable solutions. [At a lower or more concrete level] there is . . . a multitude of commitments to preferred types of instrumentation and to the ways in which accepted instruments may be legitimately employed. . . . Less local and temporary . . . are the higher level, quasi-metaphysical commitments that historical study so regularly displays. (Kuhn, 1970: pp. 40–41)

Reading the "Paths of Discovery" papers with Kuhn's framework in mind, one can identify the paradigms and associated commitments and rules with which the scientists are working, and then "plot" the course of their research as it derives from that paradigm. An illustration of this is the schematization of von Euler's work on neurotransmitters in Figure 31.2. This formulation of von Euler's work as normal puzzle-solving research could be replicated many times over for other accounts of research in this book. Similar patterns, for example, can be seen in the neuroanatomical mapping work of Brodal and Szentágothai, Weiss's study of myotopic responses, Axelrod's work on catecholamines, Bloom's investigations of synaptic transmitters in the brain, and Rushton's work on the visual physiology of light adaptation.

Kuhn's concepts also seem useful for clarifying the nature of clinical as well as laboratory research in neuroscience. Thus, for example, the study of patients with brain lesions described by Luria, Denny-Brown's investigation of reflex motor management in dystonic patients, and Jasper's and Penfield's electroencephalographic studies of the functional organization of the human brain, can all be treated as puzzle-solving researches conducted under "exemplars of past achievements" or paradigms. These clinical researches also provide examples of how discoveries and new theories emerge from anomalous findings and, particularly in Luria's paper, how paradigm competition is resolved.

The clinical research described in this volume, as well as study of a range of other clinical researches, suggests that it might be fruitful to examine in more detail how Kuhn's concepts fit the clinical-research process and whether they add new dimensions to our sociologically based understanding of this medical-scientific arena.

### Technological Factors

In most accounts of scientific research, whether by historians and philosophers of

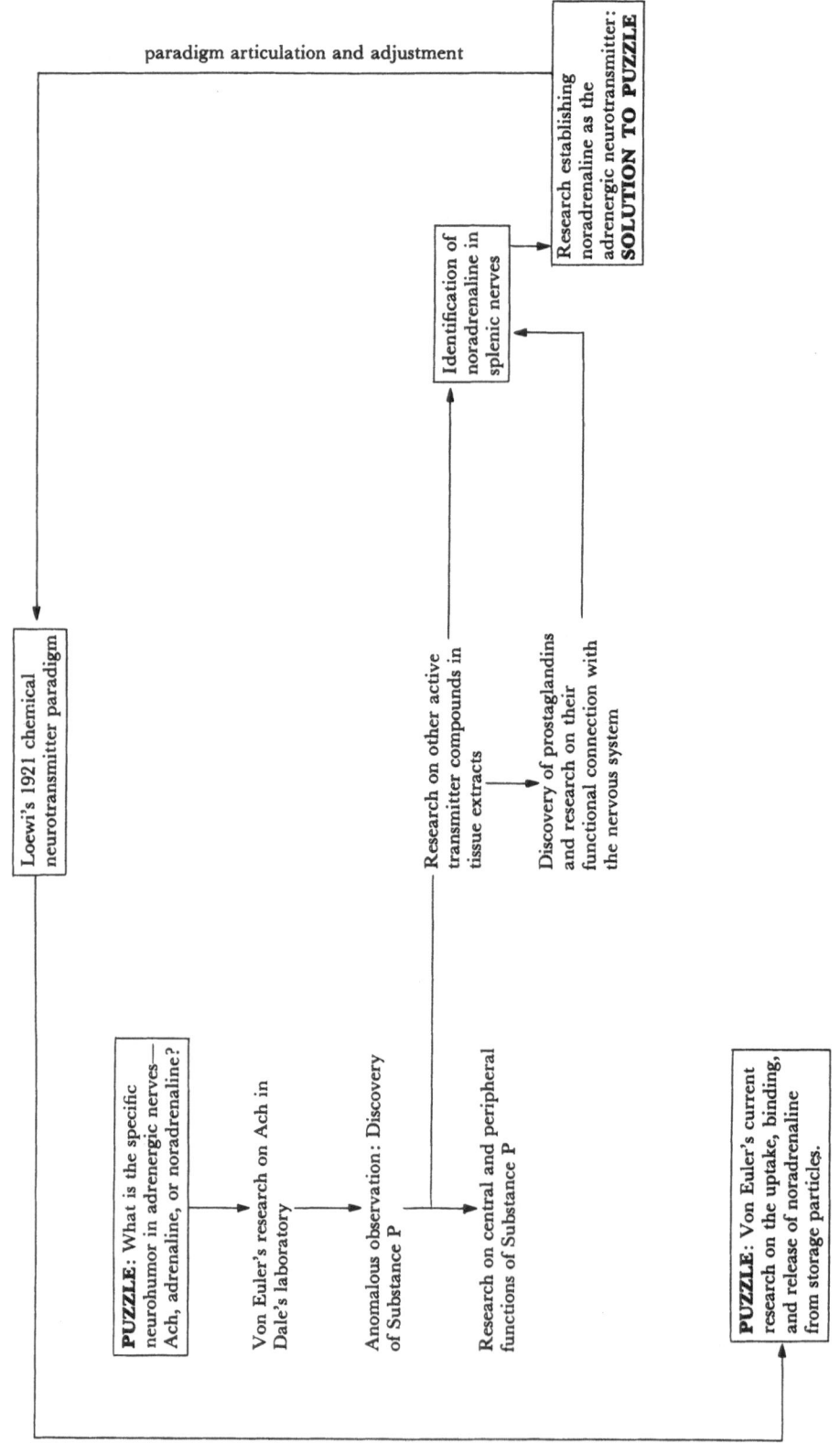

Figure 31.2    Ulf von Euler's puzzle-solving work on neurotransmitters.

science or by working scientists, the role of technique is an important theme. Kuhn refers frequently to instrumentation and techniques, the relationship between technique and paradigm, and the role of technique in normal science and discovery.

It is noteworthy that technique looms large in these neuroscientists' papers as an important determinant of what their research topics have been and how their research has proceeded. Young explores his belief that "biology evolves in many different ways, but in part following the development of technology." And Young, Granit, and others at times wonder whether, in the development and use of increasingly sophisticated techniques, such as those of single-cell experimentation, neuroscientists have become so engrossed in the pursuit of minutiae that they have lost sight of the goal of understanding the multicellular character of organized brain activity.

The development of a new instrument, research preparation, or experimental technique has often been critical to the pursuit of a given line of research, and some developments (e.g., the discovery of the EEG) have opened up whole new fields of investigation. The papers offer many illustrative examples: Bremer's isolated forebrain preparation made possible new ways of examining mechanisms regulating cortical excitability; Young's discovery of the squid giant axon and the Hodgkin-Huxley voltage-clamp technique made it possible for Cole to develop new discoveries about the paths of ions in neural membranes; the Ling-Gerard microelectrode and intracellular recording techniques set the stage for Eccles's study of synaptic physiology; new biochemical methods and concepts enabled Axelrod to develop new understandings of the biogenic amines; and, as emphasized by Bloom, Brodal, and Szentágothai, new staining and microscopic techniques have set the stage for new types of mapping and functional analysis of neuronal circuitry. Indeed, the integration of new staining and mapping techniques with new knowledge about chemical neurotransmitters has led to the emergence of a new paradigm, namely, that neural circuits (unsuspected by classical anatomical investigations) are "chemically coded" in the sense that transmission within a circuit depends upon a characteristic chemical transmitter substance.

## Competition between Paradigms

As our neuroscientists describe how they have made selections between competing paradigms, or have come to abandon one theory in favor of another, their autobiographical statements accord with Kuhn's contention that "no process yet disclosed by the historical study of scientific development at all resembles the methodological stereotype of falsification by direct comparison with nature" (Kuhn, 1970: p. 77).

For Kuhn, the process by which a scientist rejects a theory "is always based upon more than a comparison of that theory with the world. The decision to reject one paradigm is always simultaneously the decision to accept another, and the judgment leading to that decision involves the comparison of both paradigms with nature *and* with each other" (p. 77).

Examples of how the participants' accounts match this aspect of Kuhn's analysis are provided in Szentágothai's discussion of the competing reticularist and neuron doctrines in the 1930s, in Weiss's and Sperry's work on the problem of neural specificity in the context of competing plasticity and connectionist paradigms, and Luria's history of the rivalry between the narrow localizationism and equipotentiality paradigms of cerebral function. That linguistic and terminological factors complicate science is especially clear in Luria's account of how the concept of "function" as applied to a property of a tissue such as contraction (muscle) or secretion (gland) had to be painfully distinguished from such phenomena as language and locomotion that were incorrectly labeled "functions" in the early localizationism literature. That is, the paradigm had to be shifted from simplistically conceived "functions" located at particular cortical sites to the concept of complex processes brought about through the action of many different brain systems.

With respect to how long it took for scientists to acknowledge the "incorrectness" of the original facts on which ideas of classical localizationism were based, Luria makes the following Kuhnian statement:

In any deductive activity, and particularly in science, the law we can call the "law of disregard of negative information" holds good: facts that fit into a preconceived hypothesis attract attention, are singled out, and are remembered; facts that are contrary to it are disregarded, treated as "exceptions," and forgotten. (this volume, p. 339)

It may be noted in passing that this tendency toward selective attention to some facts and disregard of others is an ubiquitous feature of human affairs in general, and exceptions to it are probably more common in science than in any other class of human endeavor.

A particularly eloquent account of paradigm competition and the manner in which "theory x" was eventually accepted and "theory y" rejected, is provided by Eccles's account of his disagreement with Henry Dale about the nature of fast synaptic-transmission processes in sympathetic ganglia and neuromuscular synapses. Eccles relates how, from the early 1930s until 1951, he advocated an electrical theory of synaptic transmission, despite the "very extensive experimental evidence" for chemical transmission developed by Dale and his colleagues. Then, on a "fateful night," the results of an intracellular recording experiment that could only be explained in terms of Dale's theory forced Eccles to abandon the electrical transmission theory, "a brainchild that I had cherished for so many years."

Another example of paradigm competition is Sperry's account of his research on changes produced in the brain by learning and experience. Figure 31.3 depicts how Sperry's experiments were in part conducted to enable him to choose among competing plasticity and connectivity paradigms of the nervous system. He was testing phenomena explained by the plasticity paradigm both by "direct comparison with nature" and by comparison with the alternative connectivity paradigm.

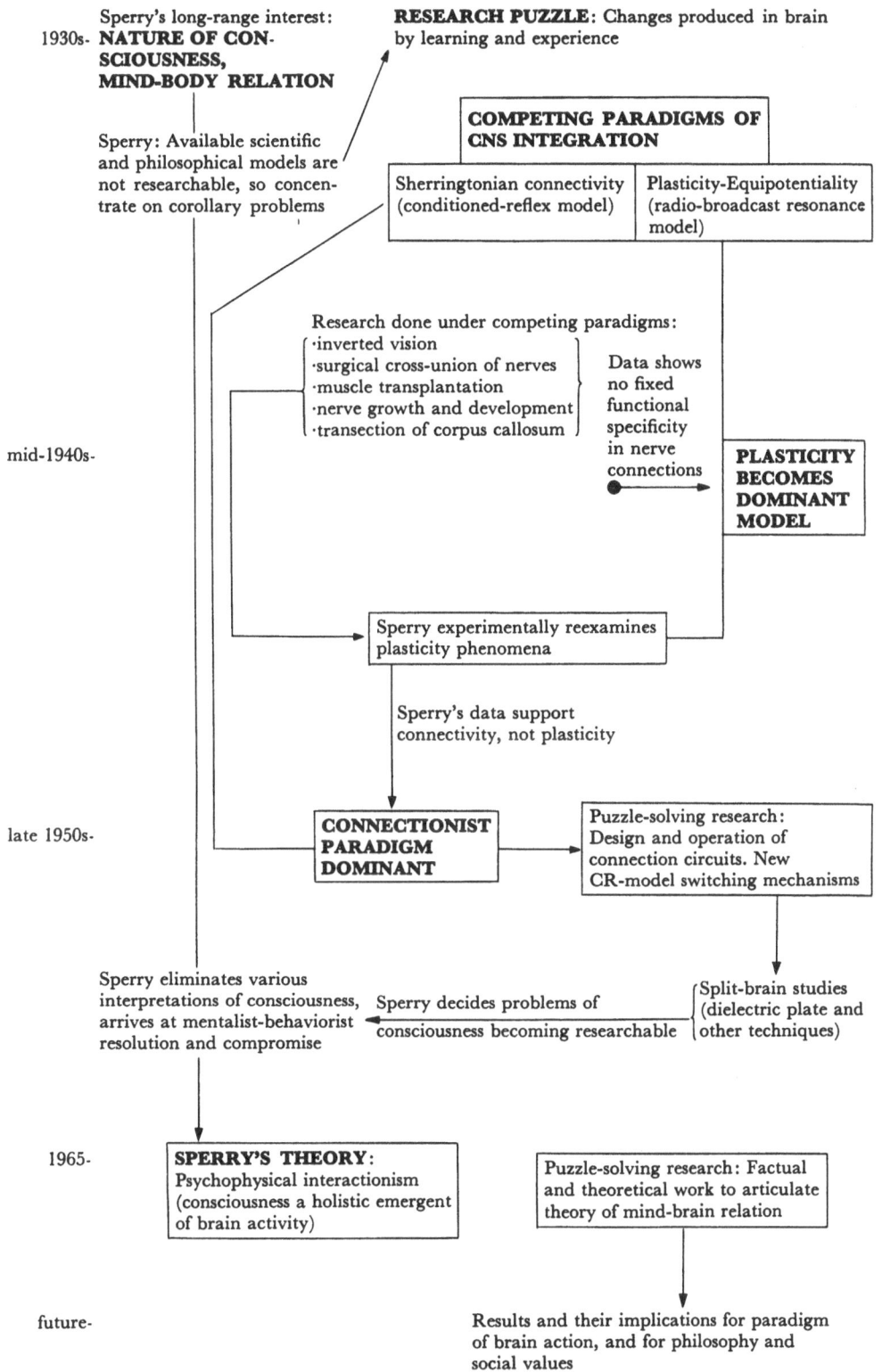

Figure 31.3   Paradigm competition and puzzle solving in Roger Sperry's work.

Figure 31.3 also illustrates another feature of science, that certain problems are "set aside on the shelf" until they are deemed researchable by the development of adequate methodologies. As Sperry tells us, his initial 1930s research interests on the nature of consciousness and the mind-brain relation were set aside until the late 1950s when he decided that his development of the split-brain and other techniques had made the problem of consciousness researchable.

## The Discovery Process and Theory Invention

According to Kuhn, normal puzzle-solving research "does not aim at novelties of fact or theory and, when successful, finds none" (Kuhn, 1970: p. 52).

As we suggested earlier, the question arises as to how well Kuhn's model of normal research in the physical sciences can be fitted to research in the life sciences, and especially in neuroscience. As we have mentioned, most of the research described by our participants corresponds to Kuhn's classification of exploratory empirical research characteristic of a science that has not yet developed mature theoretical structures and paradigms. For example, Richard Jung writes:

Most basic discoveries in neurophysiology have been made by qualitative or semiquantitative methods of exploration. In the early period when the electrophysiology of the brain and of single-fiber recording began, each experiment was a scientific adventure and chance observations were frequent. Success depended not only on chance, but also on the intuitive feeling of what could be a new scientific approach. . . . New fields could be opened up by chance observations made possible by the newly acquired tools of electrobiology. (this volume, p. 505)

The immaturity of neuroscience is complicated by another factor, namely, that it comprises a multidisciplinary attempt to explain how the nervous system mediates behavior, including the psychological and mental life of man. No one yet knows whether man's brain can understand how man's brain works, let alone which of the scientific disciplines or levels of neural organization (molecular, cellular, neural, and behavioral) might be most important for such an understanding. Within the paradigms of his own disciplinary training, the neuroscientist's research may well conform with Kuhn's statement: "Mopping-up operations are what engage most scientists throughout their careers. They constitute what I am here calling normal science" (Kuhn, 1970: p. 24). On the other hand, in the perspective of understanding the nervous system, neuroscience research may not meet Kuhn's criteria for a valid puzzle:

It is no criterion of goodness in a puzzle that its outcome be intrinsically interesting or important. On the contrary, the really pressing problems, e.g., a cure for cancer or the design of a lasting peace, are often not puzzles at all, largely because they may not have any solution. (Kuhn, 1970: p. 36)

For the neuroscientist, attempting to jump from the structured security of his own discipline to the challenging unknown of brain-behavior relationships, there is yet another profound difficulty. Neuroscience research characteristically requires experimental approaches that combine the methods and concepts of one discipline with those of another discipline. Unfortunately, disciplines don't combine in any simple additive way, but, rather, interact with each other to generate a whole new class of problems. One of us (Worden, 1966: pp. 51–54) has described elsewhere how attempts at combining behavioral and neural approaches in a study of the neural mechanisms of attention have revealed that concepts and methods adequate for research within the limits of either discipline prove unexpectedly inadequate when the two disciplines are brought together. The process is not so much one of interdisciplinary research as it is the creation of a new research discipline, and many new and complex disciplines have emerged from the interaction of the traditional sciences engaged in research on neurobehavioral problems. In his paper Julius Axelrod describes the beginnings of biochemical pharmacology; psychopharmacology, neuropharmacology, molecular neurobiology, and behavioral genetics are other examples.

These considerations suggest that neuroscience differs from the physical sciences not merely in its being in an early stage of development but also in the fact that, to understand the nervous system, the neuroscientist is forced to apply the paradigms and rules of his discipline to problems for which they were not designed and for which their relevance is by no means clear. Facing his own discipline, he may see opportunities for "mopping up"; but facing the nervous system, he sees the need to break out of disciplinary barriers in search of a new model of brain function that is not predictable from the paradigms of any present neuroscience discipline.

It remains to be seen whether, with the maturation of neuroscience, the brain-behavior problem will eventually qualify as a puzzle in Kuhn's sense. If it does not, neuroscience research will always retain an element of blind groping that does not fit the Kuhnian model of normal science. In the eyes of many, this may, perhaps, be a more seemly model of man's attempt to understand his own nature than the Kuhnian model of physical science.

In the meantime, it seems clear that the daily fare of the neuroscientist, as opposed to the practitioner of Kuhnian "normal science," consists of material more loosely related to paradigmatic expectations, often neither predicted by nor conflicting with a paradigm, and therefore neither articulating a paradigm nor initiating a phase of paradigm adjustment.

Despite these qualifications, our sample of neuroscience research can usefully be fitted to Kuhn's concepts of how discoveries and the invention of new theories emerge from normal science.

Among our sample of neuroscientists, we find several accounts of the discovery/new-theory process that are "classically Kuhnian" in content. Examples include Bremer's account of his work with the isolated-forebrain preparation; Cole's description of how the role of sodium in membrane permeability was discovered; Davis's account of early work with the EEG; the discovery of reward centers by Olds; Scharrer's finding of neurosecretory neurons; and Levi-Montalcini's account of her two-decade-long study of NGF.

Scharrer's and Levi-Montalcini's papers provide the most detailed accounts in this volume of how discoveries and new theories may emerge from a neuroscientist's normal research activities. As schematized in Figure 31.4, Berta Scharrer documents how a lengthy exploration of her husband Ernst's anomalous observation of neurosecretory neurons in 1928 led to the emergence and eventual

fall 1917- Evidence for neurohormonal activities

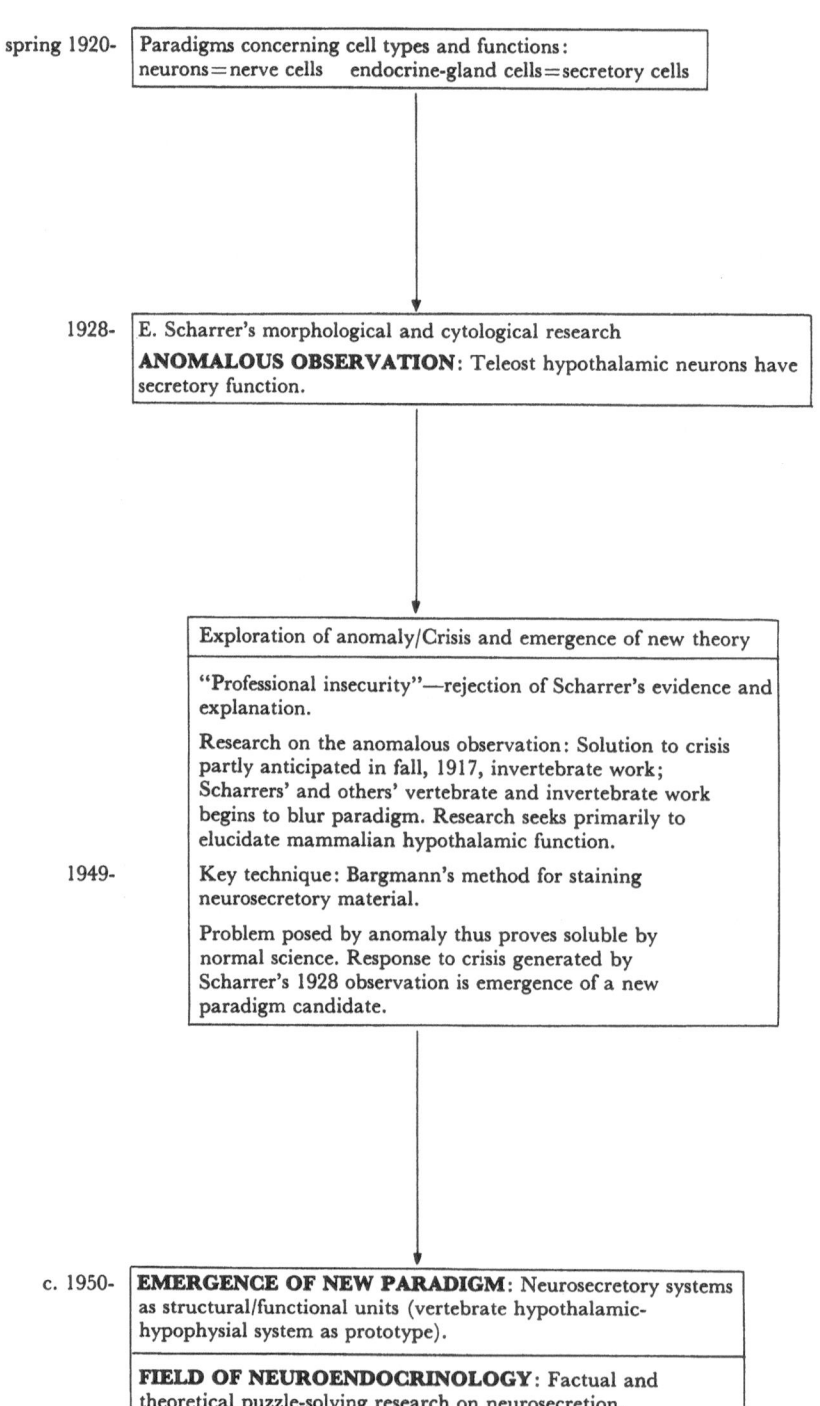

spring 1920- | Paradigms concerning cell types and functions:
neurons=nerve cells    endocrine-gland cells=secretory cells

1928- | E. Scharrer's morphological and cytological research
**ANOMALOUS OBSERVATION**: Teleost hypothalamic neurons have secretory function.

Exploration of anomaly/Crisis and emergence of new theory

"Professional insecurity"—rejection of Scharrer's evidence and explanation.

Research on the anomalous observation: Solution to crisis partly anticipated in fall, 1917, invertebrate work; Scharrers' and others' vertebrate and invertebrate work begins to blur paradigm. Research seeks primarily to elucidate mammalian hypothalamic function.

1949- | Key technique: Bargmann's method for staining neurosecretory material.

Problem posed by anomaly thus proves soluble by normal science. Response to crisis generated by Scharrer's 1928 observation is emergence of a new paradigm candidate.

c. 1950- | **EMERGENCE OF NEW PARADIGM**: Neurosecretory systems as structural/functional units (vertebrate hypothalamic-hypophysial system as prototype).

**FIELD OF NEUROENDOCRINOLOGY**: Factual and theoretical puzzle-solving research on neurosecretion.

Figure 31.4   Discovery and the emergence of a new paradigm: The case of neurosecretion.

acceptance of a new paradigm, that of neurosecretion. As seen in Figure 31.5, Rita Levi-Montalcini's story of the NGF is that of a discovery process that is not yet closed; as she writes, after twenty years of exploration, the NGF continues to "drive its hunters into new surroundings," and the NGF is still waiting to find "its place in the ever-changing game on the neuroscience chessboard."

An important point in Kuhn's analysis—that the recognition of anomaly is only the beginning of a *process* of discovery or theory invention—emerges clearly in Scharrer's and Levi-Montalcini's narratives, as well as that of the discoveries chronicled by other participants. For us, their accounts and Kuhn's concepts were further illuminated by Ragnar Granit's 1972 essay on "Discovery and understanding." Although couched in different terms, the distinction Granit makes between discovery and understanding recognizes, as does Kuhn's analysis, the processual nature of dealing with unexpected research results. Often, Granit writes, an experimenter "may not . . . understand what he has seen, though realizing that it is something quite new and probably very important" (p. 2). For Granit, this latency between discovery (defined as the perception of "something quite new") and understanding (defined as the invention of an explanatory conceptual structure) reinforces his belief that the two categories "really are different concepts and are not arbitrarily differentiated" (p. 3). One might, perhaps, equate Granit's view of discovery as a unique temporal event in which something "new" is found, with Kuhn's first stage in discovery—the perception of anomaly. Then Levi-Montalcini's account of her twenty-year "pursuit" of NGF would exemplify both Kuhn's process of discovery and Granit's thesis that "there is in discovery a quality of uniqueness tied to a particular moment in time, while understanding goes on and on from level to level of penetration and insight and thus is a process that lasts for years, in many cases for the discoverer's lifetime" (p. 3).

Granit further illustrates the difference he perceives between discovery and understanding by reference to the nature of Sir Charles Scott Sherrington's fundamental contributions to neurophysiology:

There are so many instances of discoveries having led to major advances that one is compelled to ask whether it is at all possible to make a really important contribution to experimental biology without the support of a striking discovery. Sherrington's life and work throw light on this question. Most neurophysiologists would not hesitate to call him one of the leading pioneers in their field. Yet he never made any discovery. In a systematic and skillful way he made use of known reflex types to illustrate his ideas on synaptic action and spinal cord functions. Reciprocal innervation was known before Sherrington took it up, decerebrate rigidity had been described, many other reflexes were known, inhibition had been discovered, spinal shock was familiar—at least to the group around Goltz in Strasbourg, and the general problem of muscular reception had been formulated. What Sherrington did was to supply the necessary element of "understanding," not, of course, by sitting at his writing desk, but by active experimentation around a set of gradually ripening ideas which he corrected and improved in that manner. This went on for years—a lifetime, to be precise. Ultimately a degree of conceptual accuracy was reached in his definition of synaptic excitation and inhibition that could serve as a basis for the development that has taken place in the last thirty

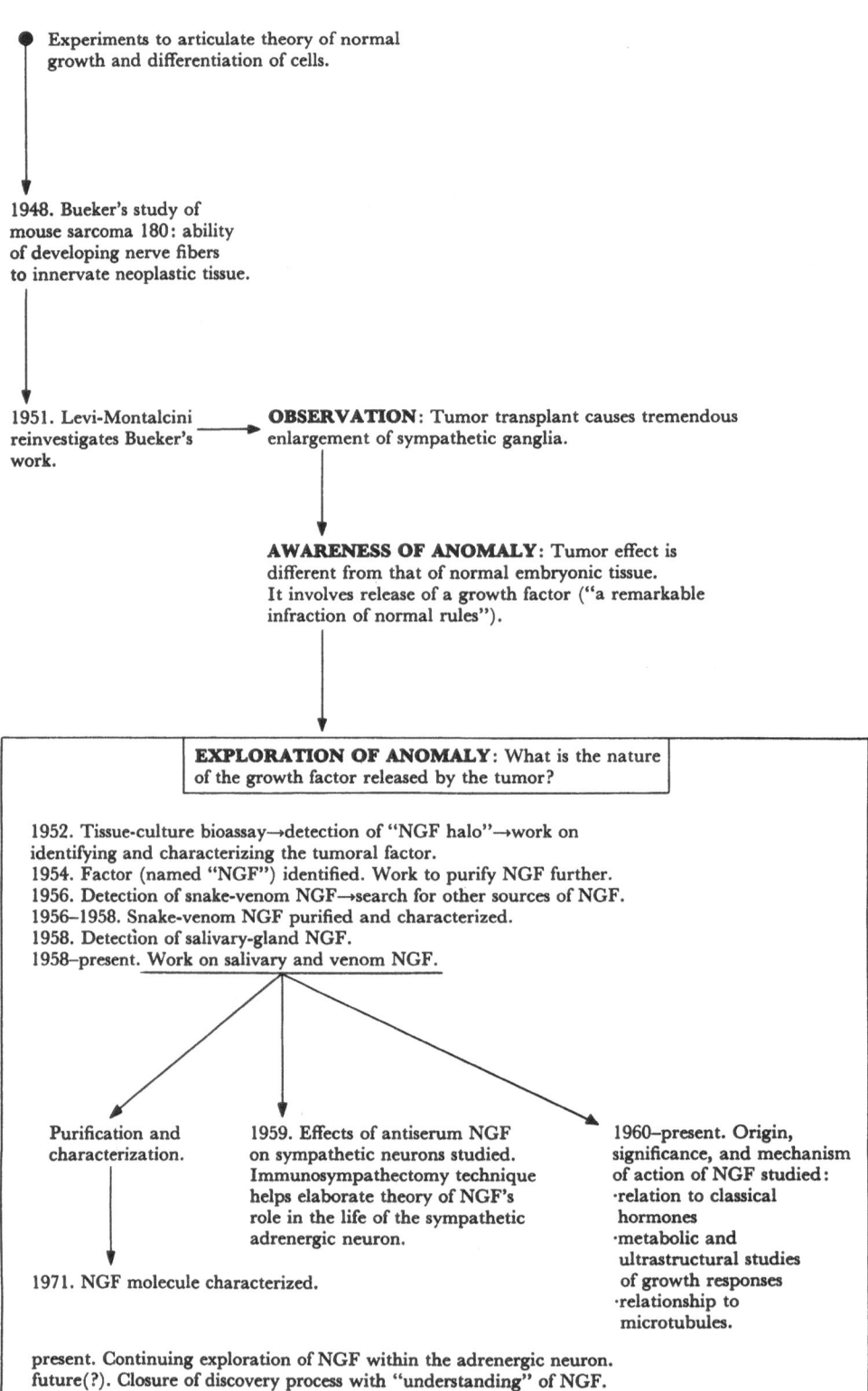

Figure 31.5 Discovery and understanding: Rita Levi-Montalcini's work on nerve growth factor.

years. His concepts are still with us, now fully incorporated in our present approach to these problems. (p. 6)

Sherrington received the Nobel Prize in 1932, specifically for his isolation and functional analysis of the single motor unit. To many, his award was long overdue. But the Nobel Prize is given for definable "discoveries," and, in support of Granit's view of Sherrington's work as providing a lifetime of understanding, it is significant that Nobel Committees in earlier years could not find a specific "discovery" by Sherrington for which they could grant him the prize (Swazey, 1969: p. 26).

To sum up, then, we have found Granit's arguments for a conceptual separation between discovery and understanding to be persuasive; persuasive as well is his contention that "understanding or insight is the real goal of our labors" (Granit, 1972: p. 10). We also find Kuhn's analysis of normal science, discovery, and theory invention to be useful for understanding the accounts of neuroscience research reported in these symposium papers. We have, however, suggested that the multidisciplinary nature of neuroscience introduces complications with regard to whether neuroscience is potentially an entity, or whether it is intrinsically a collection of sciences identifiable only on the basis of a common interest in research on the nervous system.

Finally, if one agrees with Granit that "understanding," such as that generated by Sherrington and many of our essayists, can come from what Kuhn calls normal puzzle-solving research, then this volume, in retrospect, might well have been called *The Neurosciences: Paths of Understanding*.

## References

Beveridge, W. I. B. (1957): *The Art of Scientific Investigation*. New York: Random House.

Brush, S. G. (1974): Should the history of science be rated X? *Science* 183:1164–1172.

Frank, P. G., ed. (1961): *The Validation of Scientific Theories*. New York: Collier Books.

Gillispie, C. C. (1960): *The Edge of Objectivity*. Princeton, N. J.: Princeton University Press.

Granit, R. (1972): Discovery and understanding. *Annu. Rev. Physiol.* 34:1–12.

Kuhn, T. S. (1970): *The Structure of Scientific Revolutions*. 2d ed. Chicago: The University of Chicago Press.

Lakatos, I., and Musgrave, A., eds. (1970): *Criticism and the Growth of Knowledge*. London: Cambridge University Press.

Scheffler, I. (1967): *Science and Subjectivity*. Indianapolis, Ind.: Bobbs-Merrill.

Swazey, J. P. (1969): *Reflexes and Motor Integration. Sherrington's Concept of Integrative Action*. Cambridge, Mass.: Harvard University Press.

Worden, F. G. (1966): Attention and auditory electrophysiology. *In: Progress in Physiological Psychology*. Vol. 1. Stellar, E., and Sprague, J. M., eds. New York: Academic Press.

# Name Index

# Subject Index

Abductors, *80, 83. See also* Transplant studies
Ablation studies, 329. *See also* Lesions
Acetanilide, 191
Acetylcholine, 154, 161, 182, 280, 417
Action potentials, 146–152, 312, 467
in giant fiber of squid, *18*
long-lasting, 153
nature of, 17, 21
oscilloscope records of, *147. See also* Membrane; Potentials
Adaptation, behavior as, 58–60
intercellular, 97
light-dark, 284
Adductors, *80, 83. See also* Transplant studies
Adenohypophysiotropins, 238
Adenosine monophosphate (AMP), cyclic. *See* Cyclic AMP
Adrenal cortex, 201
and induction of catecholamine biosynthetic enzyme, 196
Adrenaline, 185, 186
formation of, 201
³H-labeled, and binding sites, 197
physiological response to, 196–197
Adrenal medulla, tyrosine hydroxylase activity in, 201
Adrenergic system, 182–185, 200
and NGF, 255. *See also* Catecholamines
Affective processes, 364–370
and brain anatomy, 389
and catecholamine metabolism, 37
Afterpotentials, 473
Agnosia, 342, 347
Agraphia, 339, 342
Akinesia, 301
Albert Einstein College of Medicine, 233
Albumin–amyl acetate bilayers, 153
Alexia, 342
All-or-none law, of action potential, 17, 150, 312
Alpha rhythm, 317, 318, 409, 465, 496
arrest of, 485, 555
Amacrine cells, 25–26
American Academy of Arts and Sciences, 538
American Association of Anatomists, 522
American Physiological Society, 468, 522

Amines, biogenic, 196–197, 417, 579
indirectly acting, 185
"sympathomimetic," 192, 199. *See also specific amines;* Catecholamines; Transmitters
Amino acids, 67, 213, 418
Amnesia, 349, 418
during epileptic discharge, 444
Amphetamines, and brain-reward behavior, 382
effects of, *383*
enzyme metabolism of, 193
Amphibians, coordination in, 426
forelimb-specific coordination in, 85
myotypic correspondence in, 94
nervous system of, 427
reticulum of, 306
transplantation in, 79–82. *See also* Frog; Salamander; Transplant studies
Amplifiers, cathode-follower, 494
vacuum-tube, 407
Analgesics, 192
Analytical cytology, xviii
Analyticosynthetic activity, disturbances of, 347
Anastomosis, 89. *See also* Transplant studies
Anatomical Institute, Oslo, 124
Anatomische Gesellschaft, at Königsberg Congress, 105
Anesthesia, 555
barbital, 268
Angiography, 345
Aniline, assay for, 191
Anions, in inhibitory transmission, 169–172
Anomaly, recognition of, 585
Anoxia, 472
Anthropocentrism, 549
Antibodies
amino-acid sequences of, 67
binding sites on, *68*
combining sites of, 65
specificity of, 68
synthesis of, 66
Anticholinesterases, behavioral activating effects of, 270
Anticonnectionism, 428
Antidromic-impulse technique, 160
Antigen α, isolation of, 72–73
Antigens, 65